The Traditional Journey Continues...

The Traditional Journey Continues...

The challenge, legacy and romance of traditional bowhunting

Prepared and edited by:

M. R. James, Editor

Book editorial committee:

Roger Atwood
Glenn Hisey
John Gardner
Kevin Hisey
R. Craig Oberle
Doug Clayton
Tom Nelson
Jay St. Charles
Dirk Dieterich
Mike Schlegel
Jeff Pals
Mike Kistler

Library of Congress Control Number: 2012922224
ISBN Number: 9780961796686
Published in April, 2013

Produced and Published in the United States of America by:
The Pope and Young Club
273 Mill Creek Rd.
PO Box 548
Chatfield, MN 55923

Design and Layout by Steve Ashley

Table of Contents

Foreword

The evolution of bowhunting is a fascinating story. In a relatively short period of time, we have come from making our own simple longbows, arrows, and leather items to the full spectrum of efficient modern equipment available to today's bowhunter – from a handful of dedicated souls who struggled to exist but remained diligent to their cause, to the millions of bowhunting enthusiasts spread across this nation and throughout the world. No other group in the hunting community has grown and developed to the degree bowhunting has. Not only have hunter numbers exploded, available hunting seasons have spread rapidly across the country. Bowhunting opportunities today flourish throughout the world and are enjoyed by enthusiastic bowhunters from all walks of life.

In the words of Pope and Young Club founder, Glenn St. Charles, "*It appears that bowhunting has a bright future. However, it is not an unclouded future.*"

With growth comes new challenges for tomorrow. Numbers increase demand. Efficiency promotes success. Increased hunter numbers, coupled with higher success figures, tax the resource beyond capability. Wildlife departments struggle to keep up with demand at current economic and budgetary requirements. At some point the system will fail if we don't do something to fix the problem.

Conservationist and bowhunter Aldo Leopold once made a statement that bears as much weight today as it did years ago: "*I have the impression that the American sportsman is puzzled; he doesn't understand what's happening to him. Bigger and better gadgets are good for industry, so why not for outdoor recreation? The sporting press no longer represents sport; it has turned billboard for the gadgeteer. Wildlife administrators are too busy producing something to shoot at to worry much about the cultural value of shooting. Because everyone from Zenophon to Teddy Roosevelt said sport has value, it is assumed this value must be indestructible.*"

The sporting world as a whole is making such rapid technological advances it is difficult to keep up. The sad part of this picture is valuable time-proven principles are being left behind. The demographics of our society are changing. We are becoming an impatient hurry-up society. Success needs to come quickly and with rewards. Tempting shortcuts become the preferred way. Education often gets left behind.

As dedicated bowhunters, we are obligated to protect traditional bowhunting values. More than protect, these values must be sustained and preserved! I'm not talking about traditional equipment. I'm not referring to stickbows and wood arrows. I'm talking of important values like fair chase, ethics, sportsmanship, and challenge. The new generation of hunters needs reminded and educated on the significance and importance these elements play in the overall hunting picture. With the loss of these important values, the relationship between predator and prey is severely compromised – or even eliminated altogether. The point is our hunting passion is based on respect, honor, and love of these animals. We need to ***teach***. We need to ***preserve***.

Bowhunting is about how close…not how far. It's about facing your quarry in his element, without unfair advantage, with patience and respect. It's about being there, witnessing life in the wild, enjoying God's creation, experiencing the beauty, smelling the scents, and being one with the outdoors. It's about accepting the challenge of getting as close as possible – within an animal's living room, without detection.

The book you are holding is the second of its kind published by the Pope and Young Club. It includes listings of recorded trophies taken by bowhunters who accepted the challenge of using traditional equipment. Each recorded entry represents a successful encounter with a trophy-size animal – an encounter that undoubtedly stirred a primal reaction within the bowhunter; an adrenaline rush, the pounding heart, a roar in the ears, quickened breath, and weakening knees. It's a reaction not only tolerated, but most often sought. As bowhunters, we venture afield in hopes of experiencing yet another successful close encounter. But we also realize our unsuccessful encounters stir the emotions as well. And in the end, we discover it is indeed the little things we remember most – staring at burning campfire embers, muscles burning while packing downed meat, tolerating the miserable rainy days, relishing the camaraderie of good friends, sharing disappointment and success while we create memories.

In the end it is the little things that make our adventures meaningful and memorable. The memories we create last a lifetime!

Roger W. Atwood
P&Y President

Acknowledgements

This book represents one more step down bowhunting's long path and is another milestone in Pope and Young Club history. It is, in fact, the second *Traditional Journey* book to be published, connecting past and present in words and photographs that document individual accomplishments and collective participation in an ancient sport whose roots can be traced back in time for tens of thousands of years. And while modern man no longer relies on a bow and arrow to survive as our ancestors once did, there is little argument that reaching full draw and releasing an arrow is as necessary as breathing for those of us who proudly call ourselves bowhunters.

Certainly, modern equipment has changed our sport forever, no matter our personal choice of bow and shaft. Many of us began hunting with archery tackle when our choice was limited to recurve bows, longbows, and self bows. And for some, that choice remains unchanged. Yet regardless of the hunting bow we choose to carry afield, we can agree that it is our enduring traditional values, and the practice of legal, safe, ethical, and fair chase bowhunting, that bonds us and strengthens our resolve and common cause.

The book you hold, the words you read, reflect a collective effort by men and women who share the passion all serious archers feel. It documents and preserves forever moments in time, and its creation would not have been possible without the contributions and efforts of individuals who gave their time and talent.

Editor M. R. James, with direction and approval of the Pope and Young Club's book committee, assumed responsibility for selecting and editing the hunting adventures and other informative material featured in "The Journey Continues" section. Coordinating the entire project, Executive Secretary Kevin Hisey and the Club's staff shouldered a demanding workload that included maintaining behind-the-scenes efforts involving everything from communications to publishing responsibilities to collecting published records and ensuring their accuracy. Supportive book committee members included President Roger Atwood and the following current P&Y Board members: Glenn Hisey, Mike Schlegel, Mike Kistler, Craig Oberle, Tom Nelson, John Gardner, Doug Clayton, Jay St. Charles, Jeff Pals, and Dirk Dieterich.

Graphic designer Steve Ashley deserves special recognition for his work designing and laying out this book, coordinating its printing and timely delivery. Steve's talents have been on display in past Pope and Young Club publications for years and his latest effort is only the latest in a long line of publishing accomplishments.

Likewise, public kudos are due P&Y Headquarters' invaluable staff members, Karla Hamersma and Julie Cordes, as well as volunteer Larry Streiff. True professionals, they keep the home office functioning smoothly while allowing Glenn and Kevin Hisey time for special projects such as time-consuming book projects. Karla and Julie also assisted in proofreading duties and deserve a tip of our camo caps for their efforts.

Of course, without writers, the material printed within these pages would consist of mere facts, figures, photos, and little else. Reading "Chief" Compton's story of Ishi, Dr. Pope's account of hunting companion Art Young's memorable Alaskan adventures, Club founder Glenn St. Charles' Mad River midwinter deer hunt, and Fred Bear's Wyoming elk hunt, during which he tagged a World Record bull, allows us to share moments and adventures with men who blazed the trail we follow today.

Additionally, contemporary bowhunters and current Pope and Young Club members graciously shared their insights, hunting experiences, and advice. Mike Schlegel delves into the question,

"What is a traditional bowhunter?" Len Cardinale remembers overcoming physical and mental tribulations to achieve success on a demanding mountain goat hunt. Whitetail hunter Doug Clayton traces a bowhunting career that travels "Full Circle," while John Gardner recalls a honeymoon bowhunt for Dall's sheep with his new bride. And speaking of sheep, Marv Clyncke looks back over a long career spent hunting wild sheep and helping others achieve their dreams of collecting a trophy ram. Glenn Hisey's tips in "Basic Broadhead Collecting" provide expert advice and helpful suggestions for getting started. Jay St. Charles explores hunting for caribou in the north country barrens, while Fred Asbell explains the planning and dreaming that are such integral parts of bowhunting. Thanks to each man for his contribution to this book.

Lastly, thanks to the Boone and Crockett Club for granting permission to the Pope and Young Club to adapt, print, and use their time-tested scoring system and forms since our organization was founded in 1961.

Introduction

"I shall never forget the feeling that came over me the first morning I stepped out of camp on my first bow and arrow hunt ...

"I had a good seventy-pound bow and a quiver carrying fifteen broadhead arrows. These heads were about two inches long and one inch wide. You can well imagine how I felt to attempt to go out and kill game with, as someone put it, some sticks and a string."

Arthur H. Young

All of us who hunt or have bowhunted with traditional archery tackle don't need to imagine how Art Young felt. We *know*. We remember that feeling of excitement and anticipation because we experienced it for ourselves. We recall our own choice of hunting bow and broadhead, the self-confidence attained by long hours of drawing and releasing countless practice arrows, the eagerness of beginning our very first bowhunting adventure, and the electric rush of pulse-tingling sensory stimulation when some game animal finally stood broadside within good bow range.

It made no difference, really. Woodlot whitetail or mountain muley. Prairie pronghorn, wilderness elk, or backcountry bruin. Our first hunt and first animal are forever etched in our memory, an integral part of who we were and who we are today. What matters most is that we willingly accepted the demanding and time-consuming challenge of hunting with "some sticks and a string." We committed to what the legendary archer Howard Hill called "hunting the hard way." We recognized the need for accumulating knowledge of wildlife, honing woodscraft skills, getting close to wary game, remaining undetected, and drawing and releasing a well-aimed arrow.

This book, quite simply, is a celebration of participation in traditional bowhunting's values and principles, as well as a tribute to the effectiveness of longbows, recurves, and selfbows. It is a natural follow-up to the Pope and Young Club's 2007 publication, *A Traditional Journey*.

This book is not, however, a direct or indirect attack on compound bows or bowhunters who choose to use them. Mechanical bows have been a growing part of the contemporary archery scene since first introduced in the late 1960s. Traditional and modern hunting bows are legal and effective hunting tools readily available as a matter of personal choice. They are used efficiently season after season by bowhunters of both sexes and all ages, archers who share a common passion: *pursuing wary wild game while enjoying the highly challenging and immensely rewarding experience of fair chase bowhunting.*

The book's contents are designed to inspire and inform. Starting with President Roger Atwood's opening comments which set the tone for the Club's second traditional record book, readers curious about our sport's past can step back in time and read the story of Ishi as told by his friend and hunting companion, Will "Chief" Compton, or join Club namesake Art Young on his Alaskan bowhunts of the early 1920s, related by his friend Dr. Saxton Pope.

Want more historical hunting adventures? Read P&Y Club founder Glenn St. Charles's personal account of how he took his first-ever deer on the snowy Washington State wilderness hunt, or Fred Bear's account of how he tagged a Wyoming elk in 1952, a trophy that once ranked as the World's Record.

How about more contemporary traditional bowhunts? There's the inspirational "Hunting Hurt" story of a physically challenging mountain goat undertaken by Len Cardinale. There's John Gardner's Alaskan "Honeymoon Hunt" bowhunt for Dall's sheep, accompanied by his new outdoors-loving bride. And there's Doug Clayton's recollection of his "Coming Full Circle" from beginning archer to successful bowhunter tagging his first record book whitetail.

Want helpful advice to expand your bowhunting horizons? Turn to Glenn Hisey's step-by-step approach to "Basic Broadhead Collecting." Or the Jay St. Charles chapter on planning a successful caribou adventure, complete with tips and advice on taking these nomads of the arctic wildlands. Or Marv Clyncle's remembrances of his personal development as a successful sheep hunter, complete with hunting anecdotes and practical sheep hunting information.

Finally, veteran bowhunter Mike Schlegel tackles the topic of bowhunting with stickbows and offers answers to the question, "What Is a Traditional Bowhunter?" And Fred Asbell, a well-traveled veteran of countless hunting adventures whose membership in the Pope and Young Club dates back to the late 1960s, explains the important role "Planning and Dreaming" has in making the sport all that it can be for all serious participants.

Complementing this book's well-illustrated editorial content, of course, is the up-to-date listing of all North American big game animals entered in the Pope and Young Club's Records Program taken with longbows, recurves, and selfbows. While the first traditional record book contained some 7,755 entries, that number has grown by more than 800 during the past five years. Accompanying field photos from the Pope and Young archives span more than half a century of Club and bowhunting history.

Make no mistake, traditional bowhunting is alive and well in the twenty-first century. This book is tangible proof of that fact. And it is much more than a historical record of data; it is, in truth, a documentation of accomplishments by men, women, and youngsters who answered a siren call, accepted the daunting challenge of walking a less-traveled path, and set a standard for future generations to emulate. Such is the enduring legacy of those who hunt with mere sticks and strings.

The Journey Continues

The Story of Ishi, the Yana Indian

By W.J. "Chief" Compton

ALONG in the nineteen hundreds stockmen were puzzled by the disappearance of grain from cabins in the mountains of Tehama county, California, placed there for saddle horses used for rounding up stock in the vicinity of Deer Creek.

The thing that puzzled the cowboys the most was that the thieves never took any of the food brought to the cabins for their own use, although they often found the canned goods and other food scattered all around the place; but the rolled barley for the saddle horses was always taken whenever a cabin was broken into.

The sheepmen suffered loss at the same time by having sheep die suddenly. In making a post mortem examination they discovered obsidian arrow heads in the dead sheep. The ranchers were sure, upon this discovery, that the maurauding had been done by an Indian or Indians. But there were no wild Indians in that part of the country, or anywhere else, that they knew of.

Ishi shooting his bow near Deer Creek

For years this sort of thing had been going on when the ranchers in and around Deer Creek decided to put in a big ditch for irrigation purposes, the water to be taken out of the upper reaches of Deer Creek. In the course of time the surveyors were on the job running the line through the chaparral where the ditch was to go and it so happened one evening, at the close of the day's work, that four or five of the crew started for their camp which was on the opposite side of Deer Creek.

When they arrived at the creek bank they discovered that the creek was unusually high, and, as they had approached it at a point that was strange to them, they were discussing the feasibility of the particular spot for a crossing. They rode on horses and as they entered the water the splashing and clattering of their horses was drowned out by the roaring of the high waters of the creek.

As they followed the riffle where the water was shallowest, they were going on an angle and had just passed an elbow in the creek bank which had obscured their vision as to what was immediately beyond this projection. The riffle led them within a few feet of this shoulder. When they had advanced into the water so they could see what was behind the turn in the bank, they saw a naked Indian with a spear in his hand watching for salmon trout. He discovered them at the same instant and he ran at them, salmon spear poised to throw.

The horsemen scattered and scrambled across the creek anywhere. They rode into camp and told the others about seeing the wild Indian, but, as the days ran into weeks and no Indian was seen again, the camp commenced to doubt the story of those who claimed to have seen a naked Indian fishing with a salmon spear.

About one year after they found him fishing in the creek, they encountered him again in a most peculiar manner. Two men from this same surveying crew, one of whom was with the party when they found him the first time, were riding along on horses when the one who had been present at the first encounter suddenly exclaimed that they were in the vacinity of the spot where they discovered the wild Indian. His companion greeted this information with a laugh and said, "You fellows must have had some kind of a pipe dream about that Indian as no one except you fellows has ever seen him. Where is he now?"

Just as he asked the question an arrow zipped through the interrogator's hat. There were no more questions regarding his whereabouts. They were both of the same mind instantly: that without doubt they had found out where he was – for the moment at least. Having decided this very quickly and positively, they departed from that vicinity, pronto, in a cloud of dust, headed for camp.

Ishi and Pope hunting trip 1913

When they told the story, the hunt was on to find the bowman that had loosed the almost fatal shaft. The whole crew turned out for the hunt as they were all convinced that there were wild Indians around and uncomfortably close by. So they started to search all the big thick patches of chaparral, and they found the camp in the thickest of this growth. Just a brush wickiup, the corner posts of which where scarcely larger than a man's thumb. The roof was brush with bark laid on top. There was also a deep wide hole in the ground that they would fill with snow in the winter to furnish water for the camp.

At the time of discovery by the ditch crew only a very old woman was at home and she was in bed sick. They knew that this old woman was not the person who had sped the arrow through the hat of one of their number a few hours before, so they knew that there was at least one bowman missing; but could not know how many more. After picking up everything in sight that was loose at both ends they departed. The next day they returned, their numbers greatly augmented. But, on this second visit, they found the old woman gone. The others, they knew, had returned and carried her away.

Then the souvenir collectors fully realized that there was nothing between them and complete destruction of the camp. Joy oozed from every pore as they proceeded to demolish everything and tie in bundles the fragments for easy transportation to their homes. Where the Indians had gone they could not even guess as in their departure they left no trail.

Several months after this the butchers in a slaughter house near Oroville heard a dog barking furiously in the yard outside the building. Going outside to investigate they found a naked Indian. When asked who he was and what he wanted, they discovered that he did not understand what they were saying. They went for some Digger Indians whom they knew to do the talking but with no better results. They then tried out the linguistic powers of a Piute, but it also failed. Something had to be done if they were to find out anything about their strange visitor.

All this time the Indian thought they were planning to kill him. At this juncture someone thought of phoning the University of California at Berkeley to see if that institution could help them out of their dilemma. The university sent a Mr. Waterman, a student of dead languages.

Well, in due time, Waterman arrived, and just before the Yana tongue became a dead one as the Indian by this time was all but scared to death. Waterman tried out different Indian languages but this wild Indian knew none of them. Then he thought to try the Yana tongue on him. That brought results quickly. There was a small pine table in the room. Waterman placed his hand on this and said the word for pine in Yana. The word was "saweenie." Immediately, Ishi came alive and wanted to know of Waterman if he was a Yana, or Yahe, as he called himself. Waterman told him that he was. In less than three weeks' time, Waterman apparently was speaking Ishi's mother tongue fluently.

Arrangements were made with the University of California and Affiliated Colleges to bring Ishi to the Affiliated Colleges at San Francisco. It was there that Dr. Saxton Pope, Arthur Young, and I became acquainted with him. We found him to be a man right from the stone age. In spite of the handicap we were all in agreement that he, a savage, was the gentlest human being either of us had ever met.

Ishi skinning a deer near Deer Creek

We hunted game with him and he was always the best of companions. He was with us over five years and in that length of time we never saw him out of humor. When he hunted small game he would make calls to lure them within shooting distance that were quite effective at times. He would imitate the cry of a rabbit in distress which, he said, would bring a wild cat or lynx to him if there was one within hearing distance.

He shot up to forty yards. The bow he used would pull about fifty-five pounds. He shot the same bow for hunting and target. At target he was a poor shot. It was something foreign to his lifetime practice with the bow. His fletching job was good but he preferred the tail feathers of the turkey to the quill feathers. Why he had this preference we never found out. Probably some superstition. He made his arrow heads of obsidian – the workmanship was superb, perfect in form and extremely sharp. They would penetrate flesh as easily as the sharpest steel. In making these heads, he used a piece of buckhorn about five inches long, tapered down to about one-sixteenth of an inch across, and square in form. He held the obsidian in his hand and pressed it against the small squared end of the buckhorn and twisted. In this crude manner he chipped away the excess material, producing the most perfect stone or obsidian arrow heads I have ever seen. His bow strings were made of sinew.

His bow was made of mountain juniper if he had it. If not he would condescend to use yew. He said yew wood was only fit for the use of women. We could understand his attitude towards women, his knowledge of women being gained through association with his aunt and his sister, his uncle teaching him all the old superstitions regarding the weaker sex. But how he got the idea that juniper was better bow wood than yew we never found out. Some more supersitition, I suppose. He said to us, "Never allow a woman to touch your bow, as it will not shoot well if they do."

His uncle told him that before the white man came game was very plentiful. They hunted in large parties of twenty to fifty or more. A portion took a stand, well concealed, while the remainder of the party drove the game within shooting distance of those on stand. Of course, all this was told him by his uncle, as there were only his uncle, aunt, and sister of the Yana tribe living since Ishi was four years old. When visiting in Dr. Pope's home he would act as though he never saw Mrs. Pope at all; the idea being that it was very impolite for a man to notice another man's woman.

The thing that astonished Ishi the most of all the astonishing things he saw when he came to San Francisco was the disappearance of a window shade when it rolled up. That completely stunned him. Another thing that excited him out of his natural calm, also pleased him greatly, was to see himself in pictures. When he showed up at the slaughter house in Oroville, his hair had been burned off close to the scalp as a sign of mourning. They took several pictures of him at this time wearing short hair and always when he would see himself in one of these pictures it would please him beyond words.

Ishi enjoyed the best of health to within one year of his death, when he commenced to show signs of tuberculosis. Everything was done for him that human hands could do; but nothing could be done to arrest the disease, and bade farewell to Ishi on June the 16th, 1916. His words were "I go, you stay," and thus passed on one of the most noble of the redmen. A true friend and noble character. Everyone loved him that knew him.

We – Young, Pope, and I, saw to it that he was left alone by the religious sects, and that he was allowed to die in the faith of his people. We buried him in the Italian cemetery in San Francisco – the last of the Yana Indians.

Art Young, "Chief" Compton, and Saxton Pope

Trailblazing bowhunter Will "Chief" Comption (1862-1938), was a close friend and hunting companion of Dr. Saxton Pope, Arthur Young, and Ishi. His recollections of Ishi, the last of California's Yana Indians, were originally published in the July 1936 issue of *Ye Sylvan Archer*, a popular monthly publication printed between May 1927 and December 1941. The Pope and Young Club's highest award is named in honor of Ishi.

Alaskan Adventures
By Dr. Saxton Pope

Art Young with his Kenai Peninsula bull. Note the arrow stuck in the palm

It seems as if Fate had chosen my hunting companion, Arthur Young, to add to the honor and the legends of the bow. At any rate it fell to his lot to make trips to Alaska between the years 1922 and 1925.

He and his friend, Jack Robertson, were financed in a project to collect moving-picture scenes of the Northland.

They were instructed to show the country in all its seasonal phases, to depict the rivers, forests, glaciers, and mountains, particularly to record the summer beauties of Alaska. The animal life was to be featured in full: fish, birds, small game, caribou, mountain sheep, moose, and bear, all were to be captured on the celluloid film, and with all this a certain amount of hunting with the bow was to be included and the whole woven into a little story of adventure.

Equipped with cameras, camp outfit, and archery tackle, they sailed for Seward. From here they ventured into the wilderness as circumstances dictated. Sometimes they went by boat to Kadiac Island, sometimes to the Kenai Peninsula, or they journeyed by dog sleds and packs inland. They spent the better part of two years in this hard, exacting work, often carrying as much as a hundred pounds on their backs for many miles. Great credit must be given to

Art's partner Jack Robertson, for his energy, bravery, and fortitude. His work with the camera will make history, but for the time being we shall focus our attention on the man with the bow. Only a small portion of Young's time was devoted to hunting, the exigencies incidental to travel, and gathering animal pictures were such that archery was of secondary importance.

He hunted and shot ptarmigan, some on the wing; he added grouse and rabbit meat to the scant larder of their "go light" outfit. He shot graylings and salmon in the streams. He could easily have killed caribou because they operated close to vast herds of these foolish beasts. However, at the time it seemed that there was no hurry about the matter; they had meat in camp, and pictures were of greater interest just then. They expected to see plenty of these animals. Strangely enough the herd suddenly left the country and no further opportunity presented itself for shooting them. This was no great disappointment because the sport was too easy. What did seem worth while was the killing of the great Alaskan moose. These beasts are the largest game animals on this continent, with the exception of the almost extinct bison.

Young had his first chance at moose while on the Kenai Peninsula. Here the boys were camped and having finished his camera work Art took a day off to hunt.

In the afternoon he discovered a large old bull lying down in a burnt-over area, where approach by stealth was possible, so he began his stalk with utmost caution, paying particular attention to scent and sound. By crawling on his hands and knees he came within a hundred and fifty yards, when his progress was stopped by a fallen tree. To go around it, would expose him to vision; to climb over, would alarm the animal by snapping twigs; so Young decided to dig under. He worked with his hunting knife and hands for one hour to accomplish the operation. When he had passed this obstacle he continued his crawling till he reached a distance of sixty yards. At this stage Art called the old bull with a birch bark horn, then the moose heard him and stood up. The brush was so thick that he could not shoot immediately, but waited as the old bull circled to catch his wind and answered the challenge. When he presented a fair target at seventy yards or so, Art drove an arrow at him. It struck deep in the flank, up to the feather ranging forward. The bull was only startled a trifle and trotted off a hundred yards. Here he stopped to look and listen. Young drew his bow again, and overshooting his mark, his arrow struck one of the broad thick palms of the antlers. The point pierced the two inches of bone and wedged tight, making a sharp report as it hit. This startled the animal off at a fast trot. Young followed slowly at some distance and soon had the satisfaction of seeing the moose waver in his course and lie down. After a reasonable wait the hunter advanced to his quarry and found him dead. The triumph of such an episode is more or less mixed with misery. The pleasure undoubtedly would have been greater had some other lusty bow man been with him, but as it was he had to feast his eyes alone, moreover he had to make his way back to camp, which was some eight miles off, and night rapidly coming on.

This part of the story was thrilling to Art because he must stumble through the rough land of "little sticks" in the dark with the constant apprehension of meeting some unwelcome Alaska brown bear, which were thick there, and also the extremely unpleasant experience of running into dead trees, tripping over fallen limbs, and dropping into gullies. He reached camp ultimately, I believe. Next day he returned with his companion for meat, his antler trophy, and the picture which we present.

This bull weighed approximately sixteen hundred pounds and had a spread of sixty inches across its antlers.

Upon the second expedition a year later, Young bagged another moose. Here the arrow penetrated both sides of the chest and caused almost instant death, showing that size is not a hinderance to a quick exodus.

It is surprising even to us to see the extreme facility with which an arrow can interrupt the essential physiological process of life and destroy it. We have come to the belief that no beast is too tough or too large to be slain by an arrow. With especially constructed heads sharpened to the utmost nicety, I have shot through a double thickness of elephant hide, two inches of cardboard, a bag of shaving, and gone into an inch of wood. We feel sure that having penetrated the hide of a pachyderm his ribs can easily be severed and the heart of pulmonary cavity entered. Any considerable incision of either of these vital areas must soon cause death. And this is a field experiment which we propose to try in the near future.

There is a legitimate excuse for shooting animals such as moose, where food is a problem and the brown bears an honorable part in the episode. We feel moreover that by using the bow on this large game we are playing ultimately for game preservation. For by shaming the "mighty hunter" and his unfair methods in the use of powerful destructive agents, we feel that we help develop better sporting ethics.

It was partly on its account, and partly to answer the dare of those who have said, "You may hunt the tame bears of California and Wyoming, but you cannot fool with the Kadiac bears of Alaska with your little bow and arrow," that Young determined to go after and see if they were as fierce and invulnerable as claimed. At the present writing we who shoot the bow have slain more than a dozen bears with our shafts, but the mighty Kadiac brown grizzly has laughed at us from his frozen lair – as the literary nature fakir might say – we have been told that all that is necessary if you wish to meet a brownie, is to give him your address in Alaska and he will look you up. Also we have been told that once insulted he will tear a house down to "get even with you," – so I shook Art's hand good-bye, when he started on this Kadiac escapade, and told him to "give 'em hell."

After a long time he came back to San Francisco, and this is the story he told me – and Art has no guide in his system but is as straight as a bowstring.

"We made a false start in going after our bears. We took a boat from Seward and sailed to Seldie, then to Kenai Peninsula. Here we hunted for two solid weeks and found practically no sign of brownies.

"I decided at the end of this period to waste no more time, but to pull out of the country and sail back to Seward.

"We had but a short time to complete our picture before the last boat left the Arctic waters, but hearing of good bear signs on Kadiac Island we hit out for this place and landed in Uganik Bay. Here in the Long Arm, we found a country with many streams flowing down from the mountains which constitute this island, and much small timber in combination with open grassy glades. A type of country that is particularly suited for photographic work and bow hunting.

"After several days' exploring we discovered that the bears were catching salmon in the streams and we were successful in photographing as many as seven grizzlies at once. We took pictures of the bears wading in the water looking for fish. Usually the bear slaps the salmon out of the stream, then goes up on the bank and eats it. The "humpies" were so plentiful here, however, that they were tossed out on the bank, but not eaten, the bear preferring to capture one while in the water and then wade about on his hind legs while he held the fish in his arms and devoured it.

"We got all this and many comic antics of young bears climbing trees and playing about by using a telephoto lens. After the camera man was satisfied I proposed that we 'pull off' a 'stunt' with the bow.

"By good fortune we saw four bears coming down the mountain side to fish. They were making their way slowly through an open valley. The camera was stationed at a commanding point and I ran up a dry wash thickly grown with willow and alder to head off the bears. I was able to get within a hundred yards by use of the willow cover, then the brush became too thin to hide me, so I walked boldly out into the open to meet the bears. I practically invited them to charge since they were reputed to be so easily insulted. At first they paid little attention to me, then the two in advance sat up on their haunches in astonishment and curiosity. I approached to a distance of fifty yards, then the largest brownie began champing his jaws and growling; then he 'pinned back his ears' preparing to come at me. Just as he was about to lunge forward I shot him in the chest. The arrow went deep and stuck out a foot beyond his shoulder. He dropped on all fours and before he could make up his mind what hit him, I shot him again in the flank. This turned him and feeling himself badly wounded he wheeled about and ran. While this was going on an old female also stood in a menacing attitude, but as the wounded bear galloped past her, she came to the ground and ran diagonally from us. All of them followed suit, and as they swept out of the field of vision the wounded bear weakened and fell less than a hundred yards from the camera.

"True to his standards the camera man continued to grind out the film to the very last, so the whole picture is complete. You will see it some day for yourself and it will answer all doubts the invulnerable status of the Kadiac bears."

Young himself was not particularly elated over this conquest. He knew long ago that the Kadiac bear was no more formidable than the grizzlies we had slain and he only undertook this adventure for show purposes. Moreover though he used his heavy osage orange bow and usual broad-heads, he declares that he believes he can kill the largest bear in Alaska with a fifty pound weapon and proportionately adjusted arrows. Both Young and I are convinced of the necessity of very sharp broad-heads, and trust more to a keen blade and quick flight than to power.

During his Alaskan travels Art preferred his Osage bows to the yew. They stood being dragged over rocks and falling down mountain sides better than the

Art admires the Kadiac bear he took with a single arrow

softer yew wood. His three bows were under five feet six inches in length, short for convenience and each pulled over eighty-five pounds. The country in which he worked was so rocky that it was most disastrous on arrows, and every shot that missed meant a shattered shaft.

Possibly his roughest trip was one taken to picture mountain goats. Here a funning incident occurred. Jack and Art were stalking a herd of these wary creatures with the camera when around a point of rock the whole herd of goats appeared. Art was ahead and only just time to duck down on his hands and knees and hide his face close to the ground. He stayed so still that the entire flock passed close by him almost touching his body, while the camera man did his work from a concealed ledge higher up. Though Young counts it little to his credit, he shot one of those male goats, which was poised on so precipitous a point that it fell over and over down the mountain side and was lost as a trophy and camp meat. Humiliating as such an episode may be, it serves, however, to add a coup to the archer's count. And there we let the matter rest.

But what is of greater interest is his outwitting a Rocky Mountain Big Horn. This animal is considered the greatest game trophy in America. It is an extremely alert sheep, all eyes and wisdom. If you expose yourself but a second, though you be a mile away from the ram, probably you will be seen. And though the sheep may not move while you look at him, he is gone when you have completed your toilsome climb and peer over the last ledge of rock preparatory to shooting. Ned Frost used to say that when he hunted Big Horns he paid no attention to hearing or smell, but he was so careful about sight, that when he raised his head cautiously over a ridge to observe the sheep, he always lifed a stone and peered underneath it, or picked up a bunch of grass and gazed through it.

Most hunters are content to stalk this game within three or four hundred yards, then aim at it with telescopic sights. It is the last word in good hunting.

Stewart Edward White, the author and big game hunter, has said that the following experience is one of the finest demonstrations of stalking and understanding of animal psychology he knows.

Up near the head of Wood River, Young and his party came on a number of Big Horn Sheep and first devoted several days to film work. Then Young decided to try for a trophy with the bow. After hunting all morning he discovered with his glasses a ram a long way off.

The country was open and had no cover. The ram was resting on a ledge or rock elevated above the level of the valley. Even at a distance of half a mile it was evident to Art that the ram had seen him, so Young studied the sheep and the country carefully before deciding what plan to pursue.

From the lay of the land it was plain that no concealment was possible and no detour or ambush could be employed. The glasses showed that when the ram was a fairly old specimen and had a very sophisticated look. In fact, to Art he looked conceited and had an expression that said: "There is a man, but I am a pretty wise old sheep I know all about men; that fellow hasn't seen me yet and when he does, there is plenty of open country in back of me; my best plan is to like still and let this tenderfoot pass." So he went on ruminating and blinking at the sun.

Taking this mental attitude into consideration, Young decided that the best method of outwitting this particular sheep was to take him at his own valuation and proceed as a tenderfoot down the valley. So he walked unconcernedly along at an oblique angle to the sheep and never once taking a direct look at him. He went gaily along whistling, kicking pebbles, and swinging his bow. When he had reached a distance of two or three hundred yards the old sheep lifted up his head to see what was going on. Young paid no attention to him, though he observed him out of the corner of his eyes. So the wise old boy settled back content with his diagnosis.

Art walked along as innocently as ever. When he was a hundred and fifty yards off, the ram raised his head again and took a longer observation. He seemed to be changing his mind. Young said to himself, "He will take one more look, then he will go. Now is the time to act." So nocking an arrow on the string he ran at full speed directly at the sheep and when half way he saw the tip of his horns rise above the ledge and knew it was time to stop. He came to his shooting pose and waited, the arrow half drawn. Sure enough! Out walked the old fellow to the very edge of the parapet and gazed over. Off flew the arrow and in the twilight it was lost to vision, but he heard it strike and saw the ram wheel in flight. As it disappeared over the ridge Art followed at a run; reaching the top he peered cautiously about and saw the sheep at no great distance standing still with his legs spread wide apart. He knew by the posture that it was done for. So he went back to the valley and because of the distance from camp and the oncoming darkness he made a fire and "Siwashed it" or camped out in the open all night without blankets. In the morning he went after his trophy and found it near the spot last seen. It was a fine specimen. The arrow had pierced it from front to rear completed through and was lost; a center shot at eighty

yards; a most remarkable bit of archery and hunting stratagem. This head now decorates the dining room of the Young home in San Francisco. Unfortunately the moose antlers were cached near a river in Alaska and an unprecedented flood carried them out to sea.

While speaking of Alaskan rivers there recurs to my mind a most remarkable incident related by Young. In one picture required for their film it was necessary to show a canoe in the course of construction, the subsequent use of the vessel, and an upset in the turbulent waters of the river. To represent his bow in its canvas case, and still to spare the weapon a wetting, Young went down the river bank to pick out a stick about the same size to put in his bow case. Taking the first piece that came to hand he started to place it in the case, when struck by its smoothness he looked at it and found he had a weatherbeaten old Indian bow in his hand. It seemed like a sign, a good omen, -- for we playfully indulge in omens in these romantic adventures with the bow.

Studying this implement later I found it apparently to be a birch Urock bow, some five feet long, having nocks and a place for the usual perpendicular piece of wood bound on the handle to check the string. It would have pulled about sixty pounds, good enough for caribou hunting.

And so in brief are the adventures of Art Young in Alaska.

But who can speak of the adventures in the heart of our archer? Here is no common hunter, no insensate slayer of animals. Here we have the poet afoot, -- the archaic adventurer in modern game fields; the champion of fair play and clean sport; all that is strong and manly.

I take my hat off to Arthur Young.

Pioneer bowhunter Dr. Saxton Pope (1875-1926) published his archery classic, *Hunting with the Bow & Arrow*, in 1923. "Alaska Adventures," reprinted here, was excerpted from the second edition of that popular and inspirational book, published in 1925. Even after more than eighty years, it provides a meaningful glimpse into a time and place that still fires the imagination of 21st century bowhunters.

Mad River Yields Another Trophy

By Glenn St. Charles

Glenn St. Charles and his Mad River trophy

The deer laden cars were all stopping at the top of snowy Stevens Pass; someone off the road below, I guess, and the people strolling by our car had stopped, puzzled, beside us. You see, Hal had tied a tattered arrow across the horns of a two-point lying across our right front fender.

"Say, boys," one old fellow said, "did you really get that critter with a bow and arrow?"

"Yeah," was the reply.

"Well, I'll be darned. I heard guns were getting scarce but I didn't know it was as bad as that!"

Four of us had left Seattle about 2 a. m. Saturday for our second weekend of hunting, not knowing if we'd ever get through to the reserve. We had been snowed out the week before and, coupled with the lack of weather reports since then, it was a very skeptical but determined bunch that beat its way back over Stevens Pass that morning. The snow was deeper beside the road up to the Maverick Saddle, but a bulldozer had cleared the road in freeing some snowbound hunters and we had smooth sailing all the way.

Tink Johnson, Archie Powell, Hal Aebig, and myself found we were all alone in trying the reserve this week. All the other boys had believed it snowed up for the winter. There was twenty-six inches of snow on the level and more in the drifts. It was about 8 a. m. when we cleared a big enough patch to get a fire started and some bacon and eggs on to fry. A half hour saw us all raring to go and we hit the trails. Arch and Hal up to Maverick Peak, Tink and I down to the river. We were to follow the river about two and a half miles, then go up to McDonald Ridge, and plug the trail that Arch and Hal were driving down.

It was a cold, dead calm day with a slight overcast. Deer tracks were everywhere and it was hard to believe that a herd of sheep hadn't hit the trail ahead of us. Fresh snow had covered all tracks from the week before so we made ready for anything and everything. We followed a well-beaten trough and hadn't gone fifty yards before a buck and a doe showed themselves on the side hill toward the ridge top, too far for a shot so on we went with a little warm blood tingling through us at the sight of them. Tink turned to the right to have a look over a knoll into the Mad River, and while I waited three does and a spike trotted up the trail. We stared at each other till Tink showed up and they were soon off to parts unknown. It was just beginning to occur to me that at the sight of two different bunches I had made no effort to get ready for any possible action. This worried me a bit because I was wondering what I'd do if something really to shoot at came along!

A half mile farther along we were approaching a turn in the trail when what should pop around it but a four-point. He backtracked so fast that before we could make the thirty feet between us and the turn he was out of sight. Again I hadn't lifted my bow! Tink was behind me so I hadn't noticed his reaction. We were being as quiet as possible, not even talking while on the trail, so as to allay any chance of chasing game ahead of us. By 9:30 we had wound along the snowy river to the two and a half mile post where a trail led across the river and up the Cougar Mountain ridge on our right. Another trail led to the McDonald ridge. The latter we took.

Deer had beaten a deep path through the snow and about twenty uneventful minutes found us nearing the top. We cautiously approached it, remembering Basil Blackmore's experiences in this spot the week before when he had trouble trying to choose which of about four different bucks to shoot at. Nothing showed so we stopped for a short rest before starting back along the ridge.

The ridge leveled out in this particular spot, which prompted Tink to have a look around. I, likewise, felt like doing a little exploring farther along the ridge before starting back. We decided to meet at this spot in about half an hour. A high knoll had to be climbed first and I found my cleated rubber boots came in very handy getting over the slippery rocks. Upon reaching the top I stood awhile to look over the awe-inspiring scenery. Lake Wenathee spread out below, flanked on all sides by high snowy peaks and ridges. This view coupled with the very stillness of my surroundings was awesome.

The snow was about thirty-six inches and trees were bent under its weight. Small trees were completely covered with drifts. Without thinking much about it, I picked out a big sagging tree I'd hide behind in case a buck should happen my way; and as I could see a trail about two hundred yards ahead and about fifty yards behind it wasn't a bad spot for a stand, I thought to myself. There wasn't a bit of wind and every breath seemed to hang in the air around me. I stood there what seemed like an hour was nearer to ten minutes, getting colder and colder. The sound of shuffling one foot and then the other in the snow seemed to keep me company while in this trance, when all of a sudden I caught a slight movement up the ridge!

Two does and a buck were just sauntering out of the trees two hundred yards away! They were single file on the winding trail which gradually sloped toward me! And when the deer dipped out of sight in a little gully, I hurried over to the ambush I had previously picked out. I knelt down on one knee, threw off my hat, and waited with my bow about a quarter drawn. My ambush was about twenty-five yards from the trail, broadside to it, and only about thirty feet of the path was in view. A million things ran through my mind on what to do.

This would be my first shot at a deer if all went well. Would the deer get wind of me? Would they branch off the trail? A couple of minutes passed which helped to cool me down a bit, then a doe came into view, a two-point buck, and then the other doe! As I drew they froze in their tracks, and then came a tearing sound like a knife going into a sack of grain. The does disappeared up the ridge, the buck took two jumps ahead, stopped, looked around apparently puzzled. I had not moved a muscle after the shot. There was no extra arrow in my bow hand and none in the snow ready! Some of my plans forgotten in the excitement! After seconds of looking over his surroundings, the buck turned and walked away from me, over to the edge of the ridge.

It was then that I could see that he was hit for I could see the arrow hanging by the nock on the other side. The buck's knees seemed to sag a bit and he fell over the edge of the slope. I crept over to where I could see him. He was on his feet leaning against a tree and looking over his shoulder at me. We stood that way and stared at each other for about two minutes; then he staggered on down the slope for about thirty yards and lay down.

I then looked at my watch. It was 10:45. Tink should be waiting for me at the foot of the knoll and, with the deer down, it was my chance to go get him as I expected to need some help. I floundered down the hill as fast as three feet of snow would let me. As I neared the bottom I yelled, "Tink!" A far off answer. "Tink!" again I yelled.

A silence – then I heard, "What the h***?" as he came puffing around a boulder. "What's wrong?"

"I've got one down."

"Yeah?" he grinned. "I thought you were getting killed!"

We scrambled back to my lookout on the hill. The buck was still lying in the same spot.

We took a little time to take a few pictures and a look at our surroundings told us our troubles were really about to begin! We were almost a quarter of a mile from the saddle where the trail went down the other side to the Mad River and the quarter mile was all rocky sidehill, covered with drifts. We guessed the buck's dressed weight at one hundred seventy-five pounds. We dragged the carcass to an overhanging rock shelf which had a spot under it free of snow. We built a fire and ate our lunch and rested a bit. While Tink rigged a better carrying pole, I scouted about the rock. I found a way around this shelf and after taking our equipment to the top of the saddle and back a pretty good trail had been tramped in the snow. We dragged the buck up there a foot at a time for three and a half hours before we reached the top.

It was now 3:30. We yelled for Hal and Arch, thinking they should by this time be in the vicinity. No luck, so we started the downhill part to the Mad River trail. We practically rode the buck at this stage, reaching the trail about 4 p. m. It was then decided that Tink should go and get help and I would stay with the buck. At 5:30 Arch and Hal came steaming up the trail. Boy, was I glad to see them! They had "refreshments," dry gloves, and a good idea for sledding out the deer. Two of us pull a rope while the other steered the buck along the trail. It was 8 o'clock when we reached the campfire. Tink had steaks ready to fry, lots of firewood and the beds ready.

Between mouthfuls we talked over the day's hunt. Tink's story was soon told; Hal and Arch had run into plenty of bucks and had a couple of swell shots. Arch still wonders about missing a twenty-yard shot he had at a four-point, and Hal was thrilled to see a running buck he shot at suddenly wheel and almost run into the arrow that would have missed him by a yard. So went the day. We hit the hay to be ready for the morrow.

My buck weighed out at one hundred sixty pounds and was shot through the liver. The arrow cut half way through a rib on each side and hung by the nock.

He was shot at twenty-five yards, with a sixty-pound, four-foot eleven-inch osage takedown bow. The arrow was a twenty-five inch fir shaft, eleven-thirty secondths in diameter and weighed four hundred and forty grains.

This story about Pope and Young founder Glenn St. Charles arrowing his first deer was published in the February 1943 issue of *Ye Sylvan Archer.* He was thirty-two years old at the time. The noted bowyer helped lead the fight for archery-only seasons in the Pacific Northwest, opened Northwest Archery in suburban Seattle after World War II, and founded his North American big game archery record-keeping organization in 1961. Glenn St. Charles was elected to the Archery Hall of Fame in 1991 for his many contributions to the sport as a bowhunter, educator, and conservationist.

Wyoming Trophy Elk
By Fred Bear

Fred with his Wyoming elk that would go on to be recognized as the world's record in 1961

Up Hoback Canyon, in the Jackson Hole country of Wyoming, there is a grassy knoll some 9,000 feet high, on the north side of the mountain on which I once spent an entire day. I think I could still find that spot blindfolded, though it's been almost two years since I last saw it.

Lon Imeson of Jackson, who has been guiding in that area for many years, was ending his second season with me. He was determined now to have me bag an elk.

At three in the afternoon, on the last day of our hunt, we were lolling in the sun on top of this grassy spot just over the top of the mountain. Six or seven bull elk were bugling below us and had been since daylight that morning. While waiting for them to move up closer, we spent the day cat-napping, telling stories, and occasionally sounding our elk bugles – always getting an immediate response.

Toward the middle of the afternoon, Lon's patience had reached its limit. He sized up the situation this way: the big bull we were after was fartherest down the mountain and he had the cows safely corralled within his territory. The younger bulls, who had seemed closer to us, were hanging around the outskirts waiting for a chance to horn in on his harem.

We could plainly hear the big fellow far below us. His bugling had more volume than that of the younger bulls and was easily distinguished from theirs by the way he ended up with a series of guttural grunts – haw...haw...haw –

Packin' in

like the braying of a donkey.

There had been no rain for the past two months, and the country was bone dry. Stalking was about as quiet as an early morning milk wagon on cobblestones. There wasn't a chance of sneaking down to the herd and getting an arrow at the big bull. So we tried to sit tight until they moved up the mountain to feed on the south side where some grass still grew in spite of the drought.

It was just after three o'clock when we noticed a change. The elk were on the move, coming up the mountain to our right. Our strategy had been planned. We moved into the timber, climbing a steep ravine en route, and saw a nice young bull drinking at a trickle of stream about 250 yards away.

The elk were working up just beyond the ravine, and when we reached the top, Lon said that this looked like the place to wait. We gave them another toot on the bugle and were promptly answered as before. The big fellow, whose bugling ended with that haw, haw, haw, was still the fartherest down.

Time dragged. Every toot from our bugle brought an answer from the big bull, as far down as ever. It began to look as though he wouldn't get up to us before dark, and our nerves couldn't hold out any longer. It was finally decided that Lon would make a big circle down the mountain, get below the elk and hope to put them up to me.

For the past seven days we had been trying everything we knew that might put us within bow range of a bull elk. We had almost succeeded on three or four occasions but were defeated each time by the wind. In flat country, when the wind is blowing from the north, it blows from the north. But in these mountains, cut up with ridges and canyons, it can blow from the north and also from the south at just about the same time. For the time being, however, the wind was right, coming up the mountain and off across the ravine we had just crossed.

Lon left and I looked around for a place that offered the best cover and shooting possibilities. It suddenly occurred to me that the elk could come up through there anywhere within a distance of 200 yards. If I got myself snarled up in a blowdown or in a spot where it was too noisy to move I would simply be out of luck.

Fred receiving a congratulatory handshake from Charlie Kroll as the outfitter looks on

Just below me was an open strip covered with fairly fresh-looking grass. It offered little cover, but plenty of maneuverability. I decided to risk this and settled down beside an old log and blew my elk call.

They were coming now. No doubt about it. The answer I got meant business. He couldn't be more than 200 yards below and he made it plain that he was in no mood for foolishness. I could hear a great horning of bushes and crackling of limbs as his bugling came closer and closer.

There was a moment of doubt as to the wisdom of moving down here out of cover. Elk are big. In the rutting season they do things in a big way. The year before in this same area, I had sneaked to within sixteen steps of a bugling bull. It was too brushy for a shot, but I can still feel my spine tingling from that challenging bellow.

The log beside me shrank to the size of a twig. Even down on all fours my back stuck up above it. My bow was under my hand with an arrow on the string. My legs began to cramp just as I caught something out of the corner of my eye. Over to the right a cow elk was feeding, moving my way as unconcerned as if I were 1,000 miles from there. Just past her were two more cows and a couple of half-grown calves coming in for some of this choice grass.

Well, here we go again. The bull was not yet in sight. This herd would see or wind me and this last day would end like all the others ...

There was no way of getting out of sight without flattening out on the ground, and I couldn't shoot in that position. I tried to hold as still as possible, desperately hoping the bull would catch up with the cows that were moving closer by the minute. Twenty-five yards now, noses sniffing my way, ears pivoting front and back. Nice, sleek creatures nibbling dainties from the forest floor.

Off to my left and down the hill about sixty yards something else moved – and there he was – a great beast of light and blending into dark brown. His neck and head were black, and he carried antlers the size of a pine tree. Master of the harem and a right to be.

Lon had analyzed the problem well. He was the head bull. Those heavy shoulders and massive antlers had won him the right in combat. Some of the cows that had cast shy glances at a fringe bull perhaps knew the feel of those sharp tines, too.

The big bull was in a frenzy, grunting and bugling and ripping up a cluster of shrub maple with his antlers. He was impressive, and behind my little log I was impressed! He was in the timber, and it was too brushy to get an arrow through. Several cows were now within twenty yards and by now almost downwind. The cows were to my right and the bull to my left. Worst of all, Lon was somewhere below. Since I had moved from where he thought I would be he might come in sight at any time and put the herd to flight.

Suddenly the bull, having conquered the maples, walked straight up the mountain angling in my direction. This couldn't last. He would see me, or the wind would shift. Something was bound to go wrong – but he was still coming on. He walked out of the brush to within twenty yards and stopped, turning his head away from me. There had been no time to look at the cows for at least a full minute and I had the feeling that every one of them was looking straight at me. I couldn't risk moving my head to check. Past experience with elk had taught me that they can vanish as quickly and completely as a white-tailed deer. At the cows' first sign of alarm the bull is the first to leave. He would be two or three jumps away before I could draw my bow. It was shoot from this position, now, on hands and knees, or not at all.

Working my knees forward under my body I slowly raised the bow, horizontally, about a foot off the ground. The arrow lobbed into his midsection, and I fully expected all hell to break loose around me. The cows were behind me now, and the bull in front. The arrow struck through the back ribs, went through and lodged in a rib on the far side. The bull made a quick turn and walked back to where he had torn up the maples. He stood there, his head hanging low. It was clearly a fatal hit and he would drop on the spot in a very short time.

The cows didn't flick as much as an ear. They continued to feed without looking up even when I raised up on one elbow to watch the bull through my glasses. But suddenly they were gone in a flash, and the bull went with them. It was Lon. He came over the mountain just in time to hear the crashing of hooves and see the last of them disappear down the mountain.

His "what the blankety-blank is going on here" soon turned to a wide grin -- his "well done" grin. I'd been waiting for this for two years now and the sight of it was almost as coveted as that of the trophy I knew was lying not far over the hill.

Lon was more than a guide. He liked the spirit of bowhunting and went far beyond his job of guiding to inspire success. This was his hunting ground. He knew the elk well and credited them with a highter I.Q. than even the Rocky

Mountain bighorn sheep. Except for his persistence I might not have gone out that day. With every new sunrise the mountains had looked higher and on this, the last morning of the two years' effort, my legs were beginning to give out …

We walked to the edge of the brush and picked up the back end of my arrow. It would soon be dark, and we took up the trail without delay. We tracked him about 400 yards where he'd slid down the mountain, the great antlers caught around a tree which kept him from going farther.

Climbing the mountain and picking our way in the dark on the long ride down the other side was not difficult that night. The moon smiled in the sky, and the campfire glowed brightly. The mountains were mere molehills now.

Fred Bear's story of his Wyoming elk hunt was first published in the May, 1954 issue of *Archery Magazine*. That bull was later recognized as the bowhunting World Record in 1961 when the newly founded Pope and Young Club announced its first official list of North America's top bow-killed trophies. "Papa" Bear also served on the P&Y's Board of Directors from August 1970 to April 1978. He was elected to the Archery Hall of Fame in 1972.

What Is a Traditional Bowhunter

By Mike Schlegel

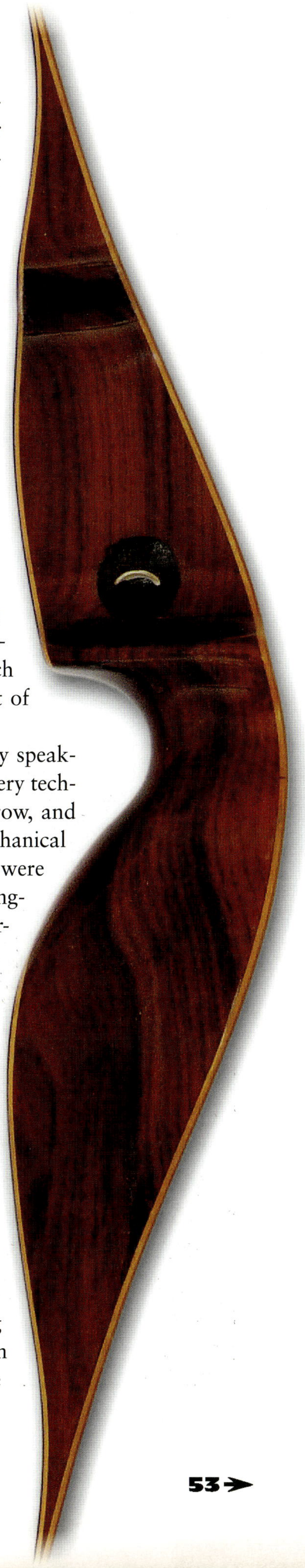

Seems to me there's an unnecessary amount of friction and stereotyping concerning "traditional" bowhunters. One of the major hunting magazines recently said that longbow shooters were "fanatical." The article accused them of trying to restrict hunting opportunity for compound bow users. Elsewhere, the most frequently used synonyms I've seen are "selfish and arrogant purists," and "ethical policemen." Currently, anyone hunting with a selfbow, longbow or recurve bow is labeled a "traditionalist." This being the case, I'd argue that all bowhunters were traditionalists prior to the 1970s, when commercial availability of the compound bow changed things forever.

However, were they truly traditionalists in the eyes of the sport's bowhunting pioneers? I think not. The vast majority did not shoot selfbows or longbows; they did not make their own bows, bowstrings, arm guards, quivers, tabs/gloves, arrows or broadheads. Rather, they took advantage of modern manufacturing and technology that blossomed after WWII. Really, since the beginning of time, man has had an I-can-make-it-better mindset, handheld killing/hunting implements included. Parallel with this is the I-want-the-newest-and-best mindset. Thus, it is no surprise that each generation moves further from the "traditional" equipment of earlier generations.

The transformation of bowhunting equipment, generally speaking, has been quite recent. Prior to the compound bow, archery technology centered on materials used in making the bow, arrow, and broadhead rather than the current emphasis on the mechanical advantage and the cadre of "necessary" accessories. Arrows were wood, fiberglass, or aluminum. Technology regarding the longbow and recurve bows focused on lamination material, fiberglass, glues, plus riser and limb design. They basically all had the same mechanical advantage; the farther the arrow is drawn, the greater the draw weight. Bow selection was based upon how much draw weight the prospective bowhunter could effectively draw while mastering the basic shooting skills. Consequently, the number of bowhunters was limited, and comparatively speaking, there were not many bowhunters nationwide.

By the mid-1950s, most states offered bow-only hunting seasons. This, coupled with the commercial availability of bows, bowhunting equipment, and more leisure time, resulted in an increased interest in bowhunting. There was skepticism among hard-core rifle hunters and many wildlife managers regarding the effectiveness of the bow and arrow for hunting big game. The Pope and Young Club was very instrumental in documenting the effectiveness of bowhunting equipment to state wildlife management agencies.

▼▼▼

Digressing a little, I feel the need to point out that initial bowhunting opportunities were very limited, predicated upon the premise that there would be few bowhunters and their success rate would be very low. Initially this proved true. However, as the availability of bowhunting equipment became widespread, the popularity of bowhunting grew and opportunity was enhanced. Increased bowhunting popularity resulted in the formation of local, state, and national archery and bowhunting organizations. Another spinoff of the increased interest in bowhunting was the formation of very successful archery and bowhunting equipment manufacturing companies, marketing campaigns, and bowhunting magazines. Consequently, pressure was applied to state wildlife management agencies for additional hunting opportunity, in terms of season length, either-sex bag limits, and species diversity.

As the various bowhunting equipment manufacturers began competing for the market, innovation and technology began to redefine bowhunting equipment and bowhunting in general. In concept, the compound bow, as a hunting weapon, surfaced in the late 1960s. Initially there was resistance to such a contraption. Enter the forward thinkers and engineers; soon the compound bow was off and running. Today the compound bow dominates the bowhunting world. Initially, the mechanical advantage of the compound bow was not that great. Today, however, it is very significant and largely responsible for the ever increasing number of bowhunters nationwide.

The following two tables display the progression of bowhunting opportunity for deer and elk hunting, plus elk bowhunter numbers and harvest data for my home state of Idaho. Table 1 is self-explanatory, displaying the number of game management units available for bow-only hunting and the number of the collective available hunting days. Table 2 tracks elk bowhunter numbers, harvest, success rates, days hunted per elk harvested, and the proportion of six-point or greater bulls in the harvest, as well as the percentage of elk the bowhunters kill in relationship to the total elk harvest by general season hunters. This data does not include controlled hunt data, as basically all bowhunting is over-the-counter. Hunter numbers peaked in 1993, the year prior to the implementation of mandatory bowhunter education. Although the days hunted per elk harvested is influenced by several variables, the elevated values for 1985, 1988, 1989, 1993, and 1997 are strongly correlated to severe conditions the previous winter. The bowhunter harvest, as a proportion of the statewide general season harvest, has increased to the mid- and high-twenty percent range.

Table 1. IDAHO DEER & ELK BOWHUNTING OPPORTUNITY, AN OVERVIEW

	DEER		ELK	
YEAR	NUMBER UNITS*	NUMBER DAYS	NUMBER UNITS*	NUMBER DAYS
1952	1	13	1	15
1955	7	130	4	44
1960	33	762	16	384
1965	26	586	16	416
1970	35	904	18	464
1975	44	1,276	29	802
1980	52	1,512	41	1,195
1985	55	1,481	40	1,232
1989	81	2,382	60	1,732
1990	79	2,412	60	1,732
1991	81A	2,447	60B	1,647
1997	79C	2,326	63D	1,636
2000	83	3,099	70	2,507
2012	101A	3,179	83E	2,400

A=4 UNITS 2 POINT AND/OR ANTLERLESS ONLY
B=5 UNITS SPIKE AND/OR ANTLERLESS ONLY
C=4 UNITS 2 POINT AND/OR ANTLERLESS ; 30 UNITS BUCK-ONLY
D=6 UNITS SPIKE AND/OR ANTLERLESS ONLY
E=11 UNITS BULL-ONLY
*SOME UNITS HAVE MULTIPLE HUNT PERIODS

▼▼▼

Prior to the compound bow, it is my contention most bowhunters were attracted to bowhunting for the challenge it afforded, plus a desire to extend their hunting time, as there were no "either-or" seasons. Bowhunter numbers remained relatively low due to the time commitment involved in becoming and staying proficient with a selfbow, longbow, and/or recurve bow. Thus, there was a mindset that centered on challenge and dedication (*challenge*) relative to the use of a short range weapon and its degree of difficulty in mastering, plus *dedication* to becoming and staying proficient with the bow and arrow. As such, shooting the bow became more than just a seasonal activity. In short, bowhunting evolved into "a way of life" for many.

Fast forwarding to 2013, the number of bowhunters nationwide is most likely at an all-time high. The recreational opportunities and equipment technology available to us are now unprecedented. Coinciding with this increase in bowhunters is the changing lifestyle and philosophy of our society. To generalize, we have evolved into an "instant self-gratification, gadget oriented" society. Understand I am not saying this is bad; however, this shift in mindset is, in my opinion, a major reason for the stereotyping and polarization within the bowhunting community.

Another major philosophical difference between a traditional bowhunter and the compound bowhunter is a concern regarding the technology of bowhunting equipment, be it bows, arrows, broadheads, or supplementary equipment, plus the addictive effect this technology will have on future hunting opportunity. This concern seems to run deeper in bowhunters that use "traditional" equipment versus those using "modern" equipment. Adding to this philosophical difference is the generational spread within the bowhunting community. Again, our society is very gadget oriented and gadget dependent. This in itself is neither good nor bad; it is just reality. In my opinion, however, at some point this ever-encroaching technology is going to result in reduced bowhunting opportunity. In 2003, I attended a planning workshop sponsored by the Idaho Department of Fish and Game. Of fifty-eight goals identified in the *"Fish and Game for Fishing and Hunting"* issue, the following goal was voted most important: *"Be more pro-active in limiting mechanical and technological advances in hunting."*

The technology issue seems to be more of an issue in the West. Many states south and east of the Rockies are trying to kill more deer by whatever means available. Meanwhile, in the western states wildlife managers are trying to balance struggling elk and mule deer populations with hunter demand to maintain over-the-counter hunting opportunities. Following are quotes from former Montana Fish, Wildlife, and Parks commissioners:

➤ "I am concerned about advancing technological aids which are designed to make the hunt easier or the kill quicker. I am concerned because hunting is not about saving time, not about killing. Instead, hunting emphasizes personal skill and perseverance while enjoying and understanding the natural world. Hunting ought to be fair competition between man and quarry, not competition between hunters. Investing in technology gives one hunter a leg up on others; soon we may all need the latest gadget, just to compete. Many hunters understand the liberal season is based upon the idea archery has little impact on game populations. If technology markedly improves archers' effectiveness, the seasons

Table 2. IDAHO ELK BOWHUNTING DATA 1975-2011

YEAR	HUNTERS*	HARVEST	% SUCCESS	DAYS HUNTED	DAYS TO HARVEST	% OF GENERAL HARVEST**
1975	3,100	100	3	NO DATA	NO DATA	1
1976	3,050	100	3	NO DATA	NO DATA	3
1977	3,100	100	3	NO DATA	NO DATA	2
1978	4,500	200	4	NO DATA	NO DATA	3
1979	6,200	170	3	38,000	224	3
1980	6,800	300	4	44,600	149	4
1981	8,100	300	4	52,600	175	3
1982	13,900	400	3	62,400	156	5
1983	8,100	300	4	53,000	177	3
1984	9,500	740	7	63,000	85	7
1985	8,200	500	6	57,000	114	5
1986	11,500	900	8	80,000	89	9
1987	14,700	1,300	9	112,800	87	12
1988	20,300	1,300	6	152,400	117	10
1989	22,900	1,500	7	167,000	111	11
1990	20,400	1,800	9	158,000	88	16
1991	21,600	1,900	9	161,900	85	15
1992	26,000	2,600	10	196,000	75	21
1993	26,700	1,900	7	194,000	102	17
1994	20,000	1,900	10	149,800	79	14
1995	21,600	1,800	8	161,400	90	15
1996	24,600	1,900	8	187,400	99	17
1997	17,000	1,200	7	131,000	109	15
1998	15,100	1,700	11	123,500	73	16
1999	15,000	1,600	11	124,000	78	18
2000	NO DATA	1,500	NO DATA	NO DATA	NO DATA	NO DATA
2001	15,100	2,100	14	105,300	50	21
2002	17,400	2,000	12	136,800	68	23
2003	17,500	2,200	13	132,200	60	20
2004	19,400	2,400	12	150,700	63	16
2005	18,300	2,600	14	143,800	55	16
2006	19,400	2,500	13	NO DATA	NO DATA	19
2007	19,600	2,600	13	155,600	60	20
2008	22,500	2,600	12	183,200	70	24
2009	19,800	2,700	14	151,000	56	26
2010	19,700	2,600	13	151,000	58	22
2011	19,800	2,500	13	153,600	61	26

*ROUNDED TO NEAREST 100 **EXCLUDES DRAW HUNT HARVEST DATA

will have to be more restrictive."

➛ "We seem to be constantly behind the technology curve and it is my desire to put rules in place to keep bowhunting primitive."

➛ "Everytime we set a season, we have a new gadget to contend with. Every time I deal with one, another one shows up. That is one of the issues we (commissioners) get tired of dealing with on a piecemeal basis."

➛ "My opinion on archery is that the generous seasons were given to bowhunters because of their primitive status. Modern bows, arrows, and sights are a far cry from where they were fifteen years ago. A line on technology must be drawn if bowhunters want to maintain liberal seasons."

▼▼▼

In spite of increases in the ranks of bowhunters, hunter numbers are declining nationwide. In addition to causing major funding issues for wildlife management agencies, this decline also tips the scale heavier to the non-hunter/anti-hunter side. Agency decision makers are not the only ones following the technological advances in hunting equipment. At some point in time the non-hunting group, as well as wildlife managers, are bound to question some of this technology. Is shooting big game at 1,000 yards really hunting? Are mechanical waterfowl decoys ethical? Do muzzleloader hunters really need a stand-alone season when elk kills over 700 yards have been documented? Unfortunately, the envelope will have been pushed too far before an unacceptable tolerance level is reached by the decision makers and the non-hunting public.

So what about my original question, "What is a traditional bowhunter?"

In my opinion, it's one's mindset regarding why one bowhunts; it is not determined by the equipment used. Rather, it is one's personal goal(s) and more importantly the challenges chosen and decisions made to achieve those goals. That being said, regardless of whether the preference is a selfbow, longbow, recurve bow or compound bow, all are bowhunters! And regardless of the bow used, there are those that push the envelope regarding fair chase and ethical behavior.

To quote Aldo Leopold, "*Ethical behavior is doing the right thing when no one else is watching – even when doing the wrong thing is legal.*" Self imposed pressure and peer pressures are major factors that contribute to poor decisions among bowhunters. All too often, in magazine or web postings, one sees the following: "I couldn't get closer than 80 yards, so I decided to take the shot." The bottom line is you and you alone know what motivates your decision making and you are responsible and accountable for those decisions.

Another Leopold quote: *"A peculiar virtue in wildlife ethics is that the hunter ordinarily has no gallery to applaud or disapprove of his conduct. Whatever his acts, they are dictated by his conscience."* Again, the decision making process is consistent, regardless of the equipment used. Your mindset governing why you bowhunt, the equipment you use, and your hunting expectations will dictate your hunting ethics. So it is time to stop the bickering between the "traditional" and compound bowhunters. We need to work together to maintain and preserve bowhunting opportunity! We need to focus on the positive aspects of the sport of hunting by promoting fair chase and ethical hunter behavior!

In the meantime, and until someone coins a better term, those choosing to hunt with a selfbow, longbow, and recurve will be labeled "traditionalists." Please know we traditional bowhunters are not selfish, arrogant, purists, or ethical police. We are simply passionate about fair chase, ethical hunter behavior, and the future and well being of bowhunting!

Idaho bowhunter Mike Schlegel is the current Pope and Young Club Conservation Chairman. He has served on the P&Y Board of Directors since 2004 and is a retired wildlife professional.

Hunting Hurt

By Len Cardinale

What a beautiful goat he is!

THE pressure is building. Shadowing my guide through the scrubby mountain brush festooned with occasional wisps of goat hair, I move as quietly as possible through the last bit of stunted cover above the rocky dropoff.

No mistakes now, buddy boy. Concentrate. Keep your eyes on Jerry. We're a two-man team. He's picking the way. He knows just what to do. Pay attention. Do the shooting. Be ready. Focus.

Motioning me to stop, Jerry Creyke creeps to the edge of the cliff and peeks over. I know Jerry feels we are above and close to the mountain goat we'd watched bed down through the spotting scope hours earlier from a mile and a half away. I am amazed that he can be so certain. I'd lost track of where the billy was hundreds of yards back.

Turning, Jerry glances at me and wordlessly indicates we need to move ahead a bit. Slowly, quietly, my heart hammering from the exertion and excitement, I follow. Then Jerry risks another careful peek. Nothing. We edge even higher. He repeats his peek-a-boo performance, backs off slowly and moves closer to the mountain crest. An August crosswind whips at us, disguising any unintentional sounds we make.

I ignore the growing tension. My mind is clear, senses sharp. I know the stalk's moment of truth is drawing nearer with each careful step we take. A lifetime of bowhunting has given me complete confidence in myself and my hunting tackle. I feel no inadequacy, only an anxiousness to discover how things are going to turn out.

We're working well together. Jerry knows it and I know it. It's as if we've been stalking together this way for years. We've blended. We've merged. We're a team, like it's supposed to be.

Again Jerry crawls to the edge, peeks over, and pushes back. Only this time he motions me to join him. I know it's time.

But still I refuse to rush it, picking my steps carefully, taking it slow and easy. Then I am kneeling beside Jerry who leans close and whispers, "He's below us to the left at about twenty-five yards. He's still bedded."

We silently swap hunched positions. I nock an arrow, ready at last. I feel the summer wind whipping at my camouflage clothing.

"Put your left foot right there," Jerry whispers, pointing. "I'll hold your belt in the back. Lean out and shoot when you're ready.

Lean out! There is nothing but empty mountain air beyond the rocky ledge. And it isn't until I stand, placing my boot exactly where Jerry had indicated, that I understand the impossibility of the situation without following my guide's instructions to the letter. The goat is bedded in a pile of loose scree beneath the ledge. There is simply no way to stand, look over the edge of the overhang, draw and shoot without leaning out into space.

Then I feel Jerry's strong hands grasp the back of my wide leather belt. He braces himself. This is a time of trust, not doubt. And my guide has my complete trust. Only a day earlier I'd frozen crossing a rock ledge above a slope that dropped hundreds of feet straight down. Seeing my plight, Jerry immediately moved to help, reassuring and steadying me. Now, without a second thought, I lean out into space. Oddly, at this particular moment I realize that my knee isn't hurting anymore.

▼▼▼

My knee. It had hurt so much in June I'd been limping around like an old man instead of a healthy fifty-year-old bowhunter who planned to work long and hard to get into top physical shape for the rigors of a high country British Columbia goat hunt.

My knee problem actually began in New Jersey in January while playing racquetball with my daughter, Maria. I'd never played before and the acute lateral movement the game demands began to cause me lingering pain after only three or four court outings. Soon the persistent pain was interfering with the normal calisthenics, running, and swimming I usually do when preparing for my bowhunts.

I've always been a strong believer in being in good physical shape to cope with the physical demands often required when seriously hunting big game animals on their home turf. I learned long ago that there is a world of difference between enjoying a bowhunt when you can handle its physical demands and the drudgery it soon becomes when you must force yourself, exhausted, through sheer willpower alone, to face the challenges of any hard hunting situation.

Initially I had hoped that sessions on the racquetball court would give me increased leg strength. Bad mistake. My doctor examined me and said I'd likely torn muscles or strained knee ligaments. He added that I'd probably need surgery to repair the damage. No way before my goat hunt! Rest and cortisone shots were my personal prescription for pre-hunt recovery. And it had worked – to a point. My knee felt fine despite days of climbing. It was the rest of me that ached after not being able to property train for this high country hunt.

Playing hurt is a choice all athletes face at some point in their careers. It comes with being totally committed to a sport, with being unwilling to give in to pain. Bowhunters are no different, really. Much of what we do is quite physical, from using our muscles to draw and shoot our bows to walking to and from our hunting area. Some hunts in rugged terrain are especially hard on our bodies. Add a nagging injury and things can get complicated. We can opt to concede defeat and bow out. Or by playing – or hunting – hurt, we can stubbornly persist in our pursuit of game despite minor aches and pains.

Of course no one should continue in the face of debilitating or life-threatening warning signs. No animal is worth a fatal heart attack or stroke. And we need to realize that there is a discernable distinction between determination and a careless disregard for personal well being. What I'm talking about here is the ability – the gritty willingness – to tolerate pain and continue an effort that may take place far from home and most likely is of particular personal value only to the hunter. Most often pride, effort, and considerable money are involved in the ultimate decision.

Speaking personally, I believe that hunting hurt is part of the total commitment we make to our sport. At times it's simply part of bowhunting. Pulled muscles, leg cramps, blisters, cuts, sprains, and strains can and do occur when we push ourselves. This is especially true as our bodies age.

Ideally, we should work into any conditioning program slowly and well in advance of any departure date, strengthening legs and lungs over the course of many months. The plan is to build endurance. And in truth our entire daily routine – diet, exercise, work, sleep – should combine to create a healthy lifestyle. Unfortunately, all too few of us stay in shape year-round. We know we should but we don't. And once we start to prepare ourselves physically for some upcoming hunt, we do what we can and hope for the best. When a minor injury enters the picture, we are faced with the decision of what to do.

In my own case an aching knee wasn't enough to stop me from heading for goat country. Any hesitation to tackle a mountain was overcome by my strong desire to bowhunt goats and sheep at least once before I go. How bad we want something has a lot to do with what we're willing to endure. And I'll admit that I was lucky. In August I first flew to Terrace, British Columbia, and then caught a float plane to the Creyke brothers' camp at Todagea Lake. My knee held up well and I finally found myself leaning out over a steep shale cliff with only my guide's grip preventing me from tumbling into space.

▼▼▼

I'm at full draw, the wind strong in my face. The billy sees me but does not scramble up and away. He simply cranes his neck, staring at me as if in disbelief that I am there – so close. I'm aware of his thick, dark horns, but I concentrate and release.

My aluminum arrow smacks the shale low, skittering harmlessly past him. He stands and looks after the clattering shaft. By the time I fit a second arrow on the string, he is looking up at me again. Jerry's grip is steady. I concentrate hard, settling on a small spot on his side. This time the arrow angles through his vitals. The goat makes his way out of his bed, heading down and away at a fast walk.

Jerry pulls me in. Goats are tough, even when fatally hit, and we watch as he makes his way some two hundred yards to our right when he beds in the center of a vast shale slide. We follow. The shifting shale underfoot makes walking difficult. We ride the loose shale down to where he is and I send an insurance arrow through his chest. Convulsively, he kicks and slides down the mountain in a small shale avalanche, coming to rest far below. Again we follow.

What a beautiful goat he is! Perhaps two hundred and fifty pounds and in prime shape, the billy wears a short summer coat and thick horns that later would score forty-six and six-eighths Pope and Young inches. I am thrilled and unashamedly give Jerry a big hug, offering him my heartfelt thanks.

We spend the next two days in camp carefully fleshing out the trophy and thoroughly salting the hide. We then cut and hang the meat properly. We are relaxed and unhurried through the process, telling and retelling the events of our hunt. There is both a mutual respect for the great animal and a sincere appreciation for a successful hunt. That night the goat meat seems unusually delicious and our camp especially happy. It is one shining moment in a bowhunting career spanning more than thirty years that I wouldn't have missed for anything in the world.

The author's memorable goat hunt took place in 1990, and this story was first published in a 1995 issue of *Bowhunter* magazine. Pope and Young Senior Member Len Cardinale ran Butts and Bows Archery in Belleville, New Jersey, for decades and has hunted extensively across North America for more than 50 years. He served on the P&Y Board of Directors (1976-1978) and was elected to the Archery Hall of Fame in 2005.

Coming Full Circle

By Doug Clayton

IT wasn't just the cool, crisp morning that had deer running everywhere; this was November 11, 1994, prime time for the whitetail deer rut in Iowa. I'd climbed into my stand a little before daylight, and as soon as I could see there'd been almost constant deer activity – does running from bucks, bucks cruising for does, and smaller bucks running from bigger bucks.

The sharp snap of a twig off to the left pulled my attention away from the eight-pointer that had just walked into range. The finger tension on the string of my Schafer Silvertip recurve bow relaxed a bit as I slowly turned my head towards the sound. "Oh, geez!" A very nice ten-point buck had just crested the ridge, following a trail that would take him to the scrape fifteen yards in front of my stand. As I watched him saunter closer, my peripheral vision detected the smaller buck still walking toward me until he disappeared somewhere beneath my stand.

With a fairly steady northwest wind, I wasn't concerned the smaller buck would scent me, but I couldn't see him so I didn't know if he had continued on or had stopped somewhere behind me. Would he alert the bigger buck that something was wrong if I started to draw my bow? The trunk of the oak tree that held my stand was too large to see around, so even trying to look where he went was useless.

As the ten-pointer came to the scrape, he raised his nose and nuzzled the small overhead branch. After a few seconds, he put his nose down and turned attention to the scrape. While watching this buck, I'd also been listening for the smaller buck somewhere behind me because I hadn't heard anything since he'd disappeared from view. With no conscious thought on my part, I started to draw my bow, focusing on a spot just behind his shoulder. An instant later the arrow was on its way!

▼▼▼

My path to that moment was anything but what you would call normal for a hunter, especially a bowhunter. To start with, I didn't come from a hunting/

fishing/outdoor family. Even with that "handicap" I always loved the outdoors, playing in the fields near where my family lived while in Michigan and then Missouri. Growing up, I spent the summer months at the family cottage in Lake Orion, Michigan, swimming in Long Lake and roaming the nearby woods. A picture from 1957 shows me with my first bow – the kind where a suction cup-tipped arrow was shot through a hole in the bow's handle.

"a little boy shooting a toy recurve bow in 1957"

Several years later my first real bow was a thirty-pound Browning Nomad that I received for Christmas. My dad and I took it to a park with an archery range in downtown St. Louis. I thought I was shooting the bow okay until a gentleman who'd been shooting near us came over and explained nicely that we'd strung the bow backwards. With the bow properly strung, that thirty pounds sure felt a lot more than that!

Not trying to date myself too much, the next bow I got was a forty-five pound Ben Pearson Cougar purchased with S&H Green Stamps. I shot this bow in our backyard on and off through high school, but that was the extent of my shooting a bow until after I graduated from college. That's when I started working for the then Iowa Conservation Commission as an Assistant Park Ranger at Viking Lake State Park in southwest Iowa.

Within a month of being hired, I purchased my first bow, a four-wheel Bear Alaskan compound. Looking back, it's probably a miracle I kept shooting a bow. I'd ordered the bow with the weight range of sixty to seventy pounds. When the bow arrived my first few attempts at pulling it back were futile, to say the least. I didn't know that I could reduce the draw weight, and as I became stronger, slowly increase it. So I never did reduce it from seventy pounds. I just kept shooting the bow until I could draw and shoot it with no problem.

Both the Park Ranger I worked with and the district forester were also bowhunters, so many lunch hours were spent shooting our bows at the park. I took my first deer, a doe, with that bow on the last day of the season in November 1979.

The following years found me shooting several other compounds, but as time went on I found myself not having as much fun as when I first started shooting. By this time I'd switched from Parks to Law Enforcement and was the State Conservation Officer (Game Warden) assigned to Pottawattamie County, living in Council Bluffs.

When I moved to Council Bluffs, I joined a recently formed local bow club called the Pottawattamie Bowhunters. Quite a few of us regularly attended club and other bow shoots in the area and afterwards would head back to our range and shoot some more. One of the members was shooting a Fred Asbell Bighorn bow. During shooting sessions, I'd first shoot an arrow with my compound and then borrow his Bighorn for a second shot at a target.

After about a year of this, I made the plunge to buy a Bighorn. In January 1989, I called up Bighorn and ordered a sixty-four inch takedown with sixty-five pound limbs, knowing the bow took about nine months to get. During this time, I was a partner in a bow shop with two other friends. Someone had left an older sixty-five pound Howard Hill longbow at the shop, and I thought I'd shoot it in the interim to learn to shoot instinctively. After about a week or so of this practice, I started wondering if I'd made a good choice in going traditional. The hand shock from that longbow was tearing up my bow arm.

Fast forward to the beginning of September, roughly two weeks before four of us were flying to Prince of Wales Island, Alaska, for a nine-day spot-and-stalk black bear hunt. That's when my Bighorn bow arrived. I opened the package with mixed emotions, assembling the bow and proceeding to our indoor range to see how bad its hand shock was. One shot. Two shots. By the third shot a very big smile came to my face. There was zero hand shock, just a smooth release with my arrow going just where I was looking!

For the next two weeks, I practiced with the Bighorn in between customers and every evening after closing the archery shop. Thinking back, I believe I was shooting close to two hundred arrows each day. When we got to Alaska, I shot a couple of practice arrows each day before heading out to hunt bruins. On the next to last day, I got my bear with a friend taping the whole episode.

Two days after returning to Iowa, I was in our archery shop taking my first practice shot when the string exploded! I knew what a brace height was and had twisted the bow's string to get it at the proper height. What I didn't know was it was a plain Dacron string, not Flemish, and you don't twist those strings. Another lesson learned.

For the next couple of years I shot my Bighorn and enjoyed every minute of it. But, like with many bowhunters, one bow is never enough, so I started looking around at others. One bow really stood out and that was the Silvertip, made by Paul Schafer. Even back in the early 1990s, Paul was a bowhunting legend, I read all the stories I could find on him.

In the summer of 1991, my wife and I headed back to Montana's Glacier

National Park for vacation. Conveniently, Paul Schafer lived not very far from where we were staying, so I gave him a call. He said to come over, mentioning he'd just gotten in some new bow brochures.

The one thing that stands out, besides seeing some of his mounts in his living room, was the price of his bow. They were just about double what most other custom bows cost. So it took some contemplation on my part, but when the next summer rolled around and I knew we were heading back to Glacier for a third year in a row, I called Paul and said I'd like to order a bow. During our conversation, I mentioned how I liked my Bighorn handle, and he said bring it over to his place so he could take measurements.

The bow was ordered in early July. Thinking it would take about nine months (like my Bighorn did), I didn't expect it until early next spring. Well, it didn't take that long at all. One day in mid-September, I was shooting my Bighorn in the backyard when a UPS truck pulled up. I didn't remember ordering anything that would require a UPS delivery until I saw the return address on the box.

I took the package into the house, assembled the bow, and went out to fling a few shafts. It was a very good thing my first couple of shots were about ten yards. I could barely hit the 3-D deer target! What it boiled down to was Schafer's bow design. With the same poundage as my Bighorn, the Silvertip was that much faster and I couldn't shoot my eighty-pound spined cedar arrows from it. Since Iowa's deer season opened in just a couple of weeks, I decided to hold off shooting the Silvertip until the following year. And after Iowa's deer season ended, I started searching for cedar arrows that would fly from my Silvertip. That was not easy since it turned out I needed ninety-pound spined arrows. Eventually, I did find a source.

Sometime in early 1993 the National Rifle Association had a white-tailed deer show that was touring the country, and one of the stops was in Omaha. Several friends and I decided to go, and after getting to the show we started walking around, checking at the various vendors. All of a sudden I heard my name being called. Searching for the source, I saw Gene Wensel standing up at his booth. After we shook hands Gene asked me if I'd heard what happened to Schafer, since he knew I shot a Silvertip. I said no and was told that Schafer had been killed in a skiing accident in January.

I didn't get a deer with Schafer's bow that fall, but during some scouting after the close of the season I came across numerous scrapes on an east-west running ridge and found a perfect spot to hang a treestand. The following year this became my favorite stand. Though I sat in another location, the new ridge produced the most action by far. And this is the stand I was in at the beginning of this story.

▼▼▼

As I released the arrow, I saw it hit just behind the buck's right front shoulder and disappear. The ten-pointer instantly took off to the east along one of the deer trails. Then I saw him take a hard left, heading down the ridge to the north and vanishing from sight. Taking a deep breath, I sat down. Several other deer in the area had bounded off, too, including the eight-point that walked under my stand and out of sight. Searching the ground through my binoculars, I could see my arrow sticking up. It was coated a bright red.

Sitting and reconstructing what had just transpired, a very big smile formed under my balaclava. While I normally wait at least thirty minutes before starting to track my deer, I knew where my arrow hit and there was absolutely no doubt in my mind the deer was already down. And it was only 9:30!

Climbing down, I walked over to the scrape, picked up my blood-soaked arrow, and started walking along the trail. The blood was easy to follow, growing in volume as I headed down the ridge. I hadn't gone much more than fifty yards before I saw him piled up next to a small tree.

Kneeling next to the buck, I ran my fingers over his hide and stared at his rack. Not until then did I see that his right ear had a large slit at the top, the tip of his left beam was missing about two inches, and a round knot about the size of a golf ball cut in half bulged out from his lower left jaw – evidence of fighting other bucks. Of all my thoughts and emotions, two were foremost in mind. One was how lucky I was to have taken such a great deer. Second was how sad it was that Paul Schafer was no longer with us.

After the required sixty-day drying period, my white-tailed deer was officially scored and entered into the Pope and Young Records at 141-3/8. Reflecting on the many roads that led to this deer, I could see a completed circle ... from a little boy shooting a toy recurve bow in 1957, to beginning to bowhunt with a compound bow, to finally coming full circle back to a recurve bow. For me it just felt right!

Doug Clayton, a P&Y Senior Member from Council Bluffs, Iowa, currently serves as a P&Y Director, is a board member of the National Bowhunter Education Foundation, and is enjoying his retirement from the Iowa Department of Natural Resources.

Basic Broadhead Collecting

By Glenn Hisey

OVERHEARD at a local bow shoot: "Didn't you shoot a Snuffer broadhead a few years ago?"

"Yeah"

"You did! Would you be willing to trade one off? I have a really neat-looking Browning Serpentine that I would be willing to trade you for one of your Snuffer heads."

"Okay. I've got one to spare."

"Oh, by the way, you still shoot a Magnus broadhead, don't you?"

"Yup, that's the one."

"I'll trade a Zwickey Eskimo for one of your Magnus heads."

▼▼▼

And just like that, it begins. A bowhunter starts a new hobby – broadhead collecting.

Collecting broadheads is fun, rewarding, and meshes well with being a bowhunter. You'll be pleasantly surprised, as it won't take long until you've assembled a collection of a dozen or more broadheads. In fact, look around at home and you may find several different kinds of broadheads that you've used at various times since first hunting with a bow.

Broadheads should be split into two major categories – glue-on heads and screw-in heads. Glue-on heads are designed to be put onto a wood shaft or a screw-in adapter with hot melt glue. Screw-in heads have an adapter with threads permanently affixed to the broadhead and are designed for use on aluminum or carbon shafts.

The older glue-on heads vary in design more than the newer screw-in heads, and in my opinion, make a more attractive display. If you're mainly interested in growing your collection in number of heads, you may want to include modern broadheads as they're easier to find.

Another avenue of collecting is fish-points and fish-heads. A wide variety of heads have been manufactured specifically for bowfishing and collecting, and many of these will make an impressive display.

▼▼▼

If you're new to broadhead collecting, I suggest two things to help you get started. First, find a mentor. Most broadhead collectors are willing to help anyone who's interested in getting started. I can remember sitting at the same table as Joe St. Charles at the Pope and Young Club Convention in Traverse City, Michigan, in 1995. Joe reached in his pocket and pulled out a Case Kiska with a .35 caliber ferrule and handed it to me. "Do you have one of these?" Joe asked. When I told him I didn't, he gave the old broadhead to me. That was my first special broadhead and it started my collection. Over the years, Lamont Granger was my mentor. I called Lamont for advice and broadhead identification many times. His knowledge and willingness to share was immeasurable. Find an experienced collector that you can call on to answer your questions and give you advice.

The second thing in getting started is to join the American Broadhead Collector's Club (ABCC). This is an organization where members are devoted to collecting broadheads and fishpoints. As a member, you receive a quarterly newsletter, an annual membership list of other collectors, an annual Official Master List of Broadheads, and an invitation to the ABCC Annual meeting. This is a good way to contact and/or meet other collectors. Call Greg Schwehr, ABCC Membership Chairman/Treasurer at (414) 463-8685 for an application form or email him at gregory.schwehr@yahoo.com.

▼▼▼

Now, a little history ...

One of the oldest commercially made broadheads was a Peck and Snyder broadhead and was listed for sale in their 1878 catalog.

Will Thompson, of the historic Thompson brothers fame, was one of the first to use a bullet jacket as a ferrule for a broadhead. This was in the late 1890s to 1900.

Captain Jack Hoefer made several different broadheads as early as 1910.

In 1923, California By-Products started manufacturing a Pope-style broadhead. They were the first widely accepted commercial head of the twentieth century. Art Young and Saxton Pope designed these heads for California By-Products. Prior to their African trip, Art Young and Saxton Pope personally made each of their own bows, arrows, and broadheads. However, they did not know what to expect in Africa. They each took six of their own bows and one hundred arrows already fletched with broadheads. In addition, they took arrow shafts, feathers and broadheads for 2,000 more arrows. They didn't have time to hand-cut and file all these broadheads. This is the underlying reason they convinced California By-Products to start making broadheads. Most were tubing ferrule, though some had a machined ferrule.

In 1926, Roy Case enters the broadhead market with a Pre-Kiska barbed head. Not too long after that, many more broadhead manufacturers produced heads to capture a share of the new market.

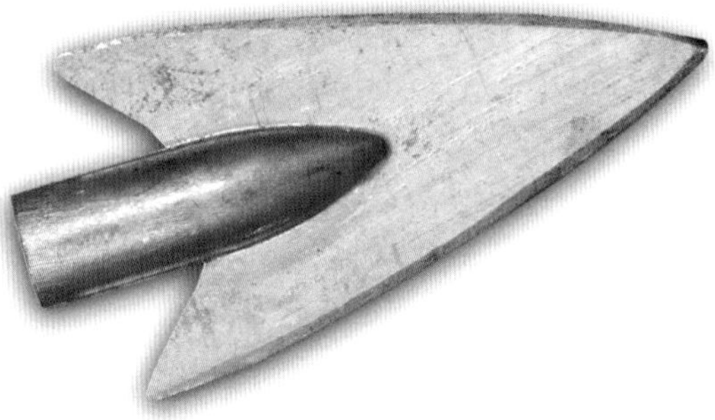

If you start with collecting glue-on broadheads, I suggest you consider the following list of 20 different manufactured heads as a goal to strive to obtain.

Ace
Bear Razorhead
Ben Pearson Skeleton Ferrule
Ben Pearson Deadhead
Ben Pearson Switchblade
Bod-Kin
Browning Serpentine
Case
Gus Adkins
Goshawk
Herters
Hilbre
Hi-Precision
Hills Hornet
Howard Hill
MA-3 Blade
Snuffer
Magnus
Wasp
Zwickey

Now, after assembling a few broadheads, I suggest that you consider building a display case to show off your collection. A display case hung in your trophy room will augment your display of mounts and your collection of bows and arrows.

▼▼▼

Here's a step-by-step instruction on building a basic display case:

Step 1 – Cut a piece of half-inch plywood to size depending on how many broadheads you have to display. For about two dozen heads, I suggest a twenty-four by twenty-four-inch size although this can be adjusted to any size you desire. This will be the back of the broadhead case.

Step 2 – Cut a piece of cloth about two inches larger each way than your piece of plywood cut in Step 1. A red or burnt orange velvet or felt cloth always makes an attractive background. Lay the cloth down on a flat surface with the face of the cloth facing down. Then lay the plywood in the center of the cloth. Pull the edges of the cloth up over the plywood and staple around the edges of the cloth.

Step 3 – Cut two pieces of pre-finished one-by-two-inch board the same length as the top and bottom of the plywood you cut for the back of the case. I prefer these boards to be oak, however, the kind of wood can be your preference.

Step 4 – Cut two more pieces of pre-finished one-by-two-inch board the length of the sides of the plywood back, less one and one-half inches.

Step 5 – The boards you cut in Steps 3 and 4 can be assembled together as a window frame with two screws at each corner. If you used oak, you'll need to pre-drill the screw holes so as not to split the wood.

Step 6 – With the window frame placed on a flat surface, lay the plywood on top of the frame with the cloth side down and fasten the plywood back to the frame around the edges with screws.

Step 7 – Next, turn the case over with the face side up. Cut pieces of three-quarter-inch screen door slat or one-quarter by three-quarter-inch molding to fit horizontally in the case. The number of rows depends on how many broadheads you have. The top row should be down at least five inches from the top frame. Each additional row should be at least five inches below the slat above it. That should allow four rows in this twenty-four by twenty-four-inch example. These slats need to be fastened to the back by short screws.

Step 8 – Install arrow clips on the slats that were put on in the previous step. These clips can be obtained from most traditional archery shops or through 3 Rivers Archery or Kustom King. If you space them one and

three-quarter inches on center, you'll get twelve on each row.

Step 9 – If you're displaying glue-on broadheads, they can be put on a short piece of wood shaft material. If you are displaying screw-on broadheads, then a short piece of aluminum shaft is needed for each head. If my math is correct, you should have spaces for forty-eight broadheads. After putting the broadheads on these shafts, place them in the arrow clips with the broadhead up. It's good to write the identification of the broadhead on each shaft.

If your collection is primarily screw-in heads, an option to Steps 7-8-9 would be to cut pieces of three-quarter by three-quarter-inch molding to fit horizontally in the case. As before, the top row should be down at least five inches from the top. Each additional row should be at least five inches below

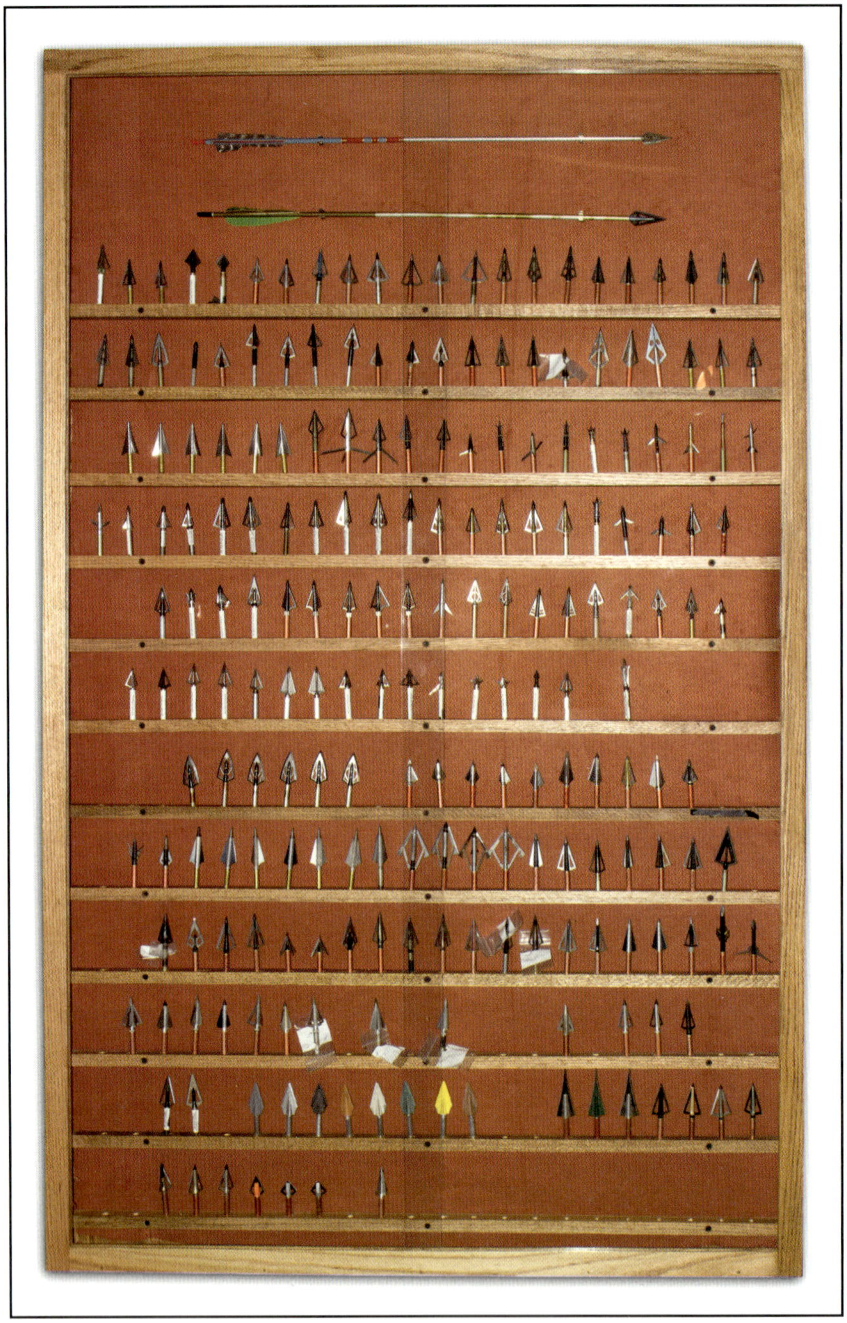

the molding above it. Before installing these moldings, drill a thirteen/sixty-fourth-inch hole vertically down through the center of the molding. These holes can be one and three-quarter-inch on center. Then, fasten these moldings to the back with short screws. The screw-in heads can be placed in these holes that you just drilled, and the glue-on heads can be set on a screw-in adapter that you can get from any archery shop. This way, you would not need the pieces of shaft or the arrow clips.

Step 10 – The cover can be plexi-glass. Cut the plexi-glass to fit the outside of the frame, and screw to the frame. Be sure to drill holes in the plexi-glass slightly larger than the screws you will be using so as not to crack the plexi-glass. The cover could be made out of a piece of window glass as well. If so, the window glass should be cut three-eighth of an inch larger both ways than the inside of the frame. Before installing the pre-finished one by two-inch board in Step 3 and Step 4, cut a quarter-inch deep groove slightly wider than the thickness of the window glass approximately one-quarter-inch in from the outside edge of the frame. That way, when the frames are in place the groove you made line up all the way around the inside of the case. Then, in Step 4, leave one side off until your case is fastened to the wall, your broadheads are clipped in place or set in place, and the window glass is slid in. At this time, the fourth side of the frame can be put in place and two screws in the end of the top and bottom will hold it in place.

I suggest you organize your collection alphabetically by the name of the manufacturer. Maybe have Ace heads first and Zwickey heads last. The heads will be easier to find if they are in a certain order. Also, you should identify your broadheads. Write the name, date and other important inf ormation on the peg. Or you could put the information on a label that is affixed to the peg so the viewer can learn the history of each broadhead. If you opt for the three-quarter-inch molding with holes drilled in it, the labels could be affixed to the molding just in front of the broadhead being described.

Now it's time to stand back and admire your very own broadhead display!

Glenn Hisey is the current Pope and Young Club First Vice President. He also serves as Director of P&Y Records and oversees the P&Y/St. Charles Museum in Chatfield, Minnesota. Glenn has been an active member of the Club's Board of Directors since 1988.

Glenn Hisey was appointed P&Y Museum Director in 2004 when the Club's museum was moved from Glenn St. Charles' Northwest Archery in Seattle, Washington, to the Club's Headquarters in Chatfield, Minnesota. Hisey knows and enjoys the history of archery and bowhunting. As Director of the P&Y Records Office, he also is very knowledgeable about the North American big game species.

Learning the collection of bows and broadheads that were moved from Seattle was a problem, however. Hisey started out shooting Herter's broadheads, then switched to Bear Razorheads and then to Zwickeys. Learning to identify all the other broadheads was a challenge. He joined the American Broadhead Collector's Club as a member. With the help of the book, Broadheads, by Wade Phillips, and the help of other collectors like Lamont Granger, Ron Sherer, Clark Gallup, and others, he has learned quite a bit. The Club's collection of broadheads in the museum is the largest collection for public viewing anywhere in the world.

The Pinnacle of Bowhunting

By Marv Clyncke

Marv with the Dall's sheep he took in 1998

USING traditional archery equipment to hunt the great rams of North America is without a doubt the most difficult challenge any bowhunter can accept. Back in 1975 when the first Pope and Young record book was published, only twenty-two rams were listed. By 2007, when the Club's *A Traditional Journey* record book was printed, not even one hundred fifty total entries had been accepted during more than three decades. And while a handful of additional sheep entries are recorded each year, it remains obvious that bighorns, desert bighorns, Dall's, and Stone's sheep are super tough species to bowhunt.

I well remember a long flight to Alaska in 1971 to bowhunt the beautiful, snowy white Dall rams. At that time maybe a dozen bowhunters – including Club namesake Art Young and the legendary Fred Bear – had tagged a Dall's sheep. In truth, very few bowhunters even contemplated pursuing these rams. But I'd dreamed of hunting Dall's and Stone's sheep since coming down with a severe case of wild sheep fever in 1960 when I first bowhunted Colorado bighorns. Yet my initial Alaska adventure was unsuccessful, in part because my guide knew nothing about hunting bows and their limitations. Additionally, I later learned the 1970s-era Alaskan sheep herds had been terribly depleted, making finding and tagging a ram even more difficult. Right then I vowed to do my research, work hard, and return to Alaska better prepared.

Even after I found an outfitter who was a Pope and Young Club Regular Member and knowledgeable about bowhunting, my 10-day Alaskan sheep hunts in 1974, 1976, and 1978 all ended with me flying home to Colorado empty handed. There I continued to draw bighorn permits, as I had throughout the 1960s when I was the only bowhunter chasing Centennial State sheep. By 1972, I'd convinced the Colorado Game and Fish Department to set aside an archery-only sheep hunting area, despite the fact the state's rifle hunting success was only ten percent. Incidentally, '73 is the same year Steve Gorr took the state's first bighorn using archery tackle. Two more rams were arrowed in 1974 and by 1977 five archery-only areas were open. Also, my long personal quest (one hundred and one days of bowhunting Colorado bighorns over a seventeen-year period) ended when I finally killed a good ram.

Bowhunting sheep was catching on at last, but the challenge of pursuing rams proved too much for many bowhunters. I can't count the number of hunters who have drawn coveted sheep permits, yet never went hunting when faced with the reality of a phycially demanding hunt with long odds against being successful. In fact, I discovered in most hunting areas only about half of permit-holders ever ventured into the typically rugged sheep country.

▼▼▼

Undoubtedly, physical and mental conditioning are the keys to preparing for a successful mountain sheep hunt. My standard advice: *Get into the best possible shape months before opening day. And don't forget to prepare mentally for what might be the most demanding and frustrating adventure any bowhunter can face.* Being unprepared is a sure-fire recipe for failure! Here's an perfect example of what I'm talking about:

Joe (not his real name) drew a sheep permit and made arrangements to arrive in Colorado a couple of days early to get acclimated to the high altitude, since most hunt areas are 10,500 feet and up. I'd located more than forty rams but opted to have Joe start hunting in an area I really liked. It contained a group of half a dozen rams. We glassed the sheep the evening before opening morning and were on the mountain at first light, primed to begin. Unfortunately, another hunter soon climbed into view and spooked the small band of sheep. "No problem," I told Joe, figuring they wouldn't go far and we could still make a stalk.

We watched two rams make their way into a steep area of slide rock where they laid down behind a large boulder. This was perfect! They were only four hundred yards above the basin floor, and I pointed Joe to a small draw east of the sheep, telling him to work his way up and get into position. I'd signal when he needed to turn toward the large rock hiding the bedded rams. I cautioned him to be extra careful and repeatedly check with me for directions. Excited, Joe took off up the steep slope, pausing every twenty yards or so to catch his breath. Meanwhile, I'm thinking to myself this simple stalk was going to result in a fifteen to twenty yard shot Honestly, I was as pumped as ol' Joe.

Finally, Joe stopped and sat down just below where I planned to signal him to turn toward the big boulder. *Good*, I thought, *he needs to catch his breath and calm down before closing in for the shot.* But minutes dragged on and Joe just continued to sit there. *What the devil was going on?* I couldn't believe my eyes when Joe finally stood up, turned back, and headed down the draw to where I waited, shocked and steaming mad. Joe sheepishly explained he was exhausted. He simply couldn't take another uphill step. *Unbelievable!* Joe was neither physically nor mentally ready to hunt sheep. He'd just blown the chance to tag a bighorn and went home deflated and defeated.

Keep Joe in mind when sweating to get in shape for sheep hunting.

▼▼▼

Other memorable sheep hunts I recall were much more satisfying. In 1993, I was finally able to go on my first Stone's sheep bowhunt with a good friend named Dave who'd bought into a good ram area near Scoop Lake in British Columbia. This ruggedly beautiful province is one of only two places with huntable Stone's sheep populations.

Marv with the first-ever Stone's sheep taken with a longbow

Due to business demands in Colorado, I only had seven days to hunt, and on the very first day I blew a shot at a good ram. Turns out that was the only chance I got, and as I always tell ram hunters, "If you miss your first shot, don't expect a second." There are exceptions but that's par for the course with sheep; hunters rarely get off more than one arrow and you have to make it count. To say I was bummed was an understatement because I figured I'd never hunt Stones again because of the expense involved. But the following June my friend Dave called when one of his hunters cancelled, asking if I could take the hunt. Ecstatic, I jumped at the chance and set aside fifteen days in mid-August.

We took horses and pack animals into the Cassiar Mountains, fulfilling one of my lifetime dreams. Stories I'd read about the B. C. sheep hunts of Fred Bear, Jack O'Connor, Warren Page, and others fired my imagination. Fact is, I'd talked with Papa Bear about his Stone's sheep adventure in 1958 when he'd become the first bowhunter to tag a Stone and claim a World's Record for that species. Fred frequently said that was one of his best hunts and best shots – ever.

I can't recall how many rams I stalked during my two weeks in the Cassiars, but I do know I never released an arrow until the last hour of the final day. That's when I stalked three rams and became the first bowhunter to take a Stone's sheep with a longbow. Talk about elation! I could have flown back to the States without a plane!

Turning my attention back to Dall's, I teamed up with hunting buddy Duke Prentup to return to Alaska in 1997. We booked the hunt with traditional bowhunter and outfitter Tom Shankster, an excellent bush pilot. That made our adventure extra special. On August 20 we flew into the Kuskokwim River country, which is upriver from where Fred Bear had tagged a Dall's ram of his own.

After rain and fog delays, we finally spotted two good rams in a steep, narrow canyon along the Post River. One of the sheep was a giant and we made an unsuccessful stalk before heading back to camp for some food and a good night's sleep. Tom, Duke, and I were back on the mountain early the following day. We tried another stalk from a different direction and this time managed to ease within a dozen yards of the white rams.

In my excitement, I flat out missed that huge ram, but seconds later Duke's arrow zipped through him. He made it no more than ten yards before tipping over a cliff that fell away several thousand feet down the rocky mountainside. Elated but concerned about the recovery, we hurried to rejoin Tom and decide how best to work our way down to Duke's ram. Three long, hard hours later the three of us stood admiring the huge sheep. Amazingly, though the fall had battered the ram's entire body, his horns were intact except for a small chunk missing from one side, proof positive that sheep horns are incredibly tough. And the tale of the tape revealed Duke's ram was forty-three and a half inches on the curl, making it the largest any of Tom's sheep hunters had tagged.

With one exceptional full-curl Dall's ram on the ground, it was my turn to try to make up for the missed shot. Early the following morning, once the fog lifted, we were back on the mountain in a spot where we'd glassed three rams in some incredibly rugged terrain the previous day. One was an exceptional full-curl ram, and it didn't take us long to spot the trio high above us. As Tom and I began our climb, Duke settled down with his binoculars to watch the unfolding stalk.

When we had closed within two hundred yards of the bedded sheep, Tom hung back as I eased closer. Finally, I nocked an arrow and cautiously peeked over the edge of the cliff, the second biggest ram was tucked in on a grassy ledge not ten yards directly below. But I couldn't see his body to take the shot! *What to do?* Mind racing, heart thudding, careful to keep the wind in my face, I slipped around and down onto the ledge. The smallest ram – a half curl – was bedded there among the boulders, but I couldn't see the giant full-curl and faced a real dilemma. One nearby ram was a little guy but the other was a legal full-curl. Both were unaware I was standing mere yards away, but there was no way to slip past for a possible shot at the largest ram. I'd likely spook both of them if I tried. Considering all the days I'd spent hunting these beautiful white sheep over the years, there was simply no way I wanted to risk blowing a perfect opportunity at near point-blank bow range, but I had to try.

Slowly, quietly, I inched past the small ram bedded mere yards away. That's when the full-curl stood up and stretched. He was maybe ten yards away. That's when I drew, releasing just as he stepped away to look over his shoulder. Unbelievably, he turned again an instant before the arrow got there, and I missed the chip shot. *Not again!* I thought. Amazingly, the sheep didn't run. The small ram rose but stood looking at me. The full-curl merely stared at where my arrow hit just in front of him. Almost instantly I had a second arrow on the string, and this time the shaft struck just behind the front shoulder He lunged ahead and launched himself into space over the end of the grassy ledge.

Even before I made it to the spot he'd disappeared, I could hear Duke shouting from far below and knew my ram was down. And there he was, lodged against a small boulder at the lip of a sheer drop-off of more than a thousand feet. I couldn't stop grinning. Two full curl Dall's rams in two days. With longbows! For once the long flight back home to Colorado wouldn't seem quite so long.

Thoughts, Tips and Tactics

As I write these words, I've been on a total of fifty-nine sheep hunts, including sixteen of my own. I've teamed up with seventeen other sheep hunters and helped them tag rams. Also, I've gladly offered where-to-hunt and how-to-hunt advice to many others who've solicited assistance. I routinely put in for desert sheep permits in three states, including Colorado, but have yet to draw a tag. But even when I can't be hunting mountain sheep, I'm spending countless hours glassing and photographing rams. So if it sounds like I'm addicted with "sheep fever," it's because I am.

Following are a few things I've learned about sheep and hunting them:

➛ Once rams are spotted during a hunt, it's usually best to watch them feeding or moving toward their daytime beds. After they're bedded down and resting, you can plan and make your stalk.

➛ Wear the best mountain boots you can find and break them in well ahead of the hunt. To avoid blisters, wear a pair of silk or poly socks against your bare feet and then slip on a pair of Smartwool socks. Be sure to trim your toenails.

➛ Layer your hunting clothes so you can shed them as the day heats up and add them when it cools down. Always carry quality lightweight raingear and plenty of water. Have a water purification filter in your pack, along with extra bowstrings, trail mix, or energy bars. Consider wearing a quality packframe for packing meat, the cape, head, and horns.

➛ Take along the best binoculars and spotting scope you can afford. Good 10x40 or 10x42s are ideal and a spotting scope is essential for long-range glassing and determining trophy quality.

➛ Always carry a camera and take plenty of photos. Sheep country is a special place and preserving memories is a great way to relive the experience.

➛ It's a good idea to have a backup bow, just in case. The typically rugged terrain can take a toll on hunting gear. Be prepared for any eventuality. I always pack a basic first aid kit and medications that include antibiotics to take in the event of illness.

➛ Finally, it's worth another mention of the need for proper physical and mental preparation. A positive attitude perfectly complements physical conditioning and lays the groundwork for sheep hunting success.

Colorado bowhunter Marv Clyncke is addicted to sheep hunting. A former Pope and Young Club Board member, he was the first bowhunter to take a Stone's ram with a longbow. Marv also was the Colorado Bowhunters Association's first Hall of Fame inductee (1974). He and wife Judy are both P&Y Senior Members and serve as bowhunting ambassadors by constantly promoting our sport.

Honeymoon Hunt

By John Gardner

Kathleen and John with John's Honeymoon Sheep

I WANTED *to cry! The legal ram I'd been watching approach finally stood only forty yards from where I crouched on the windswept Alaskan mountain. When he turned his head to stare off in the other direction, I finally rolled onto my knees, raised up, and lifted my bow to take the perfect shot. But something was terribly wrong. Both of my wait-cramped legs, which had been twisted awkwardly beneath me while the ram eased closer, were sound asleep and felt as dead as tree stumps. Worse, as blood began flowing back into my lower limbs, my head began to spin like a runaway carnival ride.*

Lightheaded, fighting a sudden wooziness, I desperately clung to consciousness until the mountain stopped moving. That's when I quickly drew, anchored, aimed instinctively, and released. My broadhead-tipped arrow cut only empty air just above the white ram's broad back. Amazingly, he didn't move, so I hurried to nock a second arrow and take a follow-up shot. Another near miss!

Meanwhile, still unalarmed by not seeing or hearing either arrow zip past, the full-curl ram simply turned and walked away toward a distant band of sheep that appeared across the open ridge. My pounding heart sank. As already mentioned, it was almost enough to make a grown man break down and start bawling. After all, my honeymoon dream hunt held the tacit promise of becoming the happiest experience of a veteran sheep hunter and his outdoors-loving bride's lifetime. Missing a trophy ram twice wasn't exactly what I had in mind.

▼▼▼

Really, it all began at a Colorado beer league hockey game when a teammate buddy invited me to join him and some other friends at a local sports bar. We'd gather to cheer on the Avalanche during an NHL playoff game, and my pal offhandedly mentioned inviting a pretty young lady that he thought I should meet. Curious, I agreed and during the subsequent gathering found myself talking to a lovely blonde woman named Kathleen. After she'd reassured me that she wasn't a vegetarian, I asked her to join me at a wildlife conservation banquet the following evening. She accepted and the rest, as they so often say, is history.

Turns out Kathleen was an active, athletic outdoors girl and an experienced mountain climber. She eagerly joined me on several sheep hunts I guided in Colorado. In truth, she'd already climbed more 14,000-foot peaks than I ever will. So sixteen months after meeting, we were making wedding and honeymoon plans that included a fishing trip to Alaska, topped off by a backpack Dall's sheep hunt. Only a week after our 2004 wedding in Silverton, Colorado, Kathleen and I were catching fish in Wasilla and preparing for flights to an Alaska Range sheep camp with veteran outfitter Tom Shankster. Talk about a great way to start our marriage!

Arriving at base camp, Tom broke the news he was temporarily a guide short due to a missed flight. That meant Paul, a rifle hunter, would be my hunting partner until the second guide arrived. Our shared guide would be Zach, an experienced goat guide from the Alaskan Peninsula. We'd be hunting a 3,000 foot tall vertical mountain towering high above us. When we gathered our gear for the climb, Zach candidly confided being a bit anxious to be guiding sheep hunters for the first time. I told him not to worry because I'd guided lots of sheep hunters and we were going to have fun on our hunt. And I meant it.

Kathleen elected to stay behind and join Tom on a flight around his hunting area. It was not only the first day of our hunt but her birthday as well. Turns out she made a good choice and enjoyed one of the most memorable flights of her life. While we made the hours-long climb, she and Tom winged around the west side of Mount McKinley, spotting lots of caribou, sheep, moose, and even one large sow grizzly with a cub.

Meanwhile, after finally topping out, we broke through the alders and followed a saddle to where we could see the backside of the mountain. Instantly we spotted two bands of rams, one to the west and another to the east. The western group, including a borderline legal ram and a no-doubt shooter, seemed to be headed our way, and so I urged Zach to sit and watch to see what they'd do.For more than an hour Zach, Paul, and I sat on the ridge in a gusting wind until the sheep dropped out of sight not far below us. Thinking they may be bedding out of the wind, we started down into a boulder field after them.

Suddenly the smaller ram popped over the edge of a pocket not twenty-five yards away. We froze. "Don't move!" I hissed as the sheep stood facing us, obviously trying to figure out what we were. Then the large ram stepped into view, but I didn't want to risk taking a front-facing shot. Slowly I raised my bow in case he turned broadside, but a ewe walked in front of the big ram and stopped, blocking any chance of a shot. The faceoff lasted for what seemed minutes until the sheep wheeled and dropped out of sight.

Zach and Kathleen taking a break from the long hours climbing

"Paul," I yelled, figuring the rams were already running well out of bow range. "Get up there and shoot!"

Paul scrambled up and dropped the big ram with a great shot, watching his thirty-seven-inch trophy tumble several hundred yards down a steep shale slope. It was 2 a.m. by the time we stumbled into base camp under heavy loads of meat topped by the ram's cape, head, and horns. One day, one legal ram. Now it was my turn.

▼▼▼

While still on the mountain, I'd spotted a small pool of water shimmering in a distant basin. This gave me an idea and the next morning I asked Tom if we could set up a spike camp there while I hunted for my sheep. "Go for it," was my outfitter's response. And we did.

Shouldering our packs shortly after noon, Kathleen, Zach, and I headed back up the mountain. Long hours later, we broke through the alders and dropped into the basin where I'd seen the water. It was 10 p.m. by the time we arrived at the small pool of crystal clear water, maybe three feet wide by ten feet long, surrounded by a patch of tall, lush grass. It was a beautiful spot. And as if an omen of good things to come, a solitary ram watched from the rocks above as we pitched our tents and cooked dinner before turning in.

John and Zach, with John's ram

The following morning Zach and I climbed the ridge to try and locate the lone ram we'd seen while setting up camp. Kathleen would head southwest and scout for sheep before circling around to meet up with us later.

As Zach and I moved along the ridge toward the east, I thought I saw a ram's horn jutting from behind a boulder below. Raising my binocs, I confirmed the horn was attached to a very good ram bedded a couple hundred yards downslope in a perfect spot for a stalk. Zach watched as I dropped low, moved up a small draw, approaching from behind with the wind in my face. I was only forty yards away when the ram suddenly stood, turned around, and plopped down facing me. Pinned in the open with my legs twisted under me, I couldn't risk moving. So for more than an hour I lay unmoving while doing my best impression of a rock. And finally, when the ram eventually rose to offer me a shot, I ended up missing twice exactly as described in the opening paragraphs.

Climbing to rejoin Zach, I was heartened to hear he'd been watching a group of five rams on the other side of the ridge. One was a real trophy with horns approaching forty inches. But rather than push things and try another stalk, I elected to patiently wait and see what would happen. Eventually the strategy paid off when the big ram worked into position below the spot where I'd missed the ram. And when he started up toward that same spot, I grabbed my bow, handed Zach the video camera, and started off down the ridge.

Getting into position just in front of the big sheep, I slipped on my headnet and shooting glove, nocked an arrow, and peeked over the ridgeline. The ram was close but now standing downwind, nose in the air. But just as he turned quartering away to run, I drew and released. My arrow merged with the big sheep as he galloped away, I could see him slowing to a walk before disappearing from sight. I scrambled after him, determined to keep him in view.

The fishing couldn't be beat!

My mind was racing. The arrow appeared to have hit his hip amd a lot of shaft showed as he ran away. Had I messed up again? And when I couldn't locate him immediately, I started to worry. Hurrying ahead, I came across a very good blood trail, which gave me renewed hope. And when I rounded a large rock outcropping, I saw him tip over and begin to roll until a horn tip dug into the sidehill dirt and stopped his skid. It was over.

A check showed my arrow actually had buried to its fletching, angling forward. As he ran, the arrow backed out and appeared to show a lack of penetration instead of a lethal hit. My lifelong dream hunt for a trophy class Dall's ram was over on only the second day of an unforgettable adventure. Although I didn't know it at the time, Zach had used my camera to videotape the stalk, shot, hit, and trailing job, giving me a special bonus that I'll always treasure.

Kathleen was nearly as excited as I was when I shared the good news back in our spike camp. After celebrating my success, we packed out the following day and caught an early flight back to Alaska's version of "civilization." More fishing, sightseeing, and fun followed as we enjoyed our honeymoon adventure before heading home to Colorado where Kathleen planned to take up bowhunting and helped make more memories in the seasons ahead.

Every guy should be so lucky to have a blind date turn out to be a dream bride who eagerly shares his outdoor adventures with each passing season.

Admittedly addicted sheep hunter John Gardner is the current Pope and Young Club Second Vice President. He and his wife, Kathleen, make their home in Durango, Colorado, where John is a master taxidermist and guides sheep and goat hunters each summer and fall.

Hunting Barren Ground Bulls

By Jay St. Charles

EACH fall the magnificent Barren Ground Caribou range mostly at or above the North American tree line. A hunter used to timber or agricultural country can be somewhat overwhelmed by the apparent lack of cover in this wide open landscape. In favorable population conditions, if the herds are present, it's not at all unusual to spot animals at the very limits of one's optics in all directions. An early discovery for the newbie can be just how fast a grazing herd moves across open land. Distances on an unfamiliar landscape can themselves be deceiving and this, combined with an unfathomably mobile quarry, often throws one's first attempts at ambush in the trash heap. Those caribou appearing an hour away are more likely no more than ten minutes out!

Low cover is available and often in abundance. For the hunter, it's a matter of being creative with what's at hand, which in itself adds to the challenge. Slight undulations of the ground, tussocks, low berry bushes, low hills, and boulders can be made to do. In this terrain, close range hunters seem to have little to their advantage, although in reality this is not true. Of major significance is the fact that a caribou's magnificent antlers are often several feet taller than their eyes and their movement – or lack thereof – telegraphing well what an individual or a small herd might be up to. We have another edge in part because the caribou are a wilderness animal, existing in a fully natural state. Caribou live in daily company with native apex predators such as wolves, grizzly bears, and wolverines. They are instinctively prepared when these dangers appear. They don't, however, seem so sure as to what to make of us.

When planning a hunt, note that the barren ground caribou herds which

Photo by Rich Bahl

Guide Morris Martin, Jay St. Charles and his 1996 Bull, MacKay Lake, NWT Canada

migrate long distances are either present or they are not. There is no more lonely landscape or a more empty hunt than to find oneself on a vast expanse of caribou habitat with no animals anywhere. Thus caribou hunting is often a hit or miss proposition. Outfitters usually have a good knowledge of when their land will receive or hold animals. Camps that have ready access to a vast amount of habitat real estate offer a huge advantage in this regard. It may not matter if huge herds are found a full twenty miles from your camp when you can get on a boat and be there in an hour. For this reason some of the best camps are often located on a large lake or navigable river. Again, it doesn't matter how many or where the caribou are if you can't get to them. I remember many years ago an airport concourse in Northern Quebec packed with departing hunters, none of whom had had a successful hunt. It was a very dark spot that afternoon.

Guides

"No more caribou. Caribou all gone!" This announcement from Tom, my Inuit guide, certainly dampened spirits just an hour into the first morning of my late September'84 hunt along Quebec'sTunulik River. Leaning a dent into my boot toe with my recurve tip, I wondered, *Did he mean they'd all gone south for the season or that there were simply none visible on the landscape at this moment?* An inch of overnight snow covered the rocky landscape, the horizon shifting back and forth from miles to less than 100 yards. Tom's sparse English and my total lack of understanding his own language made elaboration impossible. Still Tom was glassing the newly white landscape from our little promontory, so maybe there was still some hope? As days passed, I came to understand something of Tom's quirky humor, as well as his mastery of finding caribou. You've heard the story of a guide who with his naked eye could spot a good bull farther than your binoculars could reach. Well, Tom proved this to be pretty much the truth.

We bowhunters tend to be do-it-yourself hunters. This seems particularly true of those of us who hunt with a simple longbow or recurve. We've learned that it's difficult enough getting oneself within close bow range, let alone any added company. Because of local regulations requiring a guide for all non-residents hunters, universal across Canada, a caribou hunt has often been a bowhunter's first guided hunt. My '84 trip to Northern Quebec was my own first such experience.

Since then, I have enjoyed the experience of traditional bowhunts with a number of guides in caribou country, many excellent in their deep knowledge of the species and of local terrain. This has been particularly true regarding time spent with native northland people on their own traditional hunting grounds. Some stand out, such as the wonderful Jaeb family, Morris Martin, Paul MacKenzie. Lucy Chocolate, Alfred Fish. Archie, Leon, Dennis, Hardy, Peter, Patrick and others whose faces I can remember vividly.

Your new guide will be curious about your traditional equipment and what you consider your capabilities. The common question from a guide has been, "How far will you shoot?" This is a loaded question. Your answer will tell him a little of who and what you are. I've found the best answer to be a confident, "Close, very close." Words of confidence and commitment are generally always accepted and respected. Without saying much more, this relates that you are confident and committed to your craft. He'll also get the message that in this task, getting very close, he will not be of help. His real job is to help get you to where the caribou are and then safely back to camp. The large lakes and rivers of the barrens, where fickle fall weather may turn on a dime, can be treacherous and confusing. Here is where local knowledge is priceless.

Camp

Although our craft of archery dates back farther than we can see, we traditional bowhunters are a distinct minority within general bowhunting ranks. As such, just as the pioneers that went before, we have to be prepared to experience a little prejudice and curiosity from our fellow hunters. There are rifle hunters that accept the idea of the compound bow, but have mixed feelings about a longbow or recurve. There is also a segment of modern compound bowhunters who take issue with simple traditional equipment. Much of this comes from ignorance, although some is of our own making. A segment of our traditional bowhunting community fosters a strong intolerance to modern compound tackle and some of them can be quite vocal and blunt in expressing their beliefs. This does us no good. As we well know, hunting equipment is its own religion. Tolerance or lack thereof can become an issue, especially in the small confines of a wilderness outfitter's caribou lodge, and nowhere else is this tolerance more needed from all parties.

When I sense any questioning of validity or motive, I'll take that opportunity to sincerely explain, "I'm just like you. I'm hunting with this gear simply because it's what I enjoying hunting with, nothing more to it." And I may then use this opportunity to explain how my tackle is in its own way just as "dialed in" as their own. Everyone is in camp for the same reason, to have the best possible hunt. We are always our own best, or worst, ambassadors.

The Hunt

The most common successful hunt scenario is what I like to call stalking to an ambush, the point being to find a way to get in front of a group of moving caribou. To make this happen, it is essential to spot them as far out as possible. As mentioned, they are famous for covering ground very quickly, even when grazing.

As discussed, we humans are not in the caribou lexicon of normal daily events. You'll find that if you are installed on the landscape and stationary when caribou arrive, you may be able to hide in plain sight. Being there and stationary before their arrival may cause them to accept you as a part of the terrain. Numerous times I've been approached and surrounded by caribou. I've had a cow with her calf drip water on me as she drank from a pond inches from me. Once a bull decided to broom his velvet on a bush I was hiding in. In another instance, four bedded bulls I'd been crawling toward, decided to wake from their nap, stand, and feed toward me. Eventually I was surrounded on four sides, the bulls just a few feet away. With caribou, savvy hunters learn to never easily give up on a stalk, even if all appears blown.

A small herd of bachelor bulls traveling together often makes a good situation for a close encounter. Mature bulls, in particular, can be less skittish than their younger counterparts. In small groups they seem to carry an air of confidence, acting as if they believe collectively they are able to deal with whatever might arise. If they happen to have a single much younger bull with them, he may likely be a spoiler, perhaps still remembering the caution of their mother.

Sometimes you'll run across a bull traveling solo or seemingly bedded by himself. This solitary state is not natural to caribou and seeing a lone individual often indicates that this "loner" is ill or injured in some way, unable to keep up with its mates. Lone calves are also rare. For the barren ground caribou, move or be left behind is a matter of life or death.

Large herds, consisting of several hundred to several thousand individuals,

leave a lasting impression. Led by cows, it may appear to be a herd of all cows and calves with no mature bulls present. But with the faint stiringss of rut, bulls may be following directly behind in their own bachelor herd, often a mile or more behind. Staying concealed from the cows and calves, not allowing them to detect your presence, may end in a fine ambush scenario. In any situation it is always in your favor to stay concealed whether you see only cows and calves or not. Your best opportunity will always come when the caribou have no idea that you are around.

Final Thoughts

It's wise to finally invest in the best pair of binoculars you can afford. I've tried a number of combinations and personally favor a pair of high quality, lightweight 10x25s for bowhunting caribou. Such small glasses can easily be stowed in a close fitting chest pocket and then easily retrieved for quick glances. Average caribou terrain is quite open with the lighting generally good enough you'll not be hampered in any way by using this smaller glass.

Shooting will be from on or close to the ground, so it will pay to practice your shots from ground positions. As in all bowhunting, you'll benefit from identifying your shot opportunities well in advance. Having an ability to naturally perform a smooth, low-to-the-ground "snake crawl" to within a few yards of bedded or moving caribou, is well worth the practice. On a good hunt you

Minnesota bow-hunter Vince Scully and his 2007 Central Canada Bull, MacKay Lake, NWT Canada

may get to practice this skill daily!

It's common to lose a full day or more of hunting due to a change in the weather. Recognizing this probability, even under the best of circumstances with lots of caribou in an area, it's a good idea not to waste opportunity. Also, the vast open landscape allows the hunter not only to keep good visual track of the quarry, but also your hunting partner. Most guided barren ground caribou hunts involve a ratio of one guide to two hunters, or if you're on a DIY adventure you may well be hunting as part of a pair. Combine the fact that you never know what a caribou might do next, the following time-tested fair chase hunt technique may make sense. I hope it works out for you as well as it has for me and my partners. We call it "A Team - B Team" and here's how it works:

Taking turns, the "A Team" hunter assumes full and individual control over his movement and strategy in dealing with a group of caribou, just as if hunting by oneself. All "A Team" has to consider is what's in front of him. In contrast, the "B Team" hunter hangs back and observes, keeping close track of both the caribou and the "A Team" hunter and his or her movements.

At some point in time, either "A Team" is successful or the stalk is blown. This may mean that "A Team" will continue hunting, hoping for a second try at the same herd, but usually "the jig will be up," resulting in this particular herd's swift departure. During all this it will have been "B Team's" challenge to guess where these animals might be headed and set up an ambush or some other strategic shooting opportunity. Result: one bundle of caribou, two possible quality opportunities. Many, many times this plan has worked equally in favor of hunter "A" or "B."

A memorable example of this "strategy" dates back to 1992 in the company of longtime traditional bowhunter and Senior Pope and Young Member Jack Joseph of Nebraska. Jack and I were in the Northwest Territories hunting Central Canada Barren Ground caribou along a rocky shore of MacKay Lake. Taking my turn as the "A" guy, I had just completed a long, successful stalk and had started down toward my bull when I heard a voice, "If you hadn't shot him, I was going to!" It was Jack! He'd been nearby all along and I'd never noticed – nor had any caribou. Lesson? It's always good to partner with an exceptional hunter on any exceptional hunt!

Most outfitter camps in North Central Canada and Quebec are mixed camps and it's common to find traditional and compound bowhunters, muzzleloaders, and rifle hunters all sharing the same large camp. In this busy mixed atmosphere, it seems that commitment can be fleeting and I bring this up to provide a note of caution. If it is your goal to harvest a caribou with your traditional bow and arrow, then leave any rifle at home. As mentioned, the landscape that the barren ground caribou inhabit can be daunting at first glance. I've known a number of souls who have gone north with both gun and bow only to find themselves lured to an easy, but in retrospect, quite hollow success.

Remember, whatever and wherever you hunt, if the right opportunity presents itself, you will harvest game with whatever you have in your hand. It's not about the implement but rather about what's going on in the mind of the individual hunter and his or her performance as the ultimate decision maker. You cannot do better than to hunt with the style of tackle that means the most to you personally. This could be no more the truth than when on a traditional bowhunt for a magnificent quarry, conducted in that hunter's paradise, the Barrens.

Jay St. Charles, a son of Pope and Young Club Founder Glenn St. Charles, is a Senior P&Y Member from the Pacific Northwest. He is now serving his initial term on the Club's Board of Directors.

Planning and Dreaming

By G. Fred Asbell

ONE of my favorite memories is of a three-month drop camp Alaskan bowhunt that Bob Pitt and I took years ago. It was quite a deal. We found an outfitter who said we could come bowhunt for a week, or for three months, for the same price, and he didn't care which. He would furnish us a tent and grub and would move us a couple of times if necessary.

A fella can do a lot of planning and dreaming in two years...

It was a tough decision, but three months sounded like the best deal and we went with that. The price was $1000 each, and I probably need not say that $1000 was a lot more money then than it is now. It was a great deal, but it still took me two years to save the money. As we found out, the outfitter didn't know anything about bowhunting. I suppose, he figured he could drop us off and forget about us, believing that we weren't going to kill any animals to speak of using a bow and arrow. Hmmm. Well, we had two years to dream about it, and a fella can do a lot of planning and dreaming in two years.

We were pretty green back then. And truthfully, just about everyone was pretty green back then about bowhunting, particularly bowhunting in a place as exotic and remote as Alaska. We spent two years planning the trip. In looking back at it, there is no question that the planning of the adventure was one of the best parts of it. And, hey, there's nothing wrong with a three- month bowhunt for sure. But two years of anticipation, two years of dreaming, planning, and talking about it – and two years of getting out of bed every morning knowing you were one day closer to Alaska was quite a thing.

There wasn't anyone in Indiana who had ever been to Alaska bowhunting, and the only people I'd ever heard had done it were Fred Bear and Glenn St. Charles, although there were surely others. I was a member of Pope & Young Club even then and I guess I figured that gave me some sort of priority on Glenn St. Charles, because I must have called him with questions two dozen times. Glenn wasn't much of a planner, I guess, at least not like me, because about the only advice he offered was to wear wool, and to cut our pants off right at the top of our boots like the loggers did in Washington State. I knew about wool, and we tried shortening our pants to boot tops and were glad we had the time to try it before heading to Alaska because it quickly gave us very wet boots and cold legs – which wasn't to our liking.

Bob and I spent hundreds of hours together planning every detail of that trip. This was all new stuff to us and to everyone either of us knew. What arrow should we take? Someone told us that we needed at least twelve dozen arrows, and we gave a lot of thought to the cost of that before deciding that wood was the way to go. So we shot wood arrows for a year before some archery shop guy convinced us that all our arrows would warp in Alaska's wet climate and we switched to aluminum. Of course bows were a major concern, and we both jumped bow weights. Bob went from sixty-six pounds to seventy-seven pounds, and I went from fifty-five to sixty-two, and then on up to sixty-five pounds. We also made two trips to American Archery near Chicago and talked to them and shot their bows before ordering. Since the new bows required intense training, we met at the archery club at least twice a week and on weekends, shooting as many arrows as possible in preparation.

I remember Bob calling me and reading aloud a newspaper account of a lost Alaskan hunter who had frozen to death and that a simple windproof shirt/parka would have saved him, according to some authority. So for a while that winter we tested nylon wind shirts as a possibility. Bob got his wife, Barbara, to drive him down the road in zero weather while he stood up in the back of his pickup wearing a wind shirt. The results were inconclusive because his face got too cold and he had to pound on top of the cab to get his wife to pull over. After considerable discussion, we decided that a quality rain suit would serve dual purpose as a wind shirt and as a rain barrier.

We spent most of the two years trying to figure out what the correct footwear might be. We met a fellow at a sports show who had been to Alaska hunting every year for about fifteen or twenty years. He said he always wore leather boots and lost all of his toe nails each year from the wet conditions, which kept his feet perpetually wet. He said he planned each year to find a good pair of waterproof boots, but never found the time and ended up wearing the same boots each year – with the same results. We definitely didn't like the sound of that and tried many different boots including a handmade, very expensive pair, boots that were guaranteed to be waterproof, but were not. We finally came up with the perfect boot in a ten-inch ankle fit uninsulated rubber boot from L.L. Bean that worked perfectly.

And the Alaska planning tales could go on and on...

But the point I want to make is that the planning part of that hunt is what has stayed with me. Sure, the hunt itself was absolutely wonderful; an unbelievable experience. But the planning was a huge part of the enjoyment and made the hunt last for over two years, and I can never forget that. I've been a serious planner and dreamer ever since...

I love the planning part of a hunt. For me, planning a trip can be almost as good as the hunt itself. What bow am I going to shoot? What arrow? What clothes will I need? What is going to be different, and how can I get out ahead of the learning curve? Meaning that I don't want to wait until the next time I go on this hunt to know what to do, what to bring, etc. This isn't an orderly sort of thing. I don't know that I am all that orderly as a person. It's about making the "getting ready" a key part of it all. The planning is the hype that precedes the hunt itself. A well-known bowhunter of my acquaintance always made such a big deal out of being able to go on a hunt with a one-hour notice, and he loved showing up on time with a gym bag sized duffle with his clothing and gear in it. Which is, I guess, quite an accomplishment and should be applauded. That's sure not my style. The hunt is only half as much fun if I can't think about it and plan how it's going to be. Does that make me a dreamer? I guess...

Random Thoughts

Going to Alaska isn't that big a deal anymore, you say? I agree with that. Lots of bowhunters have been to Alaska. But maybe you haven't – and that's the important part, and that makes it a pretty big deal. More than likely you are going to pay more than the $1000 I paid, and that would seem to make it important that you have your fingers on and your mind around all elements of that expenditure, which is where planning becomes more than just a diversion into dreaming. Planning is important financially, and it's particularly important to your success. But, let me say it again – I love it because it expands the hunt and gives me great armchair pleasure.

It wouldn't be unusual for me to do something to make a particular hunt a little different, so maybe I'll have to plan and think about it more. Sometimes that is not even a conscious thought; it just happens. But, in truth I do something a little different on all my hunts, and I think that is simply to put a different twist to them that will require more thinking, dreaming, and planning. Maybe it's a new bow, or trying heavier arrows, or lighter ones, or a different broadhead. But – and this is the key factor – none of it is spur-of-the-moment. It's all carefully (in my mind) thought out and planned.

On a recent black bear hunt in Alberta I went with a plan to do most of my hunting on the ground by taking up a position as far away from the bait as I could and then watching it with binoculars and putting a stalk on anything desirable that showed up. Obviously, that required some planning and probably some amount of scouting after I arrived. Would I sit or stand? Would a portable tree seat be the answer? I would, of course, want to be in a place where I could watch the trails and the bait. It had to be planned as to how much cover I'd need for my stalk and so not every bait would work. Would I clear out a stalking path and go after the animal after it was on the bait or during the coming and going process? And how receptive would the guide be to my hunting in this manner?

Part of it is about adding another challenge to the hunt, and that gets intertwined with adding a twist that requires additional thinking and planning. And besides, it might be a darned good way to shoot a bait-shy bear.

Hunting with a guide is a lot different from going on your own, whether it's an elk hunt, moose hunt, or a whitetail hunt in south Georgia. But you still need to plan. Just because it's a guided hunt doesn't mean you should arrive at your outfitter's door without adequate preparation and planning. Every hunt is different, whether it's a do-it-yourself or guided one. And that's one of the pleasures of planning. When you are bowhunting on your own you know you have to figure it all out, but guided hunts require some thought, too. And there's no question that some fellas walk into them with their eyes closed. One of the problems with guided hunts is that the guide has done it in a certain way for so long that he often assumes you know what and how he does it. And truth be told, you can end up with huge disappointments on guided hunts simply because you never asked the right question about how things are gonna be done. Guided hunts are great, but they require a degree of thought and planning, too.

What Would You Do Different?

Talking to other bowhunters who have been on the same or a similar hunt is a must for me, for several reasons. I'm not talking about checking references on a guided hunt. That's stuff everyone ought to do without anyone telling them. But from the perspective of my need to plan, I want to ask, What would you do different if you could do it again?" That question almost always brings me useful information.

"I'd have brought a better sleeping bag. That weather up there was rough. You need a better tent than we had too..."

"I'd bring a bunch of hard candy...those guys were really skimpy on the sweet stuff..."

""Duct tape? Absolutely!"

"More clothes...I froze to death in those darned open boats. I had no idea there would that much of that stuff."

"I'd figure out something to get my cape and meat back...they had nothing."

"I needed waterproof boots bad. Take moleskin for sure..."

"We did a lot more glassing than I thought we would..."

"I wasn't prepared to do that much walking."

"I'd do more long-distance practice. I never saw an elk under 30 yards, and that's as far as I'm comfortable shooting."

"I'd go earlier. We went too late trying to hit the rut. There were muzzleloaders behind every tree."

And, of course, some of those comments are just how bowhunts go. But some of them could be prevented, at least improved, with a little prior knowledge and planning. On a caribou hunt to the Northwest Territories, I talked beforehand to a fella who had been on an early hunt and got a completely different set of information and suggestions than I did from a bowhunter who had been there on the last hunt, which was three weeks later and was the week I was going. Thank goodness I did that. The early hunt was full of black flies and warm weather conditions, and the other had blowing snow and very long rough boat rides requiring head to toe wind and water protection.

Equipment Lists: A Great Planning Tool

I have been making lists of what I need to take on every bowhunt for years. My lists are by animal species and location, and I have them one on top of the other on a clipboard. They include all the hunting gear, camping gear, clothing, and personal gear I need on that given hunt. I have a list for everything from sheep hunts to rabbit hunts, and there's a different list for whitetails in Alabama than for whitetails in Alberta. And usually there's a different list for a short two or three-day hunt near home over an extended stay in Nebraska. Hunting from my tipi would be completely different from staying in a cabin. I see that I have twenty-nine separate equipment lists for probably a dozen different species. There are three or four each for whitetails and mule deer. (Remember, I've been doing this bowhunting thing for 50 years!) I'll run over the list of items when I first begin planning a particular hunt, in the event that something is special. When I begin the actual packing process, which I enjoy to the maximum because it's officially the beginning of the hunt, I consider everything on the list, although I may not include it in my duffel.

Planning, thinking, and dreaming about hunting and about a specific hunt is part of the ritual of hunting. For me, it sometimes seems every bit as important as the hunt itself. Someone said that it is a ritual peculiar only to man and is not practiced by animals. I'm not sure I totally believe that, having seen sleeping dogs chasing, whining, and sometimes barking, as they relive a past chase, or dreaming of a future one. A true hunter is a breed apart, be it man or beast, and I think we all live to do it when we can, and dream and plan the rest of the time.

Fred Asbell is a successful author, bowhunter, and seminar speaker who served on the Pope and Young Board of Directors from 1978 until 2002, including an 18-year stretch as Club President. "Planning and Dreaming" originally appeared in the Dec/Jan 2010 issue of Traditional Bowhunter. Fred is also a Founder and Past President of the Compton Traditional Archers and a member of the Archery Hall of Fame.

The Records

ALASKAN BROWN BEAR

Minimum Score 20 | *Ursus arctos middendorffi* and certain related subspecies

Score	Greatest Length	Greatest Width	Sex	Area	State/ Province	Hunter's Name	Date	Rank
28 7/16	18 0/16	10 7/16	M	Ursus Cove	AK	Monty Browning	1995	1
28 0/16	17 15/16	10 1/16	M	Wide Bay	AK	Fred Bear	1960	2
27 8/16	16 6/16	11 2/16	M	Grayback Mtn.	AK	Scott Mileur	1997	3
27 5/16	16 10/16	10 11/16	M	Afognak Island	AK	Julian D. Salutregui	2008	4
27 1/16	16 15/16	10 2/16	M	Bear Bay	AK	Fred Bear	1962	5
27 1/16	16 5/16	10 12/16	M	Kodiak Island	AK	Eugene Arndt	2004	5
24 12/16	16 3/16	8 9/16	M	Port Heiden	AK	Gardner Rowell	1995	7
24 1/16	15 4/16	8 13/16	M	Togiak	AK	David "Jake" Powell	2012	8
23 6/16	14 6/16	9 0/16	M	Kodiak Island	AK	Gordon Longville	1961	9
23 4/16	14 11/16	8 9/16	M	Admiralty Island	AK	John R. Thiele	1984	10
22 15/16	14 2/16	8 13/16	F	Sheepcreek-Valdez	AK	Gerald R. Gold	1966	11
22 9/16	14 12/16	7 13/16	M	Doame River	AK	Tom Taylor	2004	12
22 8/16	15 1/16	7 7/16	M	Cinder River	AK	Ken Grosslight	1997	13
22 7/16	14 2/16	8 5/16	F	Merrill Pass	AK	Ralph Ertz	1975	14
22 6/16	14 4/16	8 2/16	M	Kodiak Island	AK	Buddy Watson	1965	15
22 5/16	14 3/16	8 2/16	M	Dice Bay	AK	Thomas M. Basch	2005	16
22 1/16	14 11/16	7 6/16	M	Talkeetna Drainage	AK	Fred Eichler	2006	17
21 10/16	13 14/16	7 12/16	F	Alaska Peninsula	AK	Bill Van Houten	1961	18
21 10/16	14 3/16	7 7/16	M	Ursus Cove	AK	Scott M. Anspaugh	2011	18
21 9/16	14 4/16	7 5/16	M	Talkeetna Drainage	AK	Fred Eichler	2011	20
21 8/16	13 11/16	7 13/16	F	Kiklukh River	AK	Ronald Rockwell	2012	21
21 7/16	13 8/16	7 15/16	F	Fidalgo Bay	AK	Joseph West	1966	22
21 7/16	13 15/16	7 8/16	M	Lynn Canal	AK	Scott J. Leslie	2001	22
21 2/16	13 14/16	7 4/16	M	Anaktuvuk River	AK	Martin Hanson	1958	24
21 0/16	13 13/16	7 3/16	M	Karluk Lake	AK	Gary G. Wall	2003	25
20 14/16	13 10/16	7 4/16	M	Chulitna River	AK	Rick D. Snell	1997	26
20 11/16	13 8/16	7 3/16	M	Yakutat	AK	Chris Parrino	2004	27
20 10/16	13 7/16	7 3/16	M	Cottonwood Bay	AK	R. Todd Beckman	2001	28
20 3/16	13 3/16	7 0/16	F	Port Heiden	AK	Rick Duggan	2001	29
20 1/16	12 14/16	7 3/16	M	Fog Lakes	AK	Dennis L. Lattery	1984	30

Hunter: James Barwick
Species: Black Bear
Score: 19 7/16
Date: 9/10/11
Location: Chilanko River, BC

BLACK BEAR

Minimum Score 18 — *Ursus americanus americanus* and certain related subspecies

Score	Greatest Length	Greatest Width	Sex	Area	State/ Province	Hunter's Name	Date	Rank
21 14/16	13 7/16	8 7/16	M	Prince of Wales Island	AK	Jim Ponciano	1990	1
21 13/16	13 8/16	8 5/16	M	Rusk County	WI	Bob Brunkow	2000	2
21 11/16	13 9/16	8 2/16	M	Nipawin	SAS	Ray Mastel	1974	3
21 11/16	13 13/16	7 14/16	M	Rusk County	WI	Gregory Baneck	1995	3
21 10/16	13 6/16	8 4/16	M	Hudson Bay	SAS	Craig Richardson	1985	5
21 9/16	13 5/16	8 4/16	M	Saddle Hills	ALB	David W. Watson	2003	6
21 7/16	13 5/16	8 2/16	M	Barron County	WI	Jeffery Tomesh	2005	7
21 7/16	13 7/16	8 0/16	M	Saline County	AR	Don Barnett	2008	7
21 6/16	13 6/16	8 0/16	M	Bayfield County	WI	Larry L. Frye	1975	9
21 6/16	13 4/16	8 2/16	M	Fannin County	GA	Stephan Patton	2009	9
21 5/16	13 6/16	7 15/16	M	Wolf Lake	ALB	Thomas G. Lester III	1998	11
21 4/16	13 5/16	7 15/16	M	Cass County	MN	Myles Keller	1980	12
21 2/16	13 2/16	8 0/16	M	Madera County	CA	Clarke Merrill	1963	13
21 2/16	13 6/16	7 12/16	M	Hubbard County	MN	Dean Como	1974	13
21 2/16	13 10/16	7 8/16	M	Langlade County	WI	Mike Steliga	1981	13
21 2/16	13 2/16	8 0/16	M	Gimli	MAN	Dave Bryce	2002	13
21 2/16	13 1/16	8 1/16	M	Suwannee River	MAN	Becky Johnston	2003	13
21 1/16	13 2/16	7 15/16	M	Quetico Provincial Park	ONT	Robert Filbrandt	1981	18
21 1/16	12 13/16	8 4/16	M	Siskiyou County	CA	Jules Pacheco	1987	18
21 0/16	12 13/16	8 3/16	M	Sioux Narrows	ONT	R. B. Cooley	1960	20
21 0/16	13 3/16	7 13/16	M	Langlade County	WI	Michael Steliga	1982	20
21 0/16	12 14/16	8 2/16	M	Lake County	MT	Colin L. Andrews	1990	20
21 0/16	12 14/16	8 2/16	M	Bay Tree	ALB	Sedgwick Bryant Loyd II	1994	20
21 0/16	13 3/16	7 13/16	M	Barron County	WI	Jeffrey P. Tomesh	1995	20
20 15/16	13 4/16	7 11/16	M	Sequoia National Forest	CA	Robert Shilling	1971	25
20 15/16	12 11/16	8 4/16	M	Routt County	CO	Mark A. Chapman	1982	25
20 15/16	13 1/16	7 14/16	M	Carrot River	SAS	Michael J. Windemuller	2011	25
20 14/16	12 8/16	8 6/16	M	Mesa County	CO	Richard A. Schreiber	1973	28
20 14/16	12 9/16	8 5/16	M	Routt County	CO	Lonny Vanatta	1984	28
20 14/16	13 2/16	7 12/16	M	Edmonton	ALB	Bruce Nederveld	1990	28
20 14/16	12 11/16	8 3/16	M	Prince of Wales Island	AK	Robert A. Meister	1992	28
20 14/16	13 2/16	7 12/16	M	Ministikwan Lake	SAS	Brent Maxwell	1992	28
20 14/16	13 2/16	7 12/16	M	Carlton County	MN	Tim Jestus	1999	28
20 13/16	13 4/16	7 9/16	M	Meadow Lake	SAS	D. Mitch Kottas	1988	34
20 13/16	13 0/16	7 13/16	M	Koochiching County	MN	Scott Schultz	2000	34
20 12/16	12 12/16	8 0/16	M	Apache County	AZ	Dr. C. G. Clare	1967	36
20 12/16	12 5/16	8 7/16	M	Emmet County	MI	Hawley H. Rhew	1974	36
20 12/16	12 12/16	8 0/16	M	Douglas County	WI	Robert J. Schmidt	1975	36
20 12/16	12 11/16	8 1/16	M	Prince of Wales Island	AK	Jack Williams	1985	36
20 12/16	12 11/16	8 1/16	M	Ketchikan	AK	Greg Munther	1991	36
20 12/16	12 14/16	7 14/16	M	Preeceville	SAS	James D. Guess	1993	36
20 12/16	12 13/16	7 15/16	M	Marquette County	MI	Darryl D. Ansel	1994	36
20 12/16	13 2/16	7 10/16	M	Apache County	AZ	Edward A. Petersen	1997	36
20 12/16	12 9/16	8 3/16	M	Taylor County	WI	Duane A. Wallner	1998	36
20 12/16	13 0/16	7 12/16	M	Prince of Wales Island	AK	Matt Meneghel	2003	36
20 12/16	12 9/16	8 3/16	M	Mistatim	SAS	Scott L. Koelzer	2005	36
20 11/16	12 13/16	7 14/16	M	Thunder Bay	ONT	Mel Johnson	1974	47
20 11/16	12 13/16	7 14/16	M	Thorne Bay	AK	Steve McCoy	1993	47
20 11/16	12 15/16	7 12/16	M	Carrot River	SAS	Robert C. Farthing	2005	47
20 11/16	13 2/16	7 9/16	M	Tioga County	PA	Marc A. Bialecki	2011	47
20 10/16	13 1/16	7 9/16	M	Rio Blanco County	CO	Frank 'Rit' Heller	1969	51
20 10/16	12 8/16	8 2/16	M	Ravalli County	MT	John C. Locke	1990	51
20 10/16	12 11/16	7 15/16	M	Nairn Township	ONT	Ron Hergott	1990	51
20 10/16	13 6/16	7 4/16	M	Hudson Bay	SAS	Dairl Hicks	1997	51
20 10/16	13 2/16	7 8/16	M	White County	GA	Duane Truslow	1999	51
20 10/16	12 14/16	7 12/16	M	Saddle Hills	ALB	Brent Watson	2005	51
20 9/16	12 12/16	7 13/16	M	Lake County	MN	Art A. Heinze	1970	57
20 9/16	12 12/16	7 13/16	M	Prince of Wales Island	AK	Roy C. Ewen	1973	57
20 9/16	12 7/16	8 2/16	M	Grand County	CO	Curt Lynn	1973	57
20 9/16	12 11/16	7 14/16	M	Aitkin County	MN	Myles Keller	1977	57
20 9/16	13 1/16	7 8/16	M	Durban	MAN	David H. Boland	1985	57
20 9/16	12 15/16	7 10/16	M	McAdam	NBW	David F. Baldwin	1988	57

BLACK BEAR

Minimum Score 18 *Continued*

Score	Greatest Length	Greatest Width	Sex	Area	State/ Province	Hunter's Name	Date	Rank
20 8/16	12 13/16	7 11/16	M	Queen Charlotte Isl.	BC	Peter Halbig	1960	63
20 8/16	12 14/16	7 10/16	M	Shawano County	WI	Bud Wiesman	1974	63
20 8/16	12 14/16	7 10/16	M	Peace River	ALB	Tom Lester	1994	63
20 8/16	12 15/16	7 9/16	M	Mistatim	SAS	Scott L. Koelzer	1998	63
20 8/16	12 11/16	7 13/16	M	Pilot Mtn.	BC	R. Newnham/L. Arnold-Smith	1999	63
20 7/16	13 9/16	6 14/16	M	Nenana	AK	Robert Dunn	1968	68
20 7/16	12 10/16	7 13/16	M	Price County	WI	Bob Eckarot	1974	68
20 7/16	12 14/16	7 9/16	M	Montezuma County	CO	Bryan C. Neeley	1974	68
20 7/16	12 7/16	8 0/16	M	Whiteshell	MAN	Ken Warkentin	1978	68
20 7/16	12 14/16	7 9/16	M	Arran	SAS	Bill Clink	1986	68
20 7/16	12 14/16	7 9/16	M	Valleyview	ALB	Stan Walchuk, Jr.	1986	68
20 7/16	12 11/16	7 12/16	M	Routt County	CO	Bill Grammer	1987	68
20 7/16	12 12/16	7 11/16	M	Iron County	WI	Brian Tessmann	1989	68
20 7/16	12 6/16	8 1/16	M	Whiteswan Lake	SAS	Terry Krahn	1997	68
20 7/16	13 0/16	7 7/16	M	Olha	MAN	Dana M. Draper	1997	68
20 7/16	12 12/16	7 11/16	M	Kitimat	BC	Harvey J. Surina	2002	68
20 7/16	12 11/16	7 12/16	M	Navajo County	AZ	Jim Willems	2002	68
20 7/16	13 1/16	7 6/16	M	Meadow Lake	SAS	Jason Toews	2009	68
20 6/16	12 11/16	7 11/16	M	Tulare County	CA	Quentin M. Boutch	1967	81
20 6/16	12 11/16	7 11/16	M	Garfield County	CO	Steve Bergman	1970	81
20 6/16	12 6/16	8 0/16	M	Saguache County	CO	Ed Wiseman	1975	81
20 6/16	12 12/16	7 10/16	M	Flower Station	ONT	Richard H. Shoup	1977	81
20 6/16	12 9/16	7 13/16	M	Ignace	ONT	Randy J. Tylke	1985	81
20 6/16	12 13/16	7 9/16	M	Wandering River	ALB	Warren Witherspoon	1988	81
20 6/16	12 13/16	7 9/16	M	Prince of Wales Island	AK	Joey Buchanan	1999	81
20 6/16	12 10/16	7 12/16	M	Lake of the Woods County	MN	Shannon E. Thomas	1999	81
20 6/16	12 11/16	7 11/16	M	Clearwater County	ID	Dennis Michael	2001	81
20 6/16	12 10/16	7 12/16	M	Rio Arriba County	NM	Jim Willems	2004	81
20 5/16	12 14/16	7 7/16	M	Shasta County	CA	Harv Ebers	1964	91
20 5/16	12 12/16	7 9/16	M	St. Louis County	MN	Jay Deones	1970	91
20 5/16	12 13/16	7 8/16	M	Buncombe County	NC	Robert T. Austin	1971	91
20 5/16	12 11/16	7 10/16	M	Lemhi County	ID	Richard R. Smith	1977	91
20 5/16	12 5/16	8 0/16	M	Meagher County	MT	Richard M. Campbell	1982	91
20 5/16	12 8/16	7 13/16	M	Archuleta County	CO	Ronald J. Murphy	1985	91
20 5/16	12 10/16	7 11/16	M	Ontonagon County	MI	Dale W. Gray	1986	91
20 5/16	12 9/16	7 12/16	M	Siskiyou County	CA	Bob Jensen	1986	91
20 5/16	12 10/16	7 11/16	M	Prince of Wales Island	AK	Larry D. Jones	1993	91
20 5/16	12 6/16	7 15/16	M	Calling Lake	ALB	Rich McGowan	1994	91
20 5/16	12 10/16	7 11/16	M	Cadillac	QUE	Ron Miller	1995	91
20 5/16	12 11/16	7 10/16	M	Prince of Wales Island	AK	Jim Bauers	1996	91
20 5/16	12 15/16	7 6/16	M	Dolores County	CO	Jerry Parker	1999	91
20 5/16	12 14/16	7 7/16	M	Koochiching County	MN	Steve Schultz	2000	91
20 5/16	12 14/16	7 7/16	M	Las Animas County	CO	Fred Eichler	2000	91
20 4/16	12 6/16	7 14/16	M	Oceana County	MI	William Benson	1967	106
20 4/16	12 11/16	7 9/16	M	Somerset County	ME	Felix Nosewicz	1968	106
20 4/16	12 10/16	7 10/16	M	Mesa County	CO	M. R. James	1971	106
20 4/16	12 12/16	7 8/16	M	Los Alamos County	NM	Kenneth A. Meyer	1971	106
20 4/16	12 12/16	7 8/16	M	Sawyer County	WI	George Geisert	1972	106
20 4/16	12 7/16	7 13/16	M	Madison County	MT	Bob Savage	1977	106
20 4/16	12 7/16	7 13/16	M	Valora	ONT	Elmer R. Luce, Jr.	1981	106
20 4/16	12 8/16	7 12/16	M	Rockingham County	VA	Charles Larry Danner	1984	106
20 4/16	12 12/16	7 8/16	M	Lanark	ONT	Ben Graham	1990	106
20 4/16	12 7/16	7 13/16	M	Madison County	MT	Larry Stackhouse	1992	106
20 4/16	12 8/16	7 12/16	M	Rocky Lake	MAN	Tim Finley	1993	106
20 4/16	12 6/16	7 14/16	M	Clarke Lake	SAS	Steve Byerly	1994	106
20 4/16	13 6/16	6 14/16	M	Athabasca River	ALB	Roger Wintle	1997	106
20 4/16	12 4/16	8 0/16	M	Koochiching County	MN	Larry M. Looman	2001	106
20 4/16	12 9/16	7 11/16	M	Kuiu Island	AK	Nathan L. Andersohn	2002	106
20 3/16	12 13/16	7 6/16	M	Prince Rupert	BC	Frank Huneck	1960	121
20 3/16	12 6/16	7 13/16	M	Ignace	ONT	Jerry Ulrich	1968	121
20 3/16	12 4/16	7 15/16	M	Adams County	ID	Joe Adams	1971	121
20 3/16	12 3/16	8 0/16	M	Boise County	ID	Mark W. Powell	1975	121

Score	Greatest Length	Greatest Width	Sex	Area	State/ Province	Hunter's Name	Date	Rank
20 3/16	12 12/16	7 7/16	M	Smoke Lake	ALB	Kenneth Szgatti	1975	121
20 3/16	12 4/16	7 15/16	M	Schefferville	QUE	Charles L. Buechel, Jr.	1987	121
20 3/16	13 0/16	7 3/16	M	Beltrami County	MN	Caio Cesolini	2000	121
20 3/16	12 7/16	7 12/16	M	Lake Wasekamio	SAS	Richard B. Dyer	2010	121
20 2/16	12 9/16	7 9/16	M	St. Louis County	MN	James Harwood	1966	129
20 2/16	12 9/16	7 9/16	M	St. Louis County	MN	Ron Johnson	1968	129
20 2/16	12 6/16	7 12/16	M	Bayfield County	WI	Clarence J. Biddle	1973	129
20 2/16	12 6/16	7 12/16	M	Iron County	WI	Chuck Ramsay	1973	129
20 2/16	12 6/16	7 12/16	M	Kuiu Island	AK	William F. Burgess	1984	129
20 2/16	12 8/16	7 10/16	M	Nipigon	ONT	Richard Scorzafava	1985	129
20 2/16	12 10/16	7 8/16	M	Latah County	ID	Mick McCullough	1990	129
20 2/16	12 10/16	7 8/16	M	Otero County	NM	Beto Gutierrez	1991	129
20 2/16	12 4/16	7 14/16	M	Square Lake	ALB	Dave Stull	1992	129
20 2/16	12 6/16	7 12/16	M	Peace River	ALB	John Lindell	1992	129
20 2/16	12 2/16	8 0/16	M	Thunder Bay	ONT	Ian Robinson	1995	129
20 2/16	12 6/16	7 12/16	M	Kuiu Island	AK	Dirk Dieterich	2002	129
20 2/16	12 13/16	7 5/16	M	Koochiching County	MN	Tom Wanke	2002	129
20 2/16	12 11/16	7 7/16	M	Ashland County	WI	Darrell "Skip" Peterson	2004	129
20 1/16	12 9/16	7 8/16	M	Shasta County	CA	Robert G. Sinclair	1967	143
20 1/16	12 5/16	7 12/16	M	Nipigon	ONT	Wilfred J. Ritchie, Jr.	1968	143
20 1/16	12 6/16	7 11/16	M	Vilas County	WI	Ben Jones	1972	143
20 1/16	12 10/16	7 7/16	M	Cowlitz County	WA	Smokey Crews	1976	143
20 1/16	12 10/16	7 7/16	M	E. Braintree	MAN	Ed Beamish	1979	143
20 1/16	12 6/16	7 11/16	M	Hudson Bay	SAS	Craig Richardson	1982	143
20 1/16	12 4/16	7 13/16	M	Lost Lake	ONT	Gunter Lemke	1985	143
20 1/16	12 10/16	7 7/16	M	Hudson Bay	SAS	Bill Zahradka	1987	143
20 1/16	12 3/16	7 14/16	M	Williams Lake	BC	Don Davidson	1989	143
20 1/16	12 9/16	7 8/16	M	Prince of Wales Island	AK	Dennis Sturgis, Jr.	1990	143
20 1/16	12 2/16	7 15/16	M	Prince of Wales Island	AK	Don Davidson	1991	143
20 1/16	12 6/16	7 11/16	M	Price County	WI	Dale S. Karch	1994	143
20 0/16	12 10/16	7 6/16	M	Oneida County	WI	Fred Felbab	1964	155
20 0/16	13 4/16	6 12/16	M	Cumberland County	TN	Louis Wix	1970	155
20 0/16	12 4/16	7 12/16	M	Shasta County	CA	Jim Dougherty	1970	155
20 0/16	12 7/16	7 9/16	M	Prince George	BC	Ron F. McKay	1980	155
20 0/16	12 6/16	7 10/16	M	Red Lake	ONT	Donald Schram	1981	155
20 0/16	12 6/16	7 10/16	M	Kenora	ONT	Steven G. Dennis	1990	155
20 0/16	12 0/16	8 0/16	M	Seibert Lake	ALB	Keith Dana	1990	155
20 0/16	12 6/16	7 10/16	M	Prince of Wales Island	AK	Connie Renfro	2001	155
20 0/16	12 9/16	7 7/16	M	Routt County	CO	Gary Wissmueller	2003	155
20 0/16	12 4/16	7 12/16	M	San Juan County	UT	Mark Turner	2006	155
20 0/16	12 11/16	7 5/16	M	Jim Lake	SAS	Michael D. Tofte	2007	155
20 0/16	12 2/16	7 14/16	M	La Ronge	SAS	Wade Schwake	2007	155
20 0/16	12 3/16	7 13/16	M	Gunnison County	CO	Joe R. Vigueria	2011	155
19 15/16	11 15/16	8 0/16	F	Iron County	WI	Robert W. Blair	1967	168
19 15/16	12 7/16	7 8/16	M	Blind River	ONT	John Lee	1973	168
19 15/16	12 0/16	7 15/16	M	Bayfield County	WI	Gary P. Kalal	1973	168
19 15/16	12 8/16	7 7/16	M	Dryden	ONT	Robert C. Kirschner	1974	168
19 15/16	12 8/16	7 7/16	M	Florence County	WI	Peter H. Kortenhorn	1981	168
19 15/16	12 0/16	7 15/16	M	Dryden	ONT	Lane Foshee	1987	168
19 15/16	12 7/16	7 8/16	M	Prince of Wales Island	AK	Chuck Lynde	1991	168
19 15/16	12 7/16	7 8/16	M	Wapawekka Lake	SAS	Douglas R. Peterson	1993	168
19 15/16	12 6/16	7 9/16	M	Chilako River	BC	Emile Matte	1997	168
19 15/16	12 8/16	7 7/16	M	Navajo County	AZ	Alfred J. Gemrich	1997	168
19 15/16	12 3/16	7 12/16	M	Alpine County	CA	Casey J. Nelson	2006	168
19 15/16	12 5/16	7 10/16	M	Wanless	MAN	Scott Rehak	2007	168
19 15/16	12 7/16	7 8/16	M	Mons Lake	ONT	Michael D. Saam	2009	168
19 14/16	12 2/16	7 12/16	M	Routt County	CO	Ronald C. Gravenkemper	1962	181
19 14/16	12 2/16	7 12/16	M	Colcord Mtn.	AZ	Hugh Pearson	1963	181
19 14/16	12 5/16	7 9/16	M	Rio Blanco County	CO	H. R. 'Dutch' Wambold	1969	181
19 14/16	12 12/16	7 2/16	M	Tulare County	CA	Ronald J. Wade	1972	181
19 14/16	12 5/16	7 9/16	M	Ear Falls	ONT	Brent Allen Poindexter	1985	181
19 14/16	12 4/16	7 10/16	M	Deep River	ONT	Rand J. Moore	1989	181

BLACK BEAR

Minimum Score 18 *Continued*

Score	Greatest Length	Greatest Width	Sex	Area	State/ Province	Hunter's Name	Date	Rank
19 14/16	12 1/16	7 13/16	M	Luce County	MI	Terry L. Cook	1989	181
19 14/16	12 5/16	7 9/16	M	Ile-a-La-Crosse	SAS	Gary Schwieters	1991	181
19 14/16	12 9/16	7 5/16	M	Pine County	MN	Kirk D. Grupa	1992	181
19 14/16	12 10/16	7 4/16	M	Riding Mtn.	MAN	Mike T. Berceau	1994	181
19 14/16	12 12/16	7 2/16	M	Washburn County	WI	Jerre Lerum	1996	181
19 14/16	12 4/16	7 10/16	M	San Juan County	NM	Jim Willems	1997	181
19 14/16	12 5/16	7 9/16	M	Seamore Canal	AK	Jerry Karsky	1998	181
19 14/16	12 9/16	7 5/16	M	Kanabec County	MN	Michael F. Hertenstein	1998	181
19 14/16	12 13/16	7 1/16	M	Lac Roger	QUE	Robert R. Agnew, Jr.	2003	181
19 14/16	12 2/16	7 12/16	M	Wabigoon	ONT	Carl Seek	2003	181
19 14/16	12 7/16	7 7/16	M	Saddle Hills	ALB	David Watson	2007	181
19 13/16	12 7/16	7 6/16	M	Shasta County	CA	Stan L. McIntyre	1968	198
19 13/16	12 7/16	7 6/16	M	Estaire	ONT	David L. Roose	1980	198
19 13/16	12 9/16	7 4/16	M	Tucker County	WV	Robert B. Golightly	1982	198
19 13/16	12 11/16	7 2/16	M	Hudson Bay	SAS	Craig Richardson	1985	198
19 13/16	12 3/16	7 10/16	M	Sioux Narrows	ONT	Phil Perry	1997	198
19 13/16	12 4/16	7 9/16	M	Jackson County	CO	Thomas G. Kelley	1997	198
19 13/16	12 10/16	7 3/16	M	Lake of the Woods County	MN	Shannon Thomas	2003	198
19 13/16	12 4/16	7 9/16	M	Sandoval County	NM	Lloyd Wood	2008	198
19 13/16	12 12/16	7 1/16	M	Onanole	MAN	Dan Stillwell	2009	198
19 12/16	11 14/16	7 14/16	M	Lake County	MT	Joe Lawrence	1966	207
19 12/16	12 2/16	7 10/16	M	Chapleau	ONT	Lawrence Gallagher	1966	207
19 12/16	12 7/16	7 5/16	M	Grand County	UT	Edmund H. Auffhammer	1968	207
19 12/16	12 7/16	7 5/16	M	Catron County	NM	Joe E. Stroube	1971	207
19 12/16	12 4/16	7 8/16	M	Itasca County	MN	James R. Kroupa	1973	207
19 12/16	12 2/16	7 10/16	M	Conejos County	CO	Michael Miller	1974	207
19 12/16	12 6/16	7 6/16	M	Bear Paw Landing	QUE	Lonnie Rumley	1988	207
19 12/16	12 4/16	7 8/16	M	Atikokan	ONT	Rick Grooms	1991	207
19 12/16	12 5/16	7 7/16	M	Lynn Lake	MAN	Ernest Gilbert	1994	207
19 12/16	12 8/16	7 4/16	M	Catron County	NM	Lew Webb, Jr.	1995	207
19 12/16	12 2/16	7 10/16	M	Longlac	ONT	Harley Krauss	1996	207
19 12/16	12 6/16	7 6/16	M	Koochiching County	MN	Jesse Turck	1996	207
19 12/16	12 9/16	7 3/16	M	Saddle Hills	ALB	Allen Avery	1997	207
19 12/16	12 6/16	7 6/16	M	Alexander Archipelago	AK	Steve Hohensee	2000	207
19 12/16	12 8/16	7 4/16	M	Koochiching County	MN	Charles W. Schultz	2010	207
19 12/16	12 3/16	7 9/16	M	Riviere Waswanipi	QUE	Ken H. Taylor	2011	207
19 11/16	12 3/16	7 8/16	M	Atikokan	ONT	Dennis Gregory	1967	223
19 11/16	12 2/16	7 9/16	M	Montezuma County	CO	Marvin Reichenau	1972	223
19 11/16	12 4/16	7 7/16	M	Larimer County	CO	Lee Kline	1974	223
19 11/16	11 15/16	7 12/16	M	Chapleau	ONT	Donald E. Meushaw	1975	223
19 11/16	12 9/16	7 2/16	M	Montezuma County	CO	Marvin Reichenau	1976	223
19 11/16	12 11/16	7 0/16	M	Boise County	ID	L. Dean Goodner	1983	223
19 11/16	12 2/16	7 9/16	M	McNulty Lake	ONT	Alexander MacPherson	1993	223
19 11/16	12 1/16	7 10/16	M	Carroll County	NH	Jack Smith	1993	223
19 11/16	12 5/16	7 6/16	M	Carrot River	SAS	Mark P. Wagner	1999	223
19 11/16	12 5/16	7 6/16	M	Grand Rapids	MAN	Robert Gagnon	2004	223
19 11/16	12 8/16	7 3/16	M	Prairie River	SAS	Ken H. Taylor	2007	223
19 10/16	12 0/16	7 10/16	M	Lincoln County	NM	David B. Terk	1964	234
19 10/16	12 6/16	7 4/16	M	Iron County	WI	William Tutt	1966	234
19 10/16	12 10/16	7 0/16	M	Madera County	CA	John D. Faulconer	1971	234
19 10/16	12 4/16	7 6/16	M	Forest County	WI	James L. Rablin	1972	234
19 10/16	12 6/16	7 4/16	M	Sandilands	MAN	Jerry Parizek	1976	234
19 10/16	12 4/16	7 6/16	M	Siskiyou County	CA	Jeff Buck	1987	234
19 10/16	12 4/16	7 6/16	M	Smeaton	SAS	Gene A. Welle	1990	234
19 10/16	12 2/16	7 8/16	M	Troy Lake	MAN	Derek McCarthy	1990	234
19 10/16	12 5/16	7 5/16	M	Canterbury	NBW	Don Rahe	1990	234
19 10/16	12 3/16	7 7/16	M	Lemhi County	ID	Michael Judas	1995	234
19 10/16	12 2/16	7 8/16	M	Los Angeles County	CA	David Whiteman	1996	234
19 10/16	12 15/16	6 11/16	M	Kelvington	SAS	Leamon Ferrell	1997	234
19 10/16	12 4/16	7 6/16	M	Wanless	MAN	Scotty A. Rehak	2001	234
19 10/16	12 10/16	7 0/16	M	Kittson County	MN	Doug Strecker	2005	234
19 10/16	12 12/16	6 14/16	M	The Pas	MAN	William Peterson	2008	234

BLACK BEAR

Minimum Score 18 *Continued*

Score	Greatest Length	Greatest Width	Sex	Area	State/ Province	Hunter's Name	Date	Rank
19 10/16	12 9/16	7 1/16	M	Swan River	MAN	Mike Mitten	2012	234
19 9/16	12 10/16	6 15/16	M	Douglas County	WI	Edwin Fitzgerald	1966	250
19 9/16	12 6/16	7 3/16	M	Grand County	CO	Judd Cooney	1967	250
19 9/16	12 0/16	7 9/16	M	Ignace	ONT	Gordon Bentley	1968	250
19 9/16	11 15/16	7 10/16	M	Idaho County	ID	Kenneth Wallenberg	1970	250
19 9/16	11 9/16	8 0/16	M	Boise County	ID	Ronald L. Sherer	1971	250
19 9/16	12 1/16	7 8/16	M	Blount County	TN	Gary Jordan	1973	250
19 9/16	12 4/16	7 5/16	M	Valley County	ID	L. Dean Goodner	1979	250
19 9/16	11 11/16	7 14/16	M	Bobcaygeon	ONT	Dale W. Gray	1982	250
19 9/16	12 5/16	7 4/16	M	Bonneville County	ID	Larry Cross	1986	250
19 9/16	12 7/16	7 2/16	M	Hudson Bay	SAS	Bill Zahradka	1987	250
19 9/16	11 11/16	7 14/16	M	Las Animas County	CO	R. L. Erdmann	1989	250
19 9/16	12 1/16	7 8/16	M	Carrot River	SAS	Quince Hale	1989	250
19 9/16	12 1/16	7 8/16	M	Boise County	ID	James L. Sullivan	1990	250
19 9/16	12 0/16	7 9/16	M	Lynn Lake	MAN	Ernest B. Gilbert	1993	250
19 9/16	12 6/16	7 3/16	M	Indian River	AK	Rick Schikora	1996	250
19 9/16	12 3/16	7 6/16	M	Weymouth	NS	Paul I. Scott	1996	250
19 9/16	11 12/16	7 13/16	M	Preissac	QUE	Paul Sedera	1997	250
19 9/16	11 15/16	7 10/16	M	Chipewyan Lake	ALB	David D. Manuszak	1998	250
19 9/16	12 2/16	7 7/16	M	Fergus County	MT	Donny Roy	1999	250
19 9/16	12 3/16	7 6/16	M	Alexander Archipelago	AK	Steve Hohensee	1999	250
19 9/16	12 3/16	7 6/16	M	Zec Capitachouane	QUE	Ken H. Taylor	2003	250
19 9/16	11 13/16	7 12/16	M	Thompson Lake	SAS	R. Charles Suttles	2005	250
19 9/16	12 3/16	7 6/16	M	Matagami	QUE	Ken H. Taylor	2007	250
19 9/16	12 5/16	7 4/16	M	Oba	ONT	Neil R. Leonard	2007	250
19 9/16	12 0/16	7 9/16	M	Okanogan County	WA	Gene A. Sieffert	2007	250
19 9/16	12 1/16	7 8/16	M	French River	ONT	Robert E. Kearney	2010	250
19 9/16	12 5/16	7 4/16	M	Chilanko River	BC	James V. Barwick	2011	250
19 8/16	12 3/16	7 5/16	M	Chapleau	ONT	Bob Sharpe	1959	277
19 8/16	12 4/16	7 4/16	M	Presque Isle County	MI	Eugene W. McKechnie	1964	277
19 8/16	12 6/16	7 2/16	M	Forest County	WI	Jerad Dittrich	1965	277
19 8/16	12 1/16	7 7/16	M	Piscataquis County	ME	James Matulis	1967	277
19 8/16	12 6/16	7 2/16	M	Crow Wing County	MN	James L. Beard	1975	277
19 8/16	12 2/16	7 6/16	M	Ear Falls	ONT	Grant Poindexter	1978	277
19 8/16	11 15/16	7 9/16	M	Madison County	MT	Shep Lantow	1982	277
19 8/16	12 4/16	7 4/16	M	Bonneville County	ID	Larry Cross	1988	277
19 8/16	12 2/16	7 6/16	M	Bigoray River	ALB	Gunter Lemke	1990	277
19 8/16	12 7/16	7 1/16	M	Mons Township	ONT	Russell Trusty	1991	277
19 8/16	12 2/16	7 6/16	M	Temiscaming	QUE	Tim Burris	1995	277
19 8/16	12 4/16	7 4/16	M	Grand Rapids	MAN	Dale Shove	1995	277
19 8/16	11 12/16	7 12/16	M	Fort Vermilion	ALB	W. C. MacCarty III, MD	1997	277
19 8/16	11 15/16	7 9/16	M	Grant County	WV	Terrence Weidman	1997	277
19 8/16	12 3/16	7 5/16	M	Buffalo Narrows	SAS	Linda Carol Graham	1998	277
19 8/16	11 15/16	7 9/16	M	Prince of Wales Island	AK	Jeff Falkner	2001	277
19 8/16	11 15/16	7 9/16	M	Ft. McMurray	ALB	Fred Eichler	2005	277
19 8/16	12 4/16	7 4/16	M	Lac du Bonnet	MAN	Douglas C. Fink	2005	277
19 8/16	12 5/16	7 3/16	M	Koochiching County	MN	Charles W. Schultz	2009	277
19 7/16	12 3/16	7 4/16	M	Kenora	ONT	Norman Pint	1964	296
19 7/16	12 5/16	7 2/16	M	Caribou County	ID	Ronald S. Curtis	1969	296
19 7/16	11 10/16	7 13/16	M	St. Louis County	MN	Don Dvoroznak	1970	296
19 7/16	11 15/16	7 8/16	M	Moffat County	CO	Louis Preba	1972	296
19 7/16	12 3/16	7 4/16	M	Garfield County	UT	Lee G. Stoddard	1975	296
19 7/16	12 7/16	7 0/16	M	Snohomish County	WA	Charles J. Bartlett	1976	296
19 7/16	12 3/16	7 4/16	M	Hudson Bay	SAS	Bill Zahradka	1986	296
19 7/16	12 3/16	7 4/16	M	Catron County	NM	Parris Nottingham	1988	296
19 7/16	12 3/16	7 4/16	M	Cadillac	QUE	Bob Drumm	1990	296
19 7/16	12 4/16	7 3/16	M	Meadow Lake	SAS	Matthew Curry	1992	296
19 7/16	11 9/16	7 14/16	M	Alpine County	CA	Kevin L. Hall	1992	296
19 7/16	12 2/16	7 5/16	M	Athabasca River	ALB	George "Matt" Potts	1994	296
19 7/16	11 13/16	7 10/16	M	Lemhi County	ID	Bruce Carlisle	1995	296
19 7/16	12 5/16	7 2/16	M	Swastika	ONT	Mike Darlak	1996	296
19 7/16	12 7/16	7 0/16	M	Mesa County	CO	Eric A. Sawyer	1996	296

BLACK BEAR

Minimum Score 18 *Continued*

Score	Greatest Length	Greatest Width	Sex	Area	State/ Province	Hunter's Name	Date	Rank
19 7/16	12 2/16	7 5/16	M	Carrot River	SAS	Mark P. Wagner	1997	296
19 7/16	12 2/16	7 5/16	M	Wabasca River	ALB	Warren Erickson	1998	296
19 7/16	12 0/16	7 7/16	M	Mistatim	SAS	Bob Morton	1999	296
19 7/16	11 15/16	7 8/16	M	Carrot River	SAS	Larry T. Fischer	2000	296
19 7/16	12 0/16	7 7/16	M	Lake Athabasca	SAS	Andrew Stanco	2002	296
19 7/16	12 3/16	7 4/16	M	Fond du Lac River	SAS	Joel M. Riotto	2004	296
19 7/16	12 5/16	7 2/16	M	Gila County	AZ	Doug Strecker	2005	296
19 7/16	12 13/16	6 10/16	M	Carrot River	SAS	Jim Kinsey	2006	296
19 7/16	12 3/16	7 4/16	M	Terrace	BC	Rick Duggan	2008	296
19 6/16	12 7/16	6 15/16	M	Shasta County	CA	L. Dale Towery	1965	320
19 6/16	11 14/16	7 8/16	M	Atikokan	ONT	Dennis Gregory	1969	320
19 6/16	11 14/16	7 8/16	M	Mesa County	CO	Charles Leidheiser	1971	320
19 6/16	12 6/16	7 0/16	F	Sawyer County	WI	Ronald Curry, Jr.	1972	320
19 6/16	12 3/16	7 3/16	M	Uncompahgre N.F.	CO	Thomas J. Hentrick	1975	320
19 6/16	12 0/16	7 6/16	M	Trinity County	CA	Robert Pearce	1986	320
19 6/16	12 0/16	7 6/16	M	Washago	ONT	James E. Doberstein	1987	320
19 6/16	12 1/16	7 5/16	M	Colfax County	NM	Daniel Hurd	1990	320
19 6/16	12 2/16	7 4/16	F	Lac La Biche	ALB	Bruce Nederveld	1991	320
19 6/16	11 12/16	7 10/16	M	Wanless	MAN	Leonard Rock	1993	320
19 6/16	12 1/16	7 5/16	M	Lynn Lake	MAN	G. Fred Asbell	1994	320
19 6/16	12 2/16	7 4/16	M	Larder Lake	ONT	Thomas R. Johnson	1994	320
19 6/16	12 6/16	7 0/16	M	Klickitat County	WA	Gary Holwegner	1995	320
19 6/16	12 2/16	7 4/16	M	Idaho County	ID	Pat Hylton	1996	320
19 6/16	12 2/16	7 4/16	M	Mariposa County	CA	Arthur M. Cain	1996	320
19 6/16	12 4/16	7 2/16	M	Big Horn County	WY	Roger Coguill	1997	320
19 6/16	12 5/16	7 1/16	M	Lynn Lake	MAN	William H. Peterson	1998	320
19 6/16	11 11/16	7 11/16	M	Lake County	MN	Kirk D. Grupa	1999	320
19 6/16	12 1/16	7 5/16	M	Boise County	ID	Anthony L. Mudd	2000	320
19 6/16	12 5/16	7 1/16	M	Athabasca River	ALB	Jack A. Berry	2002	320
19 6/16	12 1/16	7 5/16	M	Kuiu Island	AK	Stephen Tyrrell	2004	320
19 6/16	12 4/16	7 2/16	M	Tanana River	AK	Stan Parkerson	2005	320
19 6/16	12 6/16	7 0/16	M	Koochiching County	MN	Larry M. Looman	2007	320
19 6/16	12 0/16	7 6/16	M	Dipper Lake	SAS	Greg Dearth	2008	320
19 6/16	12 3/16	7 3/16	M	White County	GA	Henry Johnson	2008	320
19 6/16	12 7/16	6 15/16	M	Ste. Rose du Lac	MAN	Mark F. Pedo	2009	320
19 6/16	12 4/16	7 2/16	M	Swan River	MAN	Joe Bradley	2012	320
19 5/16	11 13/16	7 8/16	M	Clearwater County	ID	William R. Vanderhoef	1959	347
19 5/16	11 12/16	7 9/16	M	Upper Peninsula	MI	Donald Schram	1961	347
19 5/16	11 11/16	7 10/16	M	Ear Falls	ONT	Michael Mealey	1978	347
19 5/16	11 13/16	7 8/16	M	Smeaton	SAS	Gene Welle	1984	347
19 5/16	12 4/16	7 1/16	M	Cass County	MN	Brad Blanchard	1985	347
19 5/16	12 3/16	7 2/16	M	Rocky Lake	MAN	Tim Finley	1988	347
19 5/16	12 1/16	7 4/16	M	Zec Maganasipi	QUE	F.Edward Campbell	1989	347
19 5/16	12 0/16	7 5/16	M	Colfax County	NM	Stephen W. Long	1991	347
19 5/16	12 1/16	7 4/16	M	Athabasca River	ALB	Casmir S. Domurat, Jr.	1991	347
19 5/16	12 5/16	7 0/16	M	Catron County	NM	David H. Boland	1992	347
19 5/16	12 0/16	7 5/16	M	Aulneau Peninsula	ONT	Michael Koska	1993	347
19 5/16	12 5/16	7 0/16	M	Carrot River	SAS	Andrew D. Pearson	1994	347
19 5/16	12 5/16	7 0/16	M	Carrot River	SAS	Larry T. Fischer	1994	347
19 5/16	12 6/16	6 15/16	M	Aulneau Peninsula	ONT	Michael Koska	1994	347
19 5/16	12 4/16	7 1/16	M	Chapleau	ONT	Gregory James Bishop	1995	347
19 5/16	11 9/16	7 12/16	M	Cook County	MN	Mike Redig	1996	347
19 5/16	12 1/16	7 4/16	M	Whiteswan Lake	SAS	Eugene Arndt	1997	347
19 5/16	11 15/16	7 6/16	M	Carswell Lake	SAS	Dennis R. Allman	1997	347
19 5/16	11 12/16	7 9/16	M	Guthrie Lake	MAN	John Beardslee	1998	347
19 5/16	11 15/16	7 6/16	M	Zec Capitachouane	QUE	Ken H. Taylor	1998	347
19 5/16	11 11/16	7 10/16	M	Prince of Wales Island	AK	Dirk Dieterich	2001	347
19 5/16	12 5/16	7 0/16	M	Athabasca	ALB	Jay Downes, Jr.	2003	347
19 5/16	12 1/16	7 4/16	M	Cook County	MN	George Ostgarden	2003	347
19 5/16	12 6/16	6 15/16	M	Ft. McMurray	ALB	Chris Parrino	2004	347
19 5/16	12 2/16	7 3/16	M	Hodgson	MAN	Kim Engelbert	2009	347
19 5/16	12 9/16	6 12/16	M	Chippewa County	WI	Tom Herrick	2011	347

Score	Greatest Length	Greatest Width	Sex	Area	State/ Province	Hunter's Name	Date	Rank
19 4/16	12 4/16	7 0/16	M	Ashland County	WI	Herbert H. Lange	1961	373
19 4/16	12 4/16	7 0/16	M	Murphy Dome	AK	Thomas Clark	1963	373
19 4/16	11 15/16	7 5/16	M	Jackson County	OR	Bob Jacobs	1964	373
19 4/16	12 0/16	7 4/16	M	Idaho County	ID	Dick Gulman	1967	373
19 4/16	12 4/16	7 0/16	M	Bayfield County	WI	Bruce Eggenberger	1977	373
19 4/16	12 1/16	7 3/16	M	Saguache County	CO	Richard Baumfalk	1977	373
19 4/16	11 13/16	7 7/16	M	Thessalon	ONT	Robert R. Rider	1979	373
19 4/16	12 6/16	6 14/16	M	Turtle River	ONT	Al Haines	1986	373
19 4/16	11 12/16	7 8/16	M	Archuleta County	CO	Grant Adkisson	1991	373
19 4/16	12 3/16	7 1/16	M	Buffalo Head Prairie	ALB	E. Josh Isbell	1994	373
19 4/16	12 0/16	7 4/16	M	Kapuskasing	ONT	Floyd S. Kines	1995	373
19 4/16	11 15/16	7 5/16	M	Lake Besnard	SAS	Eric J. Collier	1995	373
19 4/16	12 2/16	7 2/16	M	Snow Lake	MAN	Christopher Kaforski	1996	373
19 4/16	11 15/16	7 5/16	M	Herkimer County	NY	Steve Balyszak	1997	373
19 4/16	12 3/16	7 1/16	M	Aberdeen	ONT	Gerald L. Barefield	2005	373
19 4/16	12 9/16	6 11/16	M	Carrot River	SAS	Joe Furlong	2009	373
19 4/16	11 12/16	7 8/16	M	Chilanko River	BC	Leonard Scarborough	2010	373
19 4/16	12 0/16	7 4/16	M	San Juan County	NM	Jim Willems	2010	373
19 3/16	11 12/16	7 7/16	M	Mineral County	CO	Edward Wintz	1960	391
19 3/16	12 2/16	7 1/16	M	Rio Arriba County	NM	Dan Ward	1964	391
19 3/16	11 13/16	7 6/16	M	Iron Bridge	ONT	Philip L. Hawkins	1965	391
19 3/16	11 12/16	7 7/16	M	Shasta County	CA	Michael D. Combs	1967	391
19 3/16	11 11/16	7 8/16	M	Sudbury District	ONT	Floyd Eccleston	1970	391
19 3/16	11 14/16	7 5/16	M	Sequoia National Forest	CA	Martin Szekeresh, Jr.	1973	391
19 3/16	11 11/16	7 8/16	M	Powell County	MT	Gary L. Wilson	1975	391
19 3/16	11 12/16	7 7/16	M	Grand County	CO	Jim Williams	1983	391
19 3/16	11 14/16	7 5/16	M	Kenora	ONT	Kenneth Gilb	1983	391
19 3/16	12 2/16	7 1/16	M	Durban	MAN	Bill Clink	1985	391
19 3/16	11 12/16	7 7/16	M	Park County	CO	Robert Wright	1987	391
19 3/16	12 2/16	7 1/16	M	Custer County	CO	Rod Niles	1987	391
19 3/16	11 10/16	7 9/16	M	Lac Le Truite Territory	QUE	Brian Hendricks	1989	391
19 3/16	12 1/16	7 2/16	M	Grand Rapids	MAN	James R. Kramp	1991	391
19 3/16	11 15/16	7 4/16	M	Lemhi County	ID	Gary Sims	1992	391
19 3/16	11 14/16	7 5/16	M	Wapawekka Lake	SAS	Gary Schwieters	1993	391
19 3/16	11 11/16	7 8/16	M	Athabasca	ALB	Grant Adkisson	1997	391
19 3/16	12 7/16	6 12/16	M	Green Lake	SAS	Pink Atkins	1997	391
19 3/16	12 0/16	7 3/16	M	Lemhi County	ID	Michael Judas	1999	391
19 3/16	12 0/16	7 3/16	M	Flathead County	MT	Neil L. Jacobson	2000	391
19 3/16	12 2/16	7 1/16	M	Carswell Lake	SAS	Bob Gilbert	2002	391
19 3/16	12 1/16	7 2/16	M	McAdam	NBW	Steve Senderovitz	2003	391
19 3/16	11 15/16	7 4/16	M	Gunnison County	CO	Matt Reilly	2004	391
19 3/16	11 10/16	7 9/16	M	Asotin County	WA	Kenneth Anderson	2009	391
19 2/16	11 9/16	7 9/16	M	Sudbury	ONT	Clarence Grandt	1963	415
19 2/16	11 13/16	7 5/16	M	Chapleau	ONT	Gerald E. Taft	1968	415
19 2/16	11 12/16	7 6/16	M	Blount County	TN	Don Dvoroznak	1968	415
19 2/16	11 12/16	7 6/16	M	Trinity County	CA	Fred M. Frakes	1970	415
19 2/16	11 14/16	7 4/16	M	Vilas County	WI	William L. Yessa	1971	415
19 2/16	12 3/16	6 15/16	M	Shasta County	CA	Gerald P. Doyle	1971	415
19 2/16	11 11/16	7 7/16	M	Kormak	ONT	Marvin E. Davis	1972	415
19 2/16	12 0/16	7 2/16	M	Dolores County	CO	Marvin Reichenau	1973	415
19 2/16	11 15/16	7 3/16	M	Uncompahgre N.F.	CO	Ed Bonardi	1973	415
19 2/16	12 1/16	7 1/16	M	Dryden	ONT	Ken Horton	1977	415
19 2/16	11 14/16	7 4/16	M	Skamania County	WA	John H. Wahl	1979	415
19 2/16	12 2/16	7 0/16	M	Flathead County	MT	Owen Weaver	1981	415
19 2/16	12 0/16	7 2/16	M	Durban	MAN	David H. Boland	1986	415
19 2/16	11 13/16	7 5/16	M	Rio Arriba County	NM	James H. Miller	1989	415
19 2/16	11 12/16	7 6/16	M	Oxford County	ME	Gary J. Russell	1992	415
19 2/16	11 13/16	7 5/16	M	Kelvington	SAS	Milan R. Liesener	1993	415
19 2/16	11 15/16	7 3/16	M	Remigny	QUE	Michael Bolin	1993	415
19 2/16	12 1/16	7 1/16	M	Besnard Lake	SAS	Ronald J. Collier	1994	415
19 2/16	12 0/16	7 2/16	M	High Level	ALB	Tracy Roy	1995	415
19 2/16	11 9/16	7 9/16	M	Adair Township	ONT	David G. Bockheim	1995	415

BLACK BEAR

Minimum Score 18 *Continued*

Score	Greatest Length	Greatest Width	Sex	Area	State/ Province	Hunter's Name	Date	Rank
19 2/16	12 0/16	7 2/16	M	Aroostook County	ME	Rudy Conley	1995	415
19 2/16	12 8/16	6 10/16	M	Newton County	AR	Eric R. Duncan	1996	415
19 2/16	12 4/16	6 14/16	M	Athabasca River	ALB	Henry F. Trotter III	1997	415
19 2/16	12 0/16	7 2/16	M	Carrot River	SAS	Joseph P. Furlong	2000	415
19 2/16	12 1/16	7 1/16	M	Dipper Lake	SAS	Gilbert Gresham	2001	415
19 2/16	11 13/16	7 5/16	M	Matagami	QUE	Ken H. Taylor	2008	415
19 1/16	11 1/16	8 0/16	M	Jackson County	OR	Leander Lowel	1959	441
19 1/16	11 12/16	7 5/16	M	Mineral County	CO	Ed Wintz	1959	441
19 1/16	11 15/16	7 2/16	M	Gogebic County	MI	Margaret R. Cooley	1961	441
19 1/16	11 5/16	7 12/16	M	Upper Peninsula	MI	Jerry D. Anderson	1967	441
19 1/16	11 15/16	7 2/16	M	Langlade County	WI	Roland Mantzke	1969	441
19 1/16	12 1/16	7 0/16	M	Red Lake	ONT	Don Ellett	1969	441
19 1/16	11 4/16	7 13/16	M	Vancouver Island	BC	Klaus Schultz	1974	441
19 1/16	11 15/16	7 2/16	M	Marquette County	MI	Pete Hillesheim	1974	441
19 1/16	11 10/16	7 7/16	M	Oxford County	ME	James P. Wellever	1978	441
19 1/16	11 14/16	7 3/16	M	Nestor Falls	ONT	Larry Streiff	1982	441
19 1/16	11 10/16	7 7/16	M	Inyo County	CA	Jim Voges	1988	441
19 1/16	11 14/16	7 3/16	M	Larimer County	CO	Ed Bennett	1989	441
19 1/16	11 12/16	7 5/16	M	Sandoval County	NM	Wayne C. Wendel	1991	441
19 1/16	11 15/16	7 2/16	M	Grand County	CO	Cary Laman	1992	441
19 1/16	12 2/16	6 15/16	M	Prince of Wales Island	AK	Kelly Norskog	1992	441
19 1/16	11 15/16	7 2/16	M	Prince of Wales Island	AK	Monty Moravec	1993	441
19 1/16	12 2/16	6 15/16	M	Iron County	WI	Richard M. Kanzelberger	1994	441
19 1/16	12 0/16	7 1/16	M	Fremont County	ID	Rod L. Perkins	1997	441
19 1/16	11 13/16	7 4/16	M	Wabigoon	ONT	Carl Seek	1998	441
19 1/16	12 2/16	6 15/16	M	Alexander Archipelago	AK	Steve Hohensee	1999	441
19 1/16	11 13/16	7 4/16	M	Sheridan County	WY	Randy Burtis	2002	441
19 1/16	12 1/16	7 0/16	M	Little Bear Lake	SAS	Rick Stockburger	2002	441
19 1/16	11 15/16	7 2/16	M	Ft. McMurray	ALB	Chris Parrino	2004	441
19 1/16	12 0/16	7 1/16	M	Churchill River	SAS	Mike Feldermann	2005	441
19 1/16	12 0/16	7 1/16	M	Kenora	ONT	Troy E. Brant	2006	441
19 1/16	11 15/16	7 2/16	M	Alstead Lake	SAS	Michael J. Windemuller	2009	441
19 1/16	12 4/16	6 13/16	M	Athabasca River	ALB	Joel M. Riotto	2009	441
19 1/16	11 14/16	7 3/16	M	New Hosco	QUE	Ken H. Taylor	2010	441
19 1/16	12 0/16	7 1/16	M	Swan River	MAN	Michael Casey	2012	441
19 0/16	12 6/16	6 10/16	M	Iron County	WI	Charles Kroll	1966	470
19 0/16	11 9/16	7 7/16	M	Sioux Narrows	ONT	Walter J. Sawicki	1967	470
19 0/16	11 12/16	7 4/16	M	Chapleau	ONT	Kenneth R. Larson	1968	470
19 0/16	11 13/16	7 3/16	M	Kamloops	BC	Terry J. Haines	1968	470
19 0/16	11 15/16	7 1/16	M	Ignace	ONT	Stanley Olson	1968	470
19 0/16	11 12/16	7 4/16	M	Armstrong	ONT	James Mahoney	1969	470
19 0/16	11 10/16	7 6/16	M	Kenora	ONT	Barry Englehardt	1969	470
19 0/16	11 8/16	7 8/16	M	Shasta County	CA	Patrick J. Marley	1969	470
19 0/16	11 11/16	7 5/16	M	Skamania County	WA	Dennis E. DesJardins	1970	470
19 0/16	11 7/16	7 9/16	M	Grand County	UT	C. Donald Lechner	1970	470
19 0/16	11 11/16	7 5/16	M	Ravalli County	MT	Rod Osburn	1981	470
19 0/16	11 13/16	7 3/16	M	Meadow Lake	SAS	Gary Bauer	1984	470
19 0/16	12 2/16	6 14/16	M	Ear Falls	ONT	Brent Allen Poindexter	1985	470
19 0/16	12 0/16	7 0/16	M	Smeaton	SAS	Gene Welle	1988	470
19 0/16	12 0/16	7 0/16	M	Sheridan County	WY	Larry O. Burtis	1990	470
19 0/16	11 12/16	7 4/16	M	Wallowa County	OR	Eugene Smith, Jr.	1990	470
19 0/16	11 12/16	7 4/16	M	Oxford County	ME	Gary J. Russell	1991	470
19 0/16	12 0/16	7 0/16	M	Custer County	ID	Pascal Perrin	1992	470
19 0/16	11 4/16	7 12/16	M	Wanless	MAN	Tim Finley	1992	470
19 0/16	11 14/16	7 2/16	M	Elmore County	ID	Kirk W. Reese	1994	470
19 0/16	11 7/16	7 9/16	M	Ignace	ONT	Chuck Harris	1997	470
19 0/16	12 2/16	6 14/16	M	Theodore River	AK	H. Gale McKnight	1998	470
19 0/16	11 8/16	7 8/16	M	Thompson	MAN	William F. Curlis II	1999	470
19 0/16	11 13/16	7 3/16	M	Riviere Dore	QUE	Ken H. Taylor	1999	470
19 0/16	11 15/16	7 1/16	M	Zitziana River	AK	James Jones	2000	470
19 0/16	11 13/16	7 3/16	M	Idaho County	ID	Marlon J. Clapham	2001	470
19 0/16	11 11/16	7 5/16	M	Blaine County	ID	Brian Brockette	2002	470

BLACK BEAR

Minimum Score 18 *Continued*

Score	Greatest Length	Greatest Width	Sex	Area	State/ Province	Hunter's Name	Date	Rank
19 0/16	12 0/16	7 0/16	M	High Level	ALB	John F. Harding	2002	470
19 0/16	12 4/16	6 12/16	M	Carswell Lake	SAS	Bob Gilbert	2003	470
19 0/16	12 1/16	6 15/16	M	Leaf Rapids	MAN	Shawn Sand	2005	470
19 0/16	12 3/16	6 13/16	M	Matagami	QUE	Ken H. Taylor	2006	470
19 0/16	11 12/16	7 4/16	M	Mount Arrowsmith	BC	Gregg Bekelja	2007	470
19 0/16	11 14/16	7 2/16	M	Desha County	AR	Chris Patton	2009	470
19 0/16	11 9/16	7 7/16	M	Sweet Grass County	MT	Charlie W. Green	2011	470
18 15/16	11 9/16	7 6/16	M	King George IV Lake	NFL	Frank M. Davis	1958	504
18 15/16	12 0/16	6 15/16	M	San Miguel County	NM	Dick McClain	1981	504
18 15/16	11 8/16	7 7/16	M	Poitras	ONT	Robert H. Pavlovic	1985	504
18 15/16	11 12/16	7 3/16	M	Hudson Bay	SAS	Bruce Balerud	1986	504
18 15/16	11 6/16	7 9/16	M	Elmore County	ID	John Turner	1989	504
18 15/16	11 8/16	7 7/16	M	Boise County	ID	Julian Salutregui	1989	504
18 15/16	11 13/16	7 2/16	M	Siebert Lake	ALB	Orest Popil	1992	504
18 15/16	11 13/16	7 2/16	M	Aroostook County	ME	Richard C. Tucker	1992	504
18 15/16	11 15/16	7 0/16	M	Lesser Slave Lake	ALB	Robert "Grub" Matthews	1995	504
18 15/16	11 11/16	7 4/16	M	Dryden	ONT	Al Smith	1995	504
18 15/16	12 2/16	6 13/16	M	Sublette County	WY	John Gedroez	1995	504
18 15/16	12 2/16	6 13/16	M	Wabigoon	ONT	Robert Barrie	1996	504
18 15/16	11 11/16	7 4/16	M	Clarke Lake	SAS	Gordon Braun	1997	504
18 15/16	12 3/16	6 12/16	M	Bradbury River	MAN	Larry G. Lottman	1997	504
18 15/16	11 14/16	7 1/16	M	Sheridan County	WY	Chris Boll	1998	504
18 15/16	12 3/16	6 12/16	M	Sawyer County	WI	Todd Zeuske	1998	504
18 15/16	12 2/16	6 13/16	M	Pierceland	SAS	Ray Wix	2000	504
18 15/16	11 9/16	7 6/16	M	St. John	NBW	Michael P. Mongelli	2001	504
18 15/16	11 11/16	7 4/16	M	Missoula County	MT	Matthew J. Meyer	2008	504
18 15/16	12 0/16	6 15/16	M	Bayfield County	WI	Larry Fischer	2010	504
18 15/16	12 2/16	6 13/16	M	Koochiching County	MN	Larry M. Looman	2012	504
18 14/16	11 6/16	7 8/16	M	Vermilion Bay	ONT	Wayne I. Munkel	1966	525
18 14/16	12 7/16	6 7/16	M	Vanderhoof	BC	Cecil Raphael	1967	525
18 14/16	11 2/16	7 12/16	M	Cranberry Portage	MAN	Carl Anderson	1968	525
18 14/16	11 11/16	7 3/16	M	Orleans County	VT	James Gilman	1969	525
18 14/16	11 12/16	7 2/16	M	Idaho County	ID	Peter Eremo	1970	525
18 14/16	12 3/16	6 11/16	M	Madera County	CA	Duane A. Whittle	1973	525
18 14/16	11 13/16	7 1/16	M	Lincoln County	OR	Stanley D. Miles	1974	525
18 14/16	12 2/16	6 12/16	M	Vermilion Bay	ONT	Myles Keller	1974	525
18 14/16	11 14/16	7 0/16	M	Minaki	ONT	Greg Stezenski	1975	525
18 14/16	12 0/16	6 14/16	M	Wabigoon	ONT	Keith Olson	1975	525
18 14/16	11 13/16	7 1/16	M	Prince George	BC	Jim Jackson	1976	525
18 14/16	11 9/16	7 5/16	M	Aulneau Peninsula	ONT	Michael F. Koska	1983	525
18 14/16	11 12/16	7 2/16	M	Scoop Lake	BC	Ronald Montross	1984	525
18 14/16	11 12/16	7 2/16	M	San Miguel County	NM	Dick McClain	1986	525
18 14/16	11 8/16	7 6/16	M	Ignace	ONT	Robert James Lewis	1987	525
18 14/16	12 1/16	6 13/16	M	Luce County	MI	Terry L. Cook	1988	525
18 14/16	11 12/16	7 2/16	M	Mann River	SAS	David P. Heinselman II	1989	525
18 14/16	11 13/16	7 1/16	M	Pelican Narrows	SAS	Doug Otte	1989	525
18 14/16	12 4/16	6 10/16	M	Rocky Lake	MAN	Cecil Tharp	1989	525
18 14/16	11 15/16	6 15/16	M	Sanders County	MT	Greg L. Munther	1989	525
18 14/16	11 8/16	7 6/16	M	Boise County	ID	Julian Salutrequi	1990	525
18 14/16	11 5/16	7 9/16	M	Elmore County	ID	John Turner	1990	525
18 14/16	11 7/16	7 7/16	M	Lynn Lake	MAN	Steve Gorr	1991	525
18 14/16	12 0/16	6 14/16	M	Beltrami County	MN	Brian Aune	1991	525
18 14/16	11 12/16	7 2/16	M	Monds Township	ONT	Phillip H. Fisher	1992	525
18 14/16	11 14/16	7 0/16	M	Randolph County	WV	Daniel R. Gillenwater	1992	525
18 14/16	11 12/16	7 2/16	M	Dryden	ONT	Al Forbes	1993	525
18 14/16	11 15/16	6 15/16	M	Athabasca River	ALB	William Wuerthele	1994	525
18 14/16	12 0/16	6 14/16	M	Kapuskasing	ONT	Vaughn Wright	1995	525
18 14/16	12 0/16	6 14/16	M	Oba	ONT	Robert J. Dunne, Jr.	1995	525
18 14/16	12 4/16	6 10/16	M	Athabasca River	ALB	Jewell Leadford	1996	525
18 14/16	11 14/16	7 0/16	M	Athabasca River	ALB	Roger Wintle	1997	525
18 14/16	12 0/16	6 14/16	M	Carrot River	SAS	Larry T. Fischer	1998	525
18 14/16	12 0/16	6 14/16	M	Smeaton	SAS	Gene A. Welle	1998	525

BLACK BEAR

Minimum Score 18 *Continued*

Score	Greatest Length	Greatest Width	Sex	Area	State/ Province	Hunter's Name	Date	Rank
18 14/16	11 14/16	7 0/16	M	Manitoulin Island	ONT	Warren T. Stewart	1998	525
18 14/16	12 2/16	6 12/16	M	Athabasca River	ALB	Ron Adamson	1999	525
18 14/16	11 11/16	7 3/16	M	Willow Mtn.	AK	Frankie Kish	2000	525
18 14/16	11 15/16	6 15/16	M	Bay d'Espoir	NFL	Charles W. Rehor	2002	525
18 14/16	11 12/16	7 2/16	M	Colfax County	NM	Steve E. Fegley	2003	525
18 14/16	11 9/16	7 5/16	M	Cree River	SAS	Matthew Reilly	2004	525
18 14/16	11 6/16	7 8/16	M	Fremont County	ID	Mark D. Morris	2005	525
18 14/16	11 11/16	7 3/16	M	Alstead Lake	SAS	Tim Bradley	2006	525
18 14/16	12 0/16	6 14/16	M	Teton County	WY	Ronell Skinner	2008	525
18 13/16	11 8/16	7 5/16	M	Washington County	ME	Norman Jolliffe	1973	568
18 13/16	11 6/16	7 7/16	M	Jefferson County	CO	Chuck Hutton	1974	568
18 13/16	11 8/16	7 5/16	M	Pemberton	BC	Dr. Michael R. Cummings	1976	568
18 13/16	12 0/16	6 13/16	M	The Pas	MAN	Scott Lang	1983	568
18 13/16	12 3/16	6 10/16	M	Wasilla	AK	Ted Grover	1986	568
18 13/16	11 14/16	6 15/16	M	Wawa	ONT	Dale Rohrbeck	1988	568
18 13/16	12 1/16	6 12/16	M	Pine County	MN	Arnold F. Ostgarden	1988	568
18 13/16	12 0/16	6 13/16	M	Iron County	WI	Gary G. Johnson	1988	568
18 13/16	11 7/16	7 6/16	M	Mine Centre	ONT	Willard L. Voight	1989	568
18 13/16	12 2/16	6 11/16	M	Aitkin County	MN	David Emmen	1989	568
18 13/16	12 4/16	6 9/16	M	Apache County	AZ	Stephen D. Hornady	1991	568
18 13/16	11 10/16	7 3/16	M	Kenora	ONT	Jerry Lee Andrews	1992	568
18 13/16	11 12/16	7 1/16	M	Kuiu Island	AK	Michael Speigle	1992	568
18 13/16	11 12/16	7 1/16	M	Fannin County	GA	Mac Gignilliat	1992	568
18 13/16	11 15/16	6 14/16	M	Koochiching County	MN	Scott Schultz	1993	568
18 13/16	11 9/16	7 4/16	M	Missanabie	ONT	Robert L. Busch	1995	568
18 13/16	12 0/16	6 13/16	M	Swan River	MAN	Morgan E. Lambert	1999	568
18 13/16	12 2/16	6 11/16	M	Penobscot County	ME	Daniel L. Boisvert	1999	568
18 13/16	11 13/16	7 0/16	M	Alban	ONT	Don Pedo	1999	568
18 13/16	11 15/16	6 14/16	M	House River	ALB	Tom Hentrick	2000	568
18 13/16	11 15/16	6 14/16	M	Whale Passage	AK	David Rue	2002	568
18 13/16	11 9/16	7 4/16	M	Red Lake	ONT	Andrew C. Bair	2006	568
18 13/16	11 14/16	6 15/16	M	Natrona County	WY	Randy Conner	2007	568
18 13/16	11 9/16	7 4/16	M	Lac Matagami	QUE	Ken H. Taylor	2009	568
18 13/16	11 10/16	7 3/16	M	Preissac Abitibi	QUE	Mario Pelchat	2009	568
18 13/16	11 15/16	6 14/16	M	Las Animas County	CO	James L. Anderson	2010	568
18 12/16	12 1/16	6 11/16	M	Shasta County	CA	Harv Ebers	1964	594
18 12/16	11 11/16	7 1/16	M	Carbon County	UT	Marvin Tye	1965	594
18 12/16	11 9/16	7 3/16	M	Manowam Lake	QUE	Dennis H. Driscoll	1969	594
18 12/16	11 5/16	7 7/16	M	Siskiyou County	CA	W. E. Cates	1972	594
18 12/16	12 2/16	6 10/16	M	Lesser Slave Lake	ALB	Gene Solyntjes	1978	594
18 12/16	11 3/16	7 9/16	M	Franklin County	ME	John Janelli	1980	594
18 12/16	12 3/16	6 9/16	M	Langlade County	WI	Michael Steliga	1983	594
18 12/16	11 13/16	6 15/16	M	Graham	ONT	Joe Neal Walters	1984	594
18 12/16	11 13/16	6 15/16	M	Oxford County	ME	Robert Grannis	1988	594
18 12/16	11 8/16	7 4/16	M	Gunnison County	CO	Michael K. Ward	1989	594
18 12/16	12 1/16	6 11/16	F	Rio Arriba County	NM	Kelley B. Ward	1989	594
18 12/16	11 11/16	7 1/16	M	Idaho County	ID	Dr. Andrew F. Jones	1991	594
18 12/16	12 0/16	6 12/16	M	Kootenai County	ID	Mark Jones	1992	594
18 12/16	11 8/16	7 4/16	M	Butler	ONT	Robert H. Pavlovic	1992	594
18 12/16	11 5/16	7 7/16	M	Athabasca River	ALB	William Wuerthele	1994	594
18 12/16	12 0/16	6 12/16	M	Clarke Lake	SAS	Tony Chirles	1996	594
18 12/16	11 15/16	6 13/16	M	Bradbury River	MAN	Steven W. Hartley	1997	594
18 12/16	11 12/16	7 0/16	M	Mattice	ONT	Ed Pollock	1998	594
18 12/16	12 2/16	6 10/16	M	Athabasca River	ALB	Jason Adamson	1998	594
18 12/16	11 13/16	6 15/16	M	Chipewyan Lake	ALB	David D. Manuszak	1998	594
18 12/16	12 0/16	6 12/16	M	Iron County	MI	Jeff Joseph	2003	594
18 12/16	11 12/16	7 0/16	M	Cook County	MN	Dave Vomela	2006	594
18 12/16	11 8/16	7 4/16	M	Terrace	BC	Jim Carlson	2007	594
18 12/16	11 4/16	7 8/16	M	Zec Collin	QUE	Larry G. Evans	2007	594
18 12/16	11 8/16	7 4/16	M	Brown Mtn.	AK	Mike L. Yonker	2007	594
18 12/16	11 15/16	6 13/16	M	Firebag River	ALB	Tim L. Donnelly	2008	594
18 11/16	11 12/16	6 15/16	M	Presque Isle County	MI	Herbert Miller	1957	620

BLACK BEAR

Minimum Score 18 *Continued*

Score	Greatest Length	Greatest Width	Sex	Area	State/ Province	Hunter's Name	Date	Rank
18 11/16	11 8/16	7 3/16	M	Chirvakum Creek	WA	Joe Zuend	1962	620
18 11/16	11 13/16	6 14/16	M	Iron County	MI	Donald Schram	1965	620
18 11/16	11 7/16	7 4/16	M	Shawano County	WI	Kenneth Karbon	1970	620
18 11/16	11 10/16	7 1/16	M	Lac Cayamant	QUE	Charles Shaffner	1971	620
18 11/16	11 9/16	7 2/16	M	Grand County	UT	Dennis Schoenick	1972	620
18 11/16	11 10/16	7 1/16	M	Itasca County	MN	Tom Brudeli	1981	620
18 11/16	11 5/16	7 6/16	M	Fremont County	WY	Pat Eastes	1987	620
18 11/16	11 15/16	6 12/16	M	Larimer County	CO	James Little	1989	620
18 11/16	11 10/16	7 1/16	M	Wanless	MAN	Damon Finley	1992	620
18 11/16	11 7/16	7 4/16	M	McNeil Township	ONT	Archie Mackinnon	1992	620
18 11/16	11 10/16	7 1/16	M	Houghton County	MI	Mitchell Bellamy	1992	620
18 11/16	11 5/16	7 6/16	M	Nakina	ONT	Darren W. Wilson	1993	620
18 11/16	11 15/16	6 12/16	M	High Level	ALB	Bob Hudson, Jr.	1996	620
18 11/16	11 10/16	7 1/16	M	Rapide-des-Joachims	QUE	Don Delabbio	1997	620
18 11/16	11 6/16	7 5/16	M	San Juan County	UT	David R. Lundberg	1998	620
18 11/16	12 1/16	6 10/16	M	Bayfield County	WI	Joseph C. Hinderman	1998	620
18 11/16	11 13/16	6 14/16	M	Coos County	NH	Brian J. Emerson	1998	620
18 11/16	11 11/16	7 0/16	M	Tolovana River	AK	Holland K. Sands	2000	620
18 11/16	11 8/16	7 3/16	M	Ville-Marie	QUE	Larry R. Faught	2001	620
18 11/16	11 11/16	7 0/16	M	Ft. McMurray	ALB	Doug Clayton	2003	620
18 11/16	11 14/16	6 13/16	M	Alstead Lake	SAS	Joe Bradley	2007	620
18 11/16	12 1/16	6 10/16	M	Douglas County	OR	Christopher Tipton	2007	620
18 11/16	11 11/16	7 0/16	M	Las Animas County	CO	Mark Turner	2008	620
18 10/16	11 8/16	7 2/16	M	Iron County	WI	Carl Hulbert	1963	644
18 10/16	11 8/16	7 2/16	M	Montreal River	ONT	S. Robinson/J. Beach	1966	644
18 10/16	11 7/16	7 3/16	M	Rio Grande County	CO	Rod Wintz	1967	644
18 10/16	11 11/16	6 15/16	M	Vermilion Bay	ONT	Thomas L. A. Pucci	1968	644
18 10/16	11 12/16	6 14/16	M	Idaho County	ID	Randolph Coleman	1970	644
18 10/16	12 2/16	6 8/16	M	McLeod Lake	BC	Ron McKay	1973	644
18 10/16	11 5/16	7 5/16	M	Madison County	ID	Bruce W. Baird	1974	644
18 10/16	11 0/16	7 10/16	M	Marinette County	WI	Dan Stencel	1974	644
18 10/16	11 11/16	6 15/16	M	Lewis & Clark County	MT	Scott Koelzer	1977	644
18 10/16	12 1/16	6 9/16	M	Palmer	AK	John F. Sumrall	1977	644
18 10/16	11 14/16	6 12/16	M	Pipe Lake	ONT	Richard Colby	1980	644
18 10/16	11 2/16	7 8/16	M	Judith Basin County	MT	Don Davidson	1981	644
18 10/16	11 9/16	7 1/16	M	Ear Falls	ONT	Ron Marion	1983	644
18 10/16	11 13/16	6 13/16	M	Beluga River	AK	Dennis Redden	1984	644
18 10/16	11 10/16	7 0/16	M	Fort Coulonge	QUE	F. Edward Campbell	1986	644
18 10/16	11 10/16	7 0/16	M	Sudbury	ONT	William F. Boggess	1987	644
18 10/16	11 15/16	6 11/16	M	King County	WA	David B. Young	1989	644
18 10/16	11 10/16	7 0/16	M	Blind River	ONT	Thomas M. Losiewski	1990	644
18 10/16	11 11/16	6 15/16	M	Idaho County	ID	Stephan S. Jones	1991	644
18 10/16	11 11/16	6 15/16	M	La Loche	SAS	Brian E. Ronneberg	1994	644
18 10/16	11 14/16	6 12/16	M	Clarke Lake	SAS	Ron Tandy	1995	644
18 10/16	11 10/16	7 0/16	M	Hearst	ONT	Kenyon W. Woods	1996	644
18 10/16	11 10/16	7 0/16	M	Lynn Lake	MAN	Dave Canfield	1996	644
18 10/16	11 11/16	6 15/16	M	Lemhi County	ID	Russell D. Kennedy	1996	644
18 10/16	11 13/16	6 13/16	M	Fairview	ALB	Richard Tucker	1997	644
18 10/16	11 14/16	6 12/16	M	Clearwater River	ALB	Darryl Quidort	1997	644
18 10/16	11 9/16	7 1/16	M	Riou Lake	SAS	Pat Reilly	1997	644
18 10/16	11 8/16	7 2/16	M	Caramat	ONT	Willie Runnels	2000	644
18 10/16	11 8/16	7 2/16	M	Greene County	NY	James P. Antippas	2000	644
18 10/16	11 15/16	6 11/16	M	Glaslyn	SAS	Matt Rosensweet	2001	644
18 10/16	11 12/16	6 14/16	M	Boise County	ID	Bill Watts, Jr.	2003	644
18 10/16	11 13/16	6 13/16	M	Ear Falls	ONT	Robert J. Pierson	2006	644
18 10/16	12 0/16	6 10/16	M	Fishing Lake	ALB	Scott Silverness	2007	644
18 10/16	11 4/16	7 6/16	M	Blaine County	ID	Michael J. Vigueria	2008	644
18 10/16	11 12/16	6 14/16	M	St. Louis County	MN	Timothy J. Cardinal	2009	644
18 10/16	11 6/16	7 4/16	M	Boise County	ID	James W. Lambrecht	2012	644
18 9/16	11 5/16	7 4/16	M	Chelan County	WA	Wayne Hathaway	1960	680
18 9/16	11 6/16	7 3/16	M	Olsen Bay	AK	Don Daniels	1963	680
18 9/16	11 10/16	6 15/16	M	Iron County	MI	Don Schram	1964	680

Minimum Score 18 *Continued*

Score	Greatest Length	Greatest Width	Sex	Area	State/ Province	Hunter's Name	Date	Rank
18 9/16	11 10/16	6 15/16	M	Mattawa	ONT	Dr. Max G. Menefee	1966	680
18 9/16	11 5/16	7 4/16	M	St. Louis County	MN	Ron Johnson	1967	680
18 9/16	11 5/16	7 4/16	M	Blue Jay Ridge	CA	Delbert Allmon	1972	680
18 9/16	11 12/16	6 13/16	M	Rio Arriba County	NM	Curtis W. McClahan	1973	680
18 9/16	12 0/16	6 9/16	M	Swan Hills	ALB	Gerald L. Egbert	1976	680
18 9/16	11 7/16	7 2/16	M	Franklin County	ME	John G. Morningstar	1977	680
18 9/16	11 8/16	7 1/16	M	Temiscaming	QUE	Joe G. Hopwood	1984	680
18 9/16	11 9/16	7 0/16	M	Shasta County	CA	Peter Esposito	1984	680
18 9/16	11 10/16	6 15/16	M	Pancake Bay	ONT	Brad L. Rogers	1985	680
18 9/16	11 12/16	6 13/16	M	Cass County	MN	Larry Fischer	1985	680
18 9/16	11 8/16	7 1/16	M	Temiscaming	QUE	Daniel E. Wallace	1988	680
18 9/16	11 13/16	6 12/16	M	Oneida County	WI	Tim Johnson	1988	680
18 9/16	11 10/16	6 15/16	M	Idaho County	ID	Monty Moravec	1990	680
18 9/16	11 13/16	6 12/16	M	Idaho County	ID	Monty Moravec	1991	680
18 9/16	11 11/16	6 14/16	M	Prince of Wales Island	AK	Jim Young	1993	680
18 9/16	11 11/16	6 14/16	M	Timmins	ONT	James F. Blanton	1994	680
18 9/16	11 5/16	7 4/16	M	Atikokan	ONT	Mark "Root" Gies	1994	680
18 9/16	11 9/16	7 0/16	M	Slave Lake	ALB	Gregory Noble Spickler	1995	680
18 9/16	11 11/16	6 14/16	M	Cabonga	QUE	Len Cardinale	1995	680
18 9/16	11 14/16	6 11/16	M	Shoshone County	ID	Roger Stewart	1996	680
18 9/16	11 13/16	6 12/16	M	Yavapai County	AZ	Chris J. Dunn	1997	680
18 9/16	11 12/16	6 13/16	M	Sandilands	MAN	Russell K. Mehling	1998	680
18 9/16	11 8/16	7 1/16	M	Ray River	AK	Michael Chadwick	1998	680
18 9/16	11 14/16	6 11/16	M	Mattawin	QUE	Jeffrey L. Scott	2000	680
18 9/16	11 12/16	6 13/16	M	Clisby Lake	MAN	Kim B. Engelbert	2006	680
18 9/16	11 4/16	7 5/16	M	Boise County	ID	Jeffrey Stevens	2007	680
18 9/16	11 12/16	6 13/16	M	Lemhi County	ID	R. Craig Oberle	2010	680
18 9/16	11 9/16	7 0/16	M	Blaine County	ID	Michael J. Vigueria	2011	680
18 8/16	11 8/16	7 0/16	M	Whatcom County	WA	Jack Fish	1961	711
18 8/16	11 13/16	6 11/16	M	Prince William Sound	AK	Bob Snelson	1962	711
18 8/16	11 8/16	7 0/16	M	Iron County	WI	Maynard Peck	1963	711
18 8/16	11 6/16	7 2/16	M	Clearwater County	ID	Robert J. Kreisher	1965	711
18 8/16	11 11/16	6 13/16	M	Sapawe	ONT	Dennis Gregory	1967	711
18 8/16	11 4/16	7 4/16	M	Mineral County	CO	Rod Wintz	1967	711
18 8/16	11 12/16	6 12/16	M	Siskiyou County	CA	Lyle L. Stroble	1968	711
18 8/16	11 7/16	7 1/16	M	Franklin County	ME	John Miterko	1970	711
18 8/16	12 0/16	6 8/16	M	Forest County	WI	Vilas Backhaus	1972	711
18 8/16	11 4/16	7 4/16	M	Messines	QUE	John W. Redmond	1973	711
18 8/16	11 10/16	6 14/16	M	Somerset County	ME	John D. Bonargo	1974	711
18 8/16	11 0/16	7 8/16	M	Messines	QUE	Arthur R. Litschewski	1975	711
18 8/16	11 7/16	7 1/16	M	Hopetown	ONT	Dale Bailey	1976	711
18 8/16	11 8/16	7 0/16	M	Cook County	MN	Paul Smith	1978	711
18 8/16	11 4/16	7 4/16	M	Franklin County	ME	John Janelli	1979	711
18 8/16	11 6/16	7 2/16	M	Atikokan	ONT	Larry Stewart	1982	711
18 8/16	11 10/16	6 14/16	M	Bancroft	ONT	Dean J. Farkas	1982	711
18 8/16	11 2/16	7 6/16	M	Carbon County	WY	Vaughn Cross	1984	711
18 8/16	11 10/16	6 14/16	M	Susitna River	AK	Roger Stewart	1985	711
18 8/16	11 12/16	6 12/16	M	Hartland	NBW	Frank Cinquemani	1987	711
18 8/16	11 6/16	7 2/16	M	Ravalli County	MT	John Locke	1988	711
18 8/16	11 6/16	7 2/16	M	Douglas County	OR	Tom E. Tipton	1988	711
18 8/16	11 13/16	6 11/16	M	Wapawekka Hills	SAS	Ronald J. Collier	1989	711
18 8/16	11 14/16	6 10/16	M	Doaktown	NBW	George Louis	1991	711
18 8/16	11 5/16	7 3/16	M	Custer County	ID	Darrell Nunez	1993	711
18 8/16	11 6/16	7 2/16	M	Aulneau Peninsula	ONT	Michael Koska	1993	711
18 8/16	12 0/16	6 8/16	M	Lane County	OR	Gary Nyden	1994	711
18 8/16	11 10/16	6 14/16	M	Indian River	AK	Rick Schikora	1996	711
18 8/16	11 13/16	6 11/16	M	Price County	WI	David J. Kerkove	1996	711
18 8/16	11 12/16	6 12/16	M	Shoshone County	ID	Roger Stewart	1997	711
18 8/16	11 9/16	6 15/16	M	Athabasca River	ALB	Gene Mathias	1998	711
18 8/16	11 5/16	7 3/16	M	Lynn Lake	MAN	James Thomas Schmitz	1999	711
18 8/16	11 5/16	7 3/16	M	St-Michel-des-Saints	QUE	Walt Guthrie	1999	711
18 8/16	11 4/16	7 4/16	M	Aroostook County	ME	Thomas J. Dussault	1999	711

BLACK BEAR

Minimum Score 18 *Continued*

Score	Greatest Length	Greatest Width	Sex	Area	State/ Province	Hunter's Name	Date	Rank
18 8/16	11 12/16	6 12/16	M	Aroostook County	ME	William J. Langer	2004	711
18 8/16	11 10/16	6 14/16	M	Campbell River	BC	John Duggan	2006	711
18 8/16	11 9/16	6 15/16	M	Jellicoe	ONT	Bob Kostecki	2006	711
18 8/16	11 12/16	6 12/16	M	Kenora	ONT	Thomas Filipiak	2006	711
18 8/16	11 8/16	7 0/16	M	File Lake	MAN	Randy E. Doyle	2008	711
18 8/16	11 11/16	6 13/16	M	Chugiak	AK	Rick Helton	2009	711
18 8/16	11 9/16	6 15/16	M	La Tuque	QUE	Jeff Springer	2009	711
18 7/16	11 11/16	6 12/16	M	Oneida County	WI	Tim Johnson	1967	752
18 7/16	11 5/16	7 2/16	M	Franklin County	ME	Ed Hall	1968	752
18 7/16	11 8/16	6 15/16	M	Piscataquis County	ME	John Kuhar	1975	752
18 7/16	11 13/16	6 10/16	M	Albemarle County	VA	J. C. Locke	1976	752
18 7/16	11 4/16	7 3/16	M	El Paso County	CO	Max Tallent	1981	752
18 7/16	11 10/16	6 13/16	M	Emo	ONT	Bruce Eggenberger	1986	752
18 7/16	11 11/16	6 12/16	M	Sublette County	WY	Keith Dana	1988	752
18 7/16	11 10/16	6 13/16	M	Jefferson County	WA	Wayne Haag	1989	752
18 7/16	11 13/16	6 10/16	M	Carbon County	WY	Damon Handley	1991	752
18 7/16	11 2/16	7 5/16	M	Sudbury	ONT	Phillip Wilkinson	1992	752
18 7/16	11 10/16	6 13/16	M	Fort McMurray	ALB	Jon P. Thomas	1992	752
18 7/16	11 6/16	7 1/16	M	Wallowa County	OR	Thompson Holmes	1992	752
18 7/16	11 3/16	7 4/16	M	Mountainview	ALB	Bill Van Buskirk	1995	752
18 7/16	12 1/16	6 6/16	M	Hudson Bay	SAS	Norma J. Weikert	1995	752
18 7/16	11 13/16	6 10/16	M	Delta Junction	AK	Jared Cummings	1998	752
18 7/16	11 14/16	6 9/16	M	Skamania County	WA	Lane S. Stettler	1999	752
18 7/16	11 11/16	6 12/16	M	Long Lake	MAN	Emile P. LeBlanc	2002	752
18 7/16	11 12/16	6 11/16	M	Piscataquis County	ME	Bruce Guyan	2004	752
18 7/16	11 12/16	6 11/16	M	Frobisher Lake	SAS	Kevin Dancy	2006	752
18 6/16	11 9/16	6 13/16	M	Mariposa County	CA	Douglas Walker	1965	771
18 6/16	11 8/16	6 14/16	M	Fulton County	NY	Peter Mertens	1966	771
18 6/16	11 10/16	6 12/16	M	Yancey County	NC	Jerry Rushing	1970	771
18 6/16	11 12/16	6 10/16	M	Franklin County	ME	Joe Melchiore	1970	771
18 6/16	11 8/16	6 14/16	M	Clear Creek County	CO	Jim Dougherty	1970	771
18 6/16	11 8/16	6 14/16	M	Becker County	MN	Gordon Swenson	1973	771
18 6/16	11 12/16	6 10/16	M	Flathead County	MT	Dr. Barry Wensel	1974	771
18 6/16	11 10/16	6 12/16	M	Saguache County	CO	Sandra Scheid	1975	771
18 6/16	11 5/16	7 1/16	M	Ramsey Township	ONT	Robert H. Pavlovic	1983	771
18 6/16	11 8/16	6 14/16	M	Montrose County	CO	Wayne Willis	1983	771
18 6/16	11 8/16	6 14/16	F	Spiritwood	SAS	Kent W. Brigham	1986	771
18 6/16	11 14/16	6 8/16	M	Hudson Bay	SAS	C. Randall Byers	1987	771
18 6/16	11 3/16	7 3/16	M	Wabigoon	ONT	Robert Barrie	1987	771
18 6/16	11 15/16	6 7/16	M	Hudson Bay	SAS	Kent Brandt	1988	771
18 6/16	11 1/16	7 5/16	M	Jocko River	ONT	Gary Boals	1988	771
18 6/16	11 6/16	7 0/16	M	Chibougamau	QUE	Dale R. Walburger	1988	771
18 6/16	11 1/16	7 5/16	M	Okanogan County	WA	Duane N. Fink	1988	771
18 6/16	11 11/16	6 11/16	M	Saguache County	CO	Sid Strzok	1989	771
18 6/16	11 7/16	6 15/16	M	Emo	ONT	Leo Hazelton	1989	771
18 6/16	11 7/16	6 15/16	M	Peace River	ALB	John Peruchini	1990	771
18 6/16	11 13/16	6 9/16	F	Rainy River	ONT	Michael Judas	1991	771
18 6/16	11 4/16	7 2/16	M	Colfax County	NM	Rodrigo Cruz	1991	771
18 6/16	11 11/16	6 11/16	M	Aitkin County	MN	Brad Blanchard	1994	771
18 6/16	11 9/16	6 13/16	M	Shoshone County	ID	Roger Stewart	1995	771
18 6/16	11 15/16	6 7/16	M	Aulneau Peninsula	ONT	Michael Koska	1995	771
18 6/16	11 12/16	6 10/16	M	Dryden	ONT	Rick Kroll	1995	771
18 6/16	11 7/16	6 15/16	M	Nestor Falls	ONT	Dale E. Springer	1995	771
18 6/16	11 13/16	6 9/16	M	Calling Lake	ALB	Robert Bartoshesky	1996	771
18 6/16	11 11/16	6 11/16	M	Indian River	AK	Rick Schikora	1998	771
18 6/16	11 8/16	6 14/16	M	Dryden	ONT	Alan E. Forbes	1998	771
18 6/16	11 12/16	6 10/16	M	Athabasca River	ALB	Michael Travis	1999	771
18 6/16	12 1/16	6 5/16	M	Olha	MAN	Mike Windemuller	1999	771
18 6/16	11 14/16	6 8/16	M	Carrot River	SAS	Phil Harris	1999	771
18 6/16	11 11/16	6 11/16	M	La Tuque	QUE	Thomas M. Phillips	2003	771
18 6/16	11 8/16	6 14/16	M	Shoshone County	ID	Roger Stewart	2005	771
18 6/16	12 2/16	6 4/16	M	Laurier	MAN	Robert Tastsides	2006	771

BLACK BEAR

Minimum Score 18 *Continued*

Score	Greatest Length	Greatest Width	Sex	Area	State/ Province	Hunter's Name	Date	Rank
18 6/16	11 6/16	7 0/16	M	La Tuque	QUE	Tom Phillips	2009	771
18 6/16	11 10/16	6 12/16	M	Flathead County	MT	Charles C. Stafford	2009	771
18 6/16	11 4/16	7 2/16	M	Koochiching County	MN	Charles W. Schultz	2011	771
18 6/16	11 10/16	6 12/16	M	Tanana River	AK	Steve Hohensee	2012	771
18 6/16	11 6/16	7 0/16	M	Tanana River	AK	Ted Kinney	2012	771
18 5/16	11 3/16	7 2/16	M	Iron Bridge	ONT	Philip L. Hawkins	1965	812
18 5/16	11 6/16	6 15/16	M	St. Louis County	MN	Kay A. Schuver	1969	812
18 5/16	11 9/16	6 12/16	M	Gogebic County	MI	LaVern Miller	1971	812
18 5/16	11 6/16	6 15/16	M	Florence County	WI	Jim Thimmig	1971	812
18 5/16	11 3/16	7 2/16	M	Forest County	WI	Ernie V. Hutchinson	1972	812
18 5/16	11 9/16	6 12/16	M	Park County	MT	Charles Burdette	1980	812
18 5/16	11 3/16	7 2/16	M	E. Braintree	MAN	Chester Surma	1980	812
18 5/16	11 8/16	6 13/16	M	Cold Lake	ALB	Orest Popil	1985	812
18 5/16	11 12/16	6 9/16	M	Lake County	MT	Don Davidson	1986	812
18 5/16	11 13/16	6 8/16	M	Rocky Lake	MAN	Tim Finley	1986	812
18 5/16	11 9/16	6 12/16	M	Ignace	ONT	Joel Breitung	1988	812
18 5/16	11 8/16	6 13/16	M	Boise Franc Rd.	QUE	August S. Gray	1989	812
18 5/16	11 14/16	6 7/16	M	Lemhi County	ID	Al Youman	1990	812
18 5/16	11 8/16	6 13/16	M	Atikokan	ONT	Judy Grooms	1991	812
18 5/16	11 9/16	6 12/16	M	Fort Coulonge	QUE	Russ Kay	1993	812
18 5/16	11 8/16	6 13/16	M	Sup Lake	ONT	Scott Ellery	1993	812
18 5/16	11 9/16	6 12/16	M	Smeaton	SAS	Gene A. Welle	1994	812
18 5/16	11 9/16	6 12/16	M	High Level	ALB	Donny Roy	1994	812
18 5/16	11 11/16	6 10/16	M	Smeaton	SAS	Gene Welle	1995	812
18 5/16	11 10/16	6 11/16	M	Caramat	ONT	A. Will Vander Ende	1996	812
18 5/16	11 11/16	6 10/16	M	Piscataquis County	ME	Thomas Giambrone	1996	812
18 5/16	11 8/16	6 13/16	M	Apache County	AZ	Michael J. Windemuller	1998	812
18 5/16	11 5/16	7 0/16	M	St. Mary's River	BC	Bob Wright	1999	812
18 5/16	12 0/16	6 5/16	M	Athabasca River	ALB	Ron Adamson	1999	812
18 5/16	11 9/16	6 12/16	M	Slave River	ALB	Henry Loving	1999	812
18 5/16	11 8/16	6 13/16	M	Wapawekka Lake	SAS	Gary L. Stefanovsky	2001	812
18 5/16	11 4/16	7 1/16	M	Penobscot County	ME	Galen J. Wertz	2001	812
18 5/16	11 12/16	6 9/16	F	Clark County	WI	Mark S. Biddle	2001	812
18 5/16	11 8/16	6 13/16	M	La Plata County	CO	Joseph Alan Bradley	2001	812
18 5/16	11 9/16	6 12/16	M	MacKay	ALB	David Threadgold	2002	812
18 5/16	11 9/16	6 12/16	M	Prince of Wales Island	AK	Brad Dorsing	2002	812
18 5/16	12 1/16	6 4/16	M	Duck Mtn.	MAN	Mike Dunnaway	2005	812
18 4/16	10 14/16	7 6/16	M	Latah County	ID	Don Lawrence	1962	844
18 4/16	11 7/16	6 13/16	M	Curry County	OR	Gerald Rimbey	1962	844
18 4/16	11 4/16	7 0/16	M	Shawano County	WI	Peter Erickson	1967	844
18 4/16	11 12/16	6 8/16	M	Fort Frances	ONT	Wayne Keefer	1968	844
18 4/16	11 12/16	6 8/16	M	Itasca County	MN	Lonny Herrick	1973	844
18 4/16	11 6/16	6 14/16	M	Lily Lake	BC	Stanley Moore	1979	844
18 4/16	12 0/16	6 4/16	M	Fort Assiniboine	ALB	Jim Dahlberg	1985	844
18 4/16	11 2/16	7 2/16	M	Ravalli County	MT	James Patenaude	1989	844
18 4/16	11 12/16	6 8/16	M	Rainy River	ONT	Michael Judas	1990	844
18 4/16	11 2/16	7 2/16	M	Kapuskasing	ONT	Jim Melton	1991	844
18 4/16	11 6/16	6 14/16	M	Temiscaming	QUE	Fred Wallace	1991	844
18 4/16	11 8/16	6 12/16	M	Kalkaska County	MI	David L. Roose	1991	844
18 4/16	11 5/16	6 15/16	M	Clearwater County	ID	Jim Bradford	1992	844
18 4/16	11 9/16	6 11/16	M	Pelly River	YUK	Steve Byerly	1993	844
18 4/16	11 11/16	6 9/16	M	Indian River	AK	Rick Schikora	1997	844
18 4/16	11 7/16	6 13/16	M	Dryden	ONT	Mark A. Ascheman	1997	844
18 4/16	11 4/16	7 0/16	M	La Plata County	CO	Garth Schultheis	1998	844
18 4/16	11 10/16	6 10/16	M	Fort McMurray	ALB	H. Mike Palmer	2000	844
18 4/16	11 5/16	6 15/16	M	Big Sandy Lake	SAS	John Holt	2001	844
18 4/16	11 12/16	6 8/16	M	Rappahannock County	VA	Marvin T. Breeden	2002	844
18 4/16	11 8/16	6 12/16	M	Washington County	AR	Ben Moffitt	2004	844
18 4/16	11 10/16	6 10/16	M	Alstead Lake	SAS	Joseph Bradley	2005	844
18 3/16	11 1/16	7 2/16	M	Penobscot County	ME	Charles A. Kronyak	1965	866
18 3/16	11 0/16	7 3/16	M	Custer County	ID	C. Randall Byers	1966	866
18 3/16	11 15/16	6 4/16	M	Garfield County	CO	Bob Swinehart	1967	866

BLACK BEAR

Minimum Score 18 *Continued*

Score	Greatest Length	Greatest Width	Sex	Area	State/ Province	Hunter's Name	Date	Rank
18 3/16	11 4/16	6 15/16	M	Franklin County	ME	Philip Copp	1967	866
18 3/16	11 6/16	6 13/16	M	Florence County	WI	Elaine S. Peck	1967	866
18 3/16	11 6/16	6 13/16	M	Fremont County	CO	Al Weaver	1983	866
18 3/16	11 10/16	6 9/16	M	Moose River	ONT	Bob Duncan	1984	866
18 3/16	11 6/16	6 13/16	M	King County	WA	Charles D. Singh	1986	866
18 3/16	11 9/16	6 10/16	M	Minaki	ONT	Donald Schram	1987	866
18 3/16	11 9/16	6 10/16	M	Webbwood	ONT	Stephen P. Turay	1987	866
18 3/16	11 1/16	7 2/16	M	Oxford County	ME	Jack Smith	1988	866
18 3/16	11 2/16	7 1/16	F	Prince of Wales Island	AK	Rickie D. Snell	1990	866
18 3/16	11 12/16	6 7/16	M	Bonnyville	ALB	Glen Garton	1991	866
18 3/16	11 10/16	6 9/16	M	La Ronge	SAS	Roger Wintle	1991	866
18 3/16	11 5/16	6 14/16	M	La Tuque	QUE	William R. Lewis	1991	866
18 3/16	11 10/16	6 9/16	M	Lake Besnard	SAS	Roger Wintle	1992	866
18 3/16	11 11/16	6 8/16	M	Bissett	MAN	Robert Harris	1993	866
18 3/16	11 11/16	6 8/16	M	Smeaton	SAS	Gene A. Welle	1993	866
18 3/16	11 11/16	6 8/16	M	Exshaw	ALB	Mark Wuerthele	1994	866
18 3/16	11 9/16	6 10/16	M	Prince of Wales Island	AK	Gerald L. Egbert	1994	866
18 3/16	11 14/16	6 5/16	M	Hudson Bay	SAS	William E. Lee, Jr.	1994	866
18 3/16	11 5/16	6 14/16	M	Yarmouth County	NS	Bill Terry, Sr.	1994	866
18 3/16	11 8/16	6 11/16	M	Lynn Lake	MAN	Andrew W. Szczesniak	1995	866
18 3/16	11 8/16	6 11/16	M	Idaho County	ID	Marlon J. Clapham	1998	866
18 3/16	11 8/16	6 11/16	M	Kipawa	QUE	Joseph L. Vincent	2001	866
18 3/16	12 0/16	6 3/16	M	Duck Mtns.	MAN	Royce DeCook	2005	866
18 3/16	11 2/16	7 1/16	M	Mattice	ONT	Timothy M. Schlaegel	2006	866
18 3/16	11 7/16	6 12/16	M	Perrault Falls	ONT	David G. Casady	2007	866
18 3/16	11 10/16	6 9/16	M	Prince of Wales Island	AK	Don Davidson, Jr.	2009	866
18 2/16	11 8/16	6 10/16	M	Murphy Dome	AK	Richard Cooper	1955	895
18 2/16	11 2/16	7 0/16	M	Sudbury	ONT	Floyd Eccleston	1961	895
18 2/16	11 4/16	6 14/16	M	Penobscot County	ME	Bill L. Carlos	1963	895
18 2/16	11 8/16	6 10/16	M	Penobscot County	ME	Dennis H. Driscoll	1967	895
18 2/16	11 8/16	6 10/16	M	Franklin County	ME	Kenneth Rapp	1969	895
18 2/16	11 5/16	6 13/16	M	Chapleau	ONT	Ed Helgason	1970	895
18 2/16	11 6/16	6 12/16	M	Chetwynd	BC	Lee E. Hansel	1973	895
18 2/16	11 10/16	6 8/16	M	Marinette County	WI	Jack Baxter	1976	895
18 2/16	10 14/16	7 4/16	M	Flathead County	MT	Paul P. Schafer	1976	895
18 2/16	11 8/16	6 10/16	M	Powell County	MT	Paul Brunner	1979	895
18 2/16	11 14/16	6 4/16	M	Boise County	ID	Larry Hoff	1980	895
18 2/16	11 8/16	6 10/16	M	Ignace	ONT	Elmer R. Luce, Jr.	1982	895
18 2/16	11 4/16	6 14/16	M	Fort Coulonge	QUE	Wm. Fred Stone	1985	895
18 2/16	11 15/16	6 3/16	M	Duck Mtn.	MAN	Bill Clink	1986	895
18 2/16	11 6/16	6 12/16	M	Wabigoon	ONT	Albert J. Smith	1988	895
18 2/16	11 6/16	6 12/16	M	San Miguel County	NM	Dick McClain	1989	895
18 2/16	11 2/16	7 0/16	M	Zec Restigo	QUE	Michael P. Murphy	1990	895
18 2/16	11 8/16	6 10/16	M	Stevens County	WA	Allen J. Thrush	1990	895
18 2/16	11 8/16	6 10/16	M	Lebel Township	ONT	Mike Hartling	1991	895
18 2/16	11 11/16	6 7/16	M	Grand County	CO	Barry J. Smith	1992	895
18 2/16	11 9/16	6 9/16	M	Tyson's Lake	ONT	Gary L. Seabright	1992	895
18 2/16	11 10/16	6 8/16	M	Spiritwood	SAS	Nick A. Mathews	1992	895
18 2/16	11 12/16	6 6/16	M	Bayfield County	WI	Todd W. Henck	1993	895
18 2/16	11 4/16	6 14/16	M	Rainbow Lake	ALB	Lorenzo Dow Utterback	1996	895
18 2/16	11 2/16	7 0/16	M	Fredericton	NBW	Louis Miranda III	2000	895
18 2/16	11 8/16	6 10/16	M	Chibougamau	QUE	Michael P. Kissel	2000	895
18 2/16	11 8/16	6 10/16	M	La Tuque	QUE	Leonard Billings	2003	895
18 2/16	11 10/16	6 8/16	M	Fort McMurray	ALB	H. Mike Palmer	2004	895
18 2/16	11 7/16	6 11/16	M	Lemhi County	ID	Ben L. Fahnholz	2006	895
18 2/16	11 3/16	6 15/16	M	Zama	ALB	Ron Morgan	2007	895
18 2/16	11 7/16	6 11/16	M	Las Animas County	CO	Ronald Rockwell	2008	895
18 2/16	11 8/16	6 10/16	M	Gunnison County	CO	Mike Vigueria	2011	895
18 1/16	11 8/16	6 9/16	M	Chelan County	WA	Dick Smethvrst	1965	927
18 1/16	11 4/16	6 13/16	M	Clearwater County	ID	Jess Stinichcome	1965	927
18 1/16	11 0/16	7 1/16	M	Somerset County	ME	Raymond Benedetto	1970	927
18 1/16	11 7/16	6 10/16	M	Franklin County	ME	Walter Krom	1972	927

BLACK BEAR

Minimum Score 18 — *Continued*

Score	Greatest Length	Greatest Width	Sex	Area	State/ Province	Hunter's Name	Date	Rank
18 1/16	11 3/16	6 14/16	M	Blackwater River	BC	Ron McKay	1974	927
18 1/16	10 15/16	7 2/16	M	Chapleau	ONT	Kevin E. Murphy	1975	927
18 1/16	11 9/16	6 8/16	M	Archuleta County	CO	Stephen E. Kennedy	1982	927
18 1/16	11 8/16	6 9/16	M	Dwight	ONT	Michael D. Moore	1983	927
18 1/16	11 2/16	6 15/16	M	Colfax County	NM	Dean Oatman	1983	927
18 1/16	11 9/16	6 8/16	M	McBride Lake	SAS	John Rook	1985	927
18 1/16	11 11/16	6 6/16	M	Huerfano County	CO	Jim Witcombe	1989	927
18 1/16	11 1/16	7 0/16	M	Idaho County	ID	Monty Moravec	1989	927
18 1/16	11 7/16	6 10/16	M	Nipigon	ONT	Fred W. Achilles	1990	927
18 1/16	11 7/16	6 10/16	M	Huerfano County	CO	Robert L. Beckwith	1990	927
18 1/16	11 9/16	6 8/16	M	Kelvington	SAS	Robert C. McCardell	1992	927
18 1/16	11 4/16	6 13/16	M	Searchmont	ONT	Michael J. Windemuller	1996	927
18 1/16	11 10/16	6 7/16	M	Aroostook County	ME	Mark J. Yost	1999	927
18 1/16	11 7/16	6 10/16	M	Markstay	ONT	Christopher Allen Ewald	1999	927
18 1/16	11 4/16	6 13/16	M	Lac Roger	QUE	Robert R. Agnew, Jr.	2001	927
18 1/16	11 5/16	6 12/16	M	Val-Paradis	QUE	Bryan Burkhardt	2001	927
18 1/16	11 10/16	6 7/16	M	Athabasca River	ALB	Gary Forte	2003	927
18 1/16	11 10/16	6 7/16	M	Peace River	ALB	Edward C. Evans	2004	927
18 1/16	11 0/16	7 1/16	M	Reservoir Cabonga	QUE	Tom Lagatol III	2005	927
18 1/16	11 9/16	6 8/16	M	Roscommon County	MI	Troy Spooner	2006	927
18 1/16	11 8/16	6 9/16	M	Alstead Lake	SAS	Tim Bradley	2007	927
18 1/16	10 15/16	7 2/16	M	Fort Coulonge	QUE	Walter Dixon	2007	927
18 1/16	11 11/16	6 6/16	M	Kamkota	SAS	H. Mike Palmer	2007	927
18 1/16	11 1/16	7 0/16	M	Lemhi County	ID	Ben L. Fahnholz	2009	927
18 1/16	11 9/16	6 8/16	M	Lemhi County	ID	Jeffrey Eder	2010	927
18 0/16	11 0/16	7 0/16	M	Polk County	OR	H. Dale Overholser	1959	956
18 0/16	11 9/16	6 7/16	M	Prince William Sound	AK	Bob Snelson	1962	956
18 0/16	11 12/16	6 4/16	M	Franklin County	ME	John Iannuzzo	1966	956
18 0/16	11 6/16	6 10/16	M	Jackson County	OR	Pat Mastan	1970	956
18 0/16	11 2/16	6 14/16	M	Franklin County	ME	Donald R. Pyne	1970	956
18 0/16	11 4/16	6 12/16	M		ONT	Larry Kuskie	1974	956
18 0/16	11 10/16	6 6/16	M	Franklin County	ME	Michael P. Murphy	1974	956
18 0/16	11 9/16	6 7/16	M	Marathon County	WI	Jay Schultz	1977	956
18 0/16	11 10/16	6 6/16	M	Little Susitna River	AK	Gary G. Wall	1984	956
18 0/16	11 6/16	6 10/16	M	Messines	QUE	Howard 'Butch' Malone	1984	956
18 0/16	11 2/16	6 14/16	M	Kechika Range	BC	Wade L. Carstens	1984	956
18 0/16	11 1/16	6 15/16	M	Missoula County	MT	John L. Wozniak	1986	956
18 0/16	11 4/16	6 12/16	M	Muldrew Township	ONT	Daniel P. Wieske	1987	956
18 0/16	11 11/16	6 5/16	M	Athabasca River	ALB	Grant Adkisson	1989	956
18 0/16	11 4/16	6 12/16	M	Algoma	ONT	Denis Belcourt	1989	956
18 0/16	11 9/16	6 7/16	M	Dryden	ONT	Albert J. Smith	1989	956
18 0/16	11 2/16	6 14/16	M	Rollet	QUE	Douglas E. Ott	1989	956
18 0/16	11 4/16	6 12/16	M	Beurling River	QUE	Philippe Galley	1989	956
18 0/16	11 4/16	6 12/16	M	Cranberry Portage	MAN	Bob Beardsley	1990	956
18 0/16	11 6/16	6 10/16	M	Red Lake	ONT	Fred Sprague	1990	956
18 0/16	11 10/16	6 6/16	M	Biscotasing	ONT	Everett W. Ayers	1991	956
18 0/16	11 0/16	7 0/16	M	Idaho County	ID	T. J. Conrads	1992	956
18 0/16	11 5/16	6 11/16	M	Monds Township	ONT	Bob Coyle	1995	956
18 0/16	11 11/16	6 5/16	M	Rocky Lake	MAN	Scott D. Schmidt	1995	956
18 0/16	11 7/16	6 9/16	M	Prince of Wales Island	AK	Gerald L. Egbert	1995	956
18 0/16	11 5/16	6 11/16	M	Brazeau Dam	ALB	Wayne Depperschmidt	1996	956
18 0/16	11 7/16	6 9/16	M	Hearst	ONT	Gregory D. Keeton	1996	956
18 0/16	10 12/16	7 4/16	M	Anishinabi Lake	ONT	Kim Knuti	1997	956
18 0/16	11 2/16	6 14/16	M	Kapuskasing	ONT	Mike Wiseman	1997	956
18 0/16	11 7/16	6 9/16	M	Carrot River	SAS	Joseph P. Furlong	1998	956
18 0/16	11 8/16	6 8/16	M	Iron Bridge	ALB	William G. Mason	1998	956
18 0/16	11 10/16	6 6/16	M	Rice Lake	SAS	James Frank	2005	956
18 0/16	11 9/16	6 7/16	M	Alstead Lake	SAS	Joseph Bradley	2006	956
18 0/16	11 10/16	6 6/16	M	Athabasca River	ALB	Ronald Rockwell	2006	956
18 0/16	11 7/16	6 9/16	M	Alstead Lake	SAS	Tim Bradley	2008	956
18 0/16	11 7/16	6 9/16	M	Koochiching County	MN	Charles W. Schultz	2008	956

GRIZZLY BEAR

Minimum Score 19 — *Ursus arctos horribilis*

Score	Greatest Length	Greatest Width	Sex	Area	State/ Province	Hunter's Name	Date	Rank
25 6/16	16 0/16	9 6/16	M	Anzac River	BC	Harley Tison	1972	1
25 3/16	15 13/16	9 6/16	M	Stevens Lakes	BC	Dr. Rex Hancock	1968	2
25 2/16	16 0/16	9 2/16	M	Windy	AK	Rick D. Snell	1997	3
24 15/16	15 8/16	9 7/16	M	Unalakleet	AK	Fred Eichler	2005	4
24 14/16	15 10/16	9 4/16	M	Yellowstone National Park	WY	Art Young	1920	5
24 11/16	15 2/16	9 9/16	M	Stevens Lakes	BC	Dr. R. L. Hambrick	1965	6
24 9/16	15 13/16	8 12/16	M	Bella Coola	BC	William P. Mastrangel	1956	7
24 0/16	15 0/16	9 0/16	M	Joe Poole Creek	BC	Scott L. Koelzer	2001	8
23 13/16	15 6/16	8 7/16	M	Gulkana River	AK	Art Kragness	1973	9
23 7/16	14 14/16	8 9/16	M	Kakwa River	ALB	Rick Michalski	1981	10
22 12/16	14 6/16	8 6/16	M	Brazeua River	ALB	Curt Lynn	1973	11
22 6/16	14 1/16	8 5/16	M	Joe Poole Creek	BC	Chris Parrino	2003	12
22 2/16	13 13/16	8 5/16	M	Blackstone River	YUK	Rick Duggan	2004	13
21 1/16	13 5/16	7 12/16	M	Atigun Pass	AK	J. R. Pederson	1996	14
21 0/16	13 11/16	7 5/16	M	Little Tok River	AK	Don Davidson, Jr.	1987	15
20 9/16	13 1/16	7 8/16	F	Meziadin Lake	BC	Glenn Hisey	1992	16
20 7/16	13 0/16	7 7/16	F	Kispiox River	BC	Dr. Rex Hancock	1965	17
20 2/16	13 0/16	7 2/16	F	Unalakleet River	AK	Chris Parrino	2006	18
20 1/16	13 1/16	7 0/16	M	Stevens Lakes	BC	Fred Bear	1961	19
20 0/16	12 14/16	7 2/16	F	Yellowstone National Park	WY	Saxton T. Pope	1920	20
19 14/16	12 8/16	7 6/16	F	Ptarmigan Creek	AK	Donald O. Smith	1964	21
19 12/16	12 6/16	7 6/16	F	Unalakleet	AK	Chris Parrino	2008	22
19 10/16	12 5/16	7 5/16	F	Stevens Lakes	BC	G. Fred Asbell	1969	23
19 8/16	12 8/16	7 0/16	M	Tangle Lakes	AK	John Musacchia	1972	24
19 8/16	12 10/16	6 14/16	M	Whitehorse	YUK	Scott Koelzer	1977	24
19 5/16	11 15/16	7 6/16	F	Stevens Lakes	BC	Walter Krom	1968	26
19 2/16	12 12/16	6 6/16	F	Nenana River	AK	Rick D. Snell	1993	27
19 0/16	12 10/16	6 6/16	M	Kispiox River	BC	Charles Kroll	1960	28

Hunter: Mark Turner
Species: Cougar
Score: 14 8/16
Date: 2/14/12
Location: Clark Creek County, CO

POLAR BEAR

Minimum Score 20 *Ursus maritimus*

Score	Greatest Length	Greatest Width	Sex	Area	State/ Province	Hunter's Name	Date	Rank
26 6/16	16 4/16	10 2/16	M	Cape Lisburne	AK	Richard McIntyre	1958	1
25 14/16	16 3/16	9 11/16	M	Chukchi Sea	AK	Larry Jones	1965	2
25 9/16	16 0/16	9 9/16	M	Lancaster Sound	NUN	Rick Duggan	2004	3
25 1/16	15 5/16	9 12/16	F	Baffin Island	NWT	Arthur Young	1926	4
25 0/16	15 12/16	9 4/16	M	Pond Inlet	NUN	Bryce Olson	2009	5
24 15/16	15 3/16	9 12/16	M	Taloyoak	NUN	Ricardo Longoria	1998	6
24 15/16	16 0/16	8 15/16	M	Astronomical Society Isl.	NUN	Fred Eichler	2005	6
24 9/16	16 7/16	8 2/16	M	Holman	NWT	Scott L. Koelzer	2005	8
24 5/16	15 13/16	8 8/16	M	Resolute Bay	NUN	Thomas M. Taylor	2003	9
24 4/16	15 5/16	8 15/16	F	Baffin Island	NWT	Arthur Young	1926	10
21 3/16	13 11/16	7 8/16	F	Astronomical Society Isl.	NUN	Chris Parrino	2005	11

Hunter: Scott Koelzer
Species: Polar Bear
Score: 24 9/16
Date: 5/10/05
Location: Holmen, NWT

BISON

Minimum Score 100 — *Bison bison bison* and *Bison bison athabascae*

Score	Length of R Horn	L	Circumference R of Base	L	Greatest Spread	Area	State/ Province	Hunter's Name	Date	Rank
114 4/8	18 2/8	18 1/8	13 7/8	14 2/8	29 2/8	Halfway River	BC	Ron Rockwell	2007	1
112 2/8	18 4/8	18 0/8	14 1/8	14 1/8	29 6/8	Farewell Lake	AK	George A. Moerlein	1972	2
112 0/8	18 1/8	17 7/8	13 4/8	13 6/8	27 1/8	Halfway River	BC	Tom Taylor	2002	3
112 0/8	17 2/8	17 2/8	13 0/8	13 1/8	27 1/8	Coconino County	AZ	Rick Duggan	2011	3
111 0/8	17 0/8	18 1/8	14 3/8	13 0/8	27 7/8	Teton County	WY	Luke Roush	2009	5
110 4/8	16 2/8	15 2/8	14 1/8	14 1/8	26 7/8	Garfield County	UT	Paul B. Brunner	1979	6
109 4/8	17 0/8	17 5/8	12 6/8	12 6/8	27 0/8	Teton County	WY	Tracy Villwok	2010	7
107 2/8	16 6/8	16 6/8	13 0/8	13 2/8	26 3/8	Pink Mtn.	BC	Fred Eichler	2003	8
102 4/8	15 3/8	15 3/8	12 6/8	12 6/8	24 2/8	Garfield County	UT	Dale Drilling	1991	9
102 4/8	17 1/8	16 7/8	12 2/8	12 0/8	24 2/8	Pink Mtn.	BC	Glenn E. Hisey	2000	9
100 0/8	15 5/8	15 6/8	11 5/8	11 5/8	24 4/8	Pink Mtns.	BC	Mark Turner	2010	11

Hunter: Marty Auch
Species: Pronghorn
Score: 80 0/8
Date: 9/11/10
Location: Campbell County, WY

BARREN GROUND CARIBOU

Minimum Score 325 — *Rangifer tarandus granti, Rangifer tarandus stonei* and *Rangifer tarandus arcticus*

Score	Length of R Main Beam	L	Inside Spread	Number of R Points	L	Area	State/ Province	Hunter's Name	Date	Rank
446 6/8	55 0/8	55 6/8	40 5/8	23	19	Meshik River	AK	Art Kragness	1970	1
424 0/8	52 7/8	51 1/8	41 6/8	21	12	Delta River	AK	Bill Brown	1960	2
417 0/8	50 5/8	52 3/8	33 6/8	13	17	Little Delta River	AK	Fred Bear	1959	3
414 1/8	50 5/8	50 0/8	46 7/8	12	12	Kenai Mtns.	AK	Craig E. Scarbrough	1998	4
407 4/8	54 3/8	55 3/8	51 4/8	20	14	Lake Becharof	AK	Larry Jones	1969	5
407 2/8	51 7/8	53 0/8	34 3/8	14	11	Squirrel River	AK	George A. Moerlein	1993	6
404 7/8	51 2/8	50 6/8	40 0/8	13	14	Franklin Bluffs	AK	Rickie D. Snell	1989	7
402 0/8	44 4/8	45 0/8	46 0/8	14	14	Kodiak Island	AK	Jerry Russell	2006	8
401 6/8	57 7/8	56 7/8	45 3/8	13	14	Purcell Mtn.	AK	Chris G. Sanford	1998	9
401 3/8	53 4/8	52 6/8	42 0/8	14	16	Glenn Highway	AK	Harv Ebers	1959	10
400 2/8	51 6/8	51 6/8	43 2/8	14	14	Upper Noatak River	AK	Patrick Campanella	1989	11
400 0/8	53 1/8	53 1/8	37 0/8	9	19	Lake Louise	AK	George Moerlein	1962	12
399 0/8	50 5/8	51 2/8	40 2/8	16	14	Clemmons	AK	Bob Lee	1960	13
397 4/8	46 4/8	46 3/8	34 2/8	15	15	Telaquana Lake	AK	John Moline	1971	14
396 5/8	57 7/8	55 1/8	49 4/8	13	11	Tyone Lake	AK	James Moline	1961	15
396 2/8	55 3/8	53 0/8	33 3/8	18	19	Little Delta River	AK	Keith R. Clemmons	1958	16
391 7/8	53 6/8	50 5/8	45 6/8	11	14	Glenn Highway	AK	Joe West	1965	17
391 0/8	56 0/8	53 7/8	53 0/8	12	10	Kenai	AK	David L. Hawkins	1988	18
390 5/8	53 2/8	52 7/8	42 0/8	11	10	Hook River	AK	William Elfland	1990	19
390 0/8	58 7/8	57 1/8	37 5/8	15	16	Whitefish Lake	AK	Charles C. Smith	1988	20
389 7/8	47 3/8	49 1/8	40 7/8	16	14	Mulchatna River	AK	Greg L. Munther	1987	21
388 5/8	52 0/8	53 2/8	32 6/8	18	18	Delta River	AK	Dick Bolding	1957	22
388 2/8	56 5/8	57 1/8	47 2/8	18	16	Ugashik River	AK	George Moerlein	1972	23
386 7/8	48 4/8	44 5/8	43 4/8	19	18	Little Delta	AK	Dale K. Marcy	1964	24
385 2/8	58 5/8	57 0/8	38 0/8	14	17	Talkeetna Mtns.	AK	Harvey Matz	1959	25
384 5/8	47 2/8	47 5/8	37 5/8	16	15	Talkeetna Mtns.	AK	Dr. Rex Hancock	1962	26
384 3/8	49 2/8	49 1/8	34 6/8	15	18	Alaska Peninsula	AK	Betty Gulman	1968	27
381 5/8	51 1/8	47 1/8	39 1/8	14	19	Ugashik River	AK	Stanley J. Rogers, Jr.	1974	28
381 2/8	51 4/8	53 2/8	41 1/8	11	13	Ugashik River	AK	Craig Richardson	1988	29
381 1/8	52 6/8	57 5/8	36 5/8	13	11	Lyme Village	AK	Don Nettum	1993	30
380 7/8	46 1/8	47 7/8	36 7/8	13	15	Delta Creek	AK	Wayne Trimm	1960	31
380 5/8	61 2/8	62 2/8	36 5/8	8	9	King Salmon	AK	Glenn Hisey	1984	32
380 0/8	48 1/8	49 1/8	30 2/8	14	18	Cutler River	AK	Jay Deones	1990	33
379 4/8	51 0/8	49 0/8	37 7/8	14	12	Grayling Creek	AK	Carl H. Spaeth	2002	34
378 6/8	48 2/8	52 6/8	39 5/8	10	11	Killey River	AK	Ken Wolter	2003	35
378 4/8	46 2/8	48 2/8	45 6/8	14	18	Adak Island	AK	Tom Taylor	2006	36
377 6/8	52 0/8	51 4/8	35 6/8	13	19	Dawn Lake	AK	Bob Kroll	1963	37
377 3/8	51 6/8	51 6/8	39 1/8	13	13	Cutler River	AK	Tom Kothenbeutel	1999	38
376 7/8	50 2/8	45 4/8	40 2/8	14	13	Alaska Peninsula	AK	Bob Gulman	1968	39
375 7/8	50 6/8	50 1/8	32 3/8	14	14	Cutler River	AK	Jay Deones	1997	40
374 4/8	52 5/8	48 6/8	31 0/8	11	14	Atigun River	AK	Larry Hardwig	2008	41
374 3/8	51 0/8	50 5/8	37 5/8	9	11	Kobuk River	AK	David Rue	2000	42
373 5/8	46 6/8	46 7/8	39 5/8	18	15	Devil Creek	AK	Douglas Walker	1966	43
373 2/8	57 2/8	57 0/8	34 7/8	9	8	Kilbuck Mtns.	AK	Greg Munther	1998	44
372 0/8	56 7/8	58 5/8	38 6/8	9	11	Lake Iliamna	AK	Gary Wright	1991	45
371 6/8	46 4/8	48 6/8	33 0/8	12	13	Cutler River	AK	Randy E. Doyle	1985	46
371 5/8	40 1/8	40 1/8	34 1/8	17	28	Maclaren River	AK	Dick Cooley	1962	47
371 5/8	48 5/8	49 1/8	41 6/8	20	15	Port Heiden	AK	Jim Dougherty	1968	47
371 5/8	57 1/8	58 5/8	33 3/8	12	11	Kobuk River	AK	David Rue	2000	47
371 4/8	46 4/8	45 2/8	31 2/8	11	13	Grayling Creek	AK	Carl H. Spaeth	1997	50
370 7/8	56 3/8	56 1/8	42 6/8	11	12	Arctic Coastal Plain	AK	Robin D. Johnson	1987	51
370 3/8	48 7/8	49 7/8	32 3/8	10	12	Shenjek Lake	AK	J. Keith Chastain	1984	52
370 1/8	54 0/8	52 6/8	34 5/8	10	13	North Slope	AK	Ronald L. Sherer	1983	53
370 1/8	49 6/8	49 6/8	43 4/8	9	11	Lake Clark	AK	Ron Crouch	1989	53
370 1/8	50 0/8	50 2/8	38 5/8	12	14	Keefer Creek	AK	Mark Hockenberry	1994	53
369 5/8	44 0/8	44 4/8	34 4/8	13	14	Ugashik River	AK	William J. Stonebraker	1987	56
368 7/8	52 1/8	46 3/8	36 2/8	10	10	Cheeneetnuk River	AK	Kim Wintz	1983	57
368 0/8	51 5/8	46 4/8	36 3/8	14	19	Port Heiden	AK	John E. Lawson	1970	58
367 3/8	49 4/8	50 4/8	35 4/8	8	10	Sagavanirktok River	AK	Lee S. Peterson	2006	59
366 5/8	53 7/8	53 7/8	50 3/8	8	9	Franklin Bluffs	AK	Dick Carlson	1983	60
365 6/8	48 7/8	50 4/8	38 6/8	14	12	Shotgun Hills	AK	John L. Chase	1997	61
365 4/8	47 3/8	45 7/8	27 0/8	12	12	Cutler River	AK	Larry Streiff	1990	62

BARREN GROUND CARIBOU

Minimum Score 325 *Continued*

Score	Length of R Main Beam	L	Inside Spread	Number of R Points	L	Area	State/ Province	Hunter's Name	Date	Rank
365 3/8	47 6/8	43 1/8	32 4/8	17	18	Nuna River	AK	David Rue	2009	63
364 5/8	45 6/8	45 5/8	32 7/8	14	11	Cantwell	AK	Rick D. Snell	1991	64
363 6/8	52 0/8	53 4/8	39 3/8	13	14	Dillingham	AK	Glen Shatzer	1992	65
363 2/8	53 4/8	51 7/8	36 3/8	15	13	Lake Iliamna	AK	Richard S. Kinas	1997	66
362 7/8	49 0/8	50 3/8	43 7/8	14	12	Kodiak Island	AK	William Newman	2008	67
362 4/8	44 3/8	47 2/8	34 4/8	17	18	Egegik River	AK	Walter Eslinger	1970	68
362 3/8	44 0/8	47 2/8	35 2/8	10	9	Mulchatna River	AK	Ken Conley	1994	69
362 2/8	51 1/8	51 1/8	36 0/8	14	11	Sourdough	AK	Dan Jordan	1965	70
361 5/8	54 0/8	50 1/8	34 5/8	9	11	Blackstone River	YUK	Nathan L. Andersohn	2009	71
360 7/8	51 7/8	50 2/8	41 2/8	14	16	Iliamna	AK	Scott Halbert	1992	72
360 0/8	52 3/8	52 2/8	32 6/8	8	8	Maclaren River	AK	Rick D. Snell	1994	73
358 2/8	54 7/8	57 1/8	35 5/8	13	10	Tyone Lake	AK	Jake Sonnentag	1961	74
358 2/8	53 5/8	54 5/8	41 0/8	13	13	Maclaren River	AK	George Moerlein	1963	74
358 0/8	46 1/8	45 6/8	35 3/8	14	13	Aniak Lake	AK	Kevin L. Hall	1998	76
357 7/8	48 6/8	47 2/8	35 0/8	14	14	40 Mile Range	AK	Kelsey T. Snell	2012	77
357 5/8	53 6/8	46 6/8	37 3/8	13	16	Grayling Creek	AK	Carl Spaeth	2001	78
357 3/8	48 0/8	48 3/8	39 7/8	11	13	King Salmon	AK	Gerry C. Stinski	1986	79
356 6/8	52 1/8	52 4/8	38 0/8	10	9	Lake Selby	AK	Ricky Smith	1996	80
355 7/8	54 7/8	53 7/8	26 7/8	10	10	Mulchatna River	AK	Dan L. Carroll	1987	81
355 1/8	50 1/8	50 2/8	41 5/8	11	8	Lake Iliamna	AK	Gardner Rowell	1995	82
354 0/8	44 2/8	44 3/8	34 1/8	9	8	Lime Hills	AK	Bernard G. Norton	1990	83
353 6/8	52 5/8	52 4/8	47 1/8	12	10	Ugashik Lake	AK	Scott Lang	1988	84
353 3/8	48 2/8	46 6/8	36 7/8	17	13	Caribou Creek	AK	H. R. 'Dutch' Wambold	1964	85
353 1/8	45 4/8	46 2/8	37 4/8	13	14	Richardson Highway	AK	Donald O. Smith	1963	86
352 5/8	46 5/8	48 5/8	39 1/8	11	10	Carin Mtns.	AK	Wayne Haag	1997	87
352 1/8	53 4/8	55 2/8	33 4/8	11	9	Adak Island	AK	Steve Hohensee	2005	88
351 1/8	50 6/8	49 6/8	40 2/8	13	11	Lake Clark	AK	Bob Schwanke	1988	89
350 7/8	51 3/8	49 0/8	27 2/8	20	21	Big Delta	AK	Bill Brown	1958	90
350 6/8	46 7/8	47 1/8	37 2/8	9	10	Lime Hills	AK	John Crum	1988	91
350 3/8	51 3/8	44 0/8	41 3/8	11	14	Meshik River	AK	Art Kragness	1970	92
349 5/8	50 0/8	50 0/8	46 6/8	11	8	Cutler River	AK	Jerry Notch	1997	93
349 4/8	55 7/8	57 3/8	39 2/8	9	11	Prudhoe Bay	AK	Rick Grooms	1986	94
349 4/8	46 2/8	48 0/8	34 5/8	12	13	Grayling Creek	AK	James C. Carlson	1997	94
349 4/8	52 7/8	53 2/8	41 2/8	10	11	Kogoluktuk River	AK	Steve Hohensee	1999	94
348 0/8	47 4/8	46 0/8	32 7/8	16	18	Kotzebue	AK	Carl H. Spaeth	1991	97
348 0/8	51 1/8	53 0/8	34 4/8	12	13	Taylor Mtn.	AK	Edward A. Conkell	1994	97
347 6/8	55 6/8	59 0/8	42 0/8	10	10	Stuyahok River	AK	Marlon Clapham	1989	99
347 4/8	46 2/8	45 2/8	41 7/8	12	14	Ogilvie Mtns.	YUK	Emile Gele	1965	100
347 3/8	46 7/8	52 4/8	37 0/8	9	9	Mulchatna River	AK	Ray Roussett, Jr.	1986	101
347 1/8	45 4/8	47 2/8	33 0/8	12	10	King Salmon River	AK	John "Rosey" Roseland	1991	102
346 7/8	50 4/8	50 2/8	32 0/8	10	19	Telaquana Lake	AK	Gary Wall	1974	103
346 5/8	50 0/8	48 6/8	36 5/8	10	12	Little Delta	AK	Herb Lindsay	1964	104
345 1/8	47 2/8	45 5/8	42 1/8	10	12	Mulchatna River	AK	E. Donnall Thomas, Jr.	1985	105
344 5/8	52 1/8	53 0/8	34 1/8	16	15	Squirrel River	AK	Gary Renfro	1998	106
344 2/8	46 1/8	43 6/8	27 7/8	11	13	Franklin Bluffs	AK	Matthew A. Jones	1987	107
343 6/8	52 6/8	54 1/8	30 3/8	10	9	Coleen River	AK	Marvin Whitehead	1997	108
343 1/8	51 6/8	53 2/8	38 6/8	8	13	Taylor Highway	AK	Jae Beardon	1961	109
342 5/8	47 2/8	47 1/8	38 1/8	6	9	Purcell Mtn.	AK	Chris G. Sanford	1998	110
342 5/8	47 4/8	48 4/8	30 7/8	13	10	Squirrel River	AK	Mike Brezonick	1998	110
342 3/8	51 7/8	52 1/8	34 2/8	11	12	Grayling Creek	AK	Carl H. Spaeth	2008	112
340 6/8	49 2/8	49 3/8	27 1/8	12	13	Brooks Range	AK	Roger G. Stewart	1983	113
340 0/8	54 3/8	54 3/8	27 1/8	15	13	Ambler River	AK	Alvin L. Brandenberg	2000	114
339 3/8	48 3/8	47 4/8	35 5/8	9	10	Sparvon Lake	AK	Brad Ham	1997	115
339 2/8	49 6/8	47 3/8	39 0/8	10	12	Grayling Creek	AK	Carl H. Spaeth	1997	116
339 1/8	52 5/8	53 4/8	38 4/8	10	12	Little Delta	AK	Dr. Judd Grindell	1959	117
339 0/8	45 3/8	51 5/8	34 4/8	18	13	Tyone Lake	AK	Jake Sonnentag	1963	118
338 6/8	52 6/8	49 0/8	31 6/8	13	10	Grayling Creek	AK	Carl H. Spaeth	1992	119
338 1/8	46 6/8	48 3/8	38 0/8	10	9	Ogilvie River	YUK	Terry W. Brew	2005	120
337 5/8	49 7/8	49 1/8	41 6/8	12	14	Cinder River	AK	Francis Hosch	1966	121
337 3/8	47 4/8	47 3/8	32 2/8	13	14	Maclaren River	AK	George Moerlein	1963	122
337 1/8	49 3/8	50 2/8	37 7/8	9	9	Lime Hills	AK	Clifford R. Neville, Sr.	1990	123
337 1/8	45 4/8	43 1/8	37 7/8	8	10	Nishlik Lake	AK	Robert Bartoshesky	1996	123

BARREN GROUND CARIBOU

Minimum Score 325 *Continued*

Score	Length of R Main Beam	L	Inside Spread	Number of R Points	L	Area	State/ Province	Hunter's Name	Date	Rank
336 0/8	42 7/8	45 2/8	35 0/8	13	9	Lake Iliamna	AK	Todd F. Lewis	1988	125
336 0/8	46 3/8	48 5/8	36 4/8	10	10	Mulchatna River	AK	Stan Rauch	1994	125
335 3/8	40 2/8	44 5/8	30 4/8	14	13	Godge Creek	AK	Steve Dollar	2004	127
334 1/8	45 7/8	50 0/8	35 1/8	7	10	Nikabuna Lake	AK	Timothy J. Conrads	1988	128
334 0/8	52 7/8	51 2/8	39 0/8	11	8	Wulik River	AK	Guy Leibenguth	2000	129
333 5/8	47 7/8	48 4/8	32 0/8	12	17	Happy Valley	AK	Troy Graziadei	1984	130
332 7/8	53 0/8	54 3/8	36 2/8	18	15	Denali Hwy.	AK	Junie Moll	1961	131
332 6/8	53 5/8	53 5/8	33 6/8	11	8	Dillon	YUK	Derek Duggan	2004	132
332 3/8	45 3/8	46 1/8	29 1/8	9	10	Mulchatna River	AK	Mike Barrett	1987	133
332 2/8	52 5/8	53 0/8	26 1/8	9	9	Kukaklek Lake	AK	Jay Kuhre	1989	134
332 1/8	42 1/8	44 6/8	36 5/8	10	10	Kilbuck Mtns.	AK	Greg Munther	1998	135
331 7/8	54 3/8	55 0/8	35 6/8	9	16	Port Heiden	AK	John E. Lawson	1970	136
331 5/8	45 5/8	48 1/8	35 6/8	12	11	High Lakes	AK	Doug Walker	1970	137
331 5/8	52 6/8	52 3/8	44 4/8	10	10	Dillon	YUK	Dustan Duggan	2004	137
330 5/8	51 1/8	51 5/8	35 6/8	15	10	Tangle Lakes	AK	R. Glen Williams	1966	139
330 2/8	41 4/8	40 4/8	29 4/8	11	10	Dillingham	AK	John Holt	1997	140
329 7/8	47 4/8	47 3/8	31 0/8	10	15	Dawn Lake	AK	Chuck Kroll	1963	141
329 7/8	56 4/8	53 7/8	40 3/8	9	12	Little Delta	AK	Bill Tutt	1964	141
329 7/8	52 7/8	53 6/8	32 3/8	11	13	Nuna River	AK	David Rue	2005	141
329 5/8	44 4/8	45 4/8	31 0/8	10	11	Alaska Peninsula	AK	Dave Neel	1980	144
328 4/8	42 6/8	44 3/8	31 5/8	18	12	Little Delta	AK	Roy Bryan	1964	145
328 2/8	42 0/8	39 7/8	27 2/8	8	9	Prudhoe Bay	AK	Gene Barcak	1985	146
327 6/8	39 7/8	40 0/8	37 3/8	12	11	Sagavanirktok River	AK	Roger Stewart	1989	147
327 2/8	51 3/8	51 0/8	29 4/8	10	11	Dillingham	AK	John Holt	1997	148
326 6/8	49 1/8	47 3/8	41 0/8	12	18	Anchorage	AK	Roy Bryan	1964	149
326 5/8	52 7/8	52 3/8	47 2/8	9	8	Nuna River	AK	David Rue	1998	150
326 4/8	47 7/8	49 7/8	31 7/8	10	11	Brooks Range	AK	Roger Stewart	1985	151
326 3/8	47 6/8	49 0/8	39 0/8	11	13	King Salmon	AK	Glenn Hisey	1983	152
326 3/8	51 0/8	53 1/8	41 1/8	11	10	Yellow Creek Hill	AK	Don Poole	1992	152
326 0/8	54 5/8	52 6/8	19 0/8	11	12	Squirrel River	AK	Connie Renfro	1998	154
325 3/8	46 4/8	46 5/8	34 1/8	9	10	Grayling Creek	AK	Bill Cmiel	1997	155

BARREN GROUND CARIBOU [Velvet Antlers]

Minimum Score 325 *Rangifer tarandus granti, Rangifer tarandus stonei* and *Rangifer tarandus arcticus*

Score	Length of R Main Beam	L	Inside Spread	Number of R Points	L	Area	State/ Province	Hunter's Name	Date	Rank
384 0/8	58 6/8	57 0/8	30 2/8	10	15	Bear Lake	AK	Carl H. Spaeth	1989	*
372 3/8	52 7/8	52 5/8	33 1/8	9	8	Soldotna	AK	Donny Roy	1998	*
370 7/8	47 6/8	51 4/8	34 1/8	18	16	Nushagak River	AK	Gary Potter	1998	*
357 1/8	49 3/8	49 2/8	39 3/8	12	13	Cutler River	AK	Terry Krahn	1991	*
348 0/8	44 0/8	44 0/8	40 2/8	9	9	Iliamna	AK	Jeff Barnett	1995	*
340 3/8	46 2/8	48 0/8	35 1/8	11	10	Galbraith Lake	AK	Eugene Arndt	2000	*
340 0/8	48 4/8	46 6/8	33 0/8	12	12	Dalton Highway	AK	John McCullough	1991	*
339 4/8	53 6/8	56 0/8	35 4/8	11	12	Groundhog Mtn.	AK	Steve Hohensee	1996	*
339 2/8	46 0/8	44 1/8	27 1/8	12	13	Sagavanirktok River	AK	Lee S. Peterson	2000	*
329 7/8	47 5/8	47 3/8	25 4/8	11	12	Cutler River	AK	Tom Kothenbeutel	1999	*
326 4/8	44 0/8	42 5/8	33 2/8	11	9	Kobuk River	AK	Charles C. Ellithorpe, DVM	1997	*

CENTRAL CANADA BARREN GROUND CARIBOU

Minimum Score 300 *Rangifer tarandus articus*

Score	Length of R Main Beam	Length of Main Beam L	Inside Spread	Number of R Points	Number of Points L	Area	State/ Province	Hunter's Name	Date	Rank
381 6/8	46 2/8	46 5/8	29 1/8	13	20	MacKay Lake	NWT	Tom Taylor	1995	1
374 3/8	51 1/8	51 1/8	32 6/8	15	13	Warburton Bay	NWT	Duane Hicks	1987	2
364 0/8	52 6/8	52 0/8	36 7/8	11	11	MacKay Lake	NWT	T. J. Conrads	1997	3
359 6/8	53 6/8	54 4/8	33 3/8	8	9	MacKay Lake	NWT	Tom Taylor	1995	4
358 0/8	46 4/8	46 5/8	34 6/8	14	13	MacKay Lake	NWT	Mike Wheeler	1991	5
357 4/8	51 5/8	50 5/8	30 2/8	12	12	MacKay Lake	NWT	Glenn Hisey	1997	6
353 3/8	56 3/8	54 4/8	29 2/8	10	11	Schmock Lake	MAN	Quentin R. Plett	2001	7
353 2/8	55 7/8	56 6/8	39 5/8	13	12	MacKay Lake	NWT	Tom Taylor	1995	8
348 5/8	45 6/8	44 4/8	30 3/8	14	17	Desteffany Lake	NWT	Jay Deones	1996	9
348 4/8	48 3/8	46 2/8	35 6/8	14	14	Artillery Lake	NWT	John Duggan	1999	10
343 5/8	45 2/8	44 4/8	39 4/8	15	14	Desteffany Lake	NWT	Larry D. Jones	1994	11
343 0/8	47 7/8	50 2/8	32 7/8	8	9	Nodinka Narrows	NWT	Gunter Lemke	1991	12
342 7/8	48 1/8	48 5/8	36 3/8	13	11	Humpy Lake	NWT	John Mathews	2000	13
342 4/8	50 4/8	51 1/8	26 5/8	11	15	MacKay Lake	NWT	Gerry Gwaltney	1999	14
341 4/8	52 2/8	51 7/8	31 5/8	10	10	Desteffany Lake	NWT	Kevin Hisey	1996	15
341 1/8	50 2/8	51 2/8	29 2/8	8	8	Courageous Lake	NWT	Rick C. Wilson	2000	16
340 0/8	45 0/8	44 6/8	32 6/8	13	14	Rendezvous Lake	NWT	Richard A. Hjort	1989	17
339 7/8	44 7/8	43 4/8	32 5/8	18	16	MacKay Lake	NWT	Ted Judson	2001	18
336 7/8	46 1/8	46 2/8	28 4/8	14	13	MacKay Lake	NWT	Mark Wuerthele	1995	19
336 0/8	50 0/8	49 4/8	31 0/8	11	11	MacKay Lake	NWT	William E. Terry, Sr.	1995	20
335 2/8	51 3/8	50 3/8	30 3/8	14	12	MacKay Lake	NWT	Mark Wuerthele	1995	21
334 2/8	45 2/8	43 6/8	31 6/8	10	12	MacKay Lake	NWT	Michael J. Underhill	1990	22
333 0/8	41 5/8	43 1/8	22 1/8	13	13	MacKay Lake	NWT	Jay St. Charles	1992	23
332 5/8	42 5/8	42 0/8	27 3/8	10	8	Rendezvous Lake	NWT	Dan Ermatinger	1999	24
332 3/8	42 0/8	42 0/8	39 0/8	16	19	Caribou Bay	NWT	Richard A. Case	1994	25
331 2/8	44 6/8	46 1/8	28 6/8	14	15	MacKay Lake	NWT	David Emken	1990	26
330 3/8	46 6/8	47 0/8	32 0/8	11	11	MacKay Lake	NWT	Glenn Hisey	1997	27
329 4/8	46 0/8	47 4/8	25 6/8	16	12	MacKay Lake	NWT	Nick Mathews	2001	28
328 3/8	50 2/8	51 6/8	34 2/8	13	10	MacKay Lake	NWT	Tom Vanasche	1995	29
327 7/8	49 0/8	46 5/8	30 2/8	18	14	MacKay Lake	NWT	Sheldon Showalter	1995	30
327 4/8	49 7/8	48 7/8	31 0/8	9	9	Warburton Bay	NWT	Dennis Dunn	1997	31
327 1/8	46 2/8	46 3/8	23 0/8	15	15	MacKay Lake	NWT	Don Owen	1999	32
325 5/8	48 6/8	49 0/8	19 0/8	11	11	MacKay Lake	NWT	Kevin Lyons	1997	33
324 4/8	42 5/8	43 1/8	27 4/8	13	9	Purple Cliffs	NWT	Fred Eichler	2006	34
324 2/8	49 4/8	49 7/8	34 5/8	8	8	Courageous Lake	NWT	Larry Fenton	1999	35
323 7/8	39 7/8	46 2/8	29 1/8	11	13	Desteffany Lake	NWT	Jay Deones	1996	36
322 4/8	44 6/8	43 5/8	26 4/8	13	12	MacKay Lake	NWT	Tom Taylor	1995	37
321 7/8	51 7/8	52 2/8	36 1/8	9	11	Warburton Bay	NWT	Brian Tessmann	2002	38
320 1/8	53 0/8	48 6/8	24 4/8	7	2	MacKay Lake	NWT	Nick Mathews	2001	39
319 2/8	47 6/8	46 3/8	23 6/8	12	11	MacKay Lake	NWT	Sheldon Showalter	1995	40
318 0/8	46 4/8	47 0/8	23 6/8	12	11	MacKay Lake	NWT	Doug Clayton	2001	41
317 1/8	45 5/8	45 3/8	30 2/8	11	9	Artillery Lake	NWT	Perry Oates	1999	42
316 5/8	46 6/8	48 6/8	37 0/8	7	7	MacKay Lake	NWT	Howard L. Harding	1992	43
315 5/8	50 0/8	49 4/8	28 1/8	12	11	MacKay Lake	NWT	Don Owen	1999	44
315 3/8	46 6/8	45 1/8	21 0/8	15	15	MacKay Lake	NWT	Steve Hohensee	1994	45
313 7/8	49 2/8	50 5/8	31 6/8	6	8	MacKay Lake	NWT	Wade Carstens	1993	46
313 6/8	47 2/8	49 5/8	26 0/8	11	10	Jolly River	NWT	Ty Martin	1987	47
312 2/8	47 0/8	46 0/8	38 6/8	10	10	MacKay Lake	NWT	Dennis G. Hicks	1990	48
311 3/8	43 7/8	43 1/8	27 3/8	10	10	MacKay Lake	NWT	Andy Carpenter	1997	49
311 0/8	49 1/8	51 2/8	32 5/8	13	11	MacKay Lake	NWT	Todd Szmania	2000	50
310 0/8	48 3/8	48 5/8	24 4/8	8	8	Desteffany Lake	NWT	Doug Strecker	1999	51
307 3/8	50 6/8	48 4/8	31 3/8	14	11	MacKay Lake	NWT	Warren Witherspoon	1993	52
307 2/8	48 6/8	48 6/8	35 0/8	8	9	MacKay Lake	NWT	Joseph R. St. Charles	1993	53
307 1/8	46 0/8	46 2/8	22 2/8	12	14	Artillery Lake	NWT	Rick Duggan	1999	54
306 7/8	41 6/8	41 2/8	22 6/8	8	13	Desteffany Lake	NWT	Randy E. Doyle	1996	55
304 7/8	39 4/8	38 6/8	31 2/8	11	16	MacKay Lake	NWT	Richard King	2001	56
304 6/8	45 3/8	46 0/8	25 2/8	13	11	MacKay Lake	NWT	Howard L. Harding	1992	57
303 5/8	48 5/8	47 7/8	35 5/8	8	10	Desteffany Lake	NWT	Larry Streiff	1996	58
302 6/8	44 6/8	44 2/8	29 0/8	11	9	MacKay Lake	NWT	Les Malsch	1990	59
302 1/8	50 7/8	50 7/8	30 2/8	7	10	MacKay Lake	NWT	Russ Tye	1992	60
300 6/8	40 4/8	39 7/8	30 4/8	10	15	MacKay Lake	NWT	Dale Holpainen	1992	61

CENTRAL CANADA BARREN GROUND CARIBOU [Velvet Antlers]

Minimum Score 300 *Rangifer tarandus articus*

Score	Length of R Main Beam	L	Inside Spread	Number of R Points	L	Area	State/ Province	Hunter's Name	Date	Rank
375 3/8	57 2/8	52 6/8	30 6/8	8	10	MacKay Lake	NWT	Richard Thrasher	1996	*
372 1/8	49 2/8	48 5/8	36 4/8	11	12	Little Marten Lake	NWT	Roger Stewart	1999	*
368 6/8	50 0/8	50 6/8	27 5/8	12	11	MacKay Lake	NWT	Rick Liebel	1998	*
349 7/8	52 3/8	51 5/8	30 3/8	11	12	Desteffany Lake	NWT	Neil Hamilton	2004	*
348 2/8	48 4/8	50 4/8	22 6/8	11	11	MacKay Lake	NWT	Ronald Rockwell	1996	*
344 0/8	53 2/8	54 3/8	28 5/8	10	10	MacKay Lake	NWT	Richard Thrasher	1996	*
340 5/8	42 4/8	42 7/8	39 2/8	14	13	MacKay Lake	NWT	Steve M. Martin, Jr.	2000	*
340 5/8	49 0/8	48 2/8	25 5/8	11	11	MacKay Lake	NWT	Richard A. Case	2002	*
329 5/8	49 4/8	46 5/8	30 2/8	13	13	MacKay Lake	NWT	Glenn Kuklick	1995	*
324 6/8	47 0/8	47 0/8	34 2/8	11	12	Little Marten Lake	NWT	Jeff Carlson	2004	*
324 4/8	47 3/8	45 0/8	34 4/8	6	5	MacKay Lake	NWT	Richard King	2001	*
319 7/8	49 4/8	48 4/8	37 1/8	10	8	Little Marten Lake	NWT	Roger Stewart	1999	*
317 7/8	47 1/8	44 2/8	28 7/8	8	10	MacKay Lake	NWT	Garret Decker	2001	*
317 0/8	46 3/8	46 3/8	37 7/8	9	9	Warburton Bay	NWT	Don Delabbio	1996	*
314 0/8	50 3/8	50 3/8	33 5/8	8	6	Warburton Bay	NWT	Tom Weaver	2003	*
313 3/8	50 0/8	48 0/8	22 2/8	5	7	MacKay Lake	NWT	Ronald Rockwell	1996	*
312 4/8	44 2/8	42 1/8	31 0/8	11	9	MacKay Lake	NWT	Garret Decker	2001	*

Hunter: Fred Eichler
Species: Central Canada Barren Ground Caribou
Score: 324 4/8
Date: 9/6/06
Location: Purple Cliffs, NWT

MOUNTAIN CARIBOU

Minimum Score 300 — *Rangifer tarandus osborni, Rangifer tarandus fortidens* and *Rangifer tarandus montanus*

Score	Length of R Main Beam	L	Inside Spread	Number of R Points	L	Area	State/ Province	Hunter's Name	Date	Rank
399 0/8	49 7/8	51 5/8	40 2/8	8	11	Cassiar Mtns.	BC	Joseph Hinderman	1997	1
396 4/8	44 0/8	45 1/8	42 4/8	18	15	Divide Lake	NWT	John E. Anderson	1995	2
393 2/8	50 5/8	53 2/8	41 1/8	14	10	Tuya River	BC	T. J. Conrads	2006	3
390 1/8	55 2/8	51 6/8	34 1/8	15	11	Firesteel River	BC	Melvin K. Wolf	1970	4
387 6/8	44 3/8	42 7/8	36 0/8	15	14	Cold Fish Lake	BC	Steve Gorr	1976	5
378 2/8	49 3/8	47 4/8	37 0/8	11	10	O'Grady Lake	NWT	C. Randall Byers	1988	6
374 1/8	43 6/8	47 1/8	35 0/8	16	16	Duti River	BC	Dr. Lowell Eddy	1967	7
373 3/8	47 7/8	47 3/8	30 7/8	18	15	O'Grady Lake	NWT	Randall R. Giesey	1993	8
371 7/8	48 1/8	47 1/8	39 1/8	15	13	Tatlatui Lake	BC	Larry Alma	1979	9
371 7/8	46 0/8	46 0/8	36 4/8	14	12	Divide Lake	NWT	Mike Parsons	1988	9
371 3/8	38 7/8	39 3/8	30 3/8	16	15	Thutade Lake	BC	Bob Brill	1980	11
367 4/8	47 0/8	45 4/8	29 7/8	17	13	June Lake	NWT	Nathan L. Andersohn	1995	12
365 7/8	49 0/8	49 6/8	34 0/8	12	12	O'Grady Lake	NWT	R. Brian Oates	1990	13
365 2/8	47 4/8	48 5/8	32 1/8	14	13	Horseshoe Lake	NWT	Mike Parsons	1989	14
361 5/8	41 4/8	41 6/8	31 7/8	10	10	Mountain River	NWT	Rick Duggan	2003	15
361 2/8	49 4/8	53 2/8	31 7/8	12	10	O'Grady Lake	NWT	G. Fred Asbell	1988	16
357 6/8	39 2/8	38 5/8	27 4/8	12	12	McNeil Lake	YUK	Todd Zeuske	1994	17
357 5/8	46 0/8	45 2/8	31 6/8	11	11	Caribou Pass	NWT	Duane Hicks	1996	18
356 5/8	38 7/8	40 5/8	32 0/8	15	16	Horseshoe Lake	NWT	Don Davidson	1989	19
355 4/8	48 3/8	49 2/8	29 0/8	12	10	Johiah Lake	BC	Rick Wilson	2001	20
355 1/8	57 2/8	55 7/8	43 2/8	9	11	Kawdy Plateau	BC	Harvey J. Surina	1999	21
354 3/8	41 3/8	42 4/8	28 6/8	10	10	O'Grady Lake	NWT	C. Randall Byers	1991	22
349 2/8	47 0/8	45 5/8	34 4/8	11	10	Nahanni Butte	NWT	William E. Terry, Sr.	1994	23
347 5/8	33 4/8	32 5/8	31 4/8	14	11	O'Grady Lake	NWT	Don Davidson	1991	24
342 5/8	48 1/8	48 0/8	20 1/8	12	15	Arctic Red River	NWT	Rocky Chisholm	2001	25
340 7/8	46 4/8	45 3/8	37 3/8	9	12	Horseshoe Lake	NWT	Jerry Keller	1990	26
339 6/8	42 7/8	44 6/8	20 0/8	20	18	Tatlatui Lake	BC	G. Fred Asbell	1975	27
339 5/8	44 6/8	44 5/8	24 5/8	15	13	O'Grady Lake	NWT	Mark Wuerthele	1993	28
336 3/8	45 4/8	45 0/8	31 7/8	12	9	Divide Lake	NWT	Paul Anderson	1995	29
336 1/8	42 7/8	40 5/8	36 6/8	12	15	Nacha Creek	BC	Scott L. Koelzer	1994	30
332 1/8	40 0/8	40 4/8	32 5/8	10	12	Firesteel River	BC	Larry Alma	1984	31
331 5/8	47 6/8	49 1/8	29 6/8	11	9	June Lake	NWT	Fred Eichler	2005	32
328 5/8	51 6/8	50 5/8	40 4/8	9	12	Kilgore Lake	BC	Mike Parsons	1998	33
327 0/8	41 2/8	40 1/8	31 4/8	15	15	Mountain River	NWT	Mike Barrett	1990	34
324 1/8	35 3/8	36 3/8	32 1/8	15	18	Cassiar Mtns.	BC	Mike Parsons	1994	35
324 1/8	38 5/8	40 0/8	31 5/8	10	10	Ludwig Lake	BC	Guy Leibenguth	1997	35
321 7/8	48 4/8	47 3/8	32 4/8	13	14	Kitchener Lake	BC	Stephen E. Mitchell	1970	37
320 6/8	43 6/8	44 4/8	31 7/8	11	12	Thutade Lake	BC	Jack W. Kriener	1974	38
319 3/8	45 0/8	45 2/8	40 5/8	10	12	Kitchener Lake	BC	Doug Strecker	1979	39
319 1/8	44 3/8	48 5/8	24 7/8	13	10	Cold Fish Lake	BC	Steve Gorr	1975	40
314 7/8	42 6/8	46 0/8	33 5/8	10	10	Divide Lake	NWT	Dale Drilling	1989	41
314 2/8	42 7/8	44 0/8	28 3/8	10	10	O'Grady Lake	NWT	Don Davidson, Jr.	1988	42
309 2/8	43 3/8	41 2/8	27 2/8	10	13	Thutade Lake	BC	Harold H. VanderHorst	1974	43
306 6/8	40 0/8	40 4/8	30 1/8	11	9	Chapel Lake	BC	Scott L. Koelzer	2009	44
306 2/8	43 4/8	43 2/8	29 2/8	11	10	Tatlatui Lake	BC	Robert Pitt	1975	45

MOUNTAIN CARIBOU [Velvet Antlers]

Minimum Score 300 — *Rangifer tarandus osborni, Rangifer tarandus fortidens* and *Rangifer tarandus montanus*

Score	Length of R Main Beam	L	Inside Spread	Number of R Points	L	Area	State/ Province	Hunter's Name	Date	Rank
432 2/8	46 6/8	46 3/8	43 0/8	15	15	O'Grady Lake	NWT	J. Dean Bodoh	1990	*
410 7/8	48 7/8	50 2/8	49 3/8	16	16	Arctic Red River	NWT	Rick Stockburger	2009	*
377 6/8	45 1/8	45 6/8	27 2/8	18	13	O'Grady Lake	NWT	Don Davidson	1993	*
361 3/8	45 3/8	46 4/8	35 2/8	12	11	Horseshoe Lake	NWT	Dave Brummond	1991	*
360 7/8	45 0/8	44 0/8	39 1/8	12	11	O'Grady Lake	NWT	Dan Bertalan	1993	*
353 2/8	44 7/8	46 5/8	26 0/8	14	12	O'Grady Lake	NWT	Ron Rockwell	1993	*
348 4/8	49 3/8	49 4/8	43 7/8	12	11	Mountain River	NWT	Tavis Rogers	2012	*
344 5/8	41 7/8	43 5/8	33 7/8	8	7	June Lake	NWT	Tom Vanasche	2002	*
329 2/8	43 5/8	42 0/8	31 6/8	10	12	Divide Lake	NWT	Mike Traub	1984	*
316 2/8	41 4/8	41 0/8	29 6/8	9	12	Natla River	NWT	Richard A. Case	1990	*

QUEBEC-LABRADOR CARIBOU

Minimum Score 325 — *Rangifer tarandus caboti* from Quebec and Labrador

Score	Length of R Main Beam	L	Inside Spread	Number of R Points	L	Area	State/ Province	Hunter's Name	Date	Rank
419 4/8	50 0/8	50 6/8	40 2/8	12	11	Natuak Lake	QUE	Patricio Sada Muguerza	1993	1
410 0/8	56 0/8	55 6/8	52 1/8	13	11	Schefferville	QUE	Charles L. Buechel, Jr.	1987	2
403 2/8	48 2/8	47 2/8	35 2/8	16	16	Lac Droilland	QUE	Tom Johnson	1991	3
395 6/8	59 4/8	55 4/8	47 7/8	18	16	George River	QUE	Paul Brunner	1980	4
388 2/8	46 0/8	45 4/8	42 6/8	11	11	Delay River	QUE	William G. Mason	1988	5
387 1/8	47 0/8	44 5/8	46 6/8	14	13	Weeks Lake	QUE	Donald L. Stout	1990	6
386 4/8	56 2/8	54 4/8	57 7/8	15	20	George River	QUE	John Kuhar	1972	7
385 5/8	50 2/8	52 2/8	50 7/8	16	18	Pons Island	QUE	James C. Walters	1987	8
384 4/8	46 4/8	46 5/8	42 4/8	15	16	Schefferville	QUE	Elmer R. Luce, Jr.	1987	9
384 2/8	45 2/8	46 6/8	49 0/8	19	19	Schefferville	QUE	Elmer R. Luce, Jr.	1987	10
384 0/8	46 0/8	44 3/8	43 0/8	17	13	Weymouth Inlet	QUE	Jules Pacheco	1986	11
383 7/8	47 3/8	49 5/8	46 7/8	14	17	Lake Loudin	QUE	Roger M. Schmitt	1988	12
381 4/8	51 0/8	51 6/8	41 2/8	17	19	Wedge Hill Lodge	QUE	John Janelli	1980	13
380 5/8	57 6/8	53 5/8	56 7/8	11	16	Ungava Region	QUE	Donald Schram	1982	14
379 7/8	49 0/8	50 2/8	51 2/8	15	13	Schefferville	QUE	Michael C. Dysh	1990	15
379 6/8	49 4/8	49 2/8	42 5/8	11	12	Schefferville	QUE	Ted Jaycox	1986	16
379 4/8	45 0/8	47 3/8	46 0/8	17	14	Alubiack Fiord	QUE	Tim Finley	1994	17
377 0/8	53 6/8	55 7/8	36 7/8	15	15	Delay River	QUE	Dale Underwood	1990	18
374 6/8	47 5/8	48 7/8	43 6/8	13	12	Weeks Lake	QUE	Raymond Villeneuve	1990	19
374 2/8	49 6/8	50 6/8	49 5/8	17	11	Lake Craamolet	QUE	Roland L. Kauffman	2006	20
373 3/8	53 2/8	51 1/8	40 0/8	15	13	Fort Chimo	QUE	Scott M. Showalter	1982	21
373 0/8	45 3/8	44 2/8	39 2/8	20	10	Wolf Lake	QUE	Bruce A. Hopkins	1993	22
371 2/8	46 5/8	49 5/8	38 6/8	12	10	Lac Kakiattuoluk	QUE	Ronald Rockwell	2001	23
370 7/8	46 6/8	44 5/8	41 2/8	5	6	Schefferville	QUE	Elmer R. Luce, Jr.	1986	24
370 1/8	50 3/8	49 6/8	38 3/8	11	11	Lac Merrville	QUE	Jack A. Smith	1998	25
367 0/8	47 0/8	48 0/8	44 7/8	12	12	Fort Chimo	QUE	James E. Doberstein	1989	26
364 2/8	57 7/8	58 5/8	55 2/8	13	14	Delay River	QUE	Bob Watkins	1987	27
364 2/8	51 6/8	51 4/8	44 7/8	15	12	Lac Minto	QUE	Carl Spaeth	1999	27
363 7/8	50 6/8	54 3/8	39 1/8	13	15	Lac Tasiataq	QUE	William J. Gleeson	1998	29
362 4/8	51 0/8	49 2/8	41 4/8	16	11	Lac Minto	QUE	Martin E. Cain	2004	30
362 3/8	47 6/8	48 3/8	41 4/8	13	12	Tunulik River	QUE	David J. Hell	1986	31
361 7/8	45 4/8	47 4/8	36 6/8	16	15	Musset Lake	QUE	Herman W. Kovar, Sr.	2001	32
361 3/8	47 7/8	48 6/8	41 5/8	14	14	Delay River	QUE	John Akkerman	1990	33
360 1/8	46 3/8	48 4/8	53 0/8	14	11	Schefferville	QUE	Steven H. Byerly	1989	34
359 6/8	49 6/8	50 5/8	35 4/8	13	12	Schefferville	QUE	Steven P. Salmieri	1989	35
358 5/8	54 3/8	53 3/8	41 2/8	16	15	Lac Minto	QUE	David Broyles	2006	36
358 4/8	46 5/8	46 5/8	39 0/8	13	13	Lac Minto Island	QUE	Ben Allen Graham	2000	37
358 1/8	52 1/8	50 4/8	48 0/8	14	17	Pons Island	QUE	Harold B. "Pat" Clark	1987	38
357 5/8	56 1/8	52 4/8	43 3/8	17	15	Delay River	QUE	Steve Beilgard	1989	39
357 5/8	47 3/8	49 2/8	41 5/8	17	13	Sand Lake	QUE	Dale E. Parish	2001	39
356 5/8	51 0/8	53 6/8	36 4/8	10	13	Lac Kaliattuoluk	QUE	Ivan C. James II	2001	41
356 1/8	56 0/8	55 3/8	52 2/8	6	11	George River	QUE	Royce DeCook	2002	42
356 0/8	54 3/8	54 0/8	45 6/8	10	8	Schefferville	QUE	Dr. Nicholas J. Gray	1981	43
355 1/8	49 6/8	50 2/8	39 7/8	9	9	Schefferville	QUE	Jerry Keller	1991	44
354 4/8	48 1/8	43 5/8	45 3/8	10	9	Doreen Lake	QUE	Eugene Arndt	1990	45
354 3/8	49 0/8	45 1/8	47 5/8	14	13	Lac Lefrancois	QUE	Larry Cannata	2005	46
353 6/8	46 3/8	46 1/8	38 6/8	15	13	George River	QUE	James Bailey	2003	47
353 3/8	47 0/8	48 4/8	46 1/8	10	14	Lac Minto	QUE	David Broyles	2006	48
353 1/8	47 1/8	47 3/8	39 2/8	14	11	Lac Minto	QUE	Dan Novotny	2006	49
353 0/8	45 4/8	45 3/8	30 1/8	15	18	Lac Minto	QUE	Mike Britton	2001	50
351 6/8	44 0/8	43 6/8	35 1/8	20	20	Schefferville	QUE	Steven P. Salmieri	1989	51
351 6/8	48 3/8	48 0/8	36 2/8	14	10	Barrel Lake	QUE	Thomas Egnew	1994	51
351 4/8	52 0/8	55 1/8	43 7/8	9	10	Lac Minto Island	QUE	James Carlson	1999	53
349 7/8	43 5/8	41 5/8	34 3/8	13	14	Kenny Lake	QUE	Ross Trujillo, Jr.	1991	54
349 6/8	47 0/8	44 1/8	37 4/8	20	15	George River	QUE	Charlie Kroll	1980	55
349 6/8	51 1/8	51 5/8	48 5/8	10	11	Lac Minto	QUE	Jack L. "Jackson" Ward	1997	55
349 4/8	46 5/8	46 2/8	43 2/8	17	13	Harold Lake	QUE	Don Davidson	1990	57
349 3/8	49 2/8	53 2/8	45 6/8	17	16	Harold Lake	QUE	Craig Ambos	1992	58
349 2/8	49 5/8	49 0/8	37 5/8	10	11	Fort Chimo	QUE	James E. Doberstein	1989	59
349 0/8	46 7/8	45 6/8	45 4/8	10	13	Pons River	QUE	Wayne A. Lamoreux	1991	60
347 5/8	44 6/8	45 3/8	41 7/8	13	12	Kenny Lake	QUE	Matt Lamoreux	1991	61
346 6/8	48 2/8	47 2/8	49 2/8	9	12	Delay River	QUE	Gregory J. Fries	1987	62

QUEBEC-LABRADOR CARIBOU

Minimum Score 325 *Continued*

Score	Length of R Main Beam	L	Inside Spread	Number of R Points	L	Area	State/ Province	Hunter's Name	Date	Rank
346 6/8	46 6/8	48 6/8	46 5/8	15	15	Ungava Bay	QUE	Larry Comer	1999	62
346 3/8	47 5/8	47 3/8	40 6/8	11	10	Maricourt Lake	QUE	Jerry W. Huffaker	1990	64
346 1/8	55 0/8	55 0/8	37 3/8	12	17	River Lac Cambrien	QUE	Kent W. Brigham	1987	65
345 7/8	52 6/8	54 1/8	45 2/8	11	13	Ungava Bay	QUE	John Musacchia	1978	66
345 7/8	52 3/8	51 7/8	44 2/8	13	17	Schefferville	QUE	Barry J. Smith	1988	66
345 7/8	44 6/8	43 4/8	36 0/8	17	17	Caniapiscau Lake	QUE	Walter E. Dixon	2005	66
345 6/8	52 0/8	50 2/8	33 4/8	17	14	Delay River	QUE	William Elfland	1994	69
345 6/8	49 6/8	50 3/8	43 3/8	10	10	Ungava Bay	QUE	Bruce H. Sabaini	1996	69
345 4/8	40 5/8	42 3/8	42 3/8	15	14	Lac Minto	QUE	S. Ray White	2006	71
344 5/8	46 2/8	46 1/8	41 4/8	12	12	Lac Minto	QUE	Joe P. Furlong	2001	72
344 2/8	43 2/8	43 0/8	43 2/8	13	13	Melezes River	QUE	G. Fred Asbell	1990	73
343 6/8	52 2/8	51 1/8	48 3/8	12	12	Schefferville	QUE	Gregory V. Pilot	1992	74
343 3/8	49 2/8	50 1/8	43 5/8	9	10	Ungava Bay	QUE	L. Dan Neebe	1988	75
342 7/8	53 4/8	52 6/8	39 5/8	14	14	George River	QUE	Frank Hogan	1980	76
342 5/8	47 4/8	50 3/8	46 1/8	15	10	Tuktu Camp	QUE	Martin G. Billeri	1977	77
341 5/8	53 7/8	53 5/8	41 6/8	7	11	Barrel Lake	QUE	John Hale	1994	78
341 2/8	50 4/8	49 6/8	50 2/8	15	12	Pons River	QUE	Walter F. Dotson, Jr.	1987	79
341 2/8	49 3/8	44 6/8	48 6/8	13	11	Schefferville	QUE	Robert J. Lewis	1987	79
340 3/8	39 2/8	48 7/8	31 5/8	18	15	George River	QUE	Craig Richardson	1982	81
338 4/8	45 6/8	47 1/8	39 6/8	16	15	Lac Minto	QUE	Gerald Greenleaf	2002	82
338 3/8	54 0/8	53 3/8	32 0/8	13	16	Schefferville	QUE	Kenneth C. Kaufmann	1987	83
337 7/8	44 0/8	45 3/8	42 3/8	13	11	Schefferville	QUE	Dale Drilling	1987	84
337 6/8	51 0/8	51 5/8	48 4/8	8	9	Pons River	QUE	Bradford Higson	1986	85
337 0/8	49 6/8	50 2/8	45 5/8	12	14	Melezes River	QUE	Donald L. Clark	1993	86
336 6/8	48 0/8	47 3/8	38 5/8	10	9	George River	QUE	Pat Reilly	1993	87
336 6/8	45 4/8	44 7/8	44 5/8	12	11	Lac Minto	QUE	J. Ken Martin	1995	87
336 4/8	51 3/8	52 4/8	38 6/8	10	10	Schefferville	QUE	William L. Hall	1994	89
336 0/8	42 3/8	41 7/8	45 3/8	13	13	Lake Leopard	QUE	Robert B. Seger II	1989	90
335 6/8	44 4/8	44 4/8	37 0/8	15	16	Lake Helluva	QUE	Larry E. Flowers	1999	91
335 5/8	47 3/8	46 4/8	43 6/8	14	14	Lac Boismenu	QUE	Ken H. Taylor	2007	92
335 3/8	44 0/8	45 2/8	47 0/8	13	10	Lac Minto	QUE	Art Fink	2006	93
334 7/8	53 4/8	51 3/8	47 5/8	14	11	Lac Minto	QUE	Dale Arner	2001	94
334 0/8	51 0/8	49 0/8	47 7/8	16	16	Lac Minto	QUE	Kenyon W. Woods	2001	95
333 3/8	45 3/8	48 2/8	45 3/8	12	13	Lake Loudin	QUE	Jim Gompf	1990	96
333 2/8	51 6/8	53 1/8	46 1/8	11	10	Whiskey Lake	QUE	Glenn R. Kuklick	1986	97
333 1/8	51 3/8	52 5/8	49 7/8	5	8	Schefferville	QUE	Robert James Lewis	1987	98
332 3/8	47 1/8	46 7/8	34 7/8	13	11	Lac Kakkakia	QUE	Eugene R. Thorn	2001	99
330 6/8	46 7/8	48 6/8	41 6/8	11	7	Caniapiscau River	QUE	John L. Gardner	1994	100
330 6/8	56 2/8	58 3/8	37 6/8	9	11	Schefferville	QUE	Len Rinke	2001	100
330 5/8	41 1/8	42 4/8	37 0/8	16	15	Delay River	QUE	William E. Terry, Sr.	1992	102
328 5/8	49 1/8	52 4/8	45 2/8	13	12	Pons River	QUE	Jim Ellis	1986	103
328 5/8	44 5/8	44 3/8	33 0/8	16	14	Schefferville	QUE	Kenneth C. Kaufmann	1987	103
328 4/8	44 6/8	43 0/8	39 0/8	16	16	Long Lake	QUE	David J. Stanislawski	1990	105
328 2/8	42 2/8	43 3/8	49 5/8	13	8	Delay River	QUE	August S. Gray	1988	106
326 2/8	46 1/8	45 1/8	32 7/8	13	15	Lake Leopard	QUE	Wes Seaver	1989	107
326 2/8	45 0/8	46 6/8	47 7/8	7	7	Schefferville	QUE	Jerry Parsons	1990	107
326 2/8	45 3/8	45 2/8	44 7/8	12	14	Serigny River	QUE	Greg Dearth	1999	107
325 5/8	49 4/8	48 3/8	53 0/8	8	9	Delay River	QUE	Paul Converse	1991	110
325 3/8	45 3/8	47 5/8	48 4/8	9	7	Lac Louis	QUE	Alan J. Rhinerson	1991	111
325 3/8	53 5/8	52 3/8	43 0/8	10	11	Lac Aigneau	QUE	Ken Miller	1998	111
325 1/8	51 4/8	52 4/8	49 7/8	14	9	Deception Bay	QUE	James M. Tully	1999	113

QUEBEC-LABRADOR CARIBOU [Velvet Antlers]

Minimum Score 325 *Rangifer tarandus caboti* from Quebec and Labrador

Score	Length of R Main Beam	L	Inside Spread	Number of R Points	L	Area	State/ Province	Hunter's Name	Date	Rank
357 2/8	50 2/8	50 1/8	47 5/8	8	8	George River	QUE	Herman W. Kovar, Sr.	2003	*
345 7/8	47 2/8	46 2/8	37 7/8	11	12	Delay River	QUE	John Lind	1993	*
342 4/8	49 0/8	47 2/8	42 3/8	14	15	Delay River	QUE	Wes Skakun	1993	*
341 4/8	48 6/8	53 7/8	50 1/8	10	10	Long Lake	QUE	Ted K. Jaycox	1986	*
339 7/8	49 4/8	49 0/8	45 6/8	8	12	Delay River	QUE	Duane Hicks	1993	*
339 0/8	49 0/8	47 5/8	47 1/8	9	9	Rigouville Lake	QUE	Michael E. Travis	1998	*
335 0/8	43 5/8	42 0/8	40 7/8	16	19	Lac Lefrancois	QUE	Earl O. Woods	2005	*
332 0/8	47 0/8	49 2/8	35 3/8	14	10	Jack's Lake	QUE	Roger Rothhaar	1994	*
325 5/8	44 4/8	42 6/8	35 0/8	11	11	Lac Minto	QUE	Steven Bostic	2006	*

Hunter: Gregg Bakelja
Species: Dall's Sheep
Score: 136 3/8
Date: 7/30/12
Location: Liard Mountains, NWT

WOODLAND CARIBOU

Minimum Score 220 — *Rangifer tarandus caribou* from Nova Scotia, New Brunswick, and Newfoundland

Score	Length of R Main Beam	Length of Main Beam L	Inside Spread	Number of R Points	Number of Points L	Area	State/ Province	Hunter's Name	Date	Rank
345 2/8	46 6/8	46 4/8	27 3/8	11	12	Victoria River	NFL	Dempsey Cape	1966	1
324 0/8	42 7/8	42 3/8	26 6/8	10	11	Sitdown Pond	NFL	Ed J. Bowser	1966	2
310 1/8	41 0/8	41 0/8	29 2/8	10	12	Millertown	NFL	Gerhart Huber	1966	3
299 4/8	33 3/8	38 7/8	29 4/8	8	9	Dolland Pond	NFL	Mark Turner	2006	4
284 1/8	35 2/8	36 1/8	28 5/8	12	11	Three Brook Pond	NFL	Brian Walters	2012	5
276 7/8	42 3/8	43 0/8	25 2/8	10	9	King George IV Lake	NFL	Mark McCarty	1966	6
274 7/8	38 5/8	39 4/8	29 0/8	8	8	Caribou Creek	NFL	Dan Bertalan	1992	7
267 7/8	29 3/8	31 4/8	27 4/8	12	11	Stag Lake	NFL	Richard A. Case	1993	8
262 5/8	34 5/8	37 3/8	30 6/8	9	8	Gaff Topsails	NFL	Tom Vanasche	2002	9
258 7/8	40 4/8	37 6/8	33 1/8	6	7	Lloyds River	NFL	Harold A. Hill	1964	10
258 6/8	37 6/8	38 3/8	26 7/8	7	8	Stag Lake	NFL	R. Brian Oates	1993	11
258 1/8	36 7/8	33 6/8	25 4/8	10	11	Atikonak Lake	NFL	Bill L. Carlos	1972	12
256 5/8	31 5/8	36 0/8	15 7/8	8	8	Alex Pond	NFL	Jay Deones	1993	13
256 3/8	31 7/8	33 4/8	24 4/8	8	9	Dolland Pond	NFL	Rick Duggan	2006	14
253 7/8	37 4/8	35 5/8	32 1/8	10	8	Buchans Plateau	NFL	Steve D. Munier	1990	15
249 3/8	36 5/8	37 2/8	26 4/8	9	9	Millertown	NFL	Cliff Wiseman	1962	16
248 5/8	32 3/8	30 1/8	27 2/8	6	8	Dolland Pond	NFL	Kevin Martin	2006	17
248 0/8	40 7/8	36 7/8	34 0/8	7	7	Sitdown Pond	NFL	Dr. Ed Bowser	1965	18
247 3/8	34 5/8	37 4/8	28 4/8	7	7	King George IV Lake	NFL	Bill Hirst	1966	19
245 0/8	33 2/8	31 6/8	28 1/8	8	8	Dolland Pond	NFL	Dennis Dunn	2007	20
243 6/8	34 6/8	34 6/8	27 7/8	7	6	Greys Island	NFL	Terry Krahn	1991	21
242 6/8	36 2/8	33 2/8	22 4/8	8	8	Buchans Plateau	NFL	Willard F. Collins	2000	22
242 0/8	37 1/8	36 6/8	25 7/8	8	8	Dolland Pond	NFL	Ronald Rockwell	2005	23
241 6/8	33 2/8	30 6/8	29 2/8	9	11	Buchans Plateau	NFL	Glenn Hisey	1982	24
241 2/8	33 2/8	34 6/8	25 1/8	11	11	Interior District	NFL	Bill Goff	1965	25
238 7/8	19 1/8	27 2/8	26 7/8	9	8	Stag Lake	NFL	G. Fred Asbell	1996	26
237 1/8	35 1/8	35 2/8	24 6/8	10	7	Buchans Plateau	NFL	Doug Strecker	1990	27
236 3/8	36 0/8	35 0/8	25 3/8	12	9	Princess Lake	NFL	John Musacchia	1967	28
233 2/8	38 0/8	37 0/8	27 5/8	7	4	Gull Lake	NFL	M. W. Bowser	1958	29
232 5/8	33 3/8	35 0/8	24 5/8	6	7	Deer Lake	NFL	Douglas L. Buchler	1984	30
232 1/8	30 3/8	30 7/8	25 3/8	10	8	Victoria Lake	NFL	Robert Bartoshesky	1998	31
228 1/8	33 1/8	33 0/8	27 3/8	6	8	Ocean Pond	NFL	David J. Lamoreaux	1998	32
227 7/8	36 1/8	33 0/8	27 1/8	10	9	Princess Lake	NFL	Ken Rapp	1966	33
227 7/8	32 2/8	33 0/8	26 0/8	8	7	Indian Lake	NFL	James R. Welch	2004	33
225 4/8	34 2/8	34 5/8	27 6/8	7	8	Bruce's Pond	NFL	Randy E. Doyle	1990	35
225 0/8	38 0/8	40 3/8	25 1/8	5	8	Victoria River	NFL	Clarence Bowers, Jr.	1966	36
224 1/8	36 0/8	35 0/8	27 4/8	8	9	Lloyds River	NFL	Harold A. Hill	1965	37
222 1/8	31 2/8	30 4/8	29 7/8	7	7	Greys Island	NFL	William E. Terry, Sr.	1993	38
220 1/8	33 6/8	31 3/8	24 2/8	8	8	Stony Lake	NFL	Larry Streiff	1995	39

COUGAR [Mountain Lion]

Minimum Score 13 8/16 — *Felis concolor hippolestes*

Score	Greatest Length	Greatest Width	Sex	Area	State/ Province	Hunter's Name	Date	Rank
15 8/16	8 15/16	6 9/16	M	Gila County	AZ	Stephen D. Hornady	1991	1
15 8/16	9 3/16	6 5/16	M	Lemhi County	ID	Michael Judas	1995	1
15 8/16	9 2/16	6 6/16	M	Garfield County	CO	Richard A. Mowles	1997	1
15 7/16	9 0/16	6 7/16	M	Idaho County	ID	William Egner	1972	4
15 7/16	9 2/16	6 5/16	M	Rio Arriba County	NM	Dick Ray	1985	4
15 7/16	8 15/16	6 8/16	M	Ferry County	WA	John Peruchini	1989	4
15 7/16	9 0/16	6 7/16	M	Clearwater County	ID	Dennis L. Butler	1992	4
15 6/16	9 4/16	6 2/16	M	Grand County	UT	Art Kragness	1969	8
15 6/16	9 0/16	6 6/16	M	Elmore County	ID	Kelly Dougherty	1993	8
15 5/16	8 15/16	6 6/16	M	Mesa County	CO	John Lamicq, Jr.	1969	10
15 5/16	8 15/16	6 6/16	M	Mineral County	MT	Dennis Moos	1976	10
15 5/16	9 0/16	6 5/16	M	Lindsey Lake	BC	Harvey J. Surina	1991	10
15 5/16	8 15/16	6 6/16	M	Mesa County	CO	Darryl Powell	1997	10
15 5/16	8 15/16	6 6/16	M	Tatlayoko Lake	BC	Matthew Riley	2011	10
15 4/16	9 0/16	6 4/16	M	Gila County	AZ	Dr. James L. Smith	1958	15
15 3/16	8 13/16	6 6/16	M	Ventura County	CA	Warren C. Johnston	1953	16
15 3/16	8 14/16	6 5/16	M	Huerfano County	CO	Bob Sigman	1987	16
15 3/16	9 1/16	6 2/16	M	Ravalli County	MT	Mario Locatelli	1990	16
15 3/16	8 15/16	6 4/16	M	Daggett County	UT	John Richardson	1992	16
15 3/16	9 0/16	6 3/16	M	Clear Creek County	CO	Mark Turner	1995	16
15 3/16	8 15/16	6 4/16	M	Judith Basin County	MT	John "Rosey" Roseland	1997	16
15 3/16	8 15/16	6 4/16	M	Frazier River	BC	Gene A. Welle	2000	16
15 3/16	9 2/16	6 1/16	M	Archuleta County	CO	Ernest Pinnecoose	2011	16
15 2/16	8 15/16	6 3/16	M	Rio Blanco County	CO	Leonard Cardinale	1963	24
15 2/16	9 0/16	6 2/16	M	Grand County	UT	Richard Oakleaf	1967	24
15 2/16	8 14/16	6 4/16	M	Wallowa County	OR	Terrell Buchanan	1973	24
15 2/16	8 12/16	6 6/16	M	Teton County	WY	Craig Richardson	1986	24
15 2/16	8 14/16	6 4/16	M	Conejos County	CO	Wayne Miller	1987	24
15 2/16	8 12/16	6 6/16	M	Iron County	UT	Bob Spina	1989	24
15 2/16	8 15/16	6 3/16	M	Park County	CO	Jack P. Van Vianen	1990	24
15 2/16	8 14/16	6 4/16	M	Madison County	MT	Cody Stemler	1996	24
15 2/16	8 14/16	6 4/16	M	Chain Lakes	ALB	Tom Foss	1998	24
15 1/16	8 13/16	6 4/16	M	Iron County	UT	William P. Mastrangel	1964	33
15 1/16	9 0/16	6 1/16	M	Taos County	NM	Bill Porteous	1990	33
15 0/16	8 11/16	6 5/16	M	Elko County	NV	Earl Dudley	1959	35
15 0/16	8 13/16	6 3/16	M	Utah County	UT	Richard C. Smith	1968	35
15 0/16	8 11/16	6 5/16	M	Mizzezula Mtns.	BC	Bengt G. Bjalme	1969	35
15 0/16	9 0/16	6 0/16	M	Rio Grande County	CO	Richard J. Dugas	1986	35
15 0/16	9 0/16	6 0/16	M	Elmore County	ID	Ed Strayhorn	1989	35
15 0/16	8 13/16	6 3/16	M	Montezuma County	CO	John L. Gardner	1993	35
15 0/16	8 15/16	6 1/16	M	Montezuma County	CO	Robert D. Crask	1993	35
15 0/16	8 14/16	6 2/16	M	San Juan County	UT	David A. Bronson	1997	35
15 0/16	8 12/16	6 4/16	M	South Willow Creek	ALB	Paul S. Unger	1998	35
15 0/16	8 9/16	6 7/16	M	Gallatin County	MT	Brian Koelzer	2001	35
14 15/16	8 12/16	6 3/16	M	Lincoln County	MT	Allen Apling	1959	45
14 15/16	8 11/16	6 4/16	M	Boundary County	ID	Rick Furniss	1968	45
14 15/16	8 14/16	6 1/16	M	Rio Blanco County	CO	Jack Pawlak	1971	45
14 15/16	8 14/16	6 1/16	M	Carbon County	UT	Larry Wright	1975	45
14 15/16	8 11/16	6 4/16	M	Deer Lodge County	MT	Scott Koelzer	1979	45
14 15/16	8 13/16	6 2/16	M	Elmore County	ID	Tony Mudd	1993	45
14 15/16	8 12/16	6 3/16	M	Iron County	UT	David R. Spellman	2006	45
14 14/16	8 14/16	6 0/16	M	Lincoln County	MT	Dr. B. L. Lundberg	1958	52
14 14/16	8 11/16	6 3/16	M	Flathead County	MT	Jack Whitney	1967	52
14 14/16	8 10/16	6 4/16	M	Lemhi County	ID	Ray Torrey	1969	52
14 14/16	8 10/16	6 4/16	M	Granite County	MT	John Lawler	1972	52
14 14/16	8 13/16	6 1/16	M	Elmore County	ID	Dan F. Hackney	1973	52
14 14/16	8 12/16	6 2/16	M	Madison County	MT	Cecil I. Tharp	1984	52
14 13/16	8 11/16	6 2/16	M	Chelan County	WA	Dr. R. Congdon	1951	58
14 13/16	8 11/16	6 2/16	M	Ferry County	WA	R. O. Hilderbrant	1965	58
14 13/16	8 12/16	6 1/16	F	Duchesne County	UT	Larry Jones	1967	58
14 13/16	8 13/16	6 0/16	M	Chaffee County	CO	Phillip B. Grable	1973	58
14 13/16	8 14/16	5 15/16	M	Mesa County	CO	Jim R. Lewis	1981	58

COUGAR [Mountain Lion]

Minimum Score 13 8/16 *Continued*

Score	Greatest Length	Greatest Width	Sex	Area	State/ Province	Hunter's Name	Date	Rank
14 13/16	8 11/16	6 2/16	M	San Miguel County	CO	David E. Smith	1985	58
14 13/16	8 12/16	6 1/16	M	Albany County	WY	R.D. Keeler, D.C.	1985	58
14 13/16	8 11/16	6 2/16	M	Gallatin County	MT	John Berger	1993	58
14 13/16	8 11/16	6 2/16	M	Boise County	ID	August S. Gray	1993	58
14 13/16	8 12/16	6 1/16	M	Granite County	MT	Mickey E. Lotz	1994	58
14 13/16	8 10/16	6 3/16	M	Saguache County	CO	Mike Haynes	1995	58
14 13/16	8 10/16	6 3/16	M	Okanogan County	WA	Brad Harden	2006	58
14 12/16	8 14/16	5 14/16	M	Chelan County	WA	Dr. R. Congdon	1952	70
14 12/16	8 12/16	6 0/16	M	Clallam County	WA	Lloyd Beebe	1953	70
14 12/16	8 9/16	6 3/16	M	Rio Blanco County	CO	LeRoy Wood	1965	70
14 12/16	8 12/16	6 0/16	M		CO	Clyde Hector	1967	70
14 12/16	8 13/16	5 15/16	M	Garfield County	UT	Harold Boyack	1968	70
14 12/16	8 9/16	6 3/16	M	Madison County	MT	Ken Hoehn	1985	70
14 12/16	8 12/16	6 0/16	M	Saguache County	CO	Mark Wuerthle	1988	70
14 12/16	8 8/16	6 4/16	M	Flathead County	MT	Ira S. Uradomo	1992	70
14 12/16	8 10/16	6 2/16	M	Huerfano County	CO	David Hinton	1994	70
14 12/16	8 14/16	5 14/16	M	Wayne County	UT	Robert M. Daggett	1995	70
14 12/16	8 10/16	6 2/16	M	Jefferson County	CO	Troy Cunningham	1999	70
14 12/16	8 9/16	6 3/16	M	Powell County	MT	Marlon J. Clapham	2001	70
14 12/16	8 12/16	6 0/16	M	Elmore County	ID	Dale Karch	2003	70
14 12/16	8 13/16	5 15/16	M	Clear Creek County	CO	Matt Archuleta	2004	70
14 11/16	8 10/16	6 1/16	M	Sundre	ALB	Tom Decker	1966	84
14 11/16	8 8/16	6 3/16	M	Custer County	ID	Ralph V. Pehrson	1969	84
14 11/16	8 8/16	6 3/16	M	Kettle River	BC	Irvin Plotz	1976	84
14 11/16	8 10/16	6 1/16	M	Middle Fork	ID	Robert Frank	1978	84
14 11/16	8 12/16	5 15/16	M	Utah County	UT	Fred Tarran	1982	84
14 11/16	8 8/16	6 3/16	M	Chaffee County	CO	Reggie Spiegelberg	1983	84
14 11/16	8 12/16	5 15/16	M	Teton County	WY	Joseph P. Furlong	1989	84
14 11/16	8 12/16	5 15/16	M	Montezuma County	CO	Russ Dufva	1994	84
14 11/16	8 12/16	5 15/16	M	Uintah County	UT	Scott Anspaugh	1995	84
14 11/16	8 10/16	6 1/16	M	Nye County	NV	Ricardo Longoria	1998	84
14 10/16	8 10/16	6 0/16	M	Lincoln County	MT	Dr. Lowell L. Eddy	1967	94
14 10/16	8 10/16	6 0/16	M	Idaho County	ID	C. Bruce Peeples, Jr.	1970	94
14 10/16	8 10/16	6 0/16	M	Flathead County	MT	Jerry Karsky	1976	94
14 10/16	8 10/16	6 0/16	M	Pincher Creek	ALB	Theo Mitchell	1977	94
14 10/16	8 9/16	6 1/16	M	Elmore County	ID	L. Dean Goodner	1981	94
14 10/16	8 10/16	6 0/16	M	Judith Basin County	MT	John Rosey Roseland	1989	94
14 10/16	8 10/16	6 0/16	M	Phillips County	MT	Alan Fedorenko	1993	94
14 10/16	8 11/16	5 15/16	M	Pueblo County	CO	Tommy Chambliss	1994	94
14 10/16	8 10/16	6 0/16	M	West Kettle	BC	James H. Simonds, Jr.	1995	94
14 10/16	8 9/16	6 1/16	M	Boise County	ID	Norman Henderson	1995	94
14 10/16	8 10/16	6 0/16	M	Larimer County	CO	Thomas H. Harris	1995	94
14 10/16	8 7/16	6 3/16	M	Boise County	ID	Rick Tribby	1996	94
14 10/16	8 7/16	6 3/16	M	Las Animas County	CO	Pat Powell	1998	94
14 10/16	8 10/16	6 0/16	M	Grand County	CO	Rick Duggan	2002	94
14 10/16	8 8/16	6 2/16	M	Gunnison County	CO	Matthew Reilly	2004	94
14 9/16	8 5/16	6 4/16	M	Gila County	AZ	Ben Pearson	1958	109
14 9/16	8 10/16	5 15/16	M	Idaho County	ID	Keith N. Johnson	1966	109
14 9/16	8 8/16	6 1/16	M	Lemhi County	ID	Richard E. Vail	1974	109
14 9/16	8 8/16	6 1/16	M	Grand County	UT	Henry 'Hank' Frey	1974	109
14 9/16	8 8/16	6 1/16	M	San Miguel County	NM	Richard McClain	1980	109
14 9/16	8 10/16	5 15/16	M	Rio Blanco County	CO	Don Waechtler	1984	109
14 9/16	8 10/16	5 15/16	M	Fremont County	CO	Bill Goodspeed	1986	109
14 9/16	8 9/16	6 0/16	M	Grand County	UT	Wes Walton	1987	109
14 9/16	8 8/16	6 1/16	M	Jefferson County	CO	Steve Fausel	1989	109
14 9/16	8 10/16	5 15/16	M	Kane County	UT	Jeff Buck	1989	109
14 9/16	8 8/16	6 1/16	M	Madison County	MT	Fred Richter	1991	109
14 9/16	8 9/16	6 0/16	M	Grand County	CO	Barry J. Smith	1993	109
14 9/16	8 11/16	5 14/16	M	Lemhi County	ID	Michael Judas	1996	109
14 9/16	8 10/16	5 15/16	M	Delta County	CO	Scott Hargrove	1998	109
14 9/16	8 12/16	5 13/16	M	Moffat County	CO	Mike E. Neilson	1998	109
14 9/16	8 11/16	5 14/16	M	Grand County	CO	Cary Laman	1999	109

COUGAR [Mountain Lion]

Score	Greatest Length	Greatest Width	Sex	Area	State/ Province	Hunter's Name	Date	Rank
14 9/16	8 9/16	6 0/16	M	Douglas County	CO	Turk Wendell	2000	109
14 8/16	8 9/16	5 15/16	F	Latah County	ID	Charles Kelso	1965	126
14 8/16	8 8/16	6 0/16	M	Ventura County	CA	Betty Gulman	1967	126
14 8/16	8 10/16	5 14/16	M	Catron County	NM	Ed Schaub	1970	126
14 8/16	8 9/16	5 15/16	M	Lemhi County	ID	Dr. Henry C. McDonald	1971	126
14 8/16	8 9/16	5 15/16	M	Emery County	UT	Terry Molneux	1972	126
14 8/16	8 9/16	5 15/16	M	Kane County	UT	Charles F. Maloney, Jr.	1973	126
14 8/16	8 6/16	6 2/16	M	Lincoln County	MT	Jerry Brown	1975	126
14 8/16	8 10/16	5 14/16	M	Elmore County	ID	Ronald L. Sherer	1979	126
14 8/16	8 12/16	5 12/16	M	Saguache County	CO	J. Keith Chastain	1982	126
14 8/16	8 10/16	5 14/16	M	Union County	OR	Ken Richter	1984	126
14 8/16	8 9/16	5 15/16	M	Iron County	UT	Patrick Barwick	1985	126
14 8/16	8 8/16	6 0/16	M	Lane County	OR	Larry D. Jones	1987	126
14 8/16	8 8/16	6 0/16	M	Saguache County	CO	William Larry Wray	1987	126
14 8/16	8 9/16	5 15/16	M	Missoula County	MT	Bob Lussier	1988	126
14 8/16	8 8/16	6 0/16	M	Larimer County	CO	John D. Lindell	1989	126
14 8/16	8 11/16	5 13/16	M	Judith Basin County	MT	Kelly Norskog	1990	126
14 8/16	8 10/16	5 14/16	M	Garfield County	CO	Carroll Thomas Roach	1992	126
14 8/16	8 7/16	6 1/16	M	Nakusp	BC	Dave Richardson	1992	126
14 8/16	8 10/16	5 14/16	M	Terrell County	TX	E. Josh Isbell	1992	126
14 8/16	8 8/16	6 0/16	M	Valley County	ID	Rick Addison	1994	126
14 8/16	8 8/16	6 0/16	M	Lemhi County	ID	William C. Shuster	1996	126
14 8/16	8 11/16	5 13/16	M	Garfield County	UT	Randy Forsythe	1997	126
14 8/16	8 10/16	5 14/16	M	Rio Arriba County	NM	Ryan Panzy	2002	126
14 8/16	8 8/16	6 0/16	M	Clear Creek County	CO	Mark Turner	2012	126
14 7/16	8 9/16	5 14/16	M	Elmore County	ID	William R. Vanderhoef	1966	150
14 7/16	8 11/16	5 12/16	M	Range Creek	UT	Gordy J. Longville	1967	150
14 7/16	8 6/16	6 1/16	M	White Pine County	NV	Barry L. May	1975	150
14 7/16	8 9/16	5 14/16	M	Carbon County	UT	Rick Hunckler	1977	150
14 7/16	8 9/16	5 14/16	M	Sanders County	MT	Jerry V. Finley	1983	150
14 7/16	8 8/16	5 15/16	M	Lander County	NV	David P. Lindman	1985	150
14 7/16	8 8/16	5 15/16	M	Skagit County	WA	Jerry Solie	1987	150
14 7/16	8 6/16	6 1/16	M	Idaho County	ID	Fred H. Simonton III	1992	150
14 7/16	8 9/16	5 14/16	M	Rio Blanco County	CO	Tim Cuthriell	1992	150
14 7/16	8 6/16	6 1/16	M	Rio Grande County	CO	John Olson	1992	150
14 7/16	8 9/16	5 14/16	M	Gilpin County	CO	Steve Barnhill	1992	150
14 7/16	8 10/16	5 13/16	M	Park County	CO	Ron Adamson	1993	150
14 7/16	8 6/16	6 1/16	M	Montezuma County	CO	David R. Hall	1993	150
14 7/16	8 8/16	5 15/16	M	Valley County	ID	Randy A. Reeves	1994	150
14 7/16	8 11/16	5 12/16	M	Camas County	ID	Brian Brockette	1999	150
14 7/16	8 6/16	6 1/16	M	Fremont County	CO	Robert Snell	2000	150
14 7/16	8 9/16	5 14/16	M	Fremont County	CO	Jerry Woodland	2000	150
14 6/16	8 7/16	5 15/16	M	Sequoia National Forest	CA	Douglas Walker	1960	167
14 6/16	8 7/16	5 15/16	M	Elmore County	ID	C. Randall Byers	1966	167
14 6/16	8 8/16	5 14/16	M	Missoula County	MT	John Hershey	1969	167
14 6/16	8 6/16	6 0/16	M	Falkland	BC	W. Klingsat	1974	167
14 6/16	8 7/16	5 15/16	M	Colfax County	NM	Richard A. Meyer	1974	167
14 6/16	8 8/16	5 14/16	M	Uintah County	UT	Donald Redfox	1978	167
14 6/16	8 6/16	6 0/16	M	Chaffee County	CO	Judy Clyncke	1981	167
14 6/16	8 6/16	6 0/16	M	Colfax County	NM	Joseph Wambach	1982	167
14 6/16	8 8/16	5 14/16	M	Summit County	UT	Jeffrey W. Potter	1988	167
14 6/16	8 10/16	5 12/16	M	Fremont County	CO	Daniel Daly	1988	167
14 6/16	8 6/16	6 0/16	M	Oldman River	ALB	Tom Foss	1991	167
14 6/16	8 7/16	5 15/16	M	Elmore County	ID	William MacCarty III	1991	167
14 6/16	8 7/16	5 15/16	M	Chaffee County	CO	Al Miller	1993	167
14 6/16	8 8/16	5 14/16	M	Porcupine Hills	ALB	Warren Witherspoon	1995	167
14 6/16	8 9/16	5 13/16	M	Fergus County	MT	Donny Roy	1996	167
14 6/16	8 9/16	5 13/16	M	Boise County	ID	Todd Felt	1996	167
14 6/16	8 6/16	6 0/16	M	Park County	MT	Leo P. Caito, Jr.	2002	167
14 6/16	8 5/16	6 1/16	M	Lemhi County	ID	Jeffrey A. Eder	2008	167
14 5/16	8 6/16	5 15/16	M	Esmeralda County	NV	Don Schram	1965	185
14 5/16	8 8/16	5 13/16	M	Garfield County	UT	H. R. 'Dutch' Wambold	1966	185

COUGAR [Mountain Lion]

Minimum Score 13 8/16 *Continued*

Score	Greatest Length	Greatest Width	Sex	Area	State/ Province	Hunter's Name	Date	Rank
14 5/16	8 9/16	5 12/16	M	Tatla Lake	BC	William L. Nickerson	1966	185
14 5/16	8 8/16	5 13/16	M	Garfield County	UT	Robert K. Paulson	1968	185
14 5/16	8 10/16	5 11/16	M	Carbon County	UT	M. R. James	1970	185
14 5/16	8 7/16	5 14/16	M	White Pine County	NV	James L. Beard	1975	185
14 5/16	8 8/16	5 13/16	M	Chaffee County	CO	Michael Ballard	1975	185
14 5/16	8 7/16	5 14/16	M	Las Animas County	CO	Barry L. Powell	1975	185
14 5/16	8 8/16	5 13/16	M	Beaver County	UT	Bruce Post	1975	185
14 5/16	8 9/16	5 12/16	M	Elmore County	ID	L. Dean Goodner	1978	185
14 5/16	8 7/16	5 14/16	M	Judith Basin County	MT	Don Davidson	1981	185
14 5/16	8 9/16	5 12/16	M	Dona Ana County	NM	Larry M. Sellers	1986	185
14 5/16	8 7/16	5 14/16	M	Elmore County	ID	Alfred John Gemrich	1988	185
14 5/16	8 6/16	5 15/16	M	Madison County	MT	Scott T. Smolen	1988	185
14 5/16	8 8/16	5 13/16	M	Broadwater County	MT	Mike Parsons	1990	185
14 5/16	8 9/16	5 12/16	M	Grand County	CO	Dave Parri	1994	185
14 5/16	8 9/16	5 12/16	M	Custer County	ID	Scott Brower	1994	185
14 5/16	8 8/16	5 13/16	M	Idaho County	ID	Matt March	1996	185
14 5/16	8 9/16	5 12/16	M	Moffat County	CO	Daniel L. Wells	1997	185
14 5/16	8 7/16	5 14/16	M	Mesa County	CO	Scot L. Hamilton	2005	185
14 4/16	8 8/16	5 12/16	M	Stoneman Lake	AZ	Dr. C. L. Clare	1962	205
14 4/16	8 8/16	5 12/16	M	Lincoln County	MT	Dale McNutt	1964	205
14 4/16	8 10/16	5 10/16	M	Elmore County	ID	Don Bennett	1968	205
14 4/16	8 4/16	6 0/16	M	Missoula County	MT	Tony Dumay	1968	205
14 4/16	8 6/16	5 14/16	M	Elmore County	ID	John E. Anderson	1972	205
14 4/16	8 6/16	5 14/16	M	Lemhi County	ID	Kenneth Anselmi	1972	205
14 4/16	8 6/16	5 14/16	M	Lemhi County	ID	H. R. 'Rusty' Neely	1975	205
14 4/16	8 3/16	6 1/16	M	Colfax County	NM	Stephen 'Don' Hornady	1980	205
14 4/16	8 6/16	5 14/16	M	Salt Lake County	UT	William L. Randles	1982	205
14 4/16	8 6/16	5 14/16	M	Colfax County	NM	John L. Chapman	1986	205
14 4/16	8 4/16	6 0/16	M	Nakusp	BC	Len Surina	1987	205
14 4/16	8 6/16	5 14/16	M	Ravalli County	MT	Erik "Rick" Aslesen	1988	205
14 4/16	8 8/16	5 12/16	M	Iron County	UT	Bernie E. Belfrage	1988	205
14 4/16	8 4/16	6 0/16	M	Judith Basin County	MT	Don Davidson	1992	205
14 4/16	8 8/16	5 12/16	M	Ram Creek	BC	Tom Marshall	1993	205
14 4/16	8 8/16	5 12/16	M	Washington County	UT	Pat Abalsamo	1994	205
14 4/16	8 7/16	5 13/16	M	Valley County	ID	Lewis Zane Abbott	1994	205
14 4/16	8 5/16	5 15/16	M	Platte County	WY	Mike Boughton	1996	205
14 4/16	8 8/16	5 12/16	M	Boise County	ID	Pat Lefemine	1998	205
14 4/16	8 6/16	5 14/16	M	Elmore County	ID	Joe St. Charles	1999	205
14 4/16	8 8/16	5 12/16	M	Boise County	ID	Robert P. Hudson, Jr.	1999	205
14 4/16	8 8/16	5 12/16	M	Park County	WY	Jonathan E. Rhodes	2009	205
14 4/16	8 8/16	5 12/16	M	Lincoln County	OR	Tom Nieradka	2009	205
14 4/16	8 4/16	6 0/16	M	Albany County	WY	Austin Guerin	2010	205
14 3/16	8 6/16	5 13/16	M	Nye County	NV	Dick Gulman	1968	229
14 3/16	8 7/16	5 12/16	M	Ferry County	WA	Tom Smith	1968	229
14 3/16	8 5/16	5 14/16	M	Grand County	UT	John B. Baughman	1969	229
14 3/16	8 6/16	5 13/16	M	Douglas County	NV	Bill Fuller	1972	229
14 3/16	8 7/16	5 12/16	M	Huerfano County	CO	Marvin C. Clyncke	1973	229
14 3/16	8 9/16	5 10/16	M	Juab County	UT	Samuel McCarty	1975	229
14 3/16	8 7/16	5 12/16	M	Ravalli County	MT	Jim Loughran	1990	229
14 3/16	8 14/16	5 5/16	M	Grand County	CO	Cary Laman	1991	229
14 3/16	8 5/16	5 14/16	M	Moffat County	CO	Kenny E. Leo	1996	229
14 3/16	8 8/16	5 11/16	M	Las Animas County	CO	Rick East	1997	229
14 3/16	8 4/16	5 15/16	M	Madison County	MT	Brian Koelzer	1998	229
14 2/16	8 6/16	5 12/16	M	Okanogan County	WA	Dr. Russell Congdon	1950	240
14 2/16	8 6/16	5 12/16	M	Kane County	UT	William P. Mastrangel	1957	240
14 2/16	8 4/16	5 14/16	M	Fresno County	CA	John Faulconer	1964	240
14 2/16	8 4/16	5 14/16	M	Uintah County	UT	Larry Jones	1970	240
14 2/16	8 6/16	5 12/16	M	Garfield County	CO	Phillip C. Durr	1970	240
14 2/16	8 6/16	5 12/16	M	Rio Blanco County	CO	James L. Emerson	1976	240
14 2/16	8 8/16	5 10/16	M	Ravalli County	MT	John L Wozniak	1985	240
14 2/16	8 3/16	5 15/16	M	Mesa County	CO	David A. Schroeder	1986	240
14 2/16	8 4/16	5 14/16	M	Lemhi County	ID	Kent Brandt	1988	240

COUGAR [Mountain Lion]

Minimum Score 13 8/16 *Continued*

Score	Greatest Length	Greatest Width	Sex	Area	State/ Province	Hunter's Name	Date	Rank
14 2/16	8 6/16	5 12/16	M	Park County	CO	Bryon Scott Johnson	1991	240
14 2/16	8 8/16	5 10/16	M	Wasatch County	UT	Dallas Smith	1994	240
14 2/16	8 5/16	5 13/16	M	Grand County	CO	Daniel Thomas Cresci	1995	240
14 2/16	8 4/16	5 14/16	M	Fremont County	CO	Fred Eichler	1997	240
14 2/16	8 4/16	5 14/16	M	Carbon County	MT	Mike O'Connor	1998	240
14 2/16	8 7/16	5 11/16	M	Las Animas County	CO	Fred Eichler	2000	240
14 1/16	8 14/16	5 3/16	M	Garfield County	CO	Jack Peters	1964	255
14 1/16	8 7/16	5 10/16	M	Rio Blanco County	CO	Charles Kohler	1969	255
14 1/16	8 7/16	5 10/16	M	Chaffee County	CO	Frank B. Parrish	1969	255
14 1/16	8 6/16	5 11/16	M	Elmore County	ID	Jerry E. Burt	1971	255
14 1/16	8 3/16	5 14/16	M	Mesa County	CO	Cary E. Weldon	1972	255
14 1/16	8 5/16	5 12/16	M	Carbon County	UT	David K. Elliot	1973	255
14 1/16	8 5/16	5 12/16	M	Butte County	ID	Ken Anselmi	1975	255
14 1/16	8 5/16	5 12/16	M	Rio Blanco County	CO	Chris Christian	1976	255
14 1/16	8 5/16	5 12/16	M	Wayne County	UT	C. Duane Kerr	1979	255
14 1/16	8 5/16	5 12/16	M	Cache County	UT	Val D. Larsen	1980	255
14 1/16	8 4/16	5 13/16	M	Madison County	MT	Dick Curtis	1983	255
14 1/16	8 4/16	5 13/16	M	Gilpin County	CO	Kurt W. Keskimaki	1985	255
14 1/16	8 8/16	5 9/16	M	Millard County	UT	Dave Scott	1987	255
14 1/16	8 6/16	5 11/16	M	Montrose County	CO	Joe Garvey	1988	255
14 1/16	8 5/16	5 12/16	M	Alamosa County	CO	Dan Call	1989	255
14 1/16	8 4/16	5 13/16	M	Garfield County	CO	Roger Wintle	1992	255
14 1/16	8 6/16	5 11/16	M	Park County	CO	Jack P. Van Vianen	1992	255
14 1/16	8 2/16	5 15/16	M	Sweet Grass County	MT	Bill Bryce	1998	255
14 1/16	8 4/16	5 13/16	M	Moffat County	CO	Tim M. Webster	2000	255
14 1/16	8 3/16	5 14/16	M	Madison County	MT	Michael Beutel	2003	255
14 1/16	8 5/16	5 12/16	M	Judith Basin County	MT	John "Rosey" Roseland	2007	255
14 0/16	8 4/16	5 12/16	M	Uintah County	UT	Dr. Quentin F. Mangion	1962	276
14 0/16	8 4/16	5 12/16	M	Gila County	AZ	Hugh Pearson	1963	276
14 0/16	8 4/16	5 12/16	M	Flathead County	MT	Dorn L. Brinker	1969	276
14 0/16	8 3/16	5 13/16	M	Valley County	ID	Ronald N. Kolpin	1972	276
14 0/16	8 1/16	5 15/16	M	Yavapai County	AZ	Louis A. Vohs	1973	276
14 0/16	8 5/16	5 11/16	M	Fremont County	CO	Gary Fisher	1979	276
14 0/16	8 4/16	5 12/16	M	San Juan County	UT	Gary Paluszcyk	1979	276
14 0/16	8 5/16	5 11/16	M	Madison County	MT	Leland S. Speakes, Jr.	1984	276
14 0/16	8 2/16	5 14/16	M	Madison County	MT	Carl Spaeth	1986	276
14 0/16	8 2/16	5 14/16	M	San Juan County	NM	Keith Hardy	1987	276
14 0/16	8 6/16	5 10/16	M	Judith Basin County	MT	Noel J. Poux	1988	276
14 0/16	8 4/16	5 12/16	M	Idaho County	ID	David A. Shupp	1989	276
14 0/16	8 6/16	5 10/16	M	Elmore County	ID	Julian Salutrequi	1990	276
14 0/16	8 6/16	5 10/16	M	Archuleta County	CO	Tony R. Stephens	1993	276
14 0/16	8 4/16	5 12/16	M	Valley County	ID	Kenneth Hyde	1994	276
14 0/16	8 5/16	5 11/16	M	Jefferson County	CO	Jason Adamson	1994	276
14 0/16	8 7/16	5 9/16	M	Chaffee County	CO	Ray Woods	1994	276
14 0/16	8 6/16	5 10/16	M	Longview	ALB	Lauren Hoover	1996	276
14 0/16	8 4/16	5 12/16	M	Fremont County	CO	Rick Wilson	1997	276
14 0/16	8 6/16	5 10/16	M	Clear Creek County	CO	Ron Rockwell	1999	276
14 0/16	8 5/16	5 11/16	M	Elko County	NV	Bob Ehle	2002	276
14 0/16	8 4/16	5 12/16	M	Cochrane	ALB	Dan J. Ryle, Jr.	2007	276
13 15/16	8 3/16	5 12/16	M	Gila County	AZ	Hugh Pearson	1963	298
13 15/16	8 5/16	5 10/16	M	Rio Blanco County	CO	Joel Hogan	1967	298
13 15/16	8 5/16	5 10/16	M	Lemhi County	ID	John Mascellino	1972	298
13 15/16	8 4/16	5 11/16	M	Nye County	NV	Charles Pat Walker, Jr.	1990	298
13 15/16	8 5/16	5 10/16	M	Beaverhead County	MT	Lynn Lamphiear	1991	298
13 15/16	8 2/16	5 13/16	M	Flathead County	MT	David D. Johnston	1992	298
13 14/16	8 4/16	5 10/16	M	Flathead County	MT	Jerry Almos	1970	304
13 14/16	8 1/16	5 13/16	M	Mesa County	CO	Stan Bocian	1974	304
13 14/16	8 2/16	5 12/16	M	Boise County	ID	Robert L. Bevan	1976	304
13 14/16	8 3/16	5 11/16	M	Grand County	UT	Henry 'Hank' Frey	1982	304
13 14/16	8 5/16	5 9/16	M	Chaffee County	CO	Raymond Roussett, Jr.	1985	304
13 14/16	8 5/16	5 9/16	M	Lincoln County	MT	Kenneth Mamatz	1988	304
13 14/16	8 3/16	5 11/16	M	Madison County	MT	Randall Brown	1991	304

COUGAR [Mountain Lion]

Minimum Score 13 8/16 *Continued*

Score	Greatest Length	Greatest Width	Sex	Area	State/ Province	Hunter's Name	Date	Rank
13 14/16	8 2/16	5 12/16	M	Judith Basin County	MT	John "Rosey" Roseland	1993	304
13 14/16	8 4/16	5 10/16	M	Sevier County	UT	Randy P. Forsythe	1994	304
13 14/16	8 2/16	5 12/16	M	Madison County	MT	Ken Southworth	1998	304
13 14/16	8 2/16	5 12/16	M	Summit County	CO	Bob Reedy	1998	304
13 14/16	8 3/16	5 11/16	M	Ravalli County	MT	Wendy Decker	1999	304
13 14/16	8 4/16	5 10/16	M	Elmore County	ID	Allen Clark	2008	304
13 13/16	8 2/16	5 11/16	M	Esmeralda County	NV	George Hooker	1961	317
13 13/16	7 15/16	5 14/16	M	Lemhi County	ID	Vern Herman	1969	317
13 13/16	8 3/16	5 10/16	M	Johnson County	WY	Terry Krahn	1986	317
13 13/16	8 3/16	5 10/16	M	Chaffee County	CO	David Spacek	1990	317
13 13/16	8 1/16	5 12/16	M	San Juan County	UT	Daniel Willems	1992	317
13 13/16	8 6/16	5 7/16	M	Chaffee County	CO	A. M. Salazar	1992	317
13 13/16	8 4/16	5 9/16	M	Archuleta County	CO	Grant Adkisson	1994	317
13 13/16	8 4/16	5 9/16	M	Chaffee County	CO	Paul Bohochik	1994	317
13 13/16	8 3/16	5 10/16	M	San Juan County	UT	Michael C. Parkinson	1995	317
13 13/16	8 3/16	5 10/16	M	Sanders County	MT	Wes Brogan	2001	317
13 13/16	8 4/16	5 9/16	M	Laramie County	WY	Jim Carlson	2006	317
13 13/16	8 3/16	5 10/16	M	Carbon County	WY	Michael Casey	2010	317
13 13/16	8 4/16	5 9/16	M	Washoe County	NV	Jack Spencer, Jr.	2010	317
13 12/16	8 4/16	5 8/16	M	Shasta National Forest	CA	Harv Ebers	1963	330
13 12/16	8 0/16	5 12/16	M	Ravalli County	MT	Joe Lawrence	1965	330
13 12/16	7 11/16	6 1/16	M	Garfield County	UT	Robert E. Todd	1969	330
13 12/16	8 2/16	5 10/16	M	Pima County	AZ	Sherwin Lipsitz	1976	330
13 12/16	8 3/16	5 9/16	M	Lemhi County	ID	Ray Torrey	1978	330
13 12/16	8 3/16	5 9/16	M	Rio Blanco County	CO	Tom Brakke	1990	330
13 12/16	8 3/16	5 9/16	M	Montrose County	CO	Gregory Hise	1992	330
13 12/16	8 4/16	5 8/16	M	Washington County	UT	Norman J. Roy	1993	330
13 12/16	8 1/16	5 11/16	M	Lemhi County	ID	Michael Judas	1994	330
13 12/16	8 3/16	5 9/16	M	Fremont County	CO	Ben L. Nelson	1997	330
13 11/16	8 6/16	5 5/16	M	San Juan County	UT	James Karlovec	1975	340
13 11/16	8 2/16	5 9/16	M	Carbon County	UT	Bernard R. Giacoletto	1975	340
13 11/16	8 4/16	5 7/16	M	Washington County	UT	Jules Pacheco	1989	340
13 11/16	8 1/16	5 10/16	M	Asotin County	WA	Mark Kolowith	1990	340
13 11/16	8 3/16	5 8/16	M	Fergus County	MT	Mike Bentler	1992	340
13 11/16	8 3/16	5 8/16	M	Los Alamos County	NM	Timothy F.H. Smith	1995	340
13 11/16	8 2/16	5 9/16	M	Park County	MT	Mike O'Connor	1996	340
13 11/16	8 2/16	5 9/16	M	Elmore County	ID	Tom Vanasche	1999	340
13 10/16	8 2/16	5 8/16	F	Moffat County	CO	Roland C. Gravenkemper	1959	348
13 10/16	8 0/16	5 10/16	F	Sana Arroya Canyon	UT	Edward Collins	1967	348
13 10/16	7 14/16	5 12/16	M	Fremont County	CO	Jeffrey D. McKnight	1970	348
13 10/16	7 15/16	5 11/16	M	Boise County	ID	Robert B. Braswell	1971	348
13 10/16	7 12/16	5 14/16	M	Lemhi County	ID	Richard R. Smith	1976	348
13 10/16	8 0/16	5 10/16	M	Coconino County	AZ	Les Shelton	1981	348
13 10/16	8 4/16	5 6/16	M	Colfax County	NM	I. Lionel Kelley	1986	348
13 10/16	8 2/16	5 8/16	M	Fergus County	MT	John "Rosey" Roseland	1987	348
13 10/16	8 1/16	5 9/16	M	Washington County	UT	Ken Mowerson	1988	348
13 10/16	8 3/16	5 7/16	M	Grand County	CO	Andrew C. Bair	1994	348
13 10/16	8 1/16	5 9/16	M	Eagle County	CO	Sheldon S. Showalter	1994	348
13 10/16	8 0/16	5 10/16	M	Catron County	NM	Jorge Garcia-Segovia	1995	348
13 10/16	7 15/16	5 11/16	M	Douglas County	CO	Eric H. Wolff	2011	348
13 9/16	8 0/16	5 9/16	M	Carbon County	UT	Tom Kludy	1965	361
13 9/16	8 1/16	5 8/16	M	Elmore County	ID	Larry Bergmann	1972	361
13 9/16	8 0/16	5 9/16	F	Custer County	ID	John Kuhar	1975	361
13 9/16	8 1/16	5 8/16	M	Carbon County	UT	Michael Judas	1977	361
13 9/16	7 15/16	5 10/16	M	Rio Blanco County	CO	John Horstman	1977	361
13 9/16	8 0/16	5 9/16	M	Saguache County	CO	Ed R. Wiseman	1977	361
13 9/16	7 14/16	5 11/16	M	Sanders County	MT	Fred J. Hoppe	1984	361
13 9/16	8 0/16	5 9/16	M	Sanders County	MT	Dr. Eugene T. Altiere	1985	361
13 9/16	8 1/16	5 8/16	M	Garfield County	UT	George E. Wright	1987	361
13 9/16	8 2/16	5 7/16	M	Mesa County	CO	Troy James	1991	361
13 9/16	8 2/16	5 7/16	M	Montrose County	CO	Rod Van Sickle	1997	361
13 9/16	8 0/16	5 9/16	M	Lincoln County	WY	Gary Hornberger	2009	361

COUGAR [Mountain Lion]

Minimum Score 13 8/16 *Continued*

Score	Greatest Length	Greatest Width	Sex	Area	State/ Province	Hunter's Name	Date	Rank
13 8/16	8 0/16	5 8/16	M	Uintah County	UT	Creetie Kerr	1964	373
13 8/16	8 2/16	5 6/16	M	Uintah County	UT	Dr. George A. Waldriff	1965	373
13 8/16	7 12/16	5 12/16	F	Lincoln County	MT	G. H. Malinoski	1967	373
13 8/16	8 1/16	5 7/16	M	Alamosa County	CO	Barry J. Smith	1985	373
13 8/16	7 13/16	5 11/16	M	Madison County	MT	Jim Ellis	1985	373
13 8/16	8 0/16	5 8/16	M	Fergus County	MT	John Fleharty	1988	373
13 8/16	8 0/16	5 8/16	M	Montrose County	CO	Gene Mathias	1988	373
13 8/16	8 1/16	5 7/16	M	Saguache County	CO	Michael C. Dysh	1994	373
13 8/16	7 15/16	5 9/16	M	Coconino County	AZ	Tom Egnew	1994	373
13 8/16	8 0/16	5 8/16	M	Madison County	MT	Bob Morton	1995	373
13 8/16	8 0/16	5 8/16	M	Custer County	ID	Brian Brockette	1996	373
13 8/16	8 1/16	5 7/16	F	Delta County	CO	John "Jack" Nothardt	1997	373
13 8/16	8 1/16	5 7/16	M	Montrose County	CO	Michael J. Schneider	1998	373
13 8/16	8 0/16	5 8/16	M	Fergus County	MT	Larry Jensen	1998	373
13 8/16	7 13/16	5 11/16	F	Saguache County	CO	Ron Charity	2001	373
13 8/16	8 1/16	5 7/16	M	Chaffee County	CO	Nathan L. Andersohn	2001	373
13 8/16	8 0/16	5 8/16	M	Las Animas County	CO	Fred Eichler	2010	373

Hunter: Tom Nieradka
Species: Cougar
Score: 14 4/16
Date: 8/29/09
Location: Lincoln County, OR

COLUMBIAN BLACKTAIL DEER [Typical Antlers]

Minimum Score 95 *Odocoileus hemionus columnianus*

Score	Length of R Main Beam	L	Inside Spread	Number of R Points	L	Area	State/ Province	Hunter's Name	Date	Rank
172 2/8	26 3/8	25 7/8	20 4/8	7	7	Marion County	OR	B. G. Shurtleff	1969	1
151 5/8	22 1/8	25 0/8	21 3/8	5	4	Lake County	CA	Art Young	1921	2
146 6/8	20 0/8	20 0/8	17 6/8	5	5	Marion County	OR	Jim Brackenbury	1990	3
144 5/8	21 2/8	20 3/8	16 6/8	6	7	Jackson County	OR	Leroy Bedingfield	1970	4
143 7/8	21 0/8	20 6/8	21 0/8	5	5	Shasta County	CA	Dave Swenson	1968	5
142 2/8	19 1/8	19 7/8	19 0/8	5	5	Lake County	OR	Don Chandler	1968	6
141 7/8	20 7/8	21 0/8	17 3/8	5	5	Jackson County	OR	Chester Stevenson	1917	7
140 6/8	21 3/8	20 4/8	16 2/8	3	3	Jackson County	OR	Art W. Lee	1965	8
137 6/8	22 4/8	23 3/8	26 1/8	4	4	Jackson County	OR	David Shanklin	1982	9
135 0/8	22 0/8	21 2/8	18 7/8	6	4	Jackson County	OR	Bob Staten	1964	10
134 5/8	21 0/8	16 0/8	16 2/8	5	4	Jackson County	OR	Milton L. Cady	1968	11
134 5/8	21 7/8	21 7/8	17 0/8	7	5	Trinity County	CA	Bob Auser	1981	11
133 7/8	20 6/8	20 6/8	18 7/8	4	4	Jackson County	OR	Donald R. Pritchett	1966	13
133 2/8	20 1/8	20 2/8	17 0/8	5	5	Jackson County	OR	Chester Stevenson	1921	14
133 0/8	20 5/8	21 0/8	14 6/8	5	5	Benton County	OR	Robert W. Worthean	1982	15
132 6/8	20 1/8	20 6/8	18 0/8	4	5	Jackson County	OR	Stanley Moore	1962	16
132 6/8	19 2/8	18 7/8	17 4/8	4	4	Jackson County	OR	David Shanklin	1987	16
132 3/8	18 6/8	19 4/8	16 7/8	5	5	Jackson County	OR	Donald R. Pritchett	1966	18
131 3/8	20 0/8	20 0/8	15 5/8	5	5	Jackson County	OR	Joe Williamson	1965	19
131 2/8	19 6/8	19 3/8	22 2/8	5	5	Linn County	OR	David F. Scheid	1968	20
131 2/8	19 2/8	18 3/8	18 0/8	5	5	Jackson County	OR	David Shanklin	1978	20
130 0/8	19 4/8	19 3/8	17 2/8	5	5	Clackamas County	OR	Neil M. Olsen	2007	22
129 3/8	26 1/8	25 3/8	19 2/8	5	3	Klamath County	OR	Troy Fennel	1964	23
129 1/8	20 4/8	20 4/8	16 3/8	4	4	Douglas County	OR	Tom E. Tipton	1988	24
127 1/8	18 7/8	19 5/8	16 1/8	6	5	Mendocino County	CA	James Buffum	1965	25
126 1/8	22 5/8	22 7/8	17 1/8	5	4	Douglas County	OR	Christopher Tipton	2010	26
125 6/8	19 3/8	18 5/8	14 2/8	4	4	Siskiyou County	CA	Cliff Dewell	1969	27
125 0/8	18 5/8	17 2/8	14 4/8	4	4	Jackson County	OR	Robert J. Jensen	1994	28
124 2/8	20 0/8	20 6/8	16 4/8	4	5	Lewis County	WA	Sandy Tyler	1957	29
124 2/8	18 7/8	18 1/8	16 0/8	5	5	Clackamas County	OR	Matthew Stewart	2008	29
124 0/8	19 0/8	19 2/8	19 7/8	4	4	Kitsap County	WA	Boyd Edward Shelby, Jr.	1995	31
123 4/8	21 3/8	21 7/8	16 2/8	4	4	Thurston County	WA	Al Kowalski	1978	32
123 1/8	20 0/8	20 3/8	16 1/8	5	6	Trinity County	CA	Gary Mayberry	1968	33
122 7/8	19 1/8	20 4/8	16 1/8	4	4	Trinity County	CA	Loran G. August	1981	34
122 1/8	19 5/8	19 5/8	17 4/8	5	6	Jackson County	OR	Richard G. Speer	1965	35
121 4/8	20 6/8	20 5/8	17 4/8	5	4	Lane County	OR	Bradley M. Dorsing	1991	36
121 3/8	15 5/8	15 2/8	14 0/8	5	5	Humboldt County	CA	Doug Walker	1965	37
121 0/8	18 0/8	17 0/8	14 4/8	5	5	Benton County	OR	Larry D. Jones	1966	38
120 2/8	21 2/8	20 3/8	17 1/8	5	5	Linn County	OR	Steve L. Winterstein	1991	39
120 1/8	18 3/8	18 4/8	15 5/8	5	5	Lane County	OR	Dale Drilling	1990	40
119 7/8	21 5/8	21 4/8	15 3/8	3	4	Shasta County	CA	G. Fred Asbell	2002	41
119 6/8	20 0/8	20 3/8	16 2/8	4	4	Pacific County	WA	John Higgins	1973	42
119 3/8	20 4/8	20 2/8	16 1/8	5	6	Mount Washington	BC	Harald Dittkowski	2008	43
118 6/8	20 4/8	20 2/8	14 6/8	4	5	Skamania County	WA	Frank Adkins	1967	44
118 4/8	18 1/8	18 2/8	17 0/8	5	5	Kitsap County	WA	Boyd E. Shelby, Jr.	1997	45
117 4/8	18 0/8	18 2/8	14 1/8	5	5	Benton County	OR	Chris Reed	1970	46
117 3/8	18 5/8	19 4/8	15 1/8	4	4	Jackson County	OR	Duane A. Brentano	1972	47
117 0/8	17 0/8	17 4/8	16 4/8	5	5	Lane County	OR	Brad Dorsing	1992	48
116 7/8	18 7/8	18 4/8	14 5/8	5	5	Jackson County	OR	David Shanklin	1985	49
116 5/8	19 5/8	20 3/8	12 5/8	4	5	Pacific County	WA	Smokey Crews	1969	50
116 5/8	20 4/8	21 2/8	19 1/8	4	4	Snohomish County	WA	Jerry Solie	1995	50
116 2/8	18 7/8	19 7/8	19 0/8	4	4	Jackson County	OR	Ray Gibson	1962	52
115 6/8	19 6/8	19 6/8	16 6/8	5	4	Clackamas County	OR	Jack Smith	1981	53
115 2/8	19 4/8	19 4/8	16 2/8	4	4	Humboldt County	CA	J. E. Grundman	1963	54
115 0/8	21 2/8	21 0/8	17 1/8	6	4	Marion County	OR	Chuck Lynde	1989	55
114 7/8	18 4/8	18 4/8	14 7/8	4	4	Clackamas County	OR	Dan Ellis	2006	56
113 6/8	17 3/8	15 6/8	15 0/8	5	5	Benton County	OR	Gregory M. McHuron	1966	57
113 3/8	19 2/8	18 1/8	17 1/8	4	4	Whatcom County	WA	Chad Brisky	2007	58
112 3/8	16 7/8	16 3/8	13 7/8	5	5	Snohomish County	WA	Jack Davis	1975	59
112 2/8	18 5/8	19 1/8	15 2/8	4	4	Pacific County	WA	Leon Poindexter	1968	60
112 2/8	19 2/8	19 1/8	18 0/8	4	4	Sonoma County	CA	William P. Morehead	2000	60
111 7/8	17 7/8	17 0/8	14 5/8	4	4	Pacific County	WA	Smokey Crews	1967	62

COLUMBIAN BLACKTAIL DEER [Typical Antlers]

Minimum Score 95 *Continued*

Score	Length of R Main Beam	L	Inside Spread	Number of R Points	L	Area	State/ Province	Hunter's Name	Date	Rank
111 2/8	21 1/8	20 5/8	15 6/8	3	6	Lane County	OR	Brad Dorsing	1994	63
110 7/8	18 3/8	17 0/8	16 3/8	4	4	Jackson County	OR	Dale K. Marcy	1966	64
110 6/8	21 1/8	21 4/8	22 0/8	2	2	Mendocino County	CA	Ken Bilstein	1995	65
110 4/8	17 1/8	17 3/8	13 6/8	4	4	Clark County	WA	Larry D. Nahrstedt	1968	66
110 3/8	16 1/8	15 1/8	15 7/8	5	5	Wild Deer Lake	BC	Guy Anttila	1970	67
110 3/8	19 3/8	20 0/8	13 6/8	6	5	North Vancouver	BC	Fred Day	1970	67
110 3/8	17 7/8	17 3/8	14 7/8	4	4	Mendocino County	CA	Wayne Wood	2008	67
109 6/8	21 1/8	20 7/8	16 4/8	3	4	Lewis County	WA	Barney Johnson	1974	70
109 0/8	18 0/8	18 4/8	15 4/8	4	4	Humboldt County	CA	John Roley	1996	71
108 7/8	17 6/8	19 1/8	13 7/8	5	5	Jackson County	OR	Joe Williamson	1963	72
105 6/8	19 1/8	19 6/8	19 0/8	5	3	Clackamas County	OR	Dan Ellis	2003	73
105 4/8	18 3/8	18 5/8	13 6/8	4	4	Clackamas County	OR	Daniel T. Ellis	2007	74
105 3/8	17 3/8	17 3/8	15 5/8	3	4	Coos County	OR	Gary Scorby	1971	75
105 1/8	18 6/8	19 2/8	15 7/8	3	3	Stanislaus County	CA	Harold Arnold	1970	76
104 3/8	18 5/8	18 3/8	15 3/8	4	4	Polk County	OR	Tim Nolan	1987	77
103 3/8	18 6/8	19 3/8	14 5/8	3	5	Trinity County	CA	Rick Duggan	2002	78
102 7/8	18 5/8	18 0/8	14 2/8	4	5	Lane County	OR	Larry D. Jones	1997	79
102 5/8	16 7/8	16 6/8	14 5/8	4	4	Deschutes County	OR	Steve L. Stilwell	1976	80
102 1/8	15 7/8	17 6/8	11 5/8	4	4	Douglas County	OR	Christopher Tipton	2010	81
102 0/8	17 2/8	17 0/8	13 6/8	4	4	Jackson County	OR	George Miller	1966	82
101 3/8	17 0/8	17 4/8	15 3/8	4	3	Jackson County	OR	Dr. G. Scott Jennings	1959	83
100 6/8	15 6/8	15 5/8	15 4/8	4	5	Klamath County	OR	Dr. George Miller	1964	84
99 6/8	20 6/8	20 0/8	14 0/8	2	2	Pacific County	WA	Robert A. Brown	1965	85
99 4/8	19 4/8	18 7/8	17 0/8	3	2	Solano County	CA	Brian J. Morris	2000	86
99 0/8	21 2/8	19 7/8	15 6/8	4	5	Santa Clara County	CA	Mike Walker	1979	87
99 0/8	16 2/8	16 5/8	14 6/8	4	5	Kitsap County	WA	Cecil McConnell	1986	87
98 7/8	20 2/8	19 4/8	14 1/8	3	2	Pacific County	WA	William V. Mishler	1968	89
98 7/8	17 0/8	17 2/8	13 2/8	4	6	Mt. Elphinstone	BC	Ian F. Gazeley	2002	89
98 3/8	16 1/8	16 3/8	14 3/8	4	4	Sonoma County	CA	Ray Torrey	1967	91
98 1/8	16 0/8	16 3/8	12 7/8	4	4	Rogue Unit	OR	Barbara Richardson	1964	92
98 0/8	19 7/8	19 6/8	14 6/8	4	2	Pacific County	WA	Morris Wolters	1967	93
97 7/8	16 3/8	15 5/8	16 1/8	5	5	Benton County	OR	Harold Stice	1957	94
97 5/8	17 4/8	17 6/8	13 5/8	3	3	Siskiyou County	CA	Randy Root	1987	95
97 5/8	16 4/8	16 0/8	14 3/8	3	3	Lane County	OR	Riley Savage	2007	95
97 4/8	18 6/8	18 2/8	18 6/8	3	3	Whatcom County	WA	Steve Holland	1963	97
97 1/8	16 5/8	16 7/8	13 7/8	3	3	Cowlitz County	WA	Jerry W. Adams	1994	98
97 0/8	16 7/8	16 4/8	12 0/8	4	4	Pacific County	WA	Lawrence Rogers	1972	99
96 7/8	17 2/8	17 2/8	14 7/8	4	4	Glenn County	CA	Joe Williams	1977	100
96 6/8	17 6/8	17 6/8	16 6/8	3	2	King County	WA	Tim Gordan Wallis	1995	101
96 5/8	18 3/8	18 1/8	17 1/8	4	3	Douglas County	OR	Steve Hohensee	2007	102
96 2/8	15 5/8	16 1/8	12 0/8	4	4	Marion County	OR	Roger W. Atwood	2002	103
96 1/8	15 4/8	15 0/8	13 3/8	4	4	Pacific County	WA	Lynne Sharp	1965	104
96 0/8	16 0/8	15 7/8	12 6/8	4	4	Soda Springs	OR	Harold Benson	1961	105
95 5/8	17 5/8	15 2/8	14 3/8	5	5	Siskiyou County	CA	Thomas V. Sieverding	1971	106
90 0/8	15 1/8	15 6/8	12 6/8	5	4	Douglas County	OR	Steve Hohensee	2009	107

COLUMBIAN BLACKTAIL DEER [Typical Velvet Antlers]

Minimum Score 95 *Odocoileus hemionus columnianus*

Score	Length of R Main Beam	L	Inside Spread	Number of R Points	L	Area	State/ Province	Hunter's Name	Date	Rank
125 5/8	21 5/8	21 4/8	17 3/8	4	3	Humboldt County	CA	Art Young	1918	*
113 2/8	16 2/8	16 5/8	14 0/8	4	4	Humboldt County	CA	Steven R. Davis	2005	*
111 7/8	17 2/8	18 1/8	17 5/8	5	4	Mendocino County	CA	Michael J. Schneider	2007	*
103 0/8	19 3/8	19 2/8	15 0/8	4	4	Humboldt County	CA	Randy Cooling	2006	*
102 6/8	16 7/8	17 2/8	18 1/8	4	4	Mendocino County	CA	Fred Eichler	2010	*
99 0/8	16 2/8	16 6/8	17 1/8	4	4	Mendocino County	CA	Chuck Adams	1970	*
98 7/8	17 4/8	17 5/8	14 1/8	3	3	Solano County	CA	Brian J. Morris	1998	*
96 5/8	18 2/8	18 0/8	13 3/8	3	3	Trinity County	CA	William J. Langer	2004	*

COLUMBIAN BLACKTAIL DEER [Non-Typical Antlers]

Minimum Score 115 *Odocoileus hemionus columnianus*

Score	Length of R Main Beam	L	Inside Spread	Number of R Points	L	Area	State/ Province	Hunter's Name	Date	Rank
162 5/8	23 5/8	22 2/8	21 0/8	7	7	Jackson County	OR	David Shanklin	1984	1
154 3/8	23 3/8	22 4/8	18 2/8	5	7	Jackson County	OR	David Shanklin	1983	2
152 1/8	20 2/8	19 3/8	20 1/8	5	6	Jackson County	OR	David Shanklin	1977	3
131 6/8	22 7/8	22 6/8	18 3/8	7	4	Douglas County	OR	Jerry R. De Loach	1975	4

Hunter: Mike Barrett
Species: Sitka Blacktail Deer
Score: 86 5/8
Date: 8/27/10
Location: Kodiak Island, AK

SITKA BLACKTAIL DEER [Typical Antlers]

Minimum Score 75 *Odocoileus hemionus sitkensis*

Score	Length of R Main Beam	L	Inside Spread	Number of R Points	L	Area	State/ Province	Hunter's Name	Date	Rank
116 3/8	18 4/8	18 2/8	13 1/8	5	5	Prince of Wales Island	AK	Charles Hakari	1987	1
103 1/8	17 2/8	17 3/8	15 5/8	4	4	Kodiak Island	AK	Tom Kothenbeutel	2005	2
101 6/8	15 4/8	16 2/8	13 6/8	5	5	Kodiak Island	Ak	Gary G. Wall	1986	3
99 3/8	16 5/8	16 6/8	16 1/8	4	4	Kodiak Island	AK	Todd Zeuske	1998	4
98 5/8	15 7/8	16 0/8	13 1/8	5	5	Kodiak Island	AK	Tim Moerlein	1992	5
97 5/8	16 7/8	17 2/8	16 3/8	4	4	Kodiak Island	AK	Roger Stewart	1988	6
96 6/8	15 4/8	15 1/8	14 0/8	4	5	Kodiak Island	AK	Bob Ameen	1983	7
96 3/8	16 1/8	16 6/8	15 5/8	4	4	Kodiak Island	AK	Gary G. Wall	1986	8
96 2/8	17 5/8	17 0/8	15 6/8	3	4	Kodiak Island	AK	Dick McClain	1990	9
95 6/8	17 3/8	17 2/8	14 4/8	5	4	Kodiak Island	AK	Terry Krahn	1995	10
94 2/8	15 4/8	15 1/8	14 4/8	4	4	Kodiak Island	AK	F. Dan Dinelli	1996	11
93 2/8	16 0/8	15 6/8	15 0/8	4	4	Kodiak Island	AK	John Toenes	1987	12
93 2/8	18 0/8	17 1/8	14 0/8	4	4	Kodiak Island	AK	Russell M. Kucinski	1989	12
93 1/8	17 5/8	16 6/8	15 1/8	3	3	Kodiak Island	AK	Mark Penninger	2006	14
92 3/8	17 7/8	17 1/8	16 1/8	3	3	Kodiak Island	AK	William C. Shuster	1992	15
91 6/8	15 1/8	14 7/8	13 2/8	4	4	Kodiak Island	AK	Michael J. Schneider	2004	16
91 0/8	16 0/8	15 0/8	16 7/8	4	5	Kodiak Island	AK	Tom Vanasche	2006	17
90 4/8	15 3/8	15 3/8	13 3/8	5	5	Kodiak Island	AK	Reggie Spiegelberg	1986	18
89 3/8	16 3/8	16 4/8	13 7/8	4	4	Sitkalidak Island	AK	Bill Payton	2008	19
89 0/8	16 3/8	16 4/8	18 3/8	4	3	Kodiak Island	AK	Roger Stewart	1987	20
89 0/8	15 7/8	15 0/8	13 2/8	4	4	Kodiak Island	AK	George A. Moerlein	1990	20
88 4/8	14 3/8	14 0/8	14 2/8	5	4	Afognak Island	AK	Edward L. Russell	1983	22
88 3/8	14 6/8	15 4/8	13 3/8	4	4	Kodiak Island	AK	James L. Anderson	2007	23
88 2/8	16 0/8	16 0/8	15 2/8	3	3	Kodiak Island	AK	Reggie Spiegelberg	1986	24
88 1/8	16 0/8	15 5/8	12 7/8	4	4	Kodiak Island	AK	Nathan L. Andersohn	1998	25
88 0/8	16 0/8	15 5/8	15 4/8	4	4	Kodiak Island	AK	Neil J. Russell	1993	26
87 6/8	15 0/8	15 3/8	14 2/8	5	4	Kodiak Island	AK	David Rue	2010	27
87 2/8	16 2/8	17 0/8	15 0/8	4	4	Kodiak Island	AK	Neil Russell	1996	28
86 5/8	16 1/8	15 5/8	16 3/8	4	3	Admiralty Island	AK	Charles R. Hakari	1983	29
86 5/8	17 0/8	16 7/8	15 5/8	3	4	Kodiak Island	AK	Mike Barrett	2010	29
86 2/8	16 0/8	15 1/8	14 0/8	4	4	Kodiak Island	AK	Troy Graziadei	1989	31
86 2/8	15 5/8	14 6/8	13 2/8	4	3	Deadman Bay	AK	Mark Turner	2009	31
86 1/8	15 0/8	13 7/8	12 7/8	4	4	Kodiak Island	AK	John Toenes	1987	33
86 1/8	14 6/8	15 6/8	12 5/8	3	3	Kodiak Island	AK	Elmer R. Luce, Jr.	1994	33
85 3/8	16 3/8	16 0/8	15 5/8	3	3	Afognak Island	AK	Roger Stewart	1980	35
85 1/8	16 0/8	15 4/8	15 7/8	4	3	Kodiak Island	AK	Nathan L. Andersohn	1996	36
85 0/8	13 1/8	14 4/8	15 1/8	4	4	Kodiak Island	AK	Roger Stewart	1987	37
84 7/8	16 3/8	16 1/8	15 1/8	4	5	Kodiak Island	AK	Troy A. Cunningham	2002	38
84 1/8	14 4/8	14 6/8	13 1/8	3	4	Kodiak Island	AK	Elmer R. Luce, Jr.	1994	39
83 6/8	14 1/8	14 4/8	11 6/8	4	4	Kodiak Island	AK	Russell M. Kucinski	1989	40
83 6/8	15 4/8	15 0/8	15 2/8	4	3	Kodiak Island	AK	Elmer R. Luce, Jr.	1997	40
83 6/8	14 0/8	13 7/8	14 1/8	3	4	Kodiak Island	AK	Troy A. Cunningham	2004	40
83 4/8	15 5/8	15 3/8	13 4/8	3	4	Kodiak Island	AK	Dale Holpainen	1997	43
83 1/8	15 1/8	14 1/8	13 7/8	4	3	Sitkinak Island	AK	Allen Clark	2004	44
82 6/8	15 3/8	14 0/8	12 4/8	4	4	Kodiak Island	AK	Guy Leibenguth	1998	45
82 5/8	15 1/8	14 4/8	13 5/8	4	3	Kodiak Island	AK	Greg Munther	1998	46
82 4/8	13 7/8	14 5/8	12 2/8	4	4	Kodiak Island	AK	James J. Akenson	1998	47
82 2/8	15 7/8	14 5/8	15 2/8	3	4	Kodiak Island	AK	Barry J. Smith	1996	48
82 2/8	16 1/8	15 1/8	14 2/8	3	4	Kodiak Island	AK	Greg Munther	1998	48
82 0/8	15 1/8	15 1/8	14 0/8	3	3	Kodiak Island	AK	John D. "Jack" Frost, MD	2006	50
81 6/8	14 4/8	14 5/8	13 4/8	4	4	Kodiak Island	AK	Roger Stewart	1988	51
81 3/8	15 1/8	15 7/8	14 3/8	3	5	Kodiak Island	AK	William Newman	2008	52
81 2/8	13 6/8	14 3/8	14 2/8	4	4	Kodiak Island	AK	T. J. Conrads	1996	53
81 0/8	14 7/8	15 1/8	13 6/8	3	4	Kodiak Island	AK	James J. Akenson	1998	54
80 6/8	14 2/8	13 0/8	13 0/8	4	4	Kodiak Island	AK	Roger Stewart	1988	55
80 4/8	14 2/8	15 7/8	13 2/8	4	3	Kodiak Island	AK	Steve Gorr	1989	56
80 1/8	13 4/8	13 5/8	14 7/8	4	4	Kodiak Island	AK	Ron Rockwell	2004	57
78 6/8	14 1/8	15 1/8	11 2/8	4	3	Kodiak Island	AK	Shawn McCrosky	1991	58
78 6/8	14 6/8	14 6/8	16 1/8	3	3	Glass Peninsula	AK	Marvin H. Walter	1992	58
78 5/8	15 7/8	15 7/8	19 0/8	2	3	Kodiak Island	AK	Roger Stewart	1987	60
78 5/8	14 2/8	13 2/8	13 5/8	3	3	Kodiak Island	AK	Bob Ameen	1997	60
78 2/8	14 0/8	14 2/8	14 4/8	3	4	Sitkinak Island	AK	Allen Clark	2004	62

SITKA BLACKTAIL DEER [Typical Antlers]

Minimum Score 75 *Continued*

Score	Length of R Main Beam	L	Inside Spread	Number of R Points	L	Area	State/ Province	Hunter's Name	Date	Rank
78 2/8	14 1/8	14 4/8	15 5/8	3	4	Kodiak Island	AK	Jeff Cumberworth	2006	62
78 1/8	12 2/8	12 1/8	12 1/8	4	4	Kodiak Island	AK	Tom Vanasche	2005	64
77 3/8	15 1/8	14 7/8	15 1/8	2	2	Prince of Wales Island	AK	Don Davidson	1991	65
77 2/8	13 3/8	14 2/8	16 3/8	4	3	Kodiak Island	AK	Neil Russell	1996	66
77 2/8	14 7/8	14 6/8	12 6/8	3	3	Kodiak Island	AK	William C. Shuster	1997	66
76 5/8	14 2/8	14 7/8	15 7/8	3	3	Kodiak Island	AK	David Rue	2009	68
76 0/8	15 4/8	14 6/8	14 2/8	3	3	Kodiak Island	AK	Lyle Willmarth	1986	69
75 4/8	13 1/8	12 3/8	12 0/8	3	3	Kodiak Island	AK	Mark F. Vancas	1994	70
75 2/8	13 5/8	13 1/8	12 6/8	3	4	Kodiak Island	AK	Gary G. Wall	1985	71
75 0/8	14 0/8	13 3/8	13 6/8	2	3	Sitkalidak Island	AK	Bill Payton	2008	72

Hunter: Stan Rauch
Species: Rocky Mountain Goat
Score: 42 6/8
Date: 9/11/94
Location: Summit County, CO

COUES' DEER [Typical Antlers]

Minimum Score 70 *Odocoileus virginianus couesi*

Score	Length of R Main Beam	Length of L Main Beam	Inside Spread	Number of R Points	Number of L Points	Area	State/ Province	Hunter's Name	Date	Rank
116 6/8	19 3/8	18 2/8	13 6/8	5	5	Cochise County	AZ	Ricky Carpenter	2011	1
109 1/8	17 5/8	17 3/8	12 7/8	6	7	Gila County	AZ	Dennis J. Montes	2006	2
105 6/8	16 2/8	16 6/8	14 4/8	5	4	Grant County	NM	Elmer R. Luce, Jr.	1998	3
104 2/8	18 2/8	17 7/8	15 0/8	4	4	Gila County	AZ	Larry Peterson	1978	4
103 2/8	18 1/8	19 3/8	17 2/8	4	4	Gila County	AZ	Michael P. Wanat	1997	5
102 2/8	17 2/8	17 3/8	15 7/8	5	4	Coconino County	AZ	Michael Wanat	1996	6
101 7/8	15 6/8	17 0/8	15 6/8	5	4	Grant County	NM	Peter LaScala	1989	7
100 6/8	16 6/8	16 3/8	12 6/8	4	4	Graham County	AZ	Hugh H. Hamman	1966	8
100 4/8	17 5/8	16 5/8	12 4/8	4	5	Cochise County	AZ	Dallas Scherck	1971	9
98 7/8	15 7/8	15 5/8	14 5/8	4	5	Gila County	AZ	Darryl Kessler	1978	10
98 3/8	18 2/8	18 0/8	15 1/8	4	4	Grant County	NM	Duane Beenblossom	1993	11
95 5/8	15 7/8	16 2/8	13 5/8	4	4	Santa Cruz County	AZ	Bill Krenz	1984	12
95 5/8	15 5/8	15 4/8	12 3/8	4	4	Sonora	MEX	Jim Willems	2011	12
94 7/8	17 5/8	17 5/8	12 7/8	4	4	Sonora	MEX	William J. Langer	2006	14
94 1/8	15 0/8	16 0/8	13 6/8	5	4	Greenlee County	AZ	John T. Skeen	1975	15
91 7/8	15 2/8	15 5/8	14 1/8	4	4	Cochise Coiunty	AZ	Bill VanBuskirk	2011	16
91 0/8	16 2/8	15 5/8	12 6/8	4	4	Gila County	AZ	John Radford	1972	17
90 2/8	15 5/8	16 6/8	12 6/8	4	4	Cochise County	AZ	Kim Wintz	1999	18
89 3/8	14 6/8	14 6/8	10 3/8	4	4	Grant County	NM	John L. Gardner	2005	19
88 5/8	14 2/8	15 1/8	11 3/8	4	4	Gila County	AZ	Michael P. Wanat	1995	20
88 0/8	13 7/8	13 4/8	11 2/8	4	4	Grant County	NM	Al Haines	1991	21
86 5/8	15 5/8	15 1/8	12 3/8	4	4	Santa Cruz County	AZ	David Rue	2005	22
86 5/8	17 6/8	17 1/8	10 7/8	4	4	Sonora	MEX	Dennis Dunn	2010	22
86 4/8	14 6/8	15 2/8	11 6/8	4	4	Sonora	MEX	Scott Hargrove	2011	24
85 7/8	16 1/8	15 5/8	11 5/8	3	3	Santa Cruz County	AZ	Elmer R. Luce, Jr.	2010	25
85 2/8	15 2/8	16 2/8	11 4/8	3	4	Grant County	NM	Al Haines	2001	26
84 7/8	13 0/8	12 2/8	10 3/8	4	4	Cochise County	AZ	Dave Rhodes	1976	27
84 7/8	15 5/8	15 6/8	11 5/8	4	4	Santa Cruz County	AZ	Dean Derby	2008	27
84 1/8	13 4/8	13 7/8	12 5/8	4	4	Maricopa County	AZ	Ed Matteson	1963	29
83 0/8	14 3/8	13 5/8	12 6/8	4	4	Coconino County	AZ	Elmer R. Luce, Jr.	2009	30
82 6/8	15 3/8	15 7/8	13 3/8	4	4	Coconino County	AZ	Carl Vance	1973	31
82 2/8	15 0/8	15 4/8	10 0/8	4	4	Sonora	MEX	Jim Willems	2011	32
81 2/8	12 4/8	12 1/8	13 0/8	4	4	Graham County	AZ	Bill Cross	1967	33
81 0/8	14 5/8	14 0/8	13 2/8	4	4	Grant County	NM	Bob J. Brown	1960	34
80 4/8	16 1/8	15 2/8	12 7/8	3	4	Grant County	NM	Mike Burroughs	1993	35
80 2/8	14 7/8	14 4/8	11 0/8	4	4	Cochise County	AZ	Michael L. Holm	1968	36
79 7/8	13 5/8	13 7/8	11 3/8	4	4	Grant County	NM	Elmer R. Luce, Jr.	1996	37
79 5/8	12 7/8	13 7/8	10 7/8	4	4	Grant County	NM	Mike Burroughs	1995	38
78 6/8	13 3/8	13 3/8	10 0/8	4	4	Santa Cruz County	AZ	Eric A. Sawyer	2008	39
78 5/8	13 4/8	13 2/8	9 5/8	4	4	Sonora	MEX	Dan Brockman	2010	40
77 7/8	15 3/8	14 7/8	12 5/8	3	3	Gila County	AZ	Tom Hashem	1964	41
77 6/8	14 2/8	14 1/8	9 6/8	5	5	Pima County	AZ	Richard Dawe, Jr.	1964	42
77 4/8	14 4/8	13 7/8	10 4/8	4	5	Sierra County	NM	Charles E. Franzoy	1972	43
77 3/8	14 1/8	14 0/8	11 5/8	4	4	Sonora	MEX	Rick Duggan	2003	44
77 0/8	14 1/8	14 2/8	10 6/8	4	4	Sonora	MEX	Ginger Brockman	2010	45
76 6/8	12 7/8	13 3/8	12 0/8	4	3	Cochise County	AZ	Robert G. Ables	1989	46
76 4/8	14 7/8	14 3/8	12 0/8	3	3	Gila County	AZ	Monty Dyke	1989	47
75 6/8	14 6/8	14 4/8	11 6/8	3	4	Pima County	AZ	Dalton Fisher, Jr.	2000	48
75 4/8	13 1/8	12 5/8	10 4/8	4	4	Pima County	AZ	David Huba	2010	49
75 2/8	13 6/8	13 1/8	11 4/8	4	4	Sonora	MEX	Dale Karch	2005	50
75 0/8	15 0/8	15 7/8	14 6/8	3	3	Grant County	NM	David H. Boland	1994	51
74 2/8	12 5/8	12 5/8	9 4/8	4	4	Hermosillo	MEX	Bryce H. Olson	2010	52
74 2/8	13 1/8	13 2/8	12 4/8	4	4	Cochise County	AZ	Richard King	2011	52
73 5/8	12 2/8	12 7/8	13 4/8	5	4	Hidalgo County	NM	Larry Behrends	1969	54
73 4/8	15 3/8	14 1/8	10 2/8	3	3	Sonora	MEX	Steve Miller	2009	55
73 2/8	13 0/8	13 0/8	10 6/8	4	4	Gila County	AZ	Jim Mercer	1958	56
72 4/8	11 6/8	11 6/8	11 4/8	4	4	Gila County	AZ	Gary H. Behrends	1969	57
72 0/8	13 1/8	12 6/8	10 2/8	4	3	Greenlee County	AZ	Tom Taylor	1995	58
70 7/8	13 3/8	15 0/8	12 5/8	4	4	Grant County	NM	David H. Boland	1996	59
70 3/8	15 0/8	13 4/8	10 1/8	5	4	Sonora	MEX	Richard J. Langer	2006	60
69 1/8	11 3/8	12 3/8	9 7/8	4	4	Santa Cruz County	AZ	Scott Hargrove	2008	61

COUES' DEER [Typical Velvet Antlers]

Minimum Score 70 *Odocoileus virginianus couesi*

Score	R Length of Main Beam	L Length of Main Beam	Inside Spread	R Number of Points	L Number of Points	Area	State/ Province	Hunter's Name	Date	Rank
85 5/8	13 6/8	14 1/8	14 1/8	5	4	Gila County	AZ	Fred Eichler	2009	*

COUES' DEER [Non-Typical Antlers]

Minimum Score 80 *Odocoileus virginianus couesi*

Score	R Length of Main Beam	L Length of Main Beam	Inside Spread	R Number of Points	L Number of Points	Area	State/ Province	Hunter's Name	Date	Rank
118 4/8	11 7/8	13 7/8	13 4/8	8	9	Pima County	AZ	Michael C. Dysh	2007	1

Hunter: John Romans III
Species: Pronghorn
Score: 71 0/8
Date: 8/15/12
Location: Johnson County, WY

MULE DEER [Typical Antlers]

Minimum Score 145 — *Odocoileus hemionus* and certain related subspecies

Score	Length of R Main Beam	Length of L Main Beam	Inside Spread	Number of R Points	Number of L Points	Area	State/ Province	Hunter's Name	Date	Rank
197 0/8	24 6/8	27 7/8	23 6/8	5	5	Park County	CO	Ronald E. Sniff	1969	1
192 0/8	25 6/8	26 4/8	22 0/8	7	5	Uncompahgre Plateau	CO	Donald Click	1973	2
190 1/8	25 3/8	24 6/8	25 3/8	6	5	Moffat County	CO	Glenn Pritchard	1990	3
190 0/8	25 1/8	23 4/8	24 4/8	5	5	Utah County	UT	John Edwards	1967	4
189 3/8	27 0/8	27 0/8	24 1/8	6	5	Iron County	UT	Ken Davis	1963	5
188 7/8	25 2/8	26 3/8	21 1/8	6	6	Settlement Canyon	UT	Derald R. Evans	1965	6
188 2/8	24 3/8	24 4/8	23 2/8	6	6	Mesa County	CO	John Lamicq, Jr.	1967	7
187 7/8	23 7/8	24 7/8	22 5/8	5	6	Duck Creek	UT	Gerald Clark	1966	8
187 2/8	24 3/8	22 3/8	22 2/8	5	5	Dawes County	NE	Kirk Peters	1989	9
187 1/8	26 3/8	28 1/8	28 1/8	4	4	Elmore County	ID	James H. Rainey, Sr.	1960	10
186 7/8	25 3/8	25 0/8	20 1/8	5	6	Eagle County	CO	Dr. J. D. Jones	1963	11
186 6/8	24 5/8	25 2/8	24 2/8	5	5	Cassia County	ID	Pat Miller	1965	12
186 1/8	24 1/8	24 0/8	21 3/8	5	5	Cassia County	ID	Bill Shockey	1965	13
185 5/8	24 2/8	25 4/8	27 0/8	7	6	Mesa County	CO	Art Cook	1962	14
185 0/8	26 0/8	26 2/8	26 3/8	5	6	Larimer County	CO	Don Lampert	1969	15
184 1/8	24 3/8	24 6/8	24 6/8	5	6	Delta County	CO	Scott Kolb	1963	16
184 1/8	27 5/8	27 0/8	25 3/8	5	4	Bernalillo County	NM	Gregory A. Gwash	1970	16
183 7/8	25 0/8	24 3/8	23 7/8	5	5	Moffat County	CO	Joel Hogan	1966	18
183 6/8	26 4/8	27 3/8	27 0/8	5	4	Iron County	UT	Ted Garrett	1963	19
183 5/8	26 6/8	25 7/8	25 5/8	5	5	Sanpete County	UT	Weldon Noland	1965	20
183 3/8	25 7/8	26 1/8	22 4/8	7	5	Sevier County	UT	Kyle Johnson	1970	21
183 2/8	25 5/8	26 5/8	22 7/8	6	9	Wasatch County	UT	Blake Spencer	1963	22
182 7/8	23 3/8	22 0/8	21 7/8	5	5	Caribou County	ID	Neil Dursteler	1967	23
182 6/8	22 0/8	22 2/8	20 2/8	5	7	Chelan County	WA	Glenn St. Charles	1959	24
182 5/8	23 5/8	24 4/8	25 2/8	8	6	Malheur County	OR	Jim Biigbuck Scavera	2011	25
182 4/8	25 2/8	24 2/8	25 0/8	5	5	Humboldt County	NV	George Rajnus	1985	26
181 7/8	21 7/8	21 1/8	22 6/8	5	5	Power County	ID	Austin Cummins	1970	27
181 7/8	25 6/8	24 6/8	22 5/8	7	7	Boulder County	CO	Leonard P. Quercioli	2002	27
181 5/8	24 3/8	24 0/8	21 5/8	5	5	Wakely	CO	Lynn Grace	1964	29
181 4/8	24 4/8	23 2/8	25 6/8	5	5	Garfield County	CO	Henry Wichers	1967	30
181 2/8	22 1/8	22 6/8	20 5/8	5	6	Larimer County	CO	Wayne Eberhard	1972	31
181 1/8	24 3/8	25 4/8	19 3/8	6	6	Union County	OR	Thomas Mussatto	1968	32
181 0/8	24 7/8	22 7/8	23 0/8	6	6	Rio Arriba County	NM	Charles Tapia	1965	33
180 4/8	24 1/8	25 1/8	21 6/8	4	4	Mesa County	CO	Jack Kruckenburg	1958	34
180 3/8	24 0/8	25 5/8	27 5/8	5	5	Boise County	ID	Benton K. Wetzel	1970	35
180 2/8	24 5/8	24 7/8	25 7/8	8	5	Garfield County	CO	Michael R. Allen	1967	36
180 1/8	22 0/8	22 4/8	19 3/8	5	5	Dawson County	MT	Gordon M. Quilling	1957	37
179 4/8	24 3/8	23 0/8	21 4/8	5	8	Ada County	ID	Vance Gardner	1971	38
179 4/8	22 2/8	24 3/8	19 6/8	5	5	Garfield County	CO	Ed Downard	1984	38
179 2/8	24 2/8	24 1/8	24 6/8	5	4	Illinois Creek Drainage	unk	Gordon E. Scott	1964	40
179 0/8	19 1/8	19 4/8	21 4/8	5	6	Sandoval County	NM	D. J. Heckler, Jr.	1969	41
179 0/8	22 6/8	23 0/8	20 4/8	5	5	Lincoln County	WY	Mike Barrett	1991	41
178 5/8	24 2/8	23 2/8	24 7/8	6	6	Rio Blanco County	CO	Rex Schmude	1968	43
178 4/8	24 6/8	25 0/8	23 1/8	5	5	Carter County	MT	R. C. Tucker	1958	44
178 3/8	28 3/8	27 3/8	28 3/8	5	5	Finney County	KS	Larry Ochs	1968	45
178 0/8	23 4/8	23 7/8	23 4/8	5	6	Lake County	OR	Ronald C. Halpin	1966	46
178 0/8	24 7/8	24 1/8	23 6/8	8	6	Yuma County	CO	Mike Barrett	2010	46
177 7/8	23 2/8	23 7/8	23 2/8	4	4	Routt County	CO	Richard M. Hansen	1964	48
177 6/8	22 2/8	22 4/8	24 6/8	5	6	Sublette County	WY	Dana Patrick Furgason	1992	49
177 4/8	25 6/8	26 0/8	22 4/8	5	4	Fremont County	WY	Walter Millhollin	1960	50
177 4/8	24 4/8	23 0/8	23 0/8	5	4	Ada County	ID	Neil J. Russell	1994	50
177 1/8	23 3/8	23 3/8	26 2/8	5	6	Logan County	KS	Thomas Standard	1967	52
177 1/8	23 7/8	24 4/8	22 0/8	4	6	Meagher County	MT	Mike Weitz	1981	52
177 0/8	22 0/8	23 6/8	24 4/8	6	5	White Pine County	NV	Dr. Donald Wicher	1961	54
177 0/8	24 4/8	25 5/8	21 4/8	7	6	Garfield County	CO	Steve U'Selis	1967	54
176 7/8	20 7/8	21 5/8	20 7/8	5	5	Ada County	ID	Neil Russell	1996	56
176 6/8	27 4/8	26 6/8	22 4/8	5	5	Wasatch County	UT	Doug Strecker	1987	57
176 5/8	28 0/8	27 6/8	29 7/8	6	6	Weld County	CO	Tim Bradley	1989	58
176 4/8	22 1/8	24 7/8	21 7/8	6	7	Rio Arriba County	NM	Richard W. Eustace, Jr.	2006	59
176 2/8	22 1/8	24 0/8	20 4/8	5	6	Lincoln County	NE	M. R. Buchtel	1959	60
176 1/8	24 3/8	24 3/8	23 1/8	7	6	Duchesne County	UT	M.H. 'Bill' Wilkinson, Jr.	1964	61
176 0/8	23 6/8	23 7/8	21 2/8	6	5	San Juan County	UT	Charles Farmer	1964	62

MULE DEER [Typical Antlers]

Minimum Score 145 *Continued*

Score	Length of R Main Beam	L	Inside Spread	Number of R Points	L	Area	State/ Province	Hunter's Name	Date	Rank
176 0/8	22 3/8	21 5/8	18 6/8	5	5	Coconino County	AZ	Tom Dennis	1972	62
176 0/8	22 1/8	23 4/8	21 0/8	6	6	Bingham County	ID	Doug Burkman	1994	62
175 7/8	24 5/8	24 1/8	18 0/8	6	9	Presidio County	TX	Neal Bouldin	1987	65
175 6/8	22 6/8	22 5/8	22 6/8	5	5	Uintah County	UT	Merlin L. Killpack	1957	66
175 6/8	24 4/8	24 4/8	25 4/8	6	6	Mesa County	CO	Joe Egner	1965	66
175 5/8	23 6/8	25 0/8	21 1/8	4	4	Grant County	OR	Larry Saunders	1985	68
175 3/8	21 5/8	23 1/8	19 5/8	6	7	Park County	CO	Marvin Clyncke	1981	69
175 3/8	26 5/8	26 6/8	26 3/8	4	4	Mesa County	CO	Lester R. Dayton	2003	69
175 2/8	24 1/8	24 4/8	24 4/8	7	6	Emery County	UT	Ron Myers	1958	71
175 2/8	24 6/8	25 3/8	23 6/8	6	7	Garfield County	CO	Robert Pitt	1978	71
175 1/8	21 4/8	22 7/8	21 6/8	6	6	Millard County	UT	Jason Woodland	1985	73
174 7/8	23 7/8	23 7/8	18 3/8	5	6	Delta County	CO	Paul Dickson	1969	74
174 7/8	22 5/8	21 7/8	23 0/8	5	5	Gallatin County	MT	Bob Savage	1970	74
174 7/8	24 0/8	23 4/8	23 5/8	5	6	Yuma County	CO	Mike Barrett	2008	74
174 6/8	26 2/8	26 0/8	26 2/8	5	5	Lake County	OR	Gene Lyons	1969	77
174 6/8	24 0/8	20 5/8	21 2/8	5	6	Garfield County	CO	Jewell Petz	1970	77
174 6/8	26 4/8	25 5/8	24 6/8	5	5	Lea County	NM	Eddie Carpenter	1993	77
174 6/8	24 7/8	24 6/8	20 0/8	6	5	Elbert County	CO	Tom G. Kelley	2004	77
174 5/8	25 0/8	25 0/8	23 1/8	4	4	Moffat County	CO	Roland C. Gravenkemper	1961	81
174 5/8	23 6/8	22 6/8	19 4/8	6	6	Sevier County	UT	Kenneth L. Shirley	1968	81
174 5/8	24 2/8	24 5/8	24 5/8	7	5	Sheridan County	WY	Brian Koelzer	2012	81
174 4/8	24 3/8	24 1/8	22 6/8	4	5	Mesa County	CO	Jay Verzuh	1983	84
174 2/8	22 7/8	22 6/8	21 2/8	5	5	Natrona County	WY	Pat McAteer	1982	85
174 0/8	25 0/8	25 1/8	20 6/8	5	5	Ravalli County	MT	John Schulz	1976	86
173 7/8	25 3/8	25 2/8	26 0/8	4	4	Uintah County	UT	Hal Wallentine	1962	87
173 7/8	25 4/8	25 4/8	21 0/8	5	5	Garfield County	CO	Danny C. Lloyd	1969	87
173 7/8	24 2/8	24 1/8	22 6/8	7	6	Eagle County	CO	Edward L. Berlier	1970	87
173 6/8	25 3/8	24 4/8	23 5/8	7	6	Sevier County	UT	Morris Stuart	1972	90
173 5/8	24 6/8	24 7/8	23 7/8	5	5	Morton County	ND	Pat Sullivan	1957	91
173 5/8	25 1/8	25 5/8	25 2/8	7	5	Pima County	AZ	Joe Nochta	1959	91
173 5/8	21 4/8	21 1/8	21 0/8	6	7	Chaffee County	CO	Lee Rowe	1966	91
173 5/8	23 3/8	23 7/8	24 5/8	5	5	San Juan County	UT	Robert G. Hester	1972	91
173 4/8	24 3/8	24 2/8	27 2/8	5	6	Garfield County	CO	Don Mayen	1966	95
173 4/8	22 1/8	22 1/8	20 4/8	6	7	Ada County	ID	David D. Howard	1971	95
173 4/8	23 5/8	24 4/8	21 0/8	4	6	Mesa County	CO	G. Fred Asbell	1974	95
173 3/8	22 4/8	24 1/8	24 7/8	5	5	Coconino County	AZ	Jack Richards	1958	98
173 3/8	23 0/8	23 5/8	23 5/8	5	5	Sevier County	UT	Robert J. Shumway	1969	98
173 2/8	22 6/8	23 1/8	22 6/8	6	6	Millard County	UT	George Kendall	1959	100
173 2/8	25 6/8	25 1/8	22 4/8	5	5	Dolores County	CO	Jack Acree	1969	100
173 2/8	24 0/8	24 0/8	21 4/8	5	7	Bernalillo County	NM	Rolland Hanna	1972	100
173 1/8	26 3/8	25 0/8	21 3/8	5	5	Coconino County	AZ	H. H. Harter	1961	103
173 0/8	23 4/8	24 0/8	24 1/8	5	5	Gray County	KS	Bob Barnes	1966	104
173 0/8	21 5/8	22 2/8	24 5/8	4	5	Mesa County	CO	Floyd Kendall	1966	104
172 6/8	21 5/8	21 6/8	16 7/8	6	5	Bowman County	ND	Leroy Brandenburger	1967	106
172 6/8	28 7/8	26 3/8	28 7/8	6	6	Rio Arriba County	NM	Kerino H. Revel	1968	106
172 6/8	21 7/8	22 4/8	19 4/8	5	4	Ada County	ID	Ronald B. Jones	1971	106
172 6/8	23 1/8	22 5/8	22 0/8	4	4	Rio Blanco County	CO	Brad Murray	1985	106
172 6/8	23 3/8	23 1/8	20 4/8	5	5	Franklin County	ID	Lance Henderson	1992	106
172 5/8	22 4/8	20 7/8	20 4/8	5	6	Lincoln County	WY	Mike Barrett	1990	111
172 4/8	24 3/8	22 7/8	24 3/8	6	5	Malheur County	OR	Carl R. Stone	1957	112
172 4/8	25 3/8	25 4/8	24 2/8	5	5	Owyhee County	ID	Don Rosenvall	1961	112
172 2/8	24 4/8	25 3/8	21 0/8	6	7	Rock Creek	CO	Louis Prestridge	1961	114
172 2/8	26 0/8	26 0/8	25 6/8	5	5	Wasatch County	UT	Bill Dean	1965	114
172 1/8	24 1/8	24 6/8	24 5/8	5	5	Rio Grande County	CO	Marvin Tompkins	1960	116
172 1/8	22 3/8	22 3/8	18 1/8	4	4	Lake County	OR	Ralph Hoover	1964	116
172 1/8	24 0/8	24 4/8	23 7/8	5	6	Sevier County	UT	Bob Covington	1965	116
172 1/8	21 1/8	22 2/8	20 7/8	5	5	Saguache County	CO	Rick Duggan	1984	116
172 0/8	26 1/8	26 3/8	26 1/8	4	5	Baker County	OR	James D. Hanley	1960	120
171 7/8	23 4/8	23 5/8	27 2/8	5	5	Uintah County	UT	Alvin Sisam	1966	121
171 6/8	21 4/8	23 2/8	18 0/8	5	5	Buffalo	ALB	Ron Dixon	2008	122
171 5/8	23 5/8	24 0/8	19 1/8	4	5	Sweetwater County	WY	Keith Dana	1983	123
171 2/8	23 2/8	23 6/8	20 2/8	4	5	Summit County	UT	Kent Garfield	1959	124

MULE DEER [Typical Antlers]

Minimum Score 145 *Continued*

Score	Length of R Main Beam	L	Inside Spread	Number of R Points	L	Area	State/ Province	Hunter's Name	Date	Rank
171 2/8	24 0/8	23 7/8	19 2/8	5	5	Richland County	MT	Dennis Engle	1966	124
171 1/8	28 0/8	26 4/8	22 4/8	8	7	Wasatch County	UT	Frank Snyder	1959	126
171 1/8	25 4/8	25 3/8	24 7/8	7	6	Millard County	UT	Shirley B. Pace	1962	126
171 1/8	23 3/8	23 2/8	22 7/8	5	5	Juab County	UT	Farren Anderson	1964	126
171 1/8	22 7/8	23 2/8	21 5/8	5	5	Garden County	NE	Monte Shaul	1985	126
170 7/8	21 1/8	22 7/8	19 3/8	4	4	Beaver County	UT	Richard L. Anderson	1960	130
170 7/8	23 4/8	22 1/8	19 6/8	6	5	Uncompahgre N.F.	CO	Allen G. Hughes	1974	130
170 7/8	23 3/8	22 6/8	17 6/8	5	7	Red Willow County	NE	William E. Peck, Jr.	1992	130
170 7/8	23 6/8	23 3/8	21 2/8	5	6	Sheridan County	WY	Brian Koelzer	2011	130
170 6/8	22 7/8	22 4/8	22 6/8	5	6	Chelan County	WA	Gerald Weiss	1967	134
170 6/8	25 2/8	24 5/8	25 0/8	5	5	Uintah County	UT	Eugene Damron	1974	134
170 6/8	25 4/8	25 7/8	25 4/8	6	6	Kit Carson County	CO	Larry Schaal	1988	134
170 4/8	23 2/8	23 0/8	22 0/8	5	5	Carbon County	WY	James E. Lawrence	1971	137
170 3/8	20 7/8	19 4/8	21 4/8	5	5	Ada County	ID	Jim Spearman	1970	138
170 2/8	24 7/8	23 7/8	24 2/8	6	6	Slope County	ND	Jim Peters	1954	139
170 2/8	23 1/8	23 1/8	20 1/8	5	7	Fishlake National Forest	UT	Stan Rock	1961	139
170 2/8	23 0/8	23 3/8	20 1/8	7	5	Rio Arriba County	NM	Robert H. Keadle	1966	139
170 1/8	23 3/8	24 0/8	19 7/8	5	5	Gray County	KS	Dick Masters	1968	142
170 1/8	21 6/8	24 6/8	20 7/8	7	5	Sheridan County	WY	Brian Koelzer	2008	142
170 0/8	23 3/8	22 2/8	23 6/8	7	7	Pima County	AZ	James M. Fry	1975	144
170 0/8	26 0/8	24 6/8	25 2/8	6	4	Klamath County	OR	Jeffrey A. Eder	1992	144
170 0/8	22 0/8	22 3/8	20 2/8	4	5	Gunnison County	CO	Scott Hargrove	2002	144
169 7/8	23 3/8	22 7/8	24 3/8	4	5	Caribou County	ID	Chet Hopkins	1968	147
169 6/8	23 0/8	23 3/8	23 0/8	5	5	Garfield County	CO	C. W. Gilbreath	1967	148
169 6/8	22 3/8	22 3/8	20 4/8	5	6	Camas County	ID	Jim Walters	1988	148
169 6/8	24 1/8	26 5/8	26 3/8	6	5	Yuma County	CO	Mike Barrett	2005	148
169 4/8	23 0/8	23 3/8	26 7/8	5	5	Owyhee County	ID	Dwane Marler	1955	151
169 2/8	22 3/8	21 6/8	23 4/8	6	5	Chelan County	WA	Larry Lockhart	1966	152
169 1/8	24 2/8	23 4/8	21 1/8	7	6	San Juan County	UT	Dean Wolf	1970	153
169 1/8	23 4/8	22 3/8	21 1/8	5	6	Bingham County	ID	Todd F. Lewis	1992	153
169 1/8	23 3/8	23 3/8	21 6/8	6	6	Mesa County	CO	Rick L. Cooper	2000	153
169 1/8	23 4/8	24 0/8	21 1/8	5	5	Sheridan County	WY	Scott L. Koelzer	2010	153
169 0/8	23 5/8	24 6/8	22 6/8	5	5	Fremont County	CO	Bill W. Canterbury	1975	157
169 0/8	21 6/8	21 4/8	19 7/8	7	5	Caribou County	ID	Randy K. Guinn	1994	157
169 0/8	25 0/8	23 7/8	21 4/8	5	7	Jerome County	ID	John Wells, Jr.	1997	157
168 7/8	25 2/8	25 7/8	25 3/8	6	5	Elbert County	CO	Turk Wendell	2001	160
168 6/8	23 1/8	24 2/8	20 6/8	5	4	Baker County	OR	Joe Williamsen	1959	161
168 6/8	24 7/8	24 3/8	23 4/8	5	5	Little Belt Mtns.	MT	James Ployhar	1969	161
168 6/8	21 5/8	21 5/8	17 6/8	6	5	Ravalli County	MT	Bob Brill	1977	161
168 4/8	24 6/8	24 7/8	21 6/8	5	5	Utah County	UT	Garland Bray	1970	164
168 3/8	23 5/8	23 3/8	17 5/8	5	5	Chelan County	WA	George Wells	1960	165
168 3/8	22 7/8	22 0/8	19 6/8	5	6	Mineral County	CO	Richard Kolish	1961	165
168 3/8	21 7/8	22 5/8	22 7/8	5	5	Weber County	UT	Dennis L. Shirley	1972	165
168 2/8	28 6/8	26 5/8	26 0/8	4	5	Uintah County	UT	S. K. Daniels	1908	168
168 2/8	25 0/8	24 0/8	21 7/8	5	5	Mesa County	CO	Jack Kenyon	1965	168
168 2/8	22 6/8	24 0/8	22 6/8	5	5	La Plata County	CO	Don Putterbaugh	1966	168
168 2/8	22 0/8	22 6/8	21 4/8	5	5	Mesa County	CO	Joseph Sverak	1968	168
168 2/8	23 0/8	23 3/8	22 4/8	5	5	McKenzie County	ND	Michael Tofte	2003	168
168 1/8	21 4/8	21 5/8	19 7/8	5	5	Grand County	CO	Lenard Boughton	1968	173
168 0/8	25 3/8	24 3/8	23 0/8	4	6	Uncompahgre N.F.	CO	Dick Gulman	1976	174
167 7/8	24 3/8	24 0/8	27 2/8	6	6	Grand County	UT	Bob Paulson	1967	175
167 7/8	24 6/8	25 4/8	19 7/8	5	5	Mesa County	CO	Mark Kadnuck	2006	175
167 6/8	21 4/8	20 6/8	18 5/8	5	5	Wallowa County	OR	Leonard Brooks	1967	177
167 5/8	21 2/8	22 1/8	26 0/8	5	5	Grand County	CO	Michael A. Contreras	1978	178
167 3/8	21 4/8	21 3/8	19 3/8	5	5	Owyhee County	ID	Eugene R. Mallard	1963	179
167 1/8	26 3/8	24 4/8	21 2/8	5	4	Humboldt County	NV	Jerry Stout	1965	180
167 1/8	21 1/8	19 3/8	18 2/8	6	6	Anahim Lake	BC	Guy Antilla	1965	180
167 1/8	22 0/8	21 3/8	19 3/8	5	5	Converse County	WY	Barry J. Smith	1991	180
167 0/8	23 3/8	24 2/8	22 2/8	4	4	Eagle County	CO	Arvine Routh	1965	183
167 0/8	23 2/8	23 1/8	22 1/8	7	8	Grand County	CO	Michael K. Ward	1973	183
167 0/8	23 3/8	23 7/8	19 4/8	5	5	Navajo County	AZ	Jim Willems	2007	183
166 7/8	21 2/8	22 0/8	20 5/8	5	5	Uintah County	UT	Orson Stilson	1964	186

MULE DEER [Typical Antlers]

Minimum Score 145 *Continued*

Score	Length of R Main Beam	L	Inside Spread	Number of R Points	L	Area	State/ Province	Hunter's Name	Date	Rank
166 6/8	25 1/8	25 3/8	24 4/8	5	6	Uintah County	UT	Doug Walker	1967	187
166 6/8	23 4/8	23 0/8	20 2/8	5	5	Adams County	CO	John C. Schmidt	1974	187
166 5/8	22 6/8	23 0/8	20 1/8	5	5	Baker County	OR	Lloyd V. Christensen	1959	189
166 5/8	23 4/8	23 6/8	23 5/8	5	5	Montezuma County	CO	Marvin Reichenau	1983	189
166 4/8	23 2/8	24 0/8	22 2/8	7	7	Owyhee County	ID	Bill Payne	1958	191
166 4/8	23 5/8	22 6/8	21 6/8	5	5	Owyhee County	ID	Blake Murphy	1961	191
166 4/8	23 5/8	24 5/8	18 7/8	6	5	Albany County	WY	Mark L. Nelson	2006	191
166 4/8	21 2/8	20 5/8	23 1/8	5	5	Las Animas County	CO	Fred Eichler	2010	191
166 3/8	21 7/8	22 5/8	21 0/8	6	7	Butte County	SD	L. G. Braun	1957	195
166 3/8	21 4/8	22 1/8	20 1/8	5	5	Ada County	ID	Brian Krebs	1992	195
166 2/8	25 2/8	23 1/8	18 0/8	5	5	Mesa County	CO	Lloyd Kell	1967	197
166 2/8	21 3/8	22 2/8	23 6/8	5	5	Garfield County	CO	Jim Walters	1975	197
166 2/8	19 0/8	19 4/8	16 7/8	7	5	Caribou County	ID	Randy K. Guinn	1997	197
166 2/8	24 2/8	25 5/8	19 6/8	5	5	Medicine Hat	ALB	Jerry Bowen	2004	197
166 1/8	21 4/8	22 1/8	18 1/8	5	5	Carbon County	WY	Duncan G. Weibel	1955	201
166 1/8	20 6/8	21 4/8	19 5/8	5	5	Ada County	ID	M. F. Smith	1968	201
166 0/8	20 0/8	20 7/8	19 5/8	6	5	Lincoln County	NE	M. R. Buchtel	1958	203
166 0/8	19 7/8	21 1/8	17 6/8	5	5	Mohave County	AZ	Norman J. Brown	1966	203
166 0/8	24 4/8	23 0/8	21 6/8	4	4	Garfield County	CO	Robert G. Kuper	1966	203
166 0/8	24 6/8	24 4/8	26 0/8	5	5	Cassia County	ID	Jack B. Watts	1968	203
165 7/8	23 4/8	25 0/8	25 0/8	9	6	Owyhee County	ID	Thomas Eld	1964	207
165 6/8	24 0/8	23 2/8	19 2/8	6	6	Garfield County	CO	John Richard	1972	208
165 6/8	23 6/8	23 4/8	17 2/8	5	5	Grand County	CO	Burt Thompson	1983	208
165 5/8	21 1/8	21 0/8	20 5/8	5	5	Utah County	UT	Frank Eicholt	1964	210
165 5/8	23 6/8	24 1/8	22 1/8	5	4	Sheridan County	WY	Scott L. Koelzer	2009	210
165 4/8	25 2/8	25 6/8	28 2/8	4	4	Las Animas County	CO	J. Keith Chastain	1995	212
165 3/8	27 0/8	26 0/8	19 1/8	7	5	Mesa County	CO	Kent Stumpf	1973	213
165 3/8	21 5/8	21 4/8	21 6/8	5	5	Jefferson County	CO	Steve Smith	1998	213
165 2/8	26 5/8	27 0/8	25 2/8	5	5	Mesa County	CO	Ray Carpenter	1960	215
165 2/8	20 6/8	20 1/8	18 4/8	5	5	Clear Creek County	CO	John Marolt III	1967	215
165 2/8	20 3/8	21 1/8	20 3/8	6	5	Mora County	NM	Joel P. McHorse	1975	215
165 2/8	23 5/8	24 3/8	23 6/8	5	5	Moffat County	CO	Doug Beck	2005	215
165 1/8	20 5/8	22 3/8	20 2/8	5	6	Sevier County	UT	Severin Jensen	1959	219
165 1/8	23 2/8	24 0/8	27 3/8	6	6	Douglas County	CO	Dale Slade	1967	219
165 0/8	24 3/8	25 0/8	19 7/8	5	6	Rio Arriba County	NM	David L. Chandler	1967	221
165 0/8	23 3/8	24 4/8	17 2/8	4	4	Larimer County	CO	Mitch Brown	1992	221
165 0/8	22 4/8	22 5/8	21 6/8	5	5	Scott County	KS	D. Larry Potts	1994	221
164 7/8	24 0/8	22 7/8	23 5/8	4	4	Mesa County	CO	Richard Rounds	1973	224
164 6/8	22 4/8	21 3/8	18 2/8	5	5	Skyline Drive	UT	George Heath	1964	225
164 5/8	19 5/8	20 7/8	16 6/8	6	5	Mesa County	CO	Don Walsh	1986	226
164 4/8	21 4/8	21 6/8	19 2/8	6	6	Lincoln County	WY	Mike Barrett	1989	227
164 3/8	21 7/8	23 0/8	19 7/8	5	5	Siskiyou County	CA	Dale Gatlin	1959	228
164 3/8	24 6/8	23 4/8	24 6/8	8	8	Elko County	NV	Dick Woltering	1960	228
164 3/8	22 3/8	22 3/8	23 2/8	5	5	Sheridan County	KS	Kevin J. Ryan	1974	228
164 3/8	20 7/8	20 7/8	18 3/8	5	5	Lincoln County	WY	Mike Barrett	1992	228
164 2/8	23 2/8	24 0/8	19 4/8	7	9	Garfield County	CO	D. H. Nolting	1956	232
164 2/8	22 6/8	21 6/8	22 0/8	6	5	Routt County	CO	John Hale	1975	232
164 2/8	20 0/8	21 0/8	18 5/8	6	5	Yuma County	CO	Randy Wilkins	2000	232
164 1/8	25 3/8	25 6/8	23 6/8	8	7	Owyhee County	ID	Merlie Hampton	1962	235
164 1/8	24 0/8	24 6/8	24 6/8	6	5	Summit County	CO	Russell F. Rider	1964	235
164 1/8	24 4/8	23 1/8	22 7/8	4	4	Garfield County	CO	Roy Hoff	1968	235
164 1/8	22 3/8	20 2/8	19 3/8	5	5	Yuma County	CO	Mike Barrett	2003	235
164 0/8	19 4/8	20 1/8	19 6/8	5	5	Garfield County	CO	Dr. Lowell L. Eddy	1968	239
164 0/8	20 7/8	22 0/8	23 3/8	4	4	West Desert	UT	Myron Adams	1969	239
163 7/8	20 1/8	21 3/8	17 6/8	5	6	Frontier County	NE	Keene Hueftle	1961	241
163 5/8	23 3/8	23 2/8	22 3/8	5	5	White Pine County	NV	Joe Marich	1978	242
163 4/8	23 1/8	23 2/8	20 4/8	5	5		UT	Darwin Crawford	1964	243
163 3/8	22 5/8	25 4/8	17 6/8	8	7	Coconino County	AZ	Jake Price	1963	244
163 3/8	21 6/8	22 1/8	17 5/8	5	5	Gunnison County	CO	Clark Gallup	1970	244
163 3/8	22 1/8	23 0/8	19 1/8	5	5	Harney County	OR	Gary Kephart	2009	244
163 2/8	22 6/8	22 6/8	23 4/8	5	5	Mesa County	CO	Larry D. Tillett	1972	247
163 2/8	23 4/8	22 1/8	22 2/8	5	4	Yuma County	CO	Mike Barrett	2000	247

Score	Length of R Main Beam	L	Inside Spread	Number of R Points	L	Area	State/ Province	Hunter's Name	Date	Rank
163 1/8	23 4/8	23 2/8	19 5/8	8	7	Meagher County	MT	Jim Kennedy	2009	249
163 0/8	24 2/8	23 4/8	24 2/8	5	7	La Plata County	CO	Bryan B. Owen	1964	250
162 7/8	23 6/8	23 7/8	24 1/8	5	4	Elko County	NV	Orrin M. Owens	1966	251
162 7/8	20 7/8	22 3/8	19 5/8	6	5	Cochrane	ALB	Tom Foss	1996	251
162 6/8	21 1/8	23 4/8	23 2/8	5	5	Chelan County	WA	Les Eide	1954	253
162 5/8	19 6/8	20 1/8	22 0/8	5	5	Pima County	AZ	Jerry Clarno	1978	254
162 4/8	21 3/8	21 0/8	19 2/8	5	5	Emery County	UT	Kerry Ware	1964	255
162 4/8	22 6/8	22 2/8	20 6/8	5	5	Harney County	OR	Joel Haslett	2003	255
162 3/8	25 2/8	25 0/8	22 6/8	7	7	Bear Lake County	ID	Marriner Jensen	1957	257
162 2/8	23 3/8	26 2/8	25 0/8	4	5	Owyhee County	ID	Ralph Collins	1960	258
162 2/8	20 7/8	20 2/8	22 3/8	6	5	Park County	WY	Jim Patterson	1968	258
162 1/8	23 7/8	23 6/8	18 1/8	5	5	Fishlake National Forest	UT	R. E. Kerr	1957	260
162 1/8	23 2/8	22 5/8	22 1/8	5	5	Range Creek	UT	Frank Turner	1965	260
162 1/8	22 4/8	22 5/8	17 2/8	5	6	Mesa County	CO	Joel Prickett	1973	260
162 1/8	20 3/8	22 0/8	23 0/8	7	5	Mora County	NM	Michael J. Maes	1977	260
162 0/8	22 3/8	23 1/8	21 2/8	5	5	Elko County	NV	Larry D Jones	1985	264
161 7/8	21 4/8	21 7/8	20 3/8	4	5	Rio Blanco County	CO	Douglas Kenyon	1964	265
161 7/8	22 6/8	21 6/8	21 5/8	5	5	Sevier County	UT	Clark Richards	1967	265
161 7/8	23 0/8	22 1/8	20 0/8	5	5	Rio Blanco County	CO	Leonard Conley	1973	265
161 5/8	23 5/8	22 5/8	22 6/8	6	5	Converse County	WY	Jeff Reynolds	1990	268
161 5/8	23 0/8	22 5/8	19 7/8	5	5	Sheridan County	WY	Brian Koelzer	2007	268
161 4/8	24 5/8	24 6/8	24 2/8	7	7	Salt Lake County	UT	Frank M. Davis	1957	270
161 4/8	24 4/8	24 2/8	23 6/8	4	3	Owyhee County	ID	William R. Vanderhoef	1958	270
161 4/8	22 5/8	23 5/8	19 2/8	5	5	Sheridan County	WY	Mike Barrett	1985	270
161 4/8	21 4/8	21 3/8	19 5/8	6	5	Campbell County	WY	Art Cain	1994	270
161 3/8	22 5/8	21 7/8	19 1/8	5	5	Elko County	NV	Bert W. Fox	1961	274
161 3/8	23 2/8	21 6/8	22 3/8	4	4	Box Butte County	NE	Fred H. D. Krueger	1970	274
161 3/8	22 3/8	22 4/8	18 2/8	5	6	Stanislaus County	CA	Ron Crouch	1992	274
161 2/8	21 0/8	20 6/8	16 7/8	5	6	Ada County	ID	Ed Moser	1971	277
161 2/8	23 4/8	22 3/8	17 6/8	5	5	Big Horn County	MT	Mike Barrett	1984	277
161 1/8	25 0/8	24 0/8	25 0/8	5	5	Los Alamos County	NM	J. R. McDaniels	1960	279
161 1/8	20 5/8	20 4/8	15 5/8	5	5	Garfield County	CO	John Murray	1963	279
161 1/8	22 4/8	23 0/8	21 5/8	4	4	Albany County	WY	Erik Swanson	2011	279
161 0/8	22 1/8	23 0/8	16 0/8	5	6	Niobrara County	WY	Gary Morse	1998	282
160 6/8	20 6/8	21 4/8	21 5/8	5	5	Colfax County	NM	Carl Osborne	1965	283
160 6/8	22 6/8	22 4/8	21 1/8	7	6	Chelan County	WA	Paul Cohoon	1967	283
160 6/8	21 3/8	22 2/8	26 7/8	6	6	San Juan County	UT	Dale Warren	1972	283
160 6/8	20 7/8	22 0/8	21 2/8	5	5	Ada County	ID	Greg McTee	1998	283
160 5/8	20 5/8	20 5/8	19 7/8	4	4	Humboldt County	NV	Mike Toone	1961	287
160 5/8	21 1/8	21 2/8	20 7/8	5	5	Routt County	CO	Mark Chapman	1975	287
160 5/8	21 1/8	21 6/8	17 7/8	4	4	Moffat County	CO	Glenn Pritchard	1985	287
160 5/8	24 1/8	23 5/8	19 5/8	5	5	Albany County	WY	Brent Swanson	2010	287
160 4/8	23 6/8	23 1/8	20 0/8	7	8	Elko County	NV	Robert Narrimore	1963	291
160 3/8	22 7/8	22 3/8	18 0/8	5	6	Iron County	UT	Clair Adams	1958	292
160 3/8	22 4/8	22 4/8	18 2/8	6	6	Mesa County	CO	Ed Adkins	1974	292
160 3/8	22 0/8	23 4/8	19 1/8	6	6	Jefferson County	CO	Steve Rehm	1984	292
160 2/8	19 2/8	21 2/8	19 0/8	5	5	Millard County	UT	Scott Chesley	1962	295
160 2/8	21 6/8	22 6/8	18 0/8	4	4	Sioux County	NE	Wayne Depperschmidt	1989	295
160 1/8	20 2/8	20 7/8	19 7/8	5	5	Meagher County	MT	Mickey Anderson, Jr.	1966	297
160 1/8	23 6/8	22 2/8	24 0/8	5	5	Moffat County	CO	Ron Hopkins	1970	297
160 1/8	23 1/8	23 3/8	22 0/8	6	5	Chase County	NE	William Baker	1999	297
160 0/8	22 4/8	21 7/8	20 2/8	5	5	Chaffee County	CO	Paul J. Zeisler	1964	300
160 0/8	23 0/8	22 7/8	20 3/8	6	5	Gunnison County	CO	Wayne Depperschmidt	1973	300
160 0/8	24 4/8	23 2/8	25 0/8	4	5	Coconino County	AZ	Bruce McIntyre	1979	300
160 0/8	21 7/8	22 5/8	17 6/8	5	5	Cherry County	NE	Lloyd C. Smith	1990	300
159 7/8	24 2/8	25 0/8	19 4/8	5	6	Colfax County	NM	James Kelly	1957	304
159 7/8	20 7/8	21 6/8	21 6/8	5	5	Liberty County	MT	Kenneth Aaberge	1960	304
159 7/8	22 5/8	23 0/8	18 7/8	5	5	Sanders County	MT	Walt Borgmann	1968	304
159 7/8	23 1/8	21 3/8	17 3/8	5	5	Bernalillo County	NM	Lee Burnett	1972	304
159 7/8	25 4/8	22 4/8	21 0/8	6	6	Mesa County	CO	Ronald E. Stull	1974	304
159 6/8	23 4/8	22 7/8	22 3/8	7	7	Owyhee County	ID	Gilbert Martin	1960	309
159 6/8	22 2/8	21 4/8	19 4/8	5	5	Beaver County	UT	Joe Cordonier	1974	309

MULE DEER [Typical Antlers]

Minimum Score 145 *Continued*

Score	Length of R Main Beam	L	Inside Spread	Number of R Points	L	Area	State/ Province	Hunter's Name	Date	Rank
159 6/8	20 1/8	22 1/8	19 4/8	5	5	Converse County	WY	Gene Mathias	2001	309
159 5/8	24 0/8	24 0/8	20 5/8	7	5	Johnson County	WY	Craig Grant	2005	312
159 4/8	22 2/8	23 4/8	26 1/8	5	5	Rio Arriba County	NM	Daniel G. Willems	2000	313
159 4/8	22 6/8	22 3/8	20 3/8	6	5	Campbell County	WY	Mark Johnson	2003	313
159 4/8	22 7/8	24 5/8	18 4/8	5	5	Sheridan County	WY	Mike Barrett	2009	313
159 3/8	21 1/8	20 4/8	16 5/8	4	4	Millard County	UT	Dean Todd	1956	316
159 3/8	22 4/8	22 7/8	18 4/8	5	5	Wasco County	OR	Bill Neary	1966	316
159 3/8	23 4/8	23 6/8	23 0/8	6	5	Medicine Hat	ALB	Mike Barrett	2004	316
159 2/8	23 6/8	23 0/8	24 2/8	6	6	Haakon County	SD	Matt Donnelly	2004	319
159 1/8	22 6/8	21 6/8	20 5/8	5	7	Garfield County	CO	Jack Peters	1963	320
159 0/8	24 4/8	24 3/8	22 6/8	4	4	Elko County	NV	Bill L. Conn	1971	321
159 0/8	22 3/8	21 2/8	25 2/8	5	6	Ada County	ID	Deloy Desaro	1973	321
159 0/8	22 7/8	23 3/8	19 2/8	5	5	Rio Arriba County	NM	Matthew Quintana	1994	321
158 7/8	20 1/8	19 7/8	22 1/8	5	5	Sevier County	UT	Mike Otten	1973	324
158 7/8	21 2/8	22 6/8	21 7/8	5	5	Elmore County	ID	Peter J Cintorino	1981	324
158 6/8	21 3/8	22 3/8	22 0/8	5	5	Lake County	OR	Lyle Reeder	1954	326
158 6/8	21 6/8	20 7/8	17 0/8	6	4	Stacy	MT	Dewey Olsen	1960	326
158 5/8	22 2/8	22 5/8	18 1/8	6	6	Weber County	UT	Bruce N. Moss	1974	328
158 5/8	21 3/8	21 5/8	17 7/8	5	5	Albany County	WY	Mark L. Nelson	2002	328
158 5/8	23 1/8	23 7/8	22 5/8	5	6	Yuma County	CO	Mike Barrett	2002	328
158 4/8	24 2/8	23 2/8	18 4/8	5	4	Garfield County	CO	Orvie E. Linsin	1989	331
158 3/8	22 6/8	23 6/8	20 1/8	4	5	Mesa County	CO	James C. Kennedy	1995	332
158 2/8	23 4/8	19 3/8	21 4/8	6	4	Caribou County	ID	Gary L. Vaughn	1969	333
158 2/8	23 4/8	22 1/8	20 0/8	6	7	Gosper County	NE	Johnny Hemelstrand	1972	333
158 1/8	20 7/8	20 0/8	18 3/8	5	5		UT	Gordon Young	1963	335
158 1/8	21 0/8	22 3/8	24 0/8	6	6	Rio Blanco County	CO	Jerry R. Bowen	1970	335
158 1/8	21 2/8	20 6/8	18 3/8	5	5	Mesa County	CO	David H. Boland	1978	335
158 0/8	21 4/8	21 0/8	21 4/8	7	7	Golden Valley County	ND	Bob Ross	1959	338
158 0/8	22 2/8	23 3/8	21 2/8	5	5	Owyhee County	ID	R. W. McIntire	1961	338
158 0/8	25 4/8	24 4/8	24 4/8	3	3	Garfield County	CO	Jimmy R. Speer	1970	338
158 0/8	24 7/8	24 1/8	21 0/8	6	7	Elbert County	CO	Quince Hale	1989	338
158 0/8	22 4/8	22 3/8	17 0/8	5	5	Valley County	ID	Jim Akenson	2001	338
157 7/8	23 6/8	24 0/8	21 3/8	4	5	Rio Arriba County	NM	Larry Wright	1972	343
157 6/8	23 7/8	23 4/8	23 7/8	6	5	Owyhee County	ID	Roland Duram	1964	344
157 6/8	24 6/8	24 4/8	21 0/8	5	5	Hooker County	NE	Andrew L. Glidden	2007	344
157 5/8	19 6/8	21 5/8	24 4/8	5	5	Owyhee County	ID	Fred Audette	1960	346
157 5/8	19 6/8	20 0/8	18 3/8	5	5	Montezuma County	CO	Marvin Reichenau	1973	346
157 4/8	23 1/8	23 2/8	23 2/8	5	6	Grand County	UT	Dean Caldwell	1960	348
157 4/8	22 2/8	22 4/8	19 4/8	5	5	Garfield County	CO	Bob Gulman	1966	348
157 4/8	18 5/8	19 2/8	16 0/8	4	4	Bernalillo County	NM	William R. Johnson	1969	348
157 4/8	22 4/8	21 1/8	21 0/8	5	5	Pennington County	SD	Gary Buckley	1993	348
157 4/8	21 4/8	22 0/8	18 4/8	5	5	Powell County	MT	Marlon Clapham	1995	348
157 3/8	26 2/8	26 0/8	21 2/8	7	6	Sevier County	UT	Dale Gardner	1958	353
157 3/8	21 6/8	20 0/8	20 0/8	6	6	Routt County	CO	Bing Kemp	1966	353
157 2/8	22 5/8	21 4/8	20 6/8	6	6	McKenzie County	ND	Roy Mitten	1956	355
157 2/8	21 5/8	21 4/8	16 4/8	5	5	Bernalillo County	NM	Robert F. Knight	1970	355
157 1/8	21 4/8	21 6/8	20 5/8	5	5	Clear Creek County	CO	Janet Schreur	1987	357
157 0/8	21 6/8	20 3/8	18 6/8	5	5	Lane County	KS	Dean Hamilton	1978	358
157 0/8	21 4/8	20 7/8	16 5/8	6	6	Deer Lodge County	MT	Mike Softich	1981	358
157 0/8	23 7/8	23 7/8	18 6/8	4	4	Clackamas County	OR	Thomas L. Carter	1984	358
156 7/8	24 3/8	24 0/8	23 4/8	4	4	Garfield County	CO	John Lamicq, Jr.	1966	361
156 7/8	22 1/8	21 4/8	20 7/8	4	5	Fort St. John	BC	Steven W. Hiebert	1992	361
156 6/8	23 6/8	24 2/8	20 0/8	6	5	Uintah County	UT	Terry Peck	1964	363
156 6/8	22 5/8	22 6/8	19 4/8	5	5	Mesa County	CO	Ed Meyer	1967	363
156 5/8	21 2/8	21 3/8	20 3/8	5	5	Slope County	ND	Vern R. Keim	1959	365
156 5/8	20 1/8	21 1/8	22 3/8	5	5	Rio Blanco County	CO	Jim Pickering	1966	365
156 5/8	20 0/8	21 6/8	18 6/8	6	6	Larimer County	CO	Leslie McKenzie	1970	365
156 4/8	21 7/8	22 1/8	22 0/8	5	4	Weston County	WY	Thomas L. A. Pucci	1956	368
156 4/8	20 0/8	20 0/8	19 2/8	5	5	Owyhee County	ID	Bill Leisi	1961	368
156 4/8	22 2/8	24 2/8	20 3/8	4	5	Summit County	UT	Richard Douglass	1964	368
156 4/8	23 1/8	21 4/8	20 4/8	4	4	Elko County	NV	Paul Dinan	1968	368
156 4/8	23 1/8	20 7/8	19 2/8	4	4	Chouteau County	MT	Michael R. Buesseler	1971	368

MULE DEER [Typical Antlers]

Minimum Score 145 *Continued*

Score	Length of R Main Beam	L	Inside Spread	Number of R Points	L	Area	State/ Province	Hunter's Name	Date	Rank
156 4/8	25 2/8	25 0/8	21 4/8	4	4	Powder River County	MT	Paul T. Shore	1996	368
156 3/8	21 6/8	22 6/8	25 0/8	5	6	Owyhee County	ID	Bill Kerr	1962	374
156 3/8	25 5/8	24 6/8	23 5/8	3	3	Mohave County	AZ	Bill Cross	1963	374
156 3/8	21 2/8	24 5/8	25 3/8	7	6	Lake County	OR	Orvil Winters	1965	374
156 3/8	23 5/8	21 7/8	21 7/8	4	4	Carbon County	WY	Tracy Villwok	2003	374
156 1/8	20 6/8	20 7/8	20 2/8	5	6	Millard County	UT	Jerry White	1962	378
156 1/8	22 3/8	22 3/8	20 3/8	5	5	Sheridan County	WY	Chris Boll	2009	378
156 0/8	21 4/8	21 0/8	18 2/8	5	5	Iron County	UT	Ken McKnight	1966	380
156 0/8	20 0/8	20 1/8	18 2/8	5	6	Chelan County	WA	Steve Gorr	1975	380
156 0/8	21 4/8	21 6/8	19 2/8	4	4	Powder River County	MT	Eugene Arndt	1996	380
156 0/8	22 4/8	23 3/8	19 2/8	4	4	Pennington County	SD	Barry G. Meyer	2012	380
155 7/8	24 5/8	24 5/8	25 4/8	4	4	Apache County	AZ	Bill VanBuskirk	2001	384
155 6/8	24 1/8	24 1/8	22 0/8	3	4	Owyhee County	ID	Lynn Thomas	1960	385
155 6/8	22 2/8	21 3/8	18 4/8	4	5	Garfield County	CO	Henery Jaman	1966	385
155 5/8	23 2/8	23 3/8	22 3/8	5	6	Chelan County	WA	R. F. Kelly	1960	387
155 5/8	20 7/8	20 4/8	21 6/8	5	5	Elko County	NV	Frank M. Davis	1967	387
155 5/8	24 0/8	25 6/8	22 0/8	7	5	Coconino County	AZ	Robert G. Arcieri	1977	387
155 5/8	22 7/8	24 0/8	21 5/8	4	5	Sheridan County	WY	Mike Barrett	2001	387
155 4/8	22 2/8	23 6/8	23 2/8	5	4	Millard County	UT	Shirl Pace	1966	391
155 4/8	19 4/8	21 0/8	19 6/8	5	5	Wayne County	UT	Harold Boyack	1968	391
155 3/8	21 1/8	20 5/8	19 7/8	5	5	Teller County	CO	James L. Anderson	1987	393
155 2/8	20 2/8	19 4/8	19 2/8	5	5	Owyhee County	ID	Ralph O. Collins	1957	394
155 2/8	23 6/8	24 2/8	22 4/8	6	5	Ada County	ID	Ronald K. White	1971	394
155 2/8	20 4/8	21 4/8	16 0/8	6	5	Payette County	ID	Jon Skinner	1994	394
155 2/8	22 6/8	22 1/8	17 5/8	6	6	Delta County	CO	Rick Bohl	2010	394
155 1/8	21 2/8	21 3/8	17 7/8	6	5	Sevier County	UT	Ray Shepard	1965	398
155 0/8	23 2/8	21 4/8	22 0/8	5	4	Lake County	OR	Bill Chahon	1967	399
155 0/8	23 4/8	23 0/8	22 0/8	6	5	Broadwater County	MT	Larry P. Stevens	1968	399
155 0/8	21 7/8	22 0/8	20 1/8	5	5	Mesa County	CO	Clarence Bowers, Jr.	1979	399
154 7/8	22 3/8	23 4/8	23 3/8	5	5	Dolores County	CO	Oscar A. Harden	1957	402
154 7/8	19 4/8	21 2/8	19 2/8	5	6	Millard County	UT	Dale Moore	1961	402
154 6/8	24 3/8	25 0/8	21 2/8	5	4	Hooker County	NE	Andrew L. Glidden	1999	404
154 5/8	20 3/8	21 7/8	21 7/8	5	5	Lincoln County	NV	Larry Gehre	1963	405
154 5/8	20 2/8	20 3/8	25 1/8	7	5	Converse County	WY	Ron Miller	1994	405
154 5/8	25 7/8	23 5/8	26 3/8	4	5	Empress	ALB	Tim Sailer	2003	405
154 4/8	22 3/8	22 0/8	22 2/8	5	5	Washoe County	NV	Gary Van Ness	1959	408
154 4/8	22 3/8	22 5/8	21 0/8	6	6	Routt County	CO	Douglas J. Peterson	1965	408
154 3/8	21 7/8	23 1/8	20 5/8	4	5	Pawnee County	KS	Robert E. Lagree	1970	410
154 2/8	21 7/8	21 5/8	17 4/8	8	8	Grant County	OR	Lloyd V. Christensen	1960	411
154 2/8	22 5/8	22 3/8	19 5/8	4	4	Mesa County	CO	Jimmy E. Ash	1966	411
154 2/8	22 2/8	21 2/8	21 0/8	7	8	Kirby	WY	Steve Gorr	1970	411
154 2/8	19 2/8	22 2/8	20 5/8	7	6	Las Animas County	CO	Pat Powell	2000	411
154 1/8	18 3/8	16 6/8	16 1/8	5	7	Carbon County	UT	Lieb D. Miller	1959	415
154 1/8	19 6/8	20 5/8	18 1/8	5	6	Jefferson County	OR	Doris T. Barden	1960	415
154 1/8	21 4/8	21 4/8	20 4/8	4	4	Rio Grande County	CO	Kenneth G. McCombs	1969	415
154 0/8	22 7/8	22 6/8	20 6/8	7	6	Coconino County	AZ	Stuart Diehl	1962	418
154 0/8	22 0/8	22 2/8	19 4/8	5	5	Mesa County	CO	Al Dawson	1964	418
154 0/8	22 1/8	22 1/8	18 0/8	5	4	Elmore County	ID	Timothy J. Conrads	1988	418
154 0/8	22 0/8	22 5/8	20 0/8	6	5	Sublette County	WY	Nelson J. Capestany	1991	418
154 0/8	22 2/8	23 2/8	22 3/8	6	5	San Juan County	CO	Jeff Kelly	1994	418
153 7/8	22 4/8	19 6/8	17 7/8	5	5	Uintah County	UT	Matt Brooks	1974	423
153 7/8	24 2/8	23 4/8	22 6/8	6	5	Lincoln County	CO	Michael B. Lamade	1992	423
153 6/8	23 0/8	23 6/8	24 4/8	5	5	Siskiyou County	CA	Kenneth B. Jenkins	1958	425
153 6/8	23 0/8	21 7/8	18 0/8	5	5	Dawes County	NE	William W. Plooster	1958	425
153 6/8	22 5/8	22 7/8	24 5/8	6	6	Cherry County	NE	Albert Selk	1995	425
153 6/8	23 4/8	23 4/8	20 6/8	5	5	Sheridan County	WY	Justin Ammons	2004	425
153 5/8	22 0/8	22 4/8	20 7/8	4	4	Billings County	ND	Ed Bry, Jr.	1957	429
153 5/8	21 0/8	20 7/8	17 7/8	6	7	Moffat County	CO	Zenus E. Cozart	1962	429
153 5/8	21 5/8	23 1/8	20 1/8	5	5	Grant County	WA	J. G. "Rusty" Watson	1997	429
153 4/8	22 5/8	21 7/8	22 4/8	5	5	San Juan County	UT	Roy D. Chesley	1963	432
153 4/8	23 0/8	23 0/8	17 4/8	6	5	Brown County	NE	Seth Fritzler	1965	432
153 4/8	20 1/8	22 1/8	20 2/8	6	5	Wasatch County	UT	Don Callister	1967	432

MULE DEER [Typical Antlers]

Minimum Score 145 *Continued*

Score	Length of R Main Beam	L	Inside Spread	Number of R Points	L	Area	State/ Province	Hunter's Name	Date	Rank
153 3/8	22 1/8	21 1/8	21 1/8	4	5	Garfield County	CO	John Nottingham	1974	435
153 2/8	21 0/8	21 0/8	20 0/8	5	5	Custer County	MT	Gene T. Buck	1961	436
153 2/8	21 2/8	22 1/8	16 4/8	5	5	Converse County	WY	Bill Randles	1995	436
153 2/8	20 6/8	19 5/8	20 2/8	5	5	Elbert County	CO	Tom Kelley	2002	436
153 1/8	22 0/8	22 3/8	24 0/8	5	4	Millard County	UT	Milton F. McQueary	1961	439
153 1/8	23 5/8	25 2/8	20 2/8	5	4	Rio Blanco County	CO	Thomas Nicholls	1967	439
153 1/8	21 5/8	22 3/8	22 3/8	5	5	Park County	WY	Jim Dinkins	1978	439
153 1/8	18 4/8	18 5/8	14 7/8	4	5	Meagher County	MT	Jim Kennedy	2003	439
153 0/8	24 0/8	24 0/8	21 6/8	4	4	Billings County	ND	John Holdorf	1995	443
153 0/8	23 1/8	18 7/8	19 6/8	6	6	Milk River	ALB	Tom Taylor	2002	443
153 0/8	22 6/8	22 7/8	14 6/8	7	6	Platte County	WY	Jerry Bowen	2009	443
152 7/8	20 7/8	21 7/8	20 1/8	6	6	Gove County	KS	Alan D. Beougher	1970	446
152 6/8	19 5/8	19 1/8	18 0/8	4	4	Billings County	ND	William E. Lee, Jr.	1991	447
152 6/8	20 5/8	23 0/8	23 5/8	4	4	Las Animas County	CO	Fred Eichler	2001	447
152 5/8	21 3/8	21 5/8	20 3/8	4	4	Elko County	NV	Bill Freeman	1961	449
152 5/8	22 0/8	22 3/8	22 3/8	5	5	Fremont County	WY	Gene Farley	1964	449
152 5/8	23 3/8	23 4/8	19 4/8	7	5	Valley County	ID	James J. Akenson	1984	449
152 3/8	20 6/8	20 7/8	20 2/8	5	6	Carbon County	UT	John C. Culpepper	1969	452
152 2/8	21 0/8	21 0/8	19 0/8	5	6	Grand County	UT	Roger Smith	1962	453
152 2/8	19 3/8	20 4/8	20 4/8	5	5	Elko County	NV	Jack Konvalin	1963	453
152 2/8	26 0/8	26 0/8	22 0/8	4	4	Garfield County	CO	Jim Dougherty	1968	453
152 2/8	21 5/8	21 7/8	21 5/8	5	5	Millard County	UT	David G. Snyder	1968	453
152 1/8	20 1/8	20 1/8	18 1/8	5	5	Cochise County	AZ	John Behrends	1969	457
152 1/8	20 7/8	21 2/8	18 4/8	6	5	Twin Falls County	ID	Darrell Nunez	1994	457
152 1/8	26 6/8	26 2/8	18 7/8	4	4	Rosebud County	MT	Mike Barrett	2011	457
152 0/8	20 0/8	20 6/8	17 3/8	8	7	Garfield County	UT	Bob Mackinnon	1970	460
151 7/8	22 0/8	22 2/8	20 0/8	6	8	Duchesne County	UT	Rowland S. Enomoto	1965	461
151 6/8	20 3/8	20 2/8	21 4/8	4	5	Garfield County	MT	Herman Hass	1961	462
151 6/8	18 7/8	21 7/8	21 7/8	6	6	Rio Blanco County	CO	Doug Kenyon	1967	462
151 5/8	21 5/8	20 5/8	20 5/8	5	5	Garfield County	UT	Dick Gulman	1968	464
151 5/8	20 2/8	18 7/8	18 1/8	6	6	Chaffee County	CO	Eugene K. Post	1971	464
151 4/8	20 0/8	18 3/8	21 4/8	5	5	Weld County	CO	Nathan L. Andersohn	1998	466
151 2/8	22 7/8	19 2/8	16 4/8	5	5	Chelan County	WA	G. H. Malinoski	1959	467
151 2/8	24 5/8	23 1/8	24 5/8	7	6	Bear Lake County	ID	Keith V. Hymos	1961	467
151 2/8	24 5/8	25 7/8	19 1/8	7	5	Dundy County	NE	Michael C. Dysh	1988	467
151 2/8	20 3/8	20 6/8	17 6/8	5	5	Missoula County	MT	Dan Ermatinger	2003	467
151 2/8	20 6/8	20 3/8	21 3/8	5	5	Powder River County	MT	Mike Barrett	2009	467
151 1/8	21 4/8	22 2/8	20 2/8	6	7	Butte County	SD	John Kirk	1958	472
151 0/8	22 1/8	22 0/8	20 6/8	5	7	Chelan County	WA	Deryl E. Bland	1964	473
151 0/8	18 4/8	21 2/8	17 4/8	5	5	Moffat County	CO	Wayne Liskey	1966	473
151 0/8	22 0/8	19 1/8	23 2/8	5	4	Uncompahgre N.F.	CO	Roy Miller	1972	473
150 7/8	22 3/8	23 4/8	20 0/8	6	6	Hockberry Creek	KS	Dale Redmond	1967	476
150 7/8	22 5/8	19 3/8	25 3/8	5	4	San Juan County	UT	Ken Ciarelli	1968	476
150 7/8	20 6/8	20 6/8	18 7/8	5	5	Las Animas County	CO	Jeffrey A. Falkner	1995	476
150 7/8	22 4/8	22 2/8	18 4/8	6	6	Valley County	ID	James Akenson	2007	476
150 6/8	20 5/8	20 3/8	20 5/8	5	6	Washington County	UT	Jack Richards	1960	480
150 6/8	20 2/8	20 0/8	18 2/8	5	5	Bernalillo County	NM	Lee Braudt	1968	480
150 6/8	21 2/8	23 3/8	20 0/8	4	5	Platte County	WY	Jerry Bowen	2010	480
150 4/8	22 0/8	21 6/8	22 5/8	5	5	Carbon County	UT	C. J. Coleman	1987	483
150 4/8	21 3/8	20 6/8	18 0/8	6	5	Wallowa County	OR	Billy L. Moores	1996	483
150 4/8	22 4/8	22 1/8	23 1/8	4	5	Elbert County	CO	Randy Kenderick	2011	483
150 3/8	22 6/8	22 7/8	15 5/8	4	5	Summit County	UT	Clifton Rees	1962	486
150 3/8	19 6/8	20 3/8	19 7/8	4	5	Grand County	UT	Lowell W. Dobson	1968	486
150 3/8	21 0/8	21 0/8	15 7/8	5	5	Bernalillo County	NM	Michael M. Emery	1973	486
150 3/8	19 7/8	19 4/8	23 7/8	4	4	Chouteau County	MT	Dwight P. Martin	1992	486
150 3/8	19 4/8	19 6/8	15 5/8	5	5	Custer County	CO	Logan Canterbury	2008	486
150 2/8	13 0/8	19 7/8	15 3/8	3	6	Boise County	ID	Ralph Hoobing	1964	491
150 2/8	22 1/8	22 0/8	20 4/8	6	5	Maricopa County	AZ	Mike Ottenbacher	1987	491
150 2/8	20 6/8	19 6/8	18 4/8	5	4	Caribou County	ID	Mike Barrett	2003	491
150 1/8	18 3/8	18 5/8	17 3/8	6	7	Garfield County	CO	J. B. Hogan	1961	494
150 1/8	24 7/8	26 2/8	24 0/8	5	5	Deschutes County	OR	Walter M. Graham	1963	494
150 1/8	22 3/8	22 2/8	18 3/8	5	5	Jackson County	CO	William B. Tutt	1964	494

MULE DEER [Typical Antlers]

Minimum Score 145 *Continued*

Score	Length of R Main Beam	L	Inside Spread	Number of R Points	L	Area	State/ Province	Hunter's Name	Date	Rank
150 1/8	22 6/8	22 7/8	22 3/8	4	4	Gove County	KS	Merton Ikenberry	1966	494
150 1/8	21 3/8	21 4/8	20 1/8	5	5	Washoe County	NV	Felton Hickman	1970	494
150 1/8	21 2/8	22 1/8	19 4/8	5	6	Rio Arriba County	NM	Richard W. Eustace, Jr.	1998	494
150 0/8	21 2/8	22 4/8	24 4/8	4	4	Eureka County	NV	B. Verlyn Ownes	1963	500
150 0/8	18 7/8	18 6/8	16 3/8	5	6	Garfield County	UT	Dick Gulman	1966	500
150 0/8	19 3/8	21 4/8	20 4/8	5	5	Elko County	NV	Dick Woltering	1968	500
150 0/8	22 4/8	21 7/8	18 4/8	5	4	White Pine County	NV	Scott Faiman	1987	500
150 0/8	20 5/8	19 1/8	18 0/8	6	4	Dunn County	ND	Thomas Spence	2006	500
149 7/8	21 2/8	23 0/8	20 5/8	4	4	Sevier County	UT	Milt McQueary	1964	505
149 7/8	20 1/8	20 1/8	17 3/8	5	5	Trego County	KS	Larry Pearson	1974	505
149 7/8	23 5/8	22 6/8	26 2/8	4	5	Pima County	AZ	Michael B. Cachero	1985	505
149 7/8	20 0/8	22 2/8	19 2/8	6	4	Platte County	WY	G. Fred Asbell	1988	505
149 7/8	19 7/8	18 5/8	12 7/8	5	5	Platte County	WY	Jerry Bowen	2005	505
149 6/8	21 3/8	21 1/8	19 6/8	4	4	Garfield County	CO	Skip Candahl	1966	510
149 6/8	21 0/8	21 2/8	19 6/8	5	6	Yuma County	CO	Jerry Bowen	2009	510
149 5/8	18 1/8	20 7/8	17 7/8	5	5	Rio Blanco County	CO	Joseph H. French	1972	512
149 5/8	21 2/8	20 2/8	19 1/8	6	8	Canmore	ALB	Karl Pachonik	1982	512
149 5/8	21 4/8	22 0/8	22 0/8	5	6	Sheridan County	WY	Mike Barrett	2011	512
149 4/8	22 4/8	22 4/8	19 2/8	4	4	Colfax County	NM	Ed Foster	1966	515
149 4/8	21 4/8	20 7/8	23 0/8	5	5	Delta County	CO	Larry Tiner	1986	515
149 3/8	22 4/8	21 6/8	17 1/8	4	4	Okanogan County	WA	Irl Stamps	1939	517
149 3/8	21 5/8	22 5/8	20 7/8	5	4	Mesa County	CO	John Smith	1964	517
149 3/8	22 4/8	20 4/8	19 1/8	5	7	Baker County	OR	Chuck Brackin	1964	517
149 3/8	21 1/8	20 7/8	19 5/8	5	6	Albany County	WY	Brent Swanson	2001	517
149 2/8	22 6/8	22 5/8	21 4/8	4	4	Lake County	CO	Thomas V. Sieverding	1972	521
149 2/8	25 5/8	25 2/8	25 0/8	6	4	Lincoln County	WY	Mike Barrett	1984	521
149 0/8	24 0/8	22 5/8	22 4/8	4	4	Mesa County	CO	Tom Hentrick	1974	523
149 0/8	20 7/8	20 4/8	20 2/8	5	5	Elbert County	CO	Jeff Lampe	2003	523
148 7/8	22 3/8	23 5/8	20 3/8	6	4	Hodgeman County	KS	James Wiggins	1978	525
148 7/8	21 1/8	21 4/8	19 7/8	5	5	Golden Valley County	ND	Brent Swanson	2004	525
148 7/8	20 6/8	22 4/8	19 0/8	6	6	Valley County	ID	James Akenson	2006	525
148 6/8	20 7/8	20 6/8	13 2/8	5	4	Okanogan County	WA	Dennis N. Johnson	1971	528
148 6/8	21 3/8	20 5/8	15 7/8	5	7	Lane County	KS	Vernon L. McBee	1971	528
148 6/8	21 2/8	20 0/8	17 6/8	5	5	Sioux County	NE	Jeffrey Sales	1985	528
148 6/8	21 5/8	22 6/8	21 2/8	5	5	Teton County	MT	Mark Adler	2000	528
148 5/8	21 3/8	18 7/8	23 6/8	5	5	Park County	CO	Ed Zehner	1972	532
148 5/8	19 3/8	19 2/8	18 5/8	5	5	Mineral County	CO	Ron Fief	2005	532
148 4/8	19 4/8	19 0/8	17 0/8	5	5	Boise County	ID	Floyd Audette	1964	534
148 4/8	21 7/8	22 4/8	20 4/8	5	5	Campbell County	WY	Carrol D. Wert	1979	534
148 4/8	22 0/8	21 7/8	19 6/8	5	4	San Juan County	UT	Bill Clink	1985	534
148 4/8	19 5/8	21 0/8	19 4/8	5	5	Springbank	ALB	Bry Loyd	1994	534
148 4/8	20 6/8	21 2/8	17 2/8	4	4	Lake County	OR	Tim Larocco	1996	534
148 3/8	22 5/8	23 5/8	18 5/8	5	4	Huerfano County	CO	Loren Johnson	1966	539
148 3/8	20 5/8	21 2/8	16 7/8	4	4	Chelan County	WA	Randall Moore	1998	539
148 2/8	21 0/8	23 1/8	21 1/8	4	5	Cochrane	ALB	David Richardson	1983	541
148 2/8	20 0/8	22 1/8	18 4/8	5	5	Albany County	WY	Rick L. Parish	2000	541
148 2/8	20 0/8	18 0/8	17 6/8	5	5	Montrose County	CO	Pat Reilly	2009	541
148 1/8	20 4/8	21 7/8	22 7/8	9	4	Emery County	UT	Bob Jacobsen	1961	544
148 1/8	23 0/8	22 3/8	19 5/8	4	4	Pitkin County	CO	William F. Havel	1962	544
148 1/8	21 7/8	21 7/8	21 5/8	4	5	Lake County	OR	Richard G. Speer	1964	544
148 1/8	20 0/8	21 0/8	21 0/8	4	5	Ford County	KS	Aubrey Ballard	1966	544
148 1/8	20 7/8	20 1/8	18 5/8	5	4	Grand County	CO	Judd Cooney	1969	544
148 1/8	21 4/8	22 1/8	19 3/8	5	4	Sheridan County	WY	Mike Barrett	1983	544
148 1/8	25 6/8	25 0/8	19 1/8	3	3	Yellowstone County	MT	Peter T. Weatherford	2011	544
148 0/8	21 7/8	21 4/8	21 1/8	6	6	White River N.F.	CO	Paul M. Ramsey	1959	551
148 0/8	20 5/8	19 0/8	17 4/8	4	4	Sheridan County	WY	Larry Burtis	1995	551
147 7/8	22 3/8	24 0/8	23 3/8	3	4	Millard County	UT	Milton F. McQueary	1958	553
147 7/8	21 2/8	21 0/8	18 0/8	5	6	Deschutes County	OR	Joe Reynolds	1967	553
147 7/8	21 1/8	21 3/8	18 3/8	5	5	Uintah County	UT	Brian Wagar	2000	553
147 6/8	20 0/8	20 2/8	20 2/8	5	4	Garfield County	CO	Steve Love	1972	556
147 6/8	22 5/8	22 2/8	20 5/8	5	5	Boise County	ID	Matt March, Jr.	1983	556
147 6/8	17 6/8	22 0/8	20 0/8	4	4	Malheur County	OR	Steve Savage	1985	556

MULE DEER [Typical Antlers]

Minimum Score 145 *Continued*

Score	Length of R Main Beam	L	Inside Spread	Number of R Points	L	Area	State/ Province	Hunter's Name	Date	Rank
147 6/8	19 4/8	18 7/8	17 4/8	5	5	Powder River County	MT	Mike Barrett	2006	556
147 5/8	21 6/8	21 4/8	20 1/8	5	5	Sevier County	UT	Rowland Enomoto	1963	560
147 5/8	23 0/8	23 0/8	24 2/8	5	6	San Juan County	UT	Jack Howard	1966	560
147 5/8	19 1/8	17 0/8	21 0/8	6	5	Grant County	OR	Arthur Redinger	1972	560
147 4/8	22 1/8	22 2/8	19 4/8	5	5	Bernalillo County	NM	Larry W. Johnson	1969	563
147 4/8	21 2/8	22 1/8	19 0/8	5	5	Boulder County	CO	Jack Frank	1970	563
147 4/8	23 0/8	24 0/8	24 0/8	6	6	Sweetwater County	WY	Vic Dana	1983	563
147 3/8	23 4/8	23 2/8	20 3/8	5	6	Cherry County	NE	Jack E. Joseph	1961	566
147 3/8	19 4/8	20 4/8	16 5/8	4	5	Latah County	ID	Chas. A. McDonald	1965	566
147 2/8	21 4/8	21 3/8	19 4/8	4	5	El Paso County	CO	Rick C. Wilson	1996	568
147 1/8	20 5/8	20 1/8	16 7/8	5	5	Uintah County	UT	Rolland Esterline	1967	569
147 1/8	19 6/8	23 6/8	17 0/8	6	5	Baker County	OR	Larry Garoutte	1970	569
147 1/8	22 3/8	23 0/8	17 3/8	6	5	Weld County	CO	Nathan Andersohn	2000	569
147 0/8	20 3/8	20 7/8	20 2/8	5	5	Uintah County	UT	Merlin L. Killpack	1958	572
147 0/8	25 4/8	25 0/8	23 5/8	7	5	Grand County	UT	William W. Selby	1974	572
147 0/8	22 6/8	22 3/8	20 2/8	6	6	Franklin County	ID	Dale Holpainen	1991	572
147 0/8	21 3/8	20 6/8	18 1/8	6	5	Weld County	CO	Timothy Bradley	1995	572
146 7/8	19 5/8	20 1/8	20 1/8	5	5	Grand County	UT	Norm Goodwin	1960	576
146 7/8	23 4/8	23 1/8	23 4/8	5	4	Emery County	UT	Bruce Ware	1961	576
146 7/8	21 3/8	20 7/8	21 2/8	4	4	Carbon County	WY	John Swanson	1966	576
146 7/8	19 2/8	20 0/8	18 5/8	5	5	Baker County	OR	James E. Hodson	1966	576
146 7/8	20 7/8	21 2/8	21 3/8	5	5	Meade County	SD	Kenneth McNenny	1967	576
146 7/8	21 0/8	20 2/8	18 3/8	4	4	Routt County	CO	Bob Stevens	1975	576
146 7/8	20 3/8	21 5/8	20 3/8	6	5	Sheridan County	WY	David Shoop	1980	576
146 7/8	19 7/8	20 4/8	17 3/8	5	5	Mesa County	CO	Gary L. Hoekman	1986	576
146 7/8	22 7/8	22 0/8	20 5/8	5	4	Ground Birch	BC	Tom Taylor	2005	576
146 6/8	19 6/8	20 3/8	15 7/8	4	4	Owyhee County	ID	Seneth Ward	1960	585
146 6/8	20 4/8	19 7/8	19 0/8	6	7	Elbert County	CO	Donald Ace Morgan	1983	585
146 6/8	23 3/8	24 2/8	22 2/8	4	4	Pitkin County	CO	Bill Krenz	1983	585
146 6/8	24 3/8	21 0/8	25 6/8	7	7	Weld County	CO	Monty Ace Morgan	2001	585
146 5/8	22 0/8	22 6/8	18 7/8	5	5	Chelan County	WA	Gerald King	1963	589
146 5/8	21 2/8	22 2/8	15 7/8	4	5	Lemhi County	ID	Mike Muguira	1988	589
146 4/8	23 3/8	23 0/8	21 0/8	4	5	Wayne County	UT	Harold Boyack	1968	591
146 4/8	22 3/8	22 1/8	19 6/8	5	4	Rio Grande County	CO	Jerry Woodland	1984	591
146 3/8	19 0/8	19 0/8	17 3/8	5	5	Mohave County	AZ	George Kili	1964	593
146 3/8	20 2/8	21 1/8	19 1/8	5	5	Grand County	UT	Bob Paulson	1967	593
146 3/8	24 7/8	24 6/8	24 3/8	4	4	Fall River County	SD	Michael A. Judas	1990	593
146 2/8	25 4/8	23 6/8	23 4/8	5	5	Las Animas County	CO	Tom Valamdro	1967	596
146 2/8	21 0/8	20 2/8	19 1/8	7	6	Gallatin County	MT	Scott Koelzer	1969	596
146 2/8	22 4/8	22 0/8	19 0/8	4	5	Powder River County	MT	Mike Barrett	1985	596
146 1/8	20 6/8	19 0/8	15 7/8	5	5	Ada County	ID	Jim Wenzel	1971	599
146 0/8	20 4/8	20 2/8	18 2/8	5	5	Elko County	NV	Howard Hill	1944	600
146 0/8	21 2/8	21 3/8	20 2/8	5	5	Mesa County	CO	William F. DeEsch	1966	600
145 7/8	21 4/8	20 0/8	17 0/8	6	5	Ravalli County	MT	W. "Red" Chavez	1993	602
145 6/8	17 0/8	18 3/8	16 0/8	5	5	Lake County	OR	George Rajnus	1961	603
145 6/8	22 2/8	24 3/8	21 4/8	5	3	Sevier County	UT	James R. Bell	1964	603
145 6/8	24 1/8	21 2/8	24 2/8	6	6	Chaffee County	CO	Gary Ginther	1973	603
145 6/8	22 5/8	24 2/8	17 3/8	6	5	Bernalillo County	NM	Alvin Chewiwi	1990	603
145 6/8	22 0/8	22 3/8	15 0/8	6	6	Fergus County	MT	John "Rosey" Roseland	2006	603
145 4/8	20 5/8	18 4/8	19 4/8	6	6	Boulder County	CO	Bob Byerly	1967	608
145 4/8	21 7/8	22 4/8	20 4/8	5	4	Converse County	WY	Frank N. Moore	1984	608
145 3/8	27 2/8	26 5/8	26 2/8	5	4	Moffat County	CO	Scott Showalter	1971	610
145 3/8	21 6/8	22 0/8	21 7/8	4	4	Summit County	UT	Larry Dickerson	1985	610
145 3/8	23 1/8	22 1/8	20 5/8	5	4	Weld County	CO	Nathan L. Andersohn	1996	610
145 2/8	24 0/8	24 4/8	19 6/8	4	4	Bernalillo County	NM	Robert Bulcock, Jr.	1969	613
145 1/8	23 0/8	23 3/8	22 7/8	5	4	Natrona County	WY	Bill Wade	1970	614
145 1/8	21 6/8	19 3/8	19 1/8	4	5	Eagle County	CO	Rick Duggan	1981	614
145 1/8	22 5/8	21 7/8	21 5/8	3	4	Baker County	OR	Arthur Marc Whisler	1986	614
145 0/8	22 1/8	21 6/8	17 6/8	5	5	El Paso County	CO	Thomas M. Farmer	1961	617
145 0/8	22 1/8	22 2/8	21 0/8	5	6	Trego County	KS	Don Howard	1966	617
145 0/8	17 4/8	19 0/8	19 6/8	5	5	Union County	OR	Russ Hultberg	1991	617
145 0/8	20 0/8	20 0/8	19 6/8	5	5	Albany County	WY	Brent Swanson	2004	617

MULE DEER [Typical Velvet Antlers]

Minimum Score 145 — *Odocoileus hemionus* and certain related subspecies

Score	Length of Main Beam R	L	Inside Spread	Number of Points R	L	Area	State/ Province	Hunter's Name	Date	Rank
196 5/8	26 6/8	25 7/8	25 7/8	5	5	Montrose County	CO	Dennis Garton	1970	*
193 6/8	25 5/8	25 4/8	24 6/8	4	5	Gunnison County	CO	Paul Wynes	2003	*
188 6/8	27 2/8	26 0/8	25 6/8	5	5	Alamosa County	CO	David Emken	2001	*
186 4/8	25 4/8	25 6/8	24 0/8	6	5	Grand County	UT	Perry Hansen	1982	*
185 5/8	25 5/8	23 5/8	26 6/8	5	5	Mesa County	CO	Tom Postel	1975	*
185 5/8	27 3/8	27 7/8	25 0/8	7	7	San Juan County	UT	Gary Nordgran	1993	*
185 2/8	26 3/8	26 4/8	25 3/8	6	5	Kane County	UT	Greg L. Munther	1991	*
183 6/8	25 7/8	25 5/8	29 1/8	6	5	Sanpete County	UT	Robert Rowley	1972	*
183 2/8	23 1/8	24 1/8	22 3/8	8	6	Teton County	MT	Jake Gustafson	2003	*
181 3/8	28 2/8	24 5/8	26 7/8	8	7	Sevier County	UT	Robert Cooper	1970	*
180 4/8	23 7/8	25 3/8	24 6/8	5	5	Lincoln County	NV	Troy T. Truman	1991	*
178 2/8	23 2/8	23 4/8	19 2/8	5	5	Larimer County	CO	Coy K. Glenn	1966	*
177 7/8	24 4/8	25 2/8	22 4/8	6	6	La Plata County	CO	Lew Webb	2009	*
177 4/8	23 0/8	22 7/8	19 6/8	5	5	Clear Creek County	CO	Scott George	1991	*
177 1/8	23 1/8	23 6/8	23 1/8	5	6	Grand County	UT	John E. Burton	1962	*
177 1/8	23 7/8	23 4/8	21 5/8	5	5	Converse County	WY	Erv Wagner	1995	*
176 7/8	22 7/8	23 6/8	20 7/8	5	5	Salt Lake County	UT	Darin Gardner	1999	*
175 6/8	22 2/8	23 2/8	28 3/8	5	5	Delta County	CO	Roy G. Burton	1975	*
174 2/8	21 6/8	22 1/8	19 2/8	5	5	Fremont County	WY	Dwight Sempert	1992	*
173 6/8	24 1/8	24 2/8	21 2/8	5	5	McKinley County	NM	Richard W. Eustace, Jr.	1995	*
172 2/8	24 4/8	24 3/8	21 6/8	8	9	Mesa County	CO	Franklin E. Adams	1971	*
170 7/8	24 2/8	24 6/8	26 5/8	6	5	Toole County	MT	Scott Thielmann	2008	*
170 3/8	24 3/8	23 6/8	20 5/8	5	6	Delta County	CO	T. C. Gonyo	1975	*
168 7/8	23 0/8	22 4/8	22 1/8	5	5	Wayne County	UT	Jay Clayson	1972	*
168 6/8	23 0/8	23 0/8	24 2/8	5	5	Delta County	CO	Lee McAdams	1997	*
168 6/8	22 2/8	22 5/8	19 2/8	5	5	Apex Mountain	BC	Ray Quaedvlieg	2010	*
168 5/8	24 0/8	24 4/8	21 1/8	4	5	Mesa County	CO	Bill Stonebraker	1978	*
168 4/8	23 1/8	23 4/8	24 3/8	5	5	Gunnison County	CO	Scott Hargrove	2008	*
167 1/8	23 1/8	22 6/8	21 7/8	5	6	Custer County	CO	Randall Gasch	2006	*
167 0/8	24 3/8	23 2/8	23 4/8	5	5	La Plata County	CO	Christian Dean	1998	*
166 7/8	23 1/8	23 2/8	23 0/8	6	6	Converse County	WY	Jeff Reynolds	1991	*
166 2/8	20 3/8	20 2/8	16 6/8	8	5	Albany County	WY	Brent Swanson	2005	*
166 0/8	21 0/8	22 0/8	19 4/8	5	5	Grand County	UT	Larry R. Randall	1971	*
165 3/8	21 7/8	22 3/8	19 7/8	6	5	Saguache County	CO	John Duggan	1986	*
164 3/8	21 0/8	21 6/8	17 7/8	5	5	Mesa County	CO	Marvin Cochran	1988	*
164 2/8	23 0/8	23 0/8	24 6/8	5	5	Uintah County	UT	Terry Searle	1974	*
164 0/8	22 4/8	23 2/8	21 0/8	6	6	San Juan County	UT	John Richardson	1987	*
164 0/8	21 4/8	20 2/8	21 0/8	5	5	Grand County	CO	Jim Hammer	2001	*
162 5/8	22 4/8	23 2/8	19 3/8	5	5	Iron County	UT	Troy Truce Truman	1990	*
162 4/8	22 0/8	22 3/8	19 2/8	5	7	Colfax County	NM	Mike Reid	1991	*
162 4/8	25 0/8	24 6/8	24 2/8	4	3	Moffat County	CO	Jim Willems	2000	*
161 7/8	22 0/8	21 5/8	16 5/8	5	5	Summit County	CO	G. Fred Asbell	2001	*
161 0/8	23 1/8	24 4/8	21 0/8	3	4	Albany County	WY	Nathan L. Andersohn	1987	*
160 1/8	22 3/8	19 6/8	19 7/8	5	5	Jefferson County	CO	Mike Dunn	1999	*
160 1/8	19 7/8	19 3/8	17 7/8	5	5	Summit County	CO	Tavis Rogers	2003	*
160 0/8	19 3/8	22 6/8	20 3/8	6	5	Park County	CO	John D. Hambleton	1974	*
159 6/8	23 2/8	23 4/8	20 6/8	6	4	Bonneville County	ID	Craig Keele	1975	*
159 6/8	23 2/8	23 0/8	18 3/8	5	6	Bernalillo County	NM	Lloyd Wood	2001	*
159 3/8	23 2/8	24 2/8	20 3/8	5	5	Mesa County	CO	Joe Lasch	2009	*
159 2/8	22 4/8	21 4/8	22 4/8	5	5	Harding County	SD	Mike Barrett	1976	*
158 4/8	22 7/8	20 7/8	18 6/8	5	5	Park County	CO	Marvin Clyncke	1978	*
158 1/8	23 4/8	24 5/8	21 3/8	4	5	Cache County	UT	Mike Barrett	1992	*
156 6/8	23 0/8	23 1/8	21 0/8	5	5	Moffat County	CO	Glenn Pritchard	1975	*
156 6/8	23 2/8	23 1/8	17 6/8	5	5	Elko County	NV	Rick Mason	1990	*
156 2/8	20 4/8	18 2/8	18 6/8	5	5	Saguache County	CO	Joseph R. Wilhelm	1987	*
156 0/8	20 7/8	19 3/8	16 6/8	5	5	Douglas County	CO	Reggie Spiegelberg	1990	*
154 7/8	21 5/8	21 1/8	20 5/8	5	5	Powder River County	MT	Terry Krahn	2003	*
154 5/8	23 7/8	21 1/8	22 7/8	5	5	Lassen County	CA	Robert Flores	1992	*
154 2/8	21 2/8	21 1/8	17 6/8	6	5	Garfield County	UT	Bill Van Buskirk	2008	*
154 1/8	21 2/8	21 3/8	19 5/8	5	5	Routt County	CO	Barry J. Smith	1976	*
154 1/8	22 1/8	23 4/8	20 1/8	5	4	Washington County	UT	Troy Jolley	1990	*
154 0/8	23 2/8	24 0/8	24 3/8	5	6	Mesa County	CO	Lenard Thomas	1990	*

MULE DEER [Typical Velvet Antlers]

Minimum Score 145 *Continued*

Score	Length of R Main Beam	L	Inside Spread	Number of R Points	L	Area	State/ Province	Hunter's Name	Date	Rank
153 1/8	19 5/8	19 6/8	16 6/8	6	5	Summit County	CO	Marvin Clyncke	2010	*
153 0/8	21 2/8	20 1/8	18 4/8	4	4	Carbon County	WY	Larry K. Hardwig	2011	*
152 7/8	24 1/8	23 4/8	20 3/8	4	4	Garfield County	CO	Rick C. Wilson	1971	*
151 5/8	21 3/8	22 5/8	20 1/8	4	4	Harney County	OR	Steve Savage	1980	*
150 7/8	21 2/8	19 5/8	18 1/8	4	4	Malheur County	OR	Gordon Childers	2010	*
150 1/8	22 4/8	22 3/8	15 5/8	4	5	Gilpin County	CO	Arnold Krueger	2001	*
149 5/8	21 4/8	20 2/8	16 0/8	5	4	Grand County	UT	Jay Verzuh	1985	*
149 4/8	19 7/8	19 5/8	19 2/8	5	4	Garfield County	UT	Gary McCain	2008	*
149 4/8	23 0/8	21 6/8	18 6/8	6	6	Jefferson County	CO	Randy Verfaillie	2010	*
149 3/8	21 1/8	20 0/8	23 0/8	5	5	Humboldt County	NV	Wayne Testolin	1991	*
148 1/8	19 3/8	20 2/8	16 4/8	6	5	Mesa County	CO	James L. Shanks	2001	*
148 0/8	23 2/8	22 2/8	20 3/8	4	5	Morgan County	UT	Rick Wilson	2005	*
147 7/8	20 4/8	22 1/8	17 5/8	4	5	Dawson County	MT	Dave Athas	1989	*
147 7/8	22 6/8	23 4/8	22 1/8	5	5	San Juan County	UT	Steve Stumbo	1990	*
147 4/8	21 3/8	21 4/8	18 4/8	4	4	Gunnison County	CO	Scott Hargrove	2001	*
147 0/8	23 6/8	22 7/8	26 6/8	4	5	Mesa County	CO	Jay Verzuh	1980	*
146 7/8	22 4/8	24 4/8	19 1/8	5	4	Montezuma County	CO	William C. Shuster	1979	*
146 3/8	19 6/8	20 4/8	20 7/8	6	5	Mesa County	CO	Jay Verzuh	1981	*
146 0/8	22 4/8	22 0/8	22 0/8	4	5	Catron County	NM	David Gardner	1995	*
145 3/8	16 1/8	17 7/8	17 7/8	5	5	Elko County	NV	William Newman	2010	*
145 1/8	22 5/8	21 4/8	20 7/8	5	5	Millard County	UT	David R. Scott	1964	*
145 0/8	22 2/8	22 0/8	24 5/8	4	4	Gunnison County	CO	Scott Hargrove	1995	*

Hunter: Daniel Willems
Species: Bighorn Sheep
Score: 165 7/8
Date: 9/26/12
Location: Mineral County, CO

MULE DEER [Non-Typical Antlers]

Minimum Score 170 — *Odocoileus hemionus* and certain related subspecies

Score	Length of R Main Beam	Length of L Main Beam	Inside Spread	Number of R Points	Number of L Points	Area	State/ Province	Hunter's Name	Date	Rank
274 7/8	23 7/8	26 2/8	27 0/8	23	12	Morgan County	CO	Kenneth W. Plank	1987	1
236 1/8	25 5/8	25 2/8	18 3/8	10	11	Coconino County	AZ	Stanley L. McIntyre	1965	2
234 4/8	26 4/8	27 2/8	23 7/8	10	8	Juab County	UT	Dennis M. Hickman	1972	3
232 3/8	23 2/8	21 2/8	23 0/8	12	14	Arapahoe County	CO	James P. Verney	1974	4
229 7/8	27 5/8	28 0/8	25 7/8	6	7	Lambs Canyon	UT	Lee Lindley	1942	5
226 4/8	23 5/8	27 6/8	19 2/8	9	11	Iron County	UT	Neil "Bud" Rhodes	1959	6
225 4/8	23 4/8	23 2/8	22 0/8	10	13	Herd Unit 54	UT	John C. Balch	1965	7
225 2/8	24 2/8	24 2/8	23 1/8	12	7	Garfield County	CO	Dennis Quinn	1972	8
224 6/8	25 5/8	25 5/8	22 4/8	9	9	Ferry County	WA	Romie Hilderbrant	1963	9
224 1/8	25 7/8	24 6/8	24 2/8	10	9	Washakie County	WY	William T. Ivey	1987	10
223 7/8	27 5/8	26 4/8	24 4/8	12	9	Uncompahgre Mtns.	CO	Steve Haynes	1972	11
221 3/8	22 7/8	26 4/8	18 5/8	10	10	Sheridan County	WY	Mike Barrett	2003	12
220 3/8	27 3/8	27 1/8	23 1/8	11	7	Sanpete County	UT	I. B. 'Blackie' Owen	1964	13
219 5/8	21 2/8	23 1/8	26 0/8	9	8	Mesa County	CO	Roger Lewis	1980	14
218 1/8	22 0/8	24 1/8	16 5/8	9	8	Lincoln County	WY	Mike Barrett	1987	15
217 6/8	25 5/8	24 5/8	22 5/8	12	13	White Pine County	NV	Russell Suminski	1981	16
216 2/8	21 7/8	22 5/8	19 4/8	8	9	Mohave County	AZ	William Cross	1965	17
215 5/8	21 3/8	23 0/8	16 6/8	7	10	Uintah County	UT	Jon E Bingham	1969	18
213 0/8	22 2/8	23 3/8	18 7/8	10	9	Caribou County	ID	Bob McAteer	1966	19
212 6/8	23 4/8	23 3/8	24 4/8	8	8	Ness County	KS	Ralph Stum	1966	20
212 0/8	21 0/8	20 5/8	19 5/8	10	9	Calgary	ALB	Bert Frelink	1985	21
211 1/8	23 5/8	24 4/8	20 6/8	7	7	Uintah County	UT	Vern Hatch	1962	22
211 0/8	27 2/8	26 6/8	22 3/8	7	7	Grant County	OR	Ed Woods	1957	23
210 6/8	21 5/8	20 1/8	22 7/8	8	10	Elbert County	CO	Louis Phillippe	2002	24
210 4/8	24 3/8	25 0/8	27 5/8	7	8	Montrose County	CO	Thomas Gloden	1963	25
209 6/8	26 5/8	26 2/8	26 1/8	10	8	Minidoka County	ID	Joe Dolan	1952	26
209 5/8	26 0/8	24 0/8	26 2/8	10	7	Caribou County	ID	Mike Barrett	1994	27
207 2/8	28 0/8	26 1/8	25 1/8	7	10	Elbert County	CO	Richard Boss	1999	28
207 0/8	24 2/8	23 0/8	20 6/8	7	5	Elmore County	ID	Deloy Desaro	1972	29
206 7/8	22 2/8	24 1/8	22 1/8	7	9	Lemhi County	ID	A. LaVerne Hokanson	1967	30
205 3/8	24 2/8	24 0/8	22 0/8	8	8	Grand County	UT	C. B. 'John' Olsen	1958	31
204 4/8	26 2/8	22 6/8	17 5/8	7	9	Fishlake National Forest	UT	Dick Kerr	1955	32
201 6/8	21 4/8	22 5/8	22 0/8	9	8	Caribou County	ID	Dennis Dockstader	1969	33
200 6/8	25 1/8	25 6/8	25 6/8	7	8	Lane County	KS	Matthew Shull	2003	34
200 3/8	24 4/8	24 3/8	19 4/8	8	9	Farm Creek	UT	Tex Ross	1966	35
200 2/8	27 5/8	26 2/8	19 1/8	9	8	Montezuma County	CO	Bryon C. Neeley	1971	36
199 1/8	22 6/8	23 4/8	19 1/8	9	10	Mesa County	CO	James R. Boyles	1971	37
198 5/8	21 6/8	22 6/8	23 4/8	9	8	Franklin County	KS	John R. Coblentz	1966	38
198 2/8	26 0/8	23 6/8	21 3/8	7	7	Yuma County	CO	Mike Barrett	2011	39
198 0/8	21 5/8	22 0/8	19 0/8	8	9	Routt County	CO	Bruce F. Davison	1968	40
197 5/8	22 0/8	21 1/8	20 5/8	10	8	Elmore County	ID	Jerry G. Fetters	1970	41
195 6/8	21 2/8	22 6/8	19 3/8	9	9	Weld County	CO	Densel Bolin	1974	42
195 4/8	25 2/8	23 2/8	22 3/8	9	14	Garfield County	CO	Douglas Kenyon	1964	43
195 0/8	22 2/8	22 4/8	21 0/8	5	8	Carbon County	WY	Tracy Villwok	2007	44
194 7/8	23 0/8	22 7/8	19 7/8	6	7	Finney County	KS	Jay Sloan	1967	45
194 7/8	26 5/8	25 3/8	22 3/8	6	7	Duchesne County	UT	Frank Warburton	1969	45
193 7/8	27 3/8	26 3/8	27 2/8	12	10	Yakima County	WA	James R. Lucas	1990	47
193 6/8	22 5/8	22 6/8	20 0/8	9	9	Utah County	UT	Ivan B. Henderson, Jr.	1959	48
192 4/8	24 4/8	25 0/8	21 3/8	6	7	Caribou County	ID	Mack Tigert	1985	49
192 4/8	24 2/8	22 3/8	17 7/8	6	6	Delta County	CO	Larick F. Spencer	1996	49
191 7/8	22 2/8	19 1/8	17 6/8	7	12	Bell Marsh Canyon	ID	Loren H. Dunn	1965	51
190 6/8	23 5/8	22 4/8	21 3/8	6	7	Huerfano County	CO	Ron Johnson	1970	52
190 4/8	19 4/8	20 6/8	21 4/8	6	8	La Plata County	CO	Lew Webb	2007	53
187 7/8	17 1/8	22 1/8	16 3/8	12	8	Grand County	UT	Lee Allred	1958	54
187 6/8	23 3/8	23 1/8	20 3/8	9	6	Routt County	CO	Thane Anderson	1993	55
187 3/8	25 0/8	25 1/8	21 4/8	8	7	Albany County	WY	Larry Hudson	1996	56
186 3/8	21 4/8	20 4/8	15 5/8	7	10	Sheridan County	WY	Mike Barrett	2008	57
186 2/8	23 7/8	22 3/8	22 2/8	7	7	Yuma County	CO	Jerry Bowen	2007	58
184 2/8	20 5/8	21 5/8	19 3/8	8	6	Duchesne County	UT	Dean Reynolds	1961	59
184 0/8	23 0/8	22 3/8	17 2/8	7	11	Rosebud County	MT	Rick Miller	1997	60
182 5/8	23 2/8	21 5/8	23 4/8	7	7	Sheridan County	WY	Mike Barrett	2004	61
182 3/8	24 3/8	23 2/8	24 3/8	6	6	Sheridan County	WY	Tim Goss	2003	62

MULE DEER [Non-Typical Antlers]

Minimum Score 170 *Continued*

Score	Length of R Main Beam	L	Inside Spread	Number of R Points	L	Area	State/ Province	Hunter's Name	Date	Rank
182 2/8	22 0/8	22 6/8	20 3/8	6	8	Weld County	CO	Frank Piacentino	2007	63
182 0/8	25 0/8	24 4/8	23 7/8	7	8	Kootenai County	ID	Rodney W. Willis	1977	64
181 7/8	25 0/8	23 1/8	25 2/8	6	7	Scott County	KS	John C. Walker	1996	65
181 5/8	27 0/8	24 3/8	23 4/8	7	9	Chelan County	WA	David M. Bartholemew	1967	66
181 4/8	27 5/8	26 3/8	27 4/8	5	7	Park County	CO	John Cliff	1967	67
180 4/8	21 1/8	22 5/8	21 3/8	7	5	Utah County	UT	Frank Warburton	1972	68
180 0/8	23 5/8	25 1/8	23 1/8	7	8	Lassen County	CA	Jeff Gravano	1994	69
178 5/8	25 3/8	24 6/8	31 0/8	6	7	San Juan County	UT	Guy Gates	1970	70
178 5/8	23 6/8	23 7/8	18 7/8	6	7	Sheridan County	WY	Brian F. Koelzer	2010	70
177 3/8	21 1/8	22 4/8	23 3/8	9	8	Dawson County	MT	Gerald Polesky	1970	72
176 7/8	20 5/8	22 5/8	21 2/8	7	7	Rock Creek	CO	Adolph Kuhns	1961	73
176 0/8	23 3/8	23 5/8	20 4/8	6	6	Scott County	KS	Vince Strickler	1975	74
175 3/8	20 6/8	24 2/8	29 0/8	9	9	Ness County	KS	Pete McBee	1969	75
174 2/8	23 6/8	23 1/8	22 0/8	9	10	Mesa County	CO	Art Cook	1958	76
174 1/8	25 7/8	24 5/8	18 7/8	7	5	Carbon County	UT	B. E. Epperson	1971	77
172 7/8	24 0/8	24 6/8	20 6/8	9	7	Chouteau County	MT	Mike Chouinard	1995	78
170 7/8	18 4/8	20 0/8	17 1/8	6	6	Wallowa County	OR	Wayne van Zwoll	1977	79
170 5/8	22 0/8	22 1/8	16 3/8	6	5	Lake County	OR	Bill Hendrick	1964	80
170 0/8	19 5/8	17 5/8	19 5/8	7	5	Finney County	KS	Rod Lies	1968	81
170 0/8	25 6/8	15 5/8	21 0/8	6	6	Bingham County	ID	Tim Moon	1996	81

MULE DEER [Non-Typical Velvet Antlers]

Minimum Score 170 *Odocoileus hemionus* and certain related subspecies

Score	Length of R Main Beam	L	Inside Spread	Number of R Points	L	Area	State/ Province	Hunter's Name	Date	Rank
228 6/8	25 2/8	27 2/8	19 3/8	12	8	Utah County	UT	Maurice T. Patterson	1971	*
227 3/8	26 4/8	26 5/8	20 7/8	8	7	White Pine County	NV	Tony Angelopoulos	1968	*
214 3/8	25 4/8	26 0/8	24 3/8	11	9	Caribou County	ID	Royce Brown	1984	*
207 6/8	22 5/8	21 7/8	20 3/8	7	8	Sanpete County	UT	L. Boyd Brewer	1967	*
206 5/8	26 2/8	25 6/8	21 1/8	8	7	Elko County	NV	Anthony A. Waidell	1967	*
205 4/8	23 5/8	24 1/8	18 4/8	7	8	Humboldt County	NV	Neil J. Russell	1996	*
203 3/8	23 0/8	25 1/8	25 4/8	13	8	Sheridan County	WY	Mike Barrett	1981	*
202 5/8	23 0/8	22 6/8	17 7/8	9	9	Mesa County	CO	Charles Wehmeyer	1964	*
202 3/8	23 1/8	24 5/8	18 0/8	9	8	Eagle County	CO	Rudy Meyers	1974	*
195 2/8	22 6/8	22 6/8	21 3/8	7	6	Mineral County	CO	Bob Bartoshesky	2007	*
193 4/8	23 2/8	23 5/8	21 6/8	7	6	Saguache County	CO	Barry J. Smith	1987	*
191 2/8	23 2/8	25 7/8	17 0/8	6	7	La Plata County	CO	John L. Gardner	2006	*
188 2/8	22 6/8	23 0/8	21 0/8	8	6	Jefferson County	CO	Rick Duggan	2006	*
180 6/8	27 4/8	25 3/8	21 6/8	6	8	Park County	CO	Dave Gusky	1975	*
177 6/8	25 7/8	25 5/8	22 1/8	8	6	Mesa County	CO	Thomas J. Hentrick	1972	*
175 1/8	24 5/8	23 2/8	20 2/8	7	7	Park County	CO	Marvin Clyncke	1996	*
173 5/8	25 1/8	25 3/8	22 4/8	5	6	Jefferson County	CO	Steve Smith	2000	*
172 1/8	22 7/8	23 1/8	20 7/8	6	8	Moffat County	CO	Bill Grammer	2000	*

WHITETAIL DEER [Typical Antlers]

Minimum Score 125 | *Odeocoileus virginianus* and certain related subspecies

Score	Length of R Main Beam	L	Inside Spread	Number of R Points	L	Area	State/ Province	Hunter's Name	Date	Rank
204 4/8	27 5/8	26 6/8	23 5/8	7	6	Peoria County	IL	M. J. Johnson	1965	1
197 6/8	25 6/8	26 4/8	18 6/8	7	7	Monroe County	IA	Lloyd Goad	1962	2
189 1/8	27 3/8	28 6/8	20 1/8	8	7	Kearney County	NE	Robert Vrbsky	1978	3
182 4/8	26 3/8	26 4/8	15 4/8	6	6	Jackson County	IN	Rocky Deakin	1985	4
181 7/8	25 3/8	27 3/8	21 3/8	5	5	Logan County	IL	Terry Lee Rich	1986	5
181 6/8	26 1/8	26 4/8	17 6/8	5	5	Wabasha County	MN	Lee G. Partington	1971	6
181 5/8	30 5/8	30 0/8	19 4/8	7	5	Grundy County	IL	Wesley Holm	1968	7
181 4/8	29 6/8	29 5/8	24 2/8	6	5	Keya Paha County	NE	Steve R. Pecsenye	1966	8
181 4/8	26 0/8	28 5/8	18 6/8	9	7	Fulton County	IL	Arnold Hegele	1968	8
181 2/8	26 2/8	25 2/8	22 2/8	5	6	Will County	IL	Joseph Skubisz	1996	10
180 0/8	27 2/8	24 7/8	20 7/8	7	6	Maverick County	TX	Trip du Perier	2011	11
179 4/8	28 4/8	28 5/8	16 4/8	5	5	Ellsworth County	KS	Jim Willems	2007	12
179 4/8	27 1/8	27 0/8	17 0/8	5	5	Sullivan County	NY	Domenick DeMaria	2010	12
179 3/8	26 2/8	26 5/8	17 3/8	5	5	Marshall County	SD	Phyllis Roehr	1976	14
179 1/8	25 6/8	25 6/8	16 3/8	6	5	Macoupin County	IL	William T. Wiser, Sr.	1994	15
179 1/8	25 7/8	25 6/8	17 4/8	9	6	N. Saskatchewan River	SAS	Miles Johnson	2003	15
178 3/8	25 6/8	26 4/8	21 2/8	7	8	McPherson County	KS	Larry Daniels	1967	17
177 3/8	27 1/8	26 3/8	20 3/8	6	5	Greene County	IA	Roger V. Carlson	1973	18
177 3/8	28 3/8	28 1/8	18 6/8	6	6	Wayne County	OH	Gary E. Landry	1975	18
176 6/8	25 2/8	25 0/8	23 4/8	6	6	Marshall County	KS	Ray A. Mosher	1966	20
176 2/8	28 2/8	27 5/8	21 0/8	7	5	Houston County	MN	John Zahrte	1981	21
176 2/8	26 4/8	26 5/8	20 0/8	6	8	Harvey County	KS	Davis J. Ediger	2010	21
175 7/8	27 3/8	26 4/8	24 4/8	7	6	Kandiyohi County	MN	Eldon Hauser	1969	23
175 5/8	29 0/8	28 6/8	21 3/8	4	4	Burnett County	WI	Myles Keller	1977	24
175 4/8	27 5/8	27 3/8	22 6/8	5	5	Jo Daviess County	IL	Richard McCartin	1991	25
175 2/8	26 0/8	28 3/8	21 3/8	7	5	Logan County	OH	Larry Pooler	2000	26
175 1/8	24 4/8	24 4/8	21 5/8	5	6	Marion County	IA	Gordon Hayes	1973	27
173 7/8	24 0/8	24 2/8	19 1/8	5	5	Noble County	OK	Danny McCants	1968	28
173 6/8	26 6/8	27 1/8	19 7/8	7	5	Mercer County	IL	Floyd A. Clark	1961	29
173 5/8	26 2/8	26 0/8	22 3/8	6	7	Muskingum County	OH	David R. Hatfield	1980	30
173 5/8	27 6/8	29 3/8	22 1/8	6	6	Winona County	MN	Keith Nelson	2006	30
173 4/8	25 7/8	27 0/8	23 0/8	5	5	McHenry County	IL	Gordon Sunderlage	1987	32
173 3/8	26 4/8	26 5/8	20 1/8	5	6	Warren County	IL	Larry C. Harding	1974	33
173 2/8	26 5/8	26 5/8	24 6/8	5	5	Pickaway County	OH	Tim Ritchie	1997	34
173 2/8	26 2/8	26 6/8	22 0/8	7	7	Preble County	OH	Gary H. Vest	1999	34
172 7/8	24 7/8	25 3/8	15 1/8	5	5	Clark County	OH	Jason Stull	2004	36
172 6/8	27 6/8	27 5/8	25 7/8	6	6	Mower County	MN	Art McKenzie	1972	37
172 5/8	26 7/8	27 1/8	19 2/8	5	6	Lyon County	KS	Dale Hellman	1990	38
172 2/8	26 1/8	25 3/8	19 2/8	5	6	Iowa County	IA	Ardith Lockridge	1965	39
172 2/8	27 3/8	25 1/8	18 4/8	5	5	Clinton County	IA	Robert S. Stankee	1985	39
172 1/8	27 2/8	27 5/8	18 5/8	7	7	Worcester County	MA	Paul A. Buccacio	2006	41
171 7/8	27 5/8	26 6/8	21 5/8	6	6	Lucas County	IA	Tim M. Whitlatch	1989	42
171 6/8	28 0/8	28 0/8	20 6/8	7	8	Edmonton	ALB	Warren Witherspoon	1991	43
171 6/8	25 3/8	25 7/8	18 5/8	5	6	Cookson	SAS	Bry Loyd	1999	43
171 5/8	25 0/8	25 0/8	20 3/8	6	6	Douglas County	NE	Bryan Kindler	1997	45
171 4/8	24 3/8	23 2/8	16 6/8	6	6	Ellsworth County	KS	Jim Willems	1985	46
171 3/8	22 4/8	23 7/8	24 7/8	6	6	Cass County	ND	Warren Buss	1966	47
171 1/8	26 3/8	26 4/8	20 5/8	5	5	Itasca County	MN	John Parmeter	1964	48
171 1/8	23 5/8	24 0/8	16 7/8	6	6	Piatt County	IL	Ronald E. Waugh	1971	48
171 1/8	24 3/8	23 2/8	17 4/8	8	7	Morton County	ND	Tony Schatz	1974	48
170 6/8	24 5/8	25 6/8	19 6/8	6	6	Fayette County	IA	Michael Judas	2008	51
170 4/8	25 1/8	25 4/8	20 2/8	5	4	Des Moines County	IA	Bob Fudge	1966	52
170 4/8	29 0/8	29 3/8	19 1/8	6	5	Cerro Gordo County	IA	Chuck Harris	1991	52
170 3/8	28 5/8	28 2/8	21 6/8	6	4	Hall County	NE	Gust Bergman	1965	54
170 1/8	26 1/8	27 1/8	22 1/8	5	5	Edwards County	KS	Jay Schaller	1968	55
170 1/8	28 2/8	27 4/8	22 3/8	6	5	Lake County	IL	John W. Schnider	1992	55
170 0/8	26 1/8	26 6/8	20 0/8	5	5	Scott County	KS	Monte L. Barker	1973	57
170 0/8	23 7/8	24 6/8	19 6/8	6	6	Washington County	AR	Michael Franks	2007	57
169 7/8	28 0/8	28 2/8	19 7/8	4	4	Sullivan County	NY	Domenick DeMaria	2005	59
169 6/8	24 1/8	24 7/8	20 4/8	7	6	Bearspaw	ALB	Bry Loyd	2006	60
169 5/8	24 0/8	23 4/8	22 1/8	6	7	Neosho County	KS	Jeff Friederich	1992	61
169 4/8	23 7/8	25 2/8	17 7/8	6	7	Charles Mix County	SD	Dan Carda	1974	62

WHITETAIL DEER [Typical Antlers]

Minimum Score 125 *Continued*

Score	Length of Main Beam R	Length of Main Beam L	Inside Spread	Number of Points R	Number of Points L	Area	State/ Province	Hunter's Name	Date	Rank
169 4/8	26 0/8	27 4/8	21 5/8	5	6	Knox County	IL	Don Owen	2009	62
169 3/8	23 5/8	24 5/8	19 2/8	7	7	Perry County	IN	Ronald J. Phillips	1973	64
169 2/8	26 1/8	27 2/8	21 6/8	6	6	Ashland County	OH	Darrell Huff	1985	65
169 2/8	26 7/8	26 2/8	23 2/8	5	6	Oregon County	MO	Woodrow J. Parrott	1999	65
169 1/8	28 6/8	28 2/8	19 3/8	6	5	La Salle County	IL	Dave Mrowicki	1985	67
168 6/8	27 7/8	27 0/8	16 6/8	7	7	Muskingum County	OH	Gerald Shepler	1988	68
168 5/8	27 5/8	27 1/8	20 7/8	5	5	Guernsey County	OH	Todd Lowe	1980	69
168 2/8	25 2/8	23 7/8	22 4/8	5	5	Champaign County	OH	Michael Maurice	2008	70
168 1/8	24 6/8	24 6/8	20 1/8	6	5	Murray County	MN	Marvin Brouwer	1971	71
167 6/8	26 4/8	26 2/8	19 6/8	5	5	Monona County	IA	Douglas M. Bonine	1985	72
167 6/8	27 2/8	27 1/8	21 2/8	5	5	Coffey County	KS	Edward L. Bess	1985	72
167 5/8	26 6/8	26 7/8	18 1/8	7	7	Pittsburg County	OK	Brett Foster	1994	74
167 4/8	22 4/8	23 1/8	18 4/8	5	5	Washington County	KS	Bill R. Mallean	1974	75
167 3/8	25 4/8	25 7/8	18 3/8	5	6	Douglas County	NE	Ernie Buttry	2000	76
167 2/8	29 0/8	28 5/8	21 2/8	7	5	Pepin County	WI	Terry Grotthus	2012	77
167 1/8	22 5/8	22 6/8	25 2/8	5	7	Dodge County	MN	Myles Keller	1985	78
167 0/8	26 3/8	25 7/8	18 3/8	6	6	Fountain County	IN	Steve McQueen	1991	79
166 5/8	25 6/8	26 2/8	19 7/8	6	5	St. Joseph County	MI	Donald L. Preston	2002	80
166 4/8	25 5/8	25 5/8	19 1/8	6	7	Bartholomew County	IN	Shane Sweeney	1998	81
166 3/8	26 6/8	26 4/8	17 1/8	6	6	Clarke County	IA	Dwight E. Green	1965	82
166 3/8	26 3/8	26 0/8	18 2/8	8	5	Yankton County	SD	Roger Irwin	1985	82
166 3/8	28 1/8	27 5/8	20 7/8	5	5	Van Buren County	IA	Baree Weber	2007	82
166 2/8	29 3/8	29 4/8	21 5/8	6	4	Monona County	IA	G. K. Tuttle	1967	85
166 2/8	27 4/8	27 1/8	20 2/8	9	6	Republic County	KS	Virgil Graham	1986	85
166 1/8	23 2/8	23 5/8	19 3/8	5	5	Morrison County	MN	Corey Loney	1963	87
166 1/8	25 6/8	25 2/8	21 2/8	5	4	Texas County	OK	Max Crocker	1986	87
166 0/8	23 4/8	23 5/8	18 0/8	5	5	Bon Homme County	SD	Delbert Newman	1964	89
166 0/8	27 5/8	26 7/8	23 4/8	5	5	Clinton County	IA	Loy J. Brooker	1964	89
165 7/8	23 4/8	23 4/8	16 3/8	5	5	Wapello County	IA	Richard L. Larsen	1976	91
165 7/8	23 0/8	26 0/8	18 1/8	5	5	Vermilion County	IL	Dick Bayer	1987	91
165 7/8	26 2/8	25 6/8	20 0/8	7	5	Weld County	CO	Mark Houtchens	1991	91
165 7/8	26 1/8	26 2/8	20 6/8	7	7	Woodford County	IL	Greg Gullett	1999	91
165 6/8	27 0/8	26 1/8	19 0/8	7	6	Pottawattamie County	IA	Dan Bowen	1968	95
165 6/8	27 4/8	25 4/8	26 6/8	4	5	St. Marys County	MD	John S. Hurt	2003	95
165 4/8	25 6/8	25 1/8	21 0/8	5	5	Kent County	MD	Kent Price	1962	97
165 4/8	27 6/8	27 5/8	21 0/8	6	5	Will County	IL	Donald R. Spence	1988	97
165 3/8	26 4/8	25 6/8	18 6/8	5	6	Monroe County	MO	Larry Meier	1993	99
165 3/8	25 4/8	26 5/8	15 5/8	5	4	Brown County	IN	George E. Mann	2003	99
165 3/8	26 0/8	25 4/8	18 7/8	5	5	Cherokee County	IA	Cody Brandes	2003	99
165 1/8	25 5/8	25 7/8	18 1/8	7	6	Brown County	IL	Dennis Cloninger	1997	102
165 1/8	24 7/8	25 1/8	19 1/8	5	5	Madison County	MT	James C. Kennedy	1999	102
165 0/8	25 2/8	23 5/8	19 0/8	6	6	Stony Plain	ALB	Wayne C. Prier	1983	104
164 7/8	27 5/8	27 6/8	23 1/8	7	7	Butler County	OH	Will McQueen	1989	105
164 6/8	26 5/8	26 5/8	26 4/8	8	5	Sullivan County	IN	John W. Hale	1988	106
164 5/8	25 2/8	23 5/8	18 5/8	8	6	Sumner County	KS	Archie A. Stralow	1967	107
164 5/8	26 3/8	24 6/8	20 6/8	5	6	Sedgwick County	CO	Brad Ham	1973	107
164 5/8	26 5/8	25 4/8	20 4/8	6	6	Saginaw County	MI	William J. Twarog	1990	107
164 4/8	23 5/8	23 5/8	20 7/8	6	6	Bond County	IL	Roger Munie	1987	110
164 3/8	25 4/8	25 4/8	19 6/8	8	7	Norman County	MN	Gilbert Guttormson	1953	111
164 3/8	21 4/8	21 6/8	16 5/8	6	6	Wibaux County	MT	Gerald Polesky	1959	111
164 3/8	28 4/8	28 0/8	20 4/8	7	8	Trigg County	KY	Charles Stahl	1965	111
164 3/8	25 1/8	25 2/8	18 3/8	5	5	Cottonwood County	MN	Jim Hansen	1972	111
164 3/8	26 4/8	25 5/8	17 3/8	7	8	Madison County	MT	Jim Schilke	1978	111
164 3/8	28 0/8	27 5/8	21 1/8	6	8	Fayette County	OH	Steven J. Guess	1984	111
164 2/8	27 2/8	26 2/8	21 0/8	11	7	Grundy County	IL	Jerome M. Fris	1972	117
164 1/8	26 5/8	26 3/8	18 2/8	6	5	Morrison County	MN	Lloyd Neuman	1971	118
164 1/8	25 0/8	24 5/8	22 7/8	5	5	Moultrie County	IL	Gary F. Saveley	2000	118
164 1/8	26 3/8	26 0/8	17 3/8	6	6	St. Croix County	WI	Dennis Saathoff	2004	118
164 1/8	24 4/8	24 0/8	17 1/8	5	5	Washington County	IA	Don Kieler	2007	118
164 0/8	24 3/8	25 0/8	16 6/8	5	5	Olmsted County	MN	Robert Meyer	1969	122
164 0/8	25 1/8	25 5/8	23 3/8	5	7	Morrison County	MN	Bruce Edberg	1977	122
164 0/8	25 2/8	25 3/8	15 4/8	5	5	Morrill County	NE	Tim Ray	1997	122

WHITETAIL DEER [Typical Antlers]

Minimum Score 125 *Continued*

Score	Length of Main Beam R	L	Inside Spread	Number of Points R	L	Area	State/ Province	Hunter's Name	Date	Rank
164 0/8	24 2/8	23 3/8	17 0/8	6	6	Halifax County	VA	Steven W. Blosser	2005	122
163 7/8	24 1/8	24 4/8	17 5/8	5	6	Barber County	KS	Scott Koelzer	2009	126
163 6/8	25 1/8	25 1/8	17 7/8	5	6	Spokane County	WA	Gordon Hoffnagle	2000	127
163 6/8	26 4/8	27 2/8	18 2/8	5	6	Campbell County	KY	Michael Studer	2011	127
163 5/8	25 6/8	25 4/8	21 6/8	5	5	Phillips County	KS	Bill Duncan	1969	129
163 5/8	26 5/8	26 1/8	21 3/8	6	6	Lawrence County	IL	Larry K. Karns	1975	129
163 4/8	25 4/8	25 5/8	18 0/8	5	7	Renville County	ND	Bobby Triplett	1958	131
163 3/8	24 4/8	22 7/8	22 1/8	5	5	Linn County	IA	Delmar Phillips	1960	132
163 3/8	24 3/8	23 7/8	17 5/8	8	7	Morrison County	MN	Alvin A. Diemert	1973	132
163 3/8	25 3/8	27 2/8	21 0/8	6	4	Harvey County	KS	Tim S. Ross	2001	132
163 2/8	24 6/8	26 0/8	18 6/8	5	7	Hamilton County	OH	Jay Knight	1998	135
163 1/8	24 2/8	26 0/8	16 4/8	7	7	Logan County	WV	Gilbert Sexton	1963	136
163 0/8	24 0/8	24 6/8	20 0/8	5	5	Dickinson County	IA	Harold Ehrp	1959	137
163 0/8	25 2/8	25 0/8	15 1/8	6	6	Texas County	OK	Edward F. Bryan, Jr.	1976	137
163 0/8	24 2/8	24 2/8	16 0/8	5	6	Ashland County	WI	Stanley E. Hopfensperger	1999	137
162 7/8	29 0/8	30 4/8	19 0/8	6	8	Queen Annes County	MD	L. P. Stephens, Jr.	1962	140
162 7/8	24 6/8	24 6/8	22 4/8	6	7	Cowley County	KS	Kenneth Highfill	1968	140
162 7/8	25 0/8	25 0/8	18 5/8	5	5	Baltimore County	MD	Bruce Hoover	1991	140
162 6/8	24 2/8	26 2/8	20 6/8	6	7	Rice County	MN	Ken Bakken	1957	143
162 6/8	24 6/8	25 1/8	19 4/8	5	5	Fayette County	IA	John Nagel	1969	143
162 6/8	24 7/8	23 4/8	20 2/8	6	5	Ferry County	WA	Steve Brown	1999	143
162 6/8	25 4/8	26 0/8	21 0/8	5	5	Waupaca County	WI	Edward Mertins	2002	143
162 6/8	22 5/8	23 3/8	16 4/8	6	7	Mills County	IA	John R. Horn	2008	143
162 5/8	26 2/8	24 0/8	20 7/8	8	7	Marshall County	KS	Gary W. Tobin	1966	148
162 5/8	26 4/8	27 5/8	20 3/8	4	4	Saunders County	NE	Robert Parkins	1967	148
162 5/8	26 3/8	26 7/8	17 7/8	5	7	Wyandot County	OH	David Weininger	1991	148
162 4/8	24 1/8	24 7/8	19 4/8	5	5	Champaign County	OH	Alan Shafer, Jr.	1993	151
162 4/8	26 5/8	22 5/8	24 2/8	6	7	Will County	IL	Jack Haviland	2003	151
162 4/8	23 4/8	23 4/8	18 0/8	5	5	Laurel County	KY	Jerry Oliver	2010	151
162 3/8	27 7/8	28 0/8	19 4/8	8	7	Edmonton	ALB	Dale Spooner	1992	154
162 3/8	25 5/8	26 1/8	19 3/8	6	5	Jo Daviess County	IL	Joseph C. Hinderman	2006	154
162 2/8	22 2/8	24 7/8	17 6/8	6	5	Kingsbury County	SD	Dale Peterson	1972	156
162 2/8	21 6/8	22 3/8	16 2/8	5	5	Oneida County	WI	Mark Nelis	1987	156
162 1/8	25 2/8	26 0/8	16 4/8	8	6	Monroe County	IA	Larry Whitson	1985	158
162 1/8	25 1/8	25 7/8	18 1/8	5	5	Buffalo County	WI	Terry Krahn	1994	158
162 0/8	25 0/8	27 1/8	20 4/8	8	6	Bond County	IL	Larry Nelson	1976	160
162 0/8	28 6/8	27 1/8	21 4/8	5	5	Hancock County	OH	Bradley D. DePuy	1994	160
161 7/8	23 4/8	25 4/8	18 3/8	5	5	Saunders County	NE	David Strimple	1961	162
161 7/8	24 2/8	24 1/8	20 7/8	5	5	Douglas County	NE	Noel Miller	1970	162
161 5/8	23 3/8	23 1/8	14 5/8	6	6	Dunn County	WI	Leonard Hines	1970	164
161 5/8	22 6/8	22 5/8	18 6/8	5	6	Juneau County	WI	Harlan Steindl	1971	164
161 5/8	24 0/8	24 0/8	16 3/8	5	5	E. Carroll Parish	LA	F. Lane Mitchell	1993	164
161 5/8	26 2/8	26 1/8	19 1/8	6	5	Franklin County	OH	Craig Bonham	1996	164
161 5/8	27 2/8	26 2/8	20 5/8	6	5	Clarke County	IA	Kyle R. Poindexter	2006	164
161 4/8	26 1/8	26 1/8	19 2/8	5	4	Bond County	IL	Sam White	1974	169
161 4/8	24 4/8	24 4/8	18 2/8	5	5	Hocking County	OH	Dave Hanson	1997	169
161 4/8	27 0/8	27 5/8	24 2/8	6	7	Clay County	KS	Tom E. Bowman	1997	169
161 3/8	25 0/8	23 6/8	16 6/8	8	7	Wyandotte County	KS	George F. Bigelow	1967	172
161 3/8	25 1/8	25 3/8	19 1/8	5	5	McPherson County	KS	Daniel Willems	1990	172
161 3/8	26 5/8	24 6/8	19 1/8	5	5	Jefferson County	WI	Bill Peterson	2002	172
161 3/8	28 0/8	26 6/8	21 0/8	6	6	Decatur County	KS	Floyd Boan	2008	172
161 2/8	24 6/8	24 7/8	16 7/8	6	5	Knox County	IL	Fred E. Miller	1992	176
161 1/8	26 0/8	27 1/8	18 6/8	6	5	Clark County	IN	Frank Mauk, Jr.	1966	177
161 1/8	24 2/8	24 6/8	18 0/8	5	6	Woodford County	KY	Joe Lacefield	2012	177
161 0/8	25 1/8	25 1/8	19 4/8	5	5	Geary County	KS	Kevin Foerschler	1995	179
161 0/8	23 4/8	22 3/8	19 0/8	5	6	Mountrail County	ND	Craig Richardson	2011	179
160 7/8	20 5/8	24 0/8	24 3/8	7	5	Frontier County	NE	Vernon Laverack	1959	181
160 7/8	25 5/8	27 3/8	21 4/8	5	6	Iowa County	IA	Everett Reid	1962	181
160 7/8	24 2/8	23 4/8	17 6/8	7	5	Smith County	KS	Ron Sturgeon	1965	181
160 7/8	24 4/8	24 2/8	17 1/8	5	5	Wayne County	WV	Willard Brown	1967	181
160 7/8	22 6/8	23 1/8	21 1/8	5	5	Lincoln County	MN	Bernie Ahlberg	1974	181
160 7/8	25 2/8	25 6/8	19 5/8	7	7	Rockingham County	VA	Jim Burtner	1989	181

WHITETAIL DEER [Typical Antlers]

Minimum Score 125 *Continued*

Score	Length of Main Beam R	Length of Main Beam L	Inside Spread	Number of Points R	Number of Points L	Area	State/ Province	Hunter's Name	Date	Rank
160 6/8	25 4/8	25 7/8	17 0/8	6	5	Sarpy County	NE	Lawrence A. Klabunde	1968	187
160 6/8	26 1/8	24 2/8	18 2/8	6	7	Licking County	OH	David Wamer	2002	187
160 5/8	22 3/8	23 4/8	16 4/8	7	8	Lawrence County	IL	Bob Brian	1971	189
160 5/8	25 2/8	25 3/8	14 3/8	6	5	Hubbard County	MN	Myles Keller	1982	189
160 5/8	25 5/8	25 4/8	19 1/8	7	5	Decatur County	KS	Tom Reedy	2007	189
160 5/8	24 5/8	25 0/8	20 7/8	6	5	Tallahatchie County	MS	Jim Kennedy	2010	189
160 4/8	26 6/8	25 4/8	20 7/8	7	7	Williams County	ND	John Bloom	1963	193
160 4/8	27 2/8	28 0/8	21 4/8	5	7	Lyon County	IA	Marvin H. Peterson	1970	193
160 4/8	23 1/8	22 4/8	19 0/8	6	7	Keith County	NE	Gil Wilkinson	1970	193
160 4/8	24 2/8	24 7/8	17 6/8	5	6	Dawson County	MT	Frank Legato	1978	193
160 3/8	27 2/8	26 4/8	22 2/8	7	6	Jefferson County	OH	Tom Bateman	1997	197
160 2/8	25 6/8	26 3/8	19 2/8	5	5	Nance County	NE	Ralph I. Hansen	1963	198
160 2/8	25 7/8	26 0/8	21 4/8	4	4	Cherokee County	IA	Jerry L. Smith	1969	198
160 2/8	23 7/8	23 7/8	19 7/8	6	5	Stevens County	WA	Jay Baker	1992	198
160 2/8	25 4/8	24 6/8	18 4/8	5	6	Waukesha County	WI	John Morrison	1998	198
160 2/8	24 3/8	24 4/8	21 6/8	5	5	Summit County	OH	Thomas G. Minute	1999	198
160 2/8	27 4/8	27 4/8	23 3/8	6	6	Fulton County	IL	Gary L. Olsen	2008	198
160 1/8	28 6/8	29 7/8	21 3/8	5	4	Lafayette County	WI	Jeff J. Kahle	1988	204
160 1/8	24 7/8	25 6/8	18 5/8	5	5	Clark County	MO	Tom Casey	2001	204
160 1/8	27 2/8	26 2/8	17 5/8	5	6	Russell County	KS	Glen Merica	2009	204
160 0/8	24 6/8	24 5/8	21 3/8	6	6	Worth County	IA	Terry Lynch	1972	207
160 0/8	25 4/8	26 0/8	18 6/8	5	5	Faribault County	MN	Darryl Germann	1974	207
160 0/8	27 6/8	27 7/8	23 7/8	6	7	Otero County	CO	Troy Cunningham	1995	207
160 0/8	25 4/8	25 5/8	19 6/8	6	6	Knox County	IL	Dan Hicks	2008	207
159 5/8	26 3/8	25 5/8	21 5/8	5	5	Des Moines County	IA	Richard Howard	1964	211
159 5/8	24 6/8	23 1/8	18 1/8	5	5	Madison County	IL	Barry Ash	1980	211
159 5/8	26 1/8	26 2/8	19 1/8	6	7	Ellis County	KS	Allen Landry	1986	211
159 5/8	26 2/8	25 7/8	18 0/8	7	6	Sangamon County	IL	Hobart E. Watson	1992	211
159 5/8	26 6/8	26 6/8	24 2/8	5	4	La Porte County	IN	Drake Matovich	1993	211
159 4/8	27 6/8	27 6/8	25 4/8	4	4	Blue Earth County	MN	Harold Tow	1963	216
159 4/8	22 6/8	21 5/8	17 5/8	6	5	Wabash County	IL	Tom J. McRaven	1967	216
159 4/8	24 4/8	24 6/8	18 0/8	5	6	Dade County	MO	Charles A. Myers	1985	216
159 4/8	24 6/8	23 1/8	23 4/8	5	5	Lyon County	IA	Mike Judas	1993	216
159 4/8	27 3/8	27 1/8	21 4/8	6	6	Wyandot County	OH	Mike Saam	1995	216
159 3/8	24 2/8	23 0/8	19 3/8	6	5	Delaware County	IA	Kevin Davis	2002	221
159 2/8	26 1/8	25 0/8	17 3/8	7	5	Osborne County	KS	Gary Ozias	1995	222
159 1/8	23 7/8	23 0/8	20 1/8	6	6	Lucas County	OH	Martin Higley	1962	223
159 1/8	26 2/8	25 2/8	23 5/8	4	5	Grundy County	IL	Ed Vitko, Jr.	1976	223
159 1/8	22 4/8	24 2/8	16 5/8	7	6	Clayton County	IA	Gary Troester	1978	223
159 1/8	25 1/8	25 2/8	21 7/8	5	5	Clay County	MN	Randy Nelson	1992	223
159 1/8	24 5/8	24 4/8	16 5/8	5	5	Door County	WI	Michael E. Hafenbredl	2005	223
159 0/8	25 7/8	26 1/8	17 6/8	5	5	Shelby County	IL	Gary E. Sievers	1971	228
158 7/8	24 7/8	22 4/8	20 3/8	5	6	Pope County	IL	Gary Thomas	1964	229
158 7/8	27 4/8	25 7/8	18 3/8	8	8	Blue Earth County	MN	Gordon F. Kopischke	1968	229
158 7/8	24 1/8	24 1/8	20 6/8	5	6	Putnam County	IL	David A. Heath	1975	229
158 6/8	22 5/8	22 7/8	15 6/8	7	7	Irion County	TX	John K. Watson	1977	232
158 6/8	24 4/8	25 2/8	20 3/8	5	4	Polk County	IA	Joe Crippen	1997	232
158 5/8	24 3/8	25 3/8	15 6/8	5	9	Marinette County	WI	Valerie P. Williams	1966	234
158 5/8	24 4/8	23 6/8	18 5/8	5	5	Shelby County	IL	Jim Helm	1977	234
158 5/8	27 3/8	27 6/8	21 7/8	5	5	Carroll County	IL	James C. Carlson	1999	234
158 5/8	25 6/8	25 3/8	15 3/8	5	6	Monroe County	AR	Vance Fisher	2005	234
158 4/8	23 5/8	22 3/8	20 0/8	5	6	Chautauqua County	KS	Gordon Reneberg	1965	238
158 3/8	25 3/8	26 0/8	16 4/8	7	7	Lake County	IL	James C. Carlson	1991	239
158 2/8	26 4/8	24 6/8	22 7/8	5	7	Lee County	IA	Gary Frost	1965	240
158 2/8	25 7/8	25 1/8	18 0/8	5	6	St. Marys County	MD	Pete Ropshaw	1999	240
158 1/8	25 2/8	25 4/8	16 5/8	5	4	Randolph County	IN	Ron J. Carlin	1973	242
158 1/8	26 4/8	26 4/8	18 1/8	6	6	Wyandot County	OH	Michael D Saam	1985	242
158 0/8	23 4/8	24 2/8	22 0/8	5	5	Butler County	KS	Larry D. Walker, Jr.	1996	244
157 7/8	23 0/8	25 0/8	19 5/8	5	5	Meeker County	MN	Russell T. Nelson	1974	245
157 7/8	25 0/8	26 4/8	20 5/8	5	5	Coal County	OK	R. Wayne Tarpley	2012	245
157 6/8	24 0/8	23 6/8	18 4/8	6	6	Dundy County	NE	Bryan Holley	2006	247
157 4/8	25 7/8	25 7/8	17 2/8	4	4	Marion County	IA	Charles H. Walter	1967	248

WHITETAIL DEER [Typical Antlers]

Minimum Score 125 *Continued*

Score	Length of Main Beam R	Length of Main Beam L	Inside Spread	Number of Points R	Number of Points L	Area	State/ Province	Hunter's Name	Date	Rank
157 4/8	22 7/8	23 5/8	18 0/8	5	5	Kalamazoo County	MI	Guy Stutzman	1979	248
157 3/8	23 0/8	23 2/8	22 7/8	5	5	Adams County	IL	John Musolino	1966	250
157 2/8	23 1/8	22 6/8	16 2/8	6	6	Johnson County	IL	Jim Casey	1963	251
157 2/8	22 3/8	23 2/8	16 4/8	5	5	Mountrail County	ND	Dean A. Rehak	1963	251
157 2/8	27 6/8	27 6/8	22 3/8	5	6	Jessamine County	KY	David Cartwright	1979	251
157 2/8	25 4/8	25 0/8	17 6/8	5	6	Argyle	MAN	Russ Snell	1982	251
157 2/8	25 7/8	26 7/8	17 5/8	5	6	Warren County	KY	Billy Carder	2001	251
157 0/8	25 7/8	26 0/8	16 5/8	6	7	Watonwan County	MN	Dave Ellertson	1973	256
157 0/8	25 3/8	26 0/8	15 6/8	7	8	McKenzie County	ND	Donald Olson	1974	256
156 7/8	28 7/8	26 6/8	19 7/8	5	5	Chippewa County	MN	Paul D. Lundgren	1969	258
156 7/8	26 3/8	26 1/8	20 7/8	4	4	Jo Daviess County	IL	Joseph C. Hinderman	1998	258
156 6/8	27 1/8	26 5/8	20 5/8	5	5	Lincoln County	WV	Gary Smith	1970	260
156 6/8	25 4/8	25 1/8	18 0/8	5	5	Isabella County	MI	Robert J. Warner	1993	260
156 5/8	26 1/8	25 2/8	20 2/8	6	6	Grant County	WI	Walter Edge	1957	262
156 5/8	24 6/8	26 3/8	19 5/8	5	4	Traverse County	MN	Roland L. Hausmann	1960	262
156 5/8	22 7/8	25 0/8	19 6/8	7	7	Des Moines County	IA	E. E. Smith	1965	262
156 5/8	23 7/8	24 1/8	18 7/8	7	7	Nicollet County	MN	Thomas J. Merkley	1967	262
156 5/8	27 3/8	26 2/8	21 5/8	5	5	Queen Annes County	MD	Charles Milford Squires	1969	262
156 5/8	23 0/8	23 2/8	15 4/8	6	6	Henry County	IA	Troy Alan Fenton	2000	262
156 5/8	24 0/8	24 3/8	19 5/8	5	5	Jones County	IA	Robert A. Lange	2001	262
156 4/8	26 4/8	25 7/8	22 5/8	4	5	Williamson County	IL	Roy Williams	1960	269
156 4/8	26 3/8	24 4/8	17 4/8	5	6	Forest County	WI	Daniel Radder	1968	269
156 4/8	24 1/8	23 3/8	15 0/8	6	5	Jones County	IA	Gary McCormick	1977	269
156 3/8	24 4/8	24 3/8	21 5/8	6	6	Mills County	IA	Douglas R. Roll	1986	272
156 3/8	25 0/8	24 5/8	20 5/8	5	4	Effingham County	IL	Tim Dillow	1989	272
156 2/8	26 1/8	26 4/8	21 0/8	6	6	Butler County	KS	Ralph R. Belt	1967	274
156 2/8	25 0/8	23 2/8	18 0/8	9	6	Palo Alto County	IA	Earl J. Gustafson	1972	274
156 2/8	25 1/8	26 0/8	19 4/8	5	5	Oakland County	MI	David B. Tater	1984	274
156 2/8	27 7/8	26 5/8	23 4/8	6	6	Franklin County	NE	Nathan Andersohn	1999	274
156 1/8	25 4/8	25 2/8	19 1/8	5	5	Crawford County	IL	Mickie D. Purcell	1972	278
156 1/8	25 5/8	24 6/8	22 1/8	4	4	Sumner County	KS	Ralph Shaver	1984	278
156 0/8	26 7/8	26 7/8	19 0/8	5	5	Allamakee County	IA	Gene A. Hall	1993	280
156 0/8	23 3/8	23 4/8	19 2/8	6	6	Monroe County	IA	Robert Smith	2004	280
156 0/8	23 3/8	22 4/8	19 3/8	7	5	Chase County	KS	Jim Purcaro	2005	280
155 7/8	24 7/8	24 4/8	19 7/8	5	5	Frontier County	NE	Charles Druse	1963	283
155 7/8	24 3/8	24 5/8	15 5/8	5	5	Harrison County	IA	Clarence N. Jackson, Jr.	1963	283
155 7/8	26 6/8	25 6/8	16 2/8	7	6	Anderson County	KS	Kurt A. Sayers	1990	283
155 7/8	24 7/8	24 6/8	22 5/8	5	5	Coahoma County	MS	Charles B. Neely	1994	283
155 7/8	24 4/8	23 1/8	15 5/8	6	5	Muskingum County	OH	Dan Jennings	1997	283
155 7/8	25 2/8	24 0/8	21 1/8	6	5	Buffalo County	WI	Troy R. Rauwerdink	2005	283
155 6/8	25 1/8	24 4/8	18 4/8	4	5	Madison County	NE	Dick Gambill	1967	289
155 6/8	26 0/8	26 2/8	16 6/8	5	5	Allamakee County	IA	Dan Brimeyer	1988	289
155 6/8	26 0/8	25 4/8	20 7/8	7	6	Shelby County	IL	Rick G. Cobb	2000	289
155 5/8	25 5/8	26 0/8	19 1/8	5	6	Rice County	KS	Gordon Leo Rayl	1967	292
155 5/8	22 5/8	23 0/8	18 5/8	5	5	Anne Arundel County	MD	Gene Hyatt	1976	292
155 5/8	23 7/8	22 5/8	19 1/8	7	5	Harrison County	IA	Alfred S. Foster	1978	292
155 5/8	24 5/8	23 3/8	18 7/8	5	5	Chase County	KS	Jerry Keller	1986	292
155 5/8	24 2/8	24 6/8	18 3/8	5	5	Powell County	MT	Seth Rogers	1998	292
155 5/8	24 2/8	23 4/8	16 1/8	6	5	Richland County	WI	Jack Hendricks	1999	292
155 4/8	23 2/8	25 4/8	18 6/8	7	7	Westchester County	NY	Bernard J. Crescione	1960	298
155 4/8	25 5/8	25 5/8	16 4/8	5	5	Marion County	IA	Thomas L. Tucker	1967	298
155 4/8	24 7/8	25 6/8	22 4/8	4	4	Morgan County	IL	Gerald L. Stone	1990	298
155 4/8	24 3/8	24 1/8	18 3/8	7	8	Jefferson County	OH	Michael W. Brown	1991	298
155 4/8	26 1/8	26 4/8	19 6/8	6	5	Delaware County	OH	Reno F. Carifa	2011	298
155 3/8	23 5/8	24 0/8	18 4/8	6	7	Finney County	KS	Wray Decker	1966	303
155 3/8	23 7/8	23 6/8	16 5/8	5	5	Cherry County	NE	Gary Galloway	1985	303
155 3/8	29 6/8	28 7/8	20 7/8	5	6	Highland County	OH	William E. Lee, Jr.	1987	303
155 3/8	24 6/8	24 5/8	18 1/8	5	5	Vermilion County	IL	Gary L. Wilford	1988	303
155 2/8	22 0/8	21 6/8	17 4/8	6	5	Benton County	IA	Gene Pollock	1958	307
155 2/8	25 0/8	25 5/8	18 3/8	5	6	Floyd County	IA	Richard G. Long	1967	307
155 2/8	24 7/8	24 6/8	20 2/8	5	6	Appanoose County	IA	Steven P. Salmieri	1990	307
155 2/8	24 6/8	24 7/8	17 4/8	5	5	Douglas County	WI	Kenneth W. Moen	1998	307

WHITETAIL DEER [Typical Antlers]

Minimum Score 125 *Continued*

Score	Length of Main Beam R	L	Inside Spread	Number of Points R	L	Area	State/ Province	Hunter's Name	Date	Rank
155 1/8	24 7/8	24 7/8	17 3/8	5	5	Roberts County	SD	Roland L. Hausmann	1959	311
155 1/8	25 2/8	24 7/8	19 3/8	6	6	Madison County	KY	Sonny Barker	1965	311
155 1/8	24 7/8	25 7/8	19 7/8	5	5	Calhoun County	MI	Steve D. Munier	1985	311
155 0/8	26 6/8	25 3/8	20 6/8	5	6	Polk County	IA	Dan Warren	1999	314
154 7/8	26 7/8	25 7/8	18 2/8	5	5	Murray County	MN	Craig Cohrs	1968	315
154 7/8	26 5/8	25 6/8	17 6/8	7	8	Pendleton County	KY	Thomas P. Jones	1969	315
154 7/8	24 4/8	25 7/8	18 4/8	7	6	Washtenaw County	MI	William G. Knight	1988	315
154 7/8	25 5/8	25 0/8	19 7/8	5	5	Massac County	IL	Terry B. Lewis	1990	315
154 7/8	25 2/8	25 4/8	18 5/8	7	6	Edmunds County	SD	Russell D. Leair	1992	315
154 7/8	26 0/8	26 0/8	18 3/8	5	5	Page County	IA	Bob Athen	1996	315
154 7/8	22 1/8	22 0/8	17 6/8	5	6	Custer County	MT	Scott L. Koelzer	2005	315
154 7/8	23 6/8	22 4/8	15 0/8	6	6	Cowley County	KS	Stephen C. Van Hoose	2007	315
154 7/8	25 6/8	25 5/8	17 4/8	7	8	Tuscarawas County	OH	Steve Archer	2009	315
154 6/8	24 1/8	22 5/8	17 2/8	5	5	Minnehaha County	SD	Clifford Sudenga	1962	324
154 6/8	25 1/8	21 7/8	18 4/8	6	4	Mills County	IA	Doug Roll	1985	324
154 6/8	25 6/8	26 5/8	16 2/8	9	7	Pike County	IL	Stan Chamberlain	1990	324
154 5/8	25 4/8	26 2/8	18 6/8	5	6	Jackson County	MN	Eugene C. La Maack	1955	327
154 4/8	24 0/8	24 3/8	23 2/8	5	5	Allamakee County	IA	Dayton Jones	1968	328
154 4/8	23 6/8	24 5/8	18 6/8	5	5	Davis County	IA	Ronald L. Simmons	1970	328
154 4/8	22 6/8	22 6/8	16 6/8	6	5	Fillmore County	MN	James J. Johnston	1974	328
154 4/8	24 6/8	25 0/8	18 4/8	5	5	Pittsburg County	OK	Blake Stephens	2000	328
154 3/8	23 4/8	24 3/8	18 5/8	7	7	Wythe County	VA	C. D. Tarter	1961	332
154 3/8	27 2/8	26 0/8	21 5/8	4	4	Martin County	IN	Bill Clark	1967	332
154 3/8	24 7/8	25 0/8	18 5/8	6	5	Page County	IA	Bob Athen	2008	332
154 2/8	25 4/8	27 0/8	16 6/8	6	6	Waupaca County	WI	Carl Schoenike	1948	335
154 2/8	26 2/8	26 0/8	15 4/8	4	5	Aitkin County	MN	Ervin A. Buck	1959	335
154 2/8	22 3/8	21 7/8	16 6/8	6	6	Clark County	MO	Myles Keller	1981	335
154 2/8	26 5/8	26 4/8	18 2/8	8	7	Coles County	IL	Rob King	1987	335
154 2/8	22 5/8	22 1/8	20 4/8	6	5	Laclede County	MO	Jim Atkinson	2006	335
154 1/8	24 6/8	23 7/8	15 5/8	5	5	Brown County	IN	Glen E. Parton	1970	340
154 1/8	22 6/8	24 7/8	17 5/8	5	5	Miami County	IN	Charles Wecht	1986	340
154 1/8	25 6/8	27 4/8	20 3/8	5	5	Shelby County	OH	Stephen A. Davis	1995	340
154 1/8	20 7/8	21 7/8	19 5/8	6	6	Page County	IA	Bob Athen	2004	340
154 0/8	26 5/8	26 5/8	19 1/8	4	5	Guthrie County	IA	Gordon Headlee	1977	344
154 0/8	23 0/8	22 0/8	17 4/8	5	5	Washington County	WI	Bob Lewis	2000	344
153 7/8	25 5/8	24 5/8	23 4/8	5	6	Lee County	IA	Chris Fowler	1980	346
153 7/8	24 4/8	25 4/8	19 7/8	6	6	Chase County	KS	Jerry Keller	1988	346
153 7/8	22 4/8	22 2/8	19 6/8	6	5	Decatur County	IA	Dave Sabrowsky	2011	346
153 6/8	22 1/8	22 7/8	17 0/8	5	5	Kent County	MD	S. Russell Edie	1966	349
153 6/8	24 5/8	25 1/8	18 2/8	6	5	Lee County	IL	George Nevins	1976	349
153 6/8	22 4/8	22 7/8	14 2/8	5	5	Kearney County	NE	Wayne R. Wall	1998	349
153 6/8	26 1/8	25 7/8	21 4/8	4	4	Carroll County	IL	Allen Comstock	2011	349
153 5/8	23 4/8	23 0/8	22 4/8	5	7	Harris	SAS	Garry Benson	1966	353
153 5/8	26 5/8	26 6/8	18 5/8	7	5	Hamilton County	IA	Harold Brown	1971	353
153 5/8	22 6/8	22 4/8	17 1/8	5	5	Wabasha County	MN	Tom Kothenbeutel	2007	353
153 4/8	24 0/8	23 6/8	17 4/8	5	5	Cass County	MO	Ronald Raines	2004	356
153 3/8	25 5/8	24 7/8	19 4/8	5	6	Grundy County	IL	Ed Vitko, Jr.	1972	357
153 3/8	22 7/8	22 7/8	20 6/8	7	7	Macoupin County	IL	Charles M. Woolfolk	1983	357
153 3/8	24 7/8	24 7/8	17 7/8	6	7	Calumet County	WI	Matt Fuchs	1986	357
153 3/8	25 1/8	25 2/8	19 1/8	4	4	McHenry County	IL	Roger A. Bacon	1992	357
153 3/8	23 6/8	23 4/8	19 7/8	6	6	Van Buren County	IA	Mark Story	1996	357
153 2/8	24 4/8	24 3/8	24 4/8	5	6	Burt County	NE	Harold W. Hawkins	1966	362
153 2/8	26 0/8	26 7/8	20 6/8	6	6	Winnebago County	IA	Ronald Gordon	1972	362
153 2/8	26 3/8	25 2/8	18 4/8	7	5	Butler County	KS	Larry Womack	1984	362
153 2/8	24 6/8	24 4/8	20 4/8	7	5	Edmonton	ALB	Dale Spooner	1991	362
153 2/8	25 5/8	25 7/8	18 7/8	6	5	Buffalo County	WI	Dave Hale	2006	362
153 1/8	25 2/8	25 0/8	18 5/8	4	5	Lincoln County	NE	Rich Birch	1965	367
153 1/8	25 4/8	25 2/8	19 3/8	5	5	Monroe County	NY	Dan Scorza	1990	367
153 1/8	24 0/8	24 2/8	17 7/8	5	5	Litchfield County	CT	Antonio E. Cacela	1995	367
153 0/8	25 0/8	25 2/8	17 4/8	5	5	Cambria County	PA	Andrew J. Getsy	1965	370
153 0/8	26 2/8	25 5/8	15 1/8	5	5	Saline County	NE	Don Kohout	1990	370
152 7/8	21 4/8	22 0/8	18 7/8	5	5	Polk County	WI	Wendle Johnson	1969	372

Score	Length of R Main Beam	L	Inside Spread	Number of R Points	L	Area	State/ Province	Hunter's Name	Date	Rank
152 7/8	26 0/8	25 7/8	20 2/8	6	5	Strafford County	NH	William S. Carlsen	1990	372
152 7/8	27 4/8	27 2/8	21 1/8	5	5	Lawrence County	OH	Bobby L. Willis	1991	372
152 7/8	24 3/8	24 3/8	16 6/8	6	6	Adams County	OH	Mike Hydler	2002	372
152 7/8	23 4/8	22 7/8	19 3/8	6	6	Weld County	CO	Wayne Depperschmidt	2011	372
152 6/8	25 6/8	25 1/8	21 2/8	4	5	Jefferson County	IN	Robert Schmidt	1959	377
152 6/8	23 2/8	23 6/8	18 2/8	5	6	Logan County	OH	Richard D. Fullerton	1988	377
152 6/8	26 3/8	28 3/8	22 3/8	4	5	Vermilion County	IL	John E. Fry	1989	377
152 6/8	23 4/8	23 2/8	17 7/8	6	5	Ohio County	IN	Gary Copeland	1991	377
152 6/8	24 1/8	24 7/8	18 2/8	5	5	Bear Lake	ALB	Stanley Alexander	2009	377
152 5/8	23 5/8	24 5/8	19 7/8	6	5	Ralls County	MO	Donald Curless	1959	382
152 5/8	23 2/8	23 0/8	17 2/8	5	6	Monroe County	MO	Carl T. Peak	1968	382
152 5/8	22 0/8	22 0/8	16 3/8	6	6	McPherson County	KS	Daniel G. Willems	1993	382
152 5/8	24 2/8	24 6/8	18 7/8	5	7	Adair County	MO	Anthony P. Mihalevich	2006	382
152 5/8	23 6/8	23 7/8	17 7/8	5	5	Ward County	ND	John R. Plesuk	2006	382
152 4/8	24 6/8	24 0/8	19 2/8	8	6	Hardin County	KY	Jimmy D. Neal	1973	387
152 4/8	28 7/8	26 2/8	19 2/8	5	5	Somerset County	MD	Burgess Blevins	1974	387
152 4/8	23 4/8	23 4/8	18 0/8	5	5	Harrison County	OH	Ron Fishel	1994	387
152 4/8	25 0/8	24 5/8	17 6/8	5	5	Fayette County	OH	Whitlow Wyatt	1996	387
152 3/8	24 7/8	24 7/8	18 5/8	8	6	Pope County	IL	Dr. H. Neil Becker	1968	391
152 2/8	27 6/8	27 3/8	15 5/8	6	6	Grundy County	IL	Wesley Holm	1971	392
152 2/8	23 1/8	23 1/8	18 4/8	5	6	Carroll County	IL	Allen G. Comstock	1997	392
152 2/8	24 7/8	24 1/8	19 1/8	5	6	Wabasha County	MN	Marc R. Hanson	1999	392
152 2/8	25 6/8	24 5/8	15 4/8	5	5	Black Hawk County	IA	Mike Judas	2010	392
152 1/8	21 4/8	24 2/8	21 3/8	5	4	Owen County	IN	Edward L. Armstrong	1967	396
152 1/8	25 6/8	26 1/8	20 3/8	6	4	Watonwan County	MN	David Raney	1972	396
152 1/8	24 3/8	24 0/8	20 7/8	4	5	Will County	IL	Robert L. Kamenjarin	1990	396
152 1/8	23 6/8	22 1/8	14 7/8	5	5	Waukesha County	WI	Scott A. Antczak	2011	396
152 0/8	23 0/8	23 0/8	16 0/8	5	5	Allen County	KY	Johnny Upton	1976	400
152 0/8	25 1/8	26 1/8	19 2/8	5	5	Athens County	OH	David A. Hawk	1994	400
152 0/8	23 7/8	23 7/8	18 6/8	5	6	Burleigh County	ND	William D. Helphrey	2001	400
152 0/8	21 6/8	21 5/8	15 6/8	6	7	Coles County	IL	Dan Milburn	2004	400
151 7/8	27 1/8	28 0/8	18 6/8	5	5	Morrison County	MN	Steve Smythe	1968	404
151 7/8	24 1/8	22 3/8	17 2/8	6	5	Lewis County	MO	Sam Smith	1986	404
151 7/8	26 4/8	25 4/8	20 1/8	7	6	Madison County	ID	Doug Kauer	2004	404
151 6/8	22 1/8	23 2/8	15 4/8	6	5	Flathead County	MT	Jerry D. Almos	1970	407
151 6/8	24 1/8	23 4/8	20 4/8	4	4	Bond County	IL	Allen D. Ellsworth	1987	407
151 6/8	23 2/8	23 4/8	19 4/8	6	5	Adams County	WI	Frank Knaus I	2004	407
151 5/8	26 2/8	24 7/8	20 6/8	6	6	Franklin County	KS	Kenneth Heinitz	1966	410
151 5/8	25 5/8	26 0/8	21 6/8	4	6	Jewell County	KS	Nathan Andersohn	2004	410
151 4/8	23 2/8	22 5/8	16 0/8	6	6	Gogebic County	MI	Fred Felbab	1959	412
151 4/8	23 2/8	22 7/8	16 2/8	5	5	Murray County	MN	Jim F. Wyffels	1968	412
151 4/8	23 6/8	22 5/8	17 3/8	6	5	Pittsburg County	OK	Harry Milican	1983	412
151 4/8	26 7/8	27 2/8	19 0/8	4	4	Pierce County	WI	Victor Howe	1988	412
151 4/8	23 7/8	23 2/8	16 0/8	5	5	Otero County	CO	Ron Rockwell	1990	412
151 4/8	26 3/8	26 4/8	21 1/8	7	4	Douglas County	KS	Robert Michel	1998	412
151 4/8	25 7/8	26 6/8	21 3/8	7	6	Licking County	OH	Gary Roser	2003	412
151 3/8	26 3/8	25 4/8	15 6/8	6	7	Martin County	IN	Robert E. Sloan	1968	419
151 3/8	23 2/8	23 7/8	19 1/8	5	5	Amherst County	VA	Jerry Armes	1975	419
151 3/8	26 2/8	26 3/8	18 1/8	4	4	Bayfield County	WI	Larry Fischer	1993	419
151 3/8	25 3/8	26 6/8	21 4/8	7	5	Boone County	IL	Dwight Gunderson	1994	419
151 3/8	25 5/8	25 6/8	20 3/8	4	4	Saskatoon Mtn.	ALB	Brent Watson	2011	419
151 2/8	25 5/8	25 6/8	17 2/8	4	4	Fillmore County	MN	David Carson	1967	424
151 2/8	24 1/8	23 4/8	14 1/8	7	5	Blue Earth County	MN	Rory Duetchman	1974	424
151 2/8	25 3/8	24 6/8	18 6/8	4	5	Litchfield County	CT	Donald R. Groody	1982	424
151 1/8	27 4/8	24 7/8	16 2/8	6	6	Adams County	IL	John C. Robinson	1967	427
151 1/8	22 3/8	23 4/8	18 4/8	7	5	Adair County	MO	Dr. Eddy Transano	1971	427
151 1/8	23 3/8	23 6/8	19 5/8	5	6	Okotoks	ALB	Dave Richardson	1989	427
151 1/8	23 3/8	23 1/8	17 7/8	6	7	Duval County	TX	Dean Oatman	1992	427
151 1/8	22 3/8	23 6/8	16 1/8	7	6	Chase County	KS	Jerry Keller	1999	427
151 0/8	25 0/8	26 0/8	16 6/8	5	5	Warrick County	IN	Gerald G. Taylor	1970	432
151 0/8	23 6/8	24 5/8	19 6/8	5	5	Polk County	WI	Bryan Anderson	1976	432
151 0/8	26 5/8	27 1/8	20 4/8	4	4	Appanoose County	IA	Steven P. Widmar	1986	432

WHITETAIL DEER [Typical Antlers]

Minimum Score 125 *Continued*

Score	Length of Main Beam R	Length of Main Beam L	Inside Spread	Number of Points R	Number of Points L	Area	State/ Province	Hunter's Name	Date	Rank
151 0/8	24 1/8	23 6/8	19 4/8	7	5	Clark County	OH	David L. Stull	1993	432
151 0/8	23 7/8	22 1/8	16 4/8	6	5	Dawson County	MT	Ray E. Peters	1997	432
151 0/8	21 5/8	22 0/8	16 4/8	5	5	Pike County	IL	Stan Chamberlain	1997	432
151 0/8	26 1/8	27 0/8	19 2/8	5	5	Dearborn County	IN	Dave Goodwin	2008	432
150 7/8	23 2/8	22 2/8	18 1/8	5	5	Trempealeau County	WI	Phillip Lunde	1969	439
150 7/8	23 0/8	22 2/8	20 5/8	6	6	Becker County	MN	Arnold F. Ostgarden	1987	439
150 7/8	24 4/8	23 6/8	17 0/8	5	6	Butler County	OH	Michael L. Gibbs	1992	439
150 7/8	21 4/8	21 1/8	20 1/8	7	7	Vermilion	ALB	Tim Pardely	2001	439
150 7/8	24 1/8	23 3/8	15 2/8	6	6	Dunn County	WI	Dave R. Lieffort	2004	439
150 6/8	22 3/8	22 0/8	16 4/8	5	5	Redwood County	MN	Irvin Plotz	1975	444
150 6/8	24 5/8	24 2/8	18 6/8	5	5	Butler County	MO	Steve Rueck	2001	444
150 5/8	21 0/8	22 7/8	19 3/8	6	5	Madison County	NE	Darwin Heppner	1967	446
150 5/8	22 7/8	22 6/8	22 1/8	5	6	Eaton County	MI	Greg Hoefler	1982	446
150 5/8	26 2/8	25 7/8	18 1/8	4	6	McHenry County	IL	Ken Stewart	2000	446
150 4/8	24 4/8	24 4/8	18 6/8	5	4	Scott County	IA	Howard A. Goettsch	1974	449
150 4/8	22 5/8	22 3/8	14 7/8	6	7	Wabasha County	MN	Lee Partington	1975	449
150 4/8	26 2/8	26 2/8	18 2/8	4	4	Pepin County	WI	Myles Keller	1985	449
150 4/8	23 1/8	22 5/8	17 2/8	5	5	Ravalli County	MT	Mark Moreland	1991	449
150 4/8	23 2/8	23 2/8	15 2/8	5	5	Chautauqua County	KS	Daniel L. Hubert	1997	449
150 3/8	24 5/8	24 5/8	22 5/8	4	4	Oneida County	WI	Philip Hildebrand	1967	454
150 3/8	28 0/8	27 4/8	20 5/8	4	5	Jones County	GA	Sid Hester	1969	454
150 3/8	20 4/8	21 2/8	16 7/8	5	5	Langlade County	WI	Herbert Buettner	1970	454
150 3/8	23 3/8	23 5/8	18 3/8	6	7	Graham County	KS	Russell Hull	1977	454
150 3/8	26 1/8	26 1/8	16 6/8	7	5	Jackson County	OH	Robert E. Thomas, Jr.	1987	454
150 3/8	22 3/8	22 3/8	18 1/8	6	6	Wayne County	IA	Jake Taylor	2002	454
150 3/8	26 0/8	25 0/8	17 5/8	5	8	Bremer County	IA	Dave MacDonald	2004	454
150 2/8	23 5/8	23 3/8	17 2/8	6	6	Jackson County	WI	Roger Reinart	1972	461
150 2/8	21 4/8	21 5/8	15 2/8	5	5	Bremer County	IA	Rod Heidemann	1990	461
150 2/8	24 1/8	23 7/8	19 2/8	5	5	Jefferson County	WI	Mark Zimmerman	1998	461
150 2/8	25 3/8	24 3/8	18 4/8	5	5	Jersey County	IL	Tom Price	2011	461
150 1/8	24 3/8	23 4/8	21 1/8	6	6	Luzerne County	PA	Edward Prutzman	1953	465
150 1/8	23 5/8	23 0/8	16 7/8	5	6	Johnson County	IA	Jim Keefer	1969	465
150 1/8	24 4/8	24 3/8	20 1/8	7	8	Lake County	IL	James C. Carlson	1985	465
150 1/8	26 6/8	24 0/8	20 3/8	4	6	Mercer County	OH	Hud Baker	1999	465
150 0/8	25 3/8	24 3/8	15 6/8	5	6	Adams County	WI	Daniel Becker	1957	469
150 0/8	26 6/8	27 1/8	17 6/8	5	4	Benton County	IA	Robert L. Walker	1961	469
150 0/8	26 4/8	26 4/8	18 4/8	5	5	Shelby County	IA	Einar Leistad	1970	469
150 0/8	24 0/8	23 6/8	18 1/8	5	7	Lake County	IL	James D. Maricle	1989	469
150 0/8	24 2/8	24 3/8	22 0/8	5	5	Cook County	IL	F. Dan Dinelli	1997	469
150 0/8	23 6/8	23 7/8	15 0/8	5	6	Holt County	MO	Juston W. Carr	1997	469
149 7/8	26 0/8	25 6/8	20 3/8	4	4	Pulaski County	VA	Ray S. Carter	1962	475
149 7/8	21 6/8	21 1/8	17 3/8	5	5	Adams County	IL	Lyndall W. Heyen	1967	475
149 7/8	25 0/8	25 0/8	17 5/8	5	5	Pittsburg County	OK	Doug Larimer	1985	475
149 7/8	23 1/8	22 6/8	18 5/8	6	5	Maverick County	TX	H. Mike Palmer	2005	475
149 7/8	23 5/8	23 3/8	18 3/8	6	5	Will County	IL	Thomas M. Stepanek	2005	475
149 6/8	23 4/8	23 5/8	17 1/8	7	6	Gull Lake	SAS	Keith Roney	1978	480
149 6/8	25 2/8	25 3/8	19 0/8	6	5	McHenry County	IL	Joe Florent	1995	480
149 6/8	22 7/8	23 1/8	18 5/8	6	6	Stephenson County	IL	Greg Wedig	1998	480
149 6/8	24 2/8	23 5/8	19 4/8	5	4	Bremer County	IA	David L. MacDonald	2001	480
149 6/8	22 1/8	22 5/8	18 1/8	6	6	Berks County	PA	Aaron B. Ogden	2009	480
149 6/8	26 4/8	26 4/8	17 7/8	5	5	Stearns County	MN	John E. Kline	2009	480
149 5/8	25 6/8	25 3/8	18 5/8	6	6	La Salle County	IL	Timothy L. Kakara	1975	486
149 5/8	24 7/8	25 2/8	15 4/8	7	5	Howell County	MO	Thomas E. French	1992	486
149 5/8	22 4/8	23 2/8	17 6/8	5	6	Lyon County	IA	Paul E. Blotz	1996	486
149 5/8	23 7/8	24 0/8	19 5/8	5	5	Black Hawk County	IA	Michael Judas	1999	486
149 4/8	23 3/8	26 0/8	20 2/8	5	5	Woodbury County	IA	Don Gothier	1962	490
149 4/8	25 3/8	25 2/8	18 7/8	8	8	Westchester County	NY	Steve Solomon	1963	490
149 4/8	24 2/8	23 6/8	18 6/8	5	4	Clayton County	IA	Joseph C. Hinderman	1991	490
149 4/8	25 2/8	26 0/8	16 5/8	6	6	Ross County	OH	David L. Buckler	1996	490
149 4/8	24 4/8	24 6/8	18 4/8	5	4	Logan County	CO	Janet L. George	1996	490
149 3/8	26 6/8	25 4/8	22 1/8	4	4	Westchester County	NY	George Ferber	1957	495
149 3/8	23 4/8	23 3/8	15 7/8	5	5	Madison County	KY	Robert Young	1963	495

WHITETAIL DEER [Typical Antlers]

Minimum Score 125 *Continued*

Score	Length of R Main Beam	L	Inside Spread	Number of R Points	L	Area	State/ Province	Hunter's Name	Date	Rank
149 3/8	23 4/8	24 4/8	19 3/8	5	5	Newton County	IN	Philip Kozlowski	1968	495
149 3/8	23 7/8	23 2/8	15 7/8	5	5	Buffalo County	WI	Daniel J. Brunner	1969	495
149 3/8	24 1/8	24 2/8	19 3/8	5	5	Lawrence County	OH	Ronald E. Burnette	1971	495
149 3/8	23 7/8	23 4/8	17 5/8	5	5	Waushara County	WI	Ronald Anunson	1974	495
149 3/8	24 0/8	25 2/8	17 4/8	5	4	Washington County	OH	Charles E. Vaughan	1983	495
149 3/8	26 0/8	25 5/8	19 6/8	5	6	Monona County	IA	Richard Kelly	1990	495
149 2/8	25 3/8	24 6/8	20 4/8	5	5	Westchester County	NY	Joseph H. Keeler	1957	503
149 2/8	22 0/8	22 1/8	16 6/8	5	5	Will County	IL	Daniel R. Altiery	1973	503
149 2/8	24 5/8	24 5/8	17 4/8	5	5	Montgomery County	NC	Ricky E. Jennings	1995	503
149 2/8	26 2/8	26 2/8	21 5/8	6	5	Branch County	MI	Gregory J. Raatz	1997	503
149 2/8	22 3/8	22 4/8	16 6/8	5	5	Greene County	IN	Peter M. Sartoris	2001	503
149 1/8	24 0/8	24 4/8	18 5/8	5	4	Lancaster County	NE	Bryan Bjorkman	1998	508
149 0/8	25 7/8	23 5/8	19 4/8	6	5	Mitchell County	IA	Elmer Krueger	1961	509
149 0/8	25 0/8	25 0/8	19 4/8	5	4	Black Hawk County	IA	Robert Riggle	1975	509
149 0/8	24 6/8	25 1/8	19 2/8	6	5	Warren County	IA	Grant Poindexter	1981	509
149 0/8	23 4/8	22 4/8	15 2/8	7	6	Kenosha County	WI	Alois Jeske, Sr.	1987	509
149 0/8	25 7/8	24 4/8	19 6/8	5	5	Hardin County	IA	William Stonebraker	1988	509
149 0/8	23 3/8	22 3/8	16 4/8	5	5	Duval County	TX	F. H. Becker	1997	509
148 7/8	15 4/8	26 0/8	19 4/8	5	7	Lincoln County	WI	Ronald Pond	1961	515
148 7/8	24 6/8	24 0/8	18 7/8	4	5	Grant County	SD	Larry Turbak	1964	515
148 7/8	23 1/8	23 1/8	18 4/8	6	6	Knox County	OH	John E. Bumpus	1971	515
148 7/8	24 3/8	22 7/8	20 3/8	6	6	Caledon Township	ONT	Jack Leggo	1988	515
148 7/8	24 1/8	24 6/8	19 7/8	6	6	Dane County	WI	Matthew A. Shimniok	1991	515
148 7/8	24 6/8	24 5/8	19 3/8	4	5	Hillsdale County	MI	Terry J. Gerber	1994	515
148 7/8	25 4/8	25 4/8	17 7/8	12	10	Grant County	IN	Kevin Kidwell	1997	515
148 7/8	23 4/8	24 1/8	18 1/8	5	5	Somerset County	NJ	Spiro J. Stilianessis	1998	515
148 7/8	26 0/8	26 3/8	19 6/8	4	5	Madison County	IL	Jerry Malone	2003	515
148 6/8	25 1/8	25 7/8	18 0/8	4	5	Lincoln County	MO	Robert Remmert, Sr.	1960	524
148 6/8	25 2/8	25 2/8	19 6/8	5	5	Iowa County	IA	Russ Sill	1967	524
148 6/8	23 0/8	23 0/8	18 4/8	5	5	Jefferson County	IN	Jim Coldiron	1969	524
148 6/8	21 3/8	21 1/8	19 1/8	6	6	Brown County	SD	Bill Franklin	1972	524
148 6/8	22 6/8	22 6/8	16 6/8	5	6	Reno County	KS	Davis J. Ediger	1982	524
148 6/8	25 0/8	25 0/8	22 0/8	8	5	Jasper County	IA	William E. Webster	1987	524
148 6/8	22 3/8	22 1/8	17 7/8	7	6	Winona County	MN	Tom Kothenbeutel	1988	524
148 6/8	25 0/8	25 5/8	20 6/8	4	6	Lee County	IL	Dale Dye	2004	524
148 5/8	21 7/8	22 0/8	14 7/8	5	6	Pulaski County	MO	Ron Poston	1972	532
148 5/8	23 3/8	23 6/8	16 3/8	6	6	Iroquois County	IL	Howard Brady	1994	532
148 5/8	24 3/8	24 4/8	17 2/8	5	6	Preble County	OH	Stanley Jennings	2000	532
148 5/8	24 4/8	24 2/8	20 1/8	8	5	Ogle County	IL	Michael K. Jones	2001	532
148 5/8	23 6/8	23 3/8	18 1/8	5	4	Will County	IL	Thomas S. Spence	2004	532
148 5/8	23 3/8	21 2/8	16 2/8	6	6	Carroll County	IL	Jim Carlson	2007	532
148 4/8	23 0/8	22 4/8	20 1/8	4	5	Murray County	MN	Mike Molitor	1968	538
148 4/8	26 0/8	27 5/8	19 0/8	4	4	Orange County	NC	Todd McDonald	1988	538
148 4/8	22 4/8	22 4/8	19 2/8	5	5	Baltimore County	MD	Bruce Hoover	1992	538
148 4/8	25 1/8	25 1/8	16 2/8	5	5	Lake County	IL	Jim Tahaney	1996	538
148 4/8	20 6/8	22 0/8	15 3/8	5	6	Van Buren County	IA	Joel M. Riotto	2006	538
148 3/8	25 6/8	26 7/8	22 2/8	9	5	Scott County	MN	Bob Gregory	1967	543
148 3/8	22 1/8	22 6/8	18 1/8	5	5	Morrison County	MN	Robert Redmann	1984	543
148 3/8	27 0/8	26 0/8	18 1/8	9	6	Webster County	IA	Terry Rial	1992	543
148 3/8	23 4/8	24 4/8	19 7/8	5	5	Marshall County	KS	Terry Gunn	1995	543
148 3/8	22 0/8	22 0/8	14 3/8	6	8	Jefferson County	KS	Casey Vangemern	2011	543
148 2/8	22 3/8	21 6/8	16 2/8	5	5	Walworth County	SD	Irvin Guthmiller	1956	548
148 2/8	22 7/8	21 5/8	18 2/8	5	5	Harrison County	IA	James W. Glasscock	1968	548
148 2/8	25 1/8	24 1/8	15 4/8	5	5	Woodbury County	IA	Guy Hempey	1969	548
148 2/8	26 5/8	24 3/8	19 7/8	6	8	Coshocton County	OH	Charles N. McDonald	1977	548
148 2/8	23 0/8	24 1/8	21 4/8	6	5	Clay County	MN	James F. Thompson	1987	548
148 2/8	26 0/8	26 3/8	18 3/8	4	5	Lake County	IL	Donald R. Powers	1990	548
148 2/8	24 4/8	23 5/8	16 5/8	6	5	Coles County	IL	Jim Eveland	1995	548
148 2/8	25 0/8	25 0/8	17 2/8	5	5	Pike County	IL	Don Dye, Jr.	1995	548
148 2/8	23 0/8	23 4/8	18 2/8	5	5	Houston County	GA	David Testerman	1999	548
148 2/8	25 6/8	25 6/8	18 4/8	6	6	Lake County	IL	Mark Nelsen	1999	548
148 1/8	23 0/8	22 5/8	17 2/8	5	6	Wellington County	ONT	Barry Marshall	1975	558

WHITETAIL DEER [Typical Antlers]

Minimum Score 125 *Continued*

Score	Length of Main Beam R	Length of Main Beam L	Inside Spread	Number of Points R	Number of Points L	Area	State/ Province	Hunter's Name	Date	Rank
148 1/8	20 4/8	21 3/8	17 1/8	5	5	Van Buren County	IA	David M. Ackland, Jr.	1999	558
148 1/8	24 0/8	22 3/8	18 0/8	5	7	Macon County	AL	Norman Parks	2000	558
148 1/8	23 2/8	23 0/8	16 5/8	5	5	Mountrail County	ND	James Domaskin	2000	558
148 1/8	23 2/8	22 4/8	20 5/8	5	5	Carroll County	IL	James Carlson	2002	558
148 0/8	24 5/8	25 2/8	16 2/8	5	5	McDowell County	WV	Dennis Mahaffey, Sr.	1998	563
148 0/8	22 4/8	22 7/8	17 6/8	5	5	Gage County	NE	Larry G. Lottman	2005	563
148 0/8	23 6/8	23 4/8	22 4/8	6	5	Cass County	IN	Dempsey Gross	2007	563
148 0/8	24 5/8	25 3/8	17 7/8	7	6	Dubuque County	IA	Michael Casey	2011	563
147 7/8	25 7/8	26 3/8	19 5/8	4	4	Lake County	IL	Allen G. Comstock	1990	567
147 7/8	22 6/8	23 2/8	17 5/8	5	6	Jackson County	OH	Edsel D. Duty	1993	567
147 7/8	21 5/8	20 7/8	15 3/8	5	6	Waupaca County	WI	Ronald L. Lutzewitz	1997	567
147 6/8	23 4/8	23 2/8	16 0/8	5	5	Morrison County	MN	John Zwickey	1955	570
147 6/8	23 5/8	24 3/8	16 6/8	6	7	Morrison County	MN	Allen E. Farmes	1957	570
147 6/8	25 3/8	24 5/8	19 7/8	6	5	Jo Daviess County	IL	Jerry Fritz	1971	570
147 6/8	24 1/8	25 6/8	19 0/8	5	5	Dickinson County	MI	Myles Keller	1976	570
147 6/8	24 0/8	24 2/8	17 2/8	6	5	Harlan County	NE	Ron Breitsprecher	1978	570
147 6/8	25 2/8	25 6/8	22 2/8	4	4	Delaware County	OH	Doug Fuller	1991	570
147 6/8	26 4/8	26 2/8	19 5/8	6	7	Clinton County	OH	Kevin L. Wilson	1991	570
147 6/8	24 6/8	25 0/8	20 3/8	5	6	Sherburne County	MN	Richard A. Iano	2000	570
147 6/8	26 1/8	25 3/8	20 3/8	7	5	Tallahatchie County	MS	Jim Kennedy	2009	570
147 5/8	23 7/8	23 5/8	19 1/8	5	5	Fillmore County	MN	Dale Honsey	1951	579
147 5/8	24 2/8	26 0/8	20 2/8	6	6	Nicollet County	MN	Thomas J. Merkley	1970	579
147 5/8	22 7/8	23 3/8	17 2/8	6	5	Van Buren County	IA	Elmer Luce, Jr.	1995	579
147 5/8	23 0/8	21 6/8	18 1/8	5	6	Geauga County	OH	Michael Brent	1998	579
147 4/8	22 2/8	22 6/8	18 4/8	4	4	Shawano County	WI	John Schoenike	1960	583
147 4/8	25 5/8	24 0/8	21 0/8	4	6	Montgomery County	AL	Rett Kelly	1974	583
147 4/8	23 6/8	24 4/8	17 4/8	5	5	Henry County	IA	Bruce Barrie	1990	583
147 4/8	26 5/8	27 2/8	18 3/8	5	5	McLean County	IL	David Grizzle	1997	583
147 4/8	22 3/8	21 4/8	18 1/8	6	5	Dunn County	ND	Corey Mueller	2003	583
147 4/8	24 2/8	23 7/8	18 4/8	5	5	Fond du Lac County	WI	Steve Weber	2004	583
147 3/8	25 7/8	26 0/8	18 1/8	5	5	Franklin County	VA	Mike Weaver	1978	589
147 3/8	25 6/8	26 1/8	20 1/8	5	5	Cuyahoga County	OH	Mark Johnson	1994	589
147 3/8	25 0/8	24 2/8	16 7/8	5	5	Wayne County	IN	Stephen Smitley	2000	589
147 3/8	25 7/8	25 6/8	18 1/8	4	4	Riley County	KS	John Duggan	2002	589
147 2/8	23 2/8	25 1/8	20 4/8	9	8	Clay County	IA	Uriah M. Hostetler	1964	593
147 2/8	23 2/8	22 4/8	19 2/8	5	5	Madison County	IN	Pat Moreland	1969	593
147 2/8	23 1/8	23 1/8	15 6/8	5	5	Green Lake County	WI	Don Chier	1973	593
147 2/8	21 5/8	21 5/8	18 6/8	5	5	Fergus County	MT	Mike Sweeney	1992	593
147 2/8	25 1/8	24 6/8	17 6/8	5	4	Lawrence County	MO	Jason Graff	1994	593
147 2/8	23 4/8	23 4/8	17 0/8	6	5	Lincoln County	WI	Joel Wendt	1998	593
147 2/8	21 5/8	21 7/8	16 1/8	6	7	St. Marys County	MD	Pete Ropshaw	2008	593
147 1/8	25 1/8	25 3/8	16 7/8	5	5	Delaware County	IA	Blair Berens	1963	600
147 1/8	24 7/8	24 4/8	18 4/8	7	7	Red Willow County	NE	Gary Ginther	1967	600
147 1/8	22 6/8	23 1/8	17 5/8	5	6	Iron County	WI	Dr. C. J. Rainaldo	1967	600
147 1/8	21 0/8	21 5/8	16 3/8	6	6	Vilas County	WI	Anthony J. Sahulcik, Jr.	1969	600
147 1/8	25 0/8	24 0/8	18 3/8	4	4	Dodge County	MN	Clark Gallup	1974	600
147 1/8	24 5/8	24 5/8	18 2/8	7	5	Burleigh County	ND	Chuck Welch	1986	600
147 1/8	23 3/8	24 0/8	19 5/8	5	4	Lake County	IL	Robert A. Turner	1989	600
147 1/8	20 7/8	21 7/8	15 7/8	6	4	Dubuque County	IA	Joseph Hinderman	2007	600
147 0/8	24 6/8	25 5/8	17 4/8	5	7	Dodge County	WI	Alex B. Feucht	1963	608
147 0/8	28 5/8	26 2/8	19 0/8	5	4	Fulton County	IL	Bernard Smith	1974	608
147 0/8	24 3/8	24 7/8	19 2/8	7	7	Dodge County	NE	Donald W. Robinson	1974	608
147 0/8	23 5/8	23 5/8	20 0/8	5	5	Allamakee County	IA	Don Kieler	1979	608
147 0/8	25 1/8	25 2/8	19 6/8	7	6	Charles County	MD	David G. Wilson	1981	608
147 0/8	22 5/8	21 3/8	16 4/8	5	5	Ogle County	IL	Roy Nichols	1996	608
147 0/8	23 1/8	23 4/8	15 2/8	6	6	Des Moines County	IA	Robert J. Lewis	1996	608
147 0/8	21 4/8	21 5/8	15 3/8	6	7	Doniphan County	KS	Stephen D. Wolfram	1997	608
147 0/8	24 5/8	24 4/8	16 2/8	5	5	Ross County	OH	John M. Hinkle, Jr.	1999	608
147 0/8	22 1/8	22 1/8	16 0/8	5	5	Mountrail County	ND	Craig Richardson	2004	608
146 7/8	24 0/8	24 3/8	22 1/8	5	7	Cass County	ND	Duane H. Olsen	1959	618
146 7/8	21 5/8	21 0/8	16 3/8	5	5	Buffalo County	NE	Dwight Bond	1970	618
146 7/8	22 2/8	21 2/8	19 7/8	5	5	Waushara County	WI	Norman A. Moss	1976	618

WHITETAIL DEER [Typical Antlers]

Minimum Score 125 *Continued*

Score	Length of R Main Beam	L	Inside Spread	Number of R Points	L	Area	State/ Province	Hunter's Name	Date	Rank
146 7/8	23 4/8	24 4/8	17 1/8	5	6	Augusta Township	ONT	Henry P. Bouchard	1991	618
146 7/8	24 1/8	24 4/8	16 7/8	5	5	Hardin County	IA	William Stonebraker	1994	618
146 7/8	25 3/8	25 3/8	16 2/8	5	5	Winona County	MN	Jeffrey R. Fechner	2000	618
146 6/8	23 6/8	24 1/8	17 0/8	5	4	Allegany County	MD	James Wilhelm	1961	624
146 6/8	23 3/8	23 1/8	17 6/8	7	8	Roberts County	SD	Robert Hendren	1967	624
146 6/8	24 6/8	27 0/8	20 2/8	5	4	Neosho County	KS	Carl Walker	1968	624
146 6/8	26 1/8	25 3/8	18 1/8	4	5	Lee County	IA	Jim Bohnenkamp	1970	624
146 6/8	25 6/8	25 5/8	18 0/8	5	5	Shelby County	IL	Gary E. Sievers	1972	624
146 6/8	20 2/8	20 4/8	15 1/8	6	5	Pulaski County	GA	Dan B. Clifton	1989	624
146 6/8	21 7/8	22 2/8	17 0/8	5	5	Sterling County	TX	Mike Belanger	1992	624
146 6/8	23 5/8	23 2/8	18 4/8	5	4	Breckinridge County	KY	R. Ray Wix	1993	624
146 6/8	20 5/8	21 2/8	17 0/8	5	5	Sheridan County	WY	George K. Warner	1994	624
146 6/8	25 6/8	24 7/8	18 0/8	5	5	Hillsdale County	MI	Gale E. Pauken	1996	624
146 6/8	21 0/8	25 0/8	21 6/8	5	6	Page County	IA	Bob Athen	2006	624
146 6/8	24 7/8	24 5/8	16 7/8	5	6	Pierce County	WI	Nate Gruber	2008	624
146 5/8	22 5/8	22 5/8	18 7/8	4	3	Brown County	SD	Donald Grote	1963	636
146 5/8	23 2/8	24 1/8	17 3/8	5	6	Buckingham County	VA	Larry D. Baker	1974	636
146 5/8	24 7/8	24 1/8	17 7/8	5	4	Hartford County	CT	Peter J.M. Kiendzoir	1986	636
146 5/8	25 5/8	23 6/8	18 1/8	6	5	Ontario County	NY	Robert D. Koutras	1992	636
146 5/8	22 6/8	23 0/8	18 3/8	5	5	Trempealeau County	WI	David Mikrut	1994	636
146 5/8	22 5/8	21 3/8	16 2/8	6	7	Van Buren County	IA	Russ Miller	1996	636
146 5/8	25 4/8	26 3/8	22 2/8	5	6	Montgomery County	OH	Dale Shepard	1997	636
146 5/8	24 5/8	24 1/8	18 3/8	4	4	Fayette County	WV	David Kinningham	1998	636
146 5/8	22 3/8	22 4/8	16 7/8	5	6	Bayfield County	WI	Jason Yarwood	2003	636
146 5/8	21 2/8	21 7/8	16 3/8	5	5	Monroe County	OH	Bob Cover	2005	636
146 4/8	22 0/8	22 1/8	17 4/8	5	5	Wood County	WI	George Davis	1967	646
146 4/8	24 4/8	24 0/8	15 0/8	5	5	Amherst County	VA	Garry B. Pruitt	1972	646
146 4/8	22 0/8	23 3/8	18 0/8	5	5	Vigo County	IN	Richard E. Smith	1973	646
146 4/8	23 6/8	23 5/8	19 2/8	4	5	Vilas County	WI	Eugene A. Pribek	1994	646
146 4/8	23 4/8	23 6/8	17 6/8	6	8	Muskingum County	OH	Bryan Cooper	1997	646
146 4/8	25 0/8	26 1/8	19 4/8	4	4	Frederick County	MD	Edward Reilly	2000	646
146 4/8	24 4/8	24 4/8	21 0/8	4	5	Richland County	OH	Keith D. Levendorf	2003	646
146 4/8	22 6/8	22 2/8	20 3/8	6	5	Marquette County	WI	Arlyn Johnson	2009	646
146 3/8	23 5/8	23 1/8	27 1/8	5	5	De Kalb County	IN	Stanley Bremer	1969	654
146 3/8	21 4/8	22 2/8	15 4/8	6	5	Oakland County	MI	Dwight D. Paslean	1997	654
146 3/8	23 1/8	23 0/8	16 3/8	6	5	Brown County	IN	Gordon Braun	2006	654
146 2/8	24 3/8	24 1/8	18 2/8	5	5	Stearns County	MN	Mike Beuning	1945	657
146 2/8	24 6/8	24 1/8	18 2/8	5	5	Butler County	KS	David R. Rogers	1976	657
146 2/8	22 6/8	22 0/8	17 5/8	5	6	Dodge County	MN	David Lyke	1985	657
146 2/8	24 3/8	24 5/8	19 1/8	5	6	Redwood County	MN	R. Tetrick/M. Tetrick	1986	657
146 2/8	24 1/8	25 6/8	20 6/8	6	5	Lake County	IL	Kenneth D. Staples	1986	657
146 2/8	23 0/8	23 4/8	15 2/8	6	7	Ferry County	WA	Antonio P. Bamba	1998	657
146 2/8	23 4/8	23 7/8	18 4/8	4	4	Trumbull County	OH	Michael Mace	1999	657
146 1/8	24 1/8	25 0/8	18 3/8	5	5	Lamar County	GA	Joe A. Medcalf	1977	664
146 1/8	25 0/8	24 0/8	22 3/8	5	6	Queen Annes County	MD	Ross F. Mills	1990	664
146 1/8	23 5/8	24 0/8	17 5/8	5	5	Dane County	WI	Todd R. Zeuske	1999	664
146 1/8	26 4/8	25 6/8	18 2/8	6	5	Lake County	IL	Carl H. Spaeth	2002	664
146 0/8	22 1/8	22 2/8	19 2/8	6	6	Richland County	MT	James L. Kelly	1958	668
146 0/8	21 1/8	20 6/8	17 5/8	5	6	Cass County	IN	William D. Finks	1967	668
146 0/8	24 0/8	25 0/8	18 4/8	5	5	Will County	IL	Terry Marcukaitis	1971	668
146 0/8	25 0/8	24 0/8	22 0/8	5	5	Winona County	MN	Henry Scharmack,Jr.	1973	668
146 0/8	20 5/8	20 5/8	14 3/8	7	8	Murray County	MN	John Stenke	1973	668
146 0/8	21 7/8	23 1/8	16 4/8	5	6	Plymouth County	IA	Robert J. Pierson	2001	668
146 0/8	23 0/8	24 1/8	21 1/8	5	6	Jo Daviess County	IL	Mike Feldermann	2011	668
145 7/8	25 1/8	25 0/8	16 3/8	4	5	Wood County	WI	Laddimere Beranek	1959	675
145 7/8	25 4/8	24 4/8	19 1/8	4	5	Greeley County	NE	Bill W. Surface	1962	675
145 7/8	22 4/8	22 7/8	20 7/8	4	8	Lafayette County	WI	James Goetzke	1972	675
145 7/8	25 1/8	26 3/8	22 1/8	3	3	Steuben County	IN	Jim Loughran	1992	675
145 6/8	20 0/8	20 5/8	17 0/8	5	6	Okmulgee County	OK	Pat Giulioli	1973	679
145 6/8	22 3/8	23 2/8	16 6/8	4	5	Jefferson County	KS	John Welborn	1981	679
145 6/8	23 3/8	24 0/8	18 0/8	7	6	Van Buren County	IA	Don Kieler	1987	679
145 6/8	23 4/8	24 0/8	17 0/8	6	5	Jasper County	IL	Elmer R. Luce, Jr.	1991	679

Minimum Score 125 *Continued*

Score	Length of Main Beam R	Length of Main Beam L	Inside Spread	Number of Points R	Number of Points L	Area	State/ Province	Hunter's Name	Date	Rank
145 6/8	21 4/8	21 4/8	17 2/8	5	5	Plymouth County	IA	Cash N. Howe	1991	679
145 6/8	24 7/8	24 2/8	18 2/8	4	4	Franklin County	OH	Gregory J. Rustemeyer	1995	679
145 6/8	23 4/8	22 5/8	20 1/8	5	6	Ashtabula County	OH	Larry M. Knapp	2004	679
145 6/8	25 4/8	25 6/8	16 2/8	8	5	Vinton County	OH	Matt Bethel	2009	679
145 6/8	22 3/8	22 2/8	16 5/8	5	6	Knox County	IL	Don Owen	2009	679
145 5/8	25 1/8	24 4/8	19 7/8	6	5	Waseca County	MN	Robert Barrie	1971	688
145 5/8	22 5/8	21 5/8	17 4/8	6	6	Moody County	SD	Harvey R. Benton	1976	688
145 5/8	24 2/8	25 3/8	19 7/8	4	4	McHenry County	IL	Joe Roos	1991	688
145 5/8	25 0/8	25 4/8	18 2/8	5	5	Fulton County	IL	Patrick Cebuhar	1994	688
145 5/8	24 1/8	24 2/8	15 7/8	5	5	Codington County	SD	Troy A. Richardson	1995	688
145 5/8	24 4/8	23 5/8	18 3/8	5	5	Van Buren County	IA	Christopher W. Fechner	2004	688
145 4/8	23 3/8	24 0/8	17 2/8	6	7	Morton County	ND	Eddy Wallery	1959	694
145 4/8	25 2/8	24 1/8	18 6/8	5	5	Waupaca County	WI	Craig Shambeau	1968	694
145 4/8	25 5/8	24 4/8	19 0/8	4	5	Monroe County	WI	Larry Arentz	1977	694
145 4/8	23 2/8	22 0/8	18 2/8	7	6	Mower County	MN	Walter E. Bauer	1977	694
145 4/8	24 4/8	24 6/8	17 2/8	4	4	Ravalli County	MT	Chris Landstrom	1988	694
145 4/8	22 5/8	23 4/8	16 3/8	5	6	Appanoose County	IA	Steven P. Salmieri	1991	694
145 4/8	25 6/8	25 5/8	16 2/8	5	5	Meade County	KY	Aaron H. Pierce	1992	694
145 4/8	23 4/8	23 4/8	16 0/8	5	6	Chase County	KS	Tony Cassity	1997	694
145 4/8	25 0/8	24 3/8	18 2/8	5	5	Decatur County	IA	Royce DeCook	2005	694
145 4/8	25 4/8	25 0/8	14 6/8	6	5	Weld County	CO	Nathan Andersohn	2005	694
145 4/8	22 5/8	23 6/8	16 6/8	5	5	Pulaski County	MO	Aaron M. C. Cisco	2010	694
145 3/8	24 3/8	25 1/8	20 5/8	5	5	Lyon County	KS	Edward Bess	1967	705
145 3/8	22 4/8	22 5/8	18 1/8	5	5	Morrison County	MN	Gerald A. Young	1971	705
145 3/8	23 4/8	24 5/8	16 3/8	7	5	Montgomery County	MD	James C. Dalrymple, Jr.	1988	705
145 2/8	21 6/8	21 3/8	18 2/8	5	5	Des Moines County	IA	Gary Biles	1973	708
145 2/8	22 5/8	22 7/8	17 2/8	5	5	Yankton County	SD	Gordon Orton	1976	708
145 2/8	23 7/8	24 4/8	17 5/8	5	5	Livingston County	MI	Alan K. Newberry	1979	708
145 2/8	23 5/8	24 7/8	18 4/8	6	5	Lincoln County	MT	Michael F. Shepard	1987	708
145 2/8	23 1/8	23 6/8	16 3/8	6	4	Lawrence County	IL	Lary Caddell	1988	708
145 2/8	24 2/8	23 5/8	16 2/8	5	6	Wood County	OH	George Scott Burwell	2010	708
145 1/8	21 4/8	21 4/8	19 7/8	5	5	Valley County	MT	Gene Henck	1993	714
145 1/8	22 5/8	22 7/8	18 3/8	5	5	Delaware County	IA	Kelly Salow	1996	714
145 1/8	26 2/8	27 1/8	20 4/8	4	5	Des Moines County	IA	Jeff Wilson	1997	714
145 1/8	22 2/8	22 7/8	18 3/8	7	6	Poplarfield	MAN	David A. Shupp	2000	714
145 1/8	23 2/8	23 0/8	15 0/8	4	6	Comanche County	KS	Fred Eichler	2005	714
145 1/8	21 7/8	22 3/8	16 4/8	7	6	Walsh County	ND	Gary Monsebroten	2006	714
145 0/8	22 5/8	21 2/8	14 2/8	6	6	Washington County	IA	Doron Whitlock	1966	720
145 0/8	24 0/8	24 5/8	20 0/8	4	4	Grant County	SD	Kevin Bronson	1973	720
145 0/8	21 5/8	22 4/8	15 4/8	5	5	Jones County	IA	Jim H. Dougherty	1979	720
145 0/8	23 0/8	23 4/8	19 1/8	9	5	St. Joseph County	IN	Monty Layne	1984	720
144 7/8	23 0/8	23 4/8	19 7/8	6	6	Wyandotte County	KS	George F. Bigelow	1966	724
144 7/8	21 7/8	22 3/8	17 1/8	5	6	Sussex County	VA	Alvin D. Skinner	1972	724
144 7/8	20 2/8	20 1/8	15 2/8	6	5	Richland County	MT	Garth N. Kallevig	1980	724
144 7/8	21 7/8	21 2/8	18 7/8	6	5	Texas County	OK	Max Crocker	1985	724
144 7/8	22 4/8	21 4/8	16 6/8	6	5	Lake County	IL	Carl H. Spaeth	1987	724
144 7/8	24 2/8	24 6/8	20 1/8	5	7	Sangamon County	IL	Jonathan P. Smith	2001	724
144 7/8	22 1/8	21 5/8	16 7/8	7	5	Summit County	OH	Dave Juchnowski	2002	724
144 7/8	23 3/8	25 0/8	17 4/8	6	6	Daviess County	MO	Brett R. DeGraaf	2003	724
144 7/8	23 1/8	23 2/8	16 3/8	6	4	Page County	IA	Bob Athen	2006	724
144 7/8	26 0/8	26 0/8	18 7/8	5	4	Jo Daviess County	IL	Jim Carlson	2006	724
144 6/8	24 1/8	25 4/8	18 0/8	4	5	Hancock County	MS	Alan J. Guess	1987	734
144 6/8	24 5/8	24 6/8	18 6/8	5	5	Fayette County	IA	Mike Judas	1999	734
144 6/8	23 7/8	22 4/8	16 5/8	6	6	Fulton County	OH	W. Tennyson Krugh	1999	734
144 5/8	24 0/8	24 4/8	21 3/8	4	4	Spink County	SD	Jerald Shantz	1959	737
144 5/8	22 1/8	22 1/8	15 7/8	5	5	Rock County	WI	T. Lawrence Hesgard	1966	737
144 5/8	23 6/8	23 6/8	19 7/8	4	4	Geauga County	OH	Rudy Grecar	1970	737
144 5/8	24 7/8	26 3/8	21 7/8	5	6	Valley County	MT	Leith Wimmer	1971	737
144 5/8	24 2/8	23 4/8	15 1/8	5	6	Traill County	ND	Arlin Ingebretson	1983	737
144 5/8	22 0/8	23 2/8	19 1/8	5	5	Lake County	IL	Carl Spaeth	2004	737
144 5/8	22 7/8	24 0/8	19 4/8	4	5	McKenzie County	ND	Michael Tofte	2004	737
144 5/8	23 4/8	23 0/8	16 0/8	6	6	Fayette County	IA	Mike Judas	2006	737

Score	Length of R Main Beam	L	Inside Spread	Number of R Points	L	Area	State/ Province	Hunter's Name	Date	Rank
144 4/8	22 6/8	22 6/8	16 3/8	5	7	Hall County	NE	Verne Skow	1958	745
144 4/8	23 2/8	23 6/8	16 3/8	5	6	Day County	SD	John E. Sigdestad	1963	745
144 4/8	25 7/8	24 3/8	20 5/8	5	5	Winneshiek County	IA	Wayne Lamoreux	1993	745
144 4/8	23 6/8	23 5/8	20 1/8	6	5	Kankakee County	IL	Rick Renzi	1994	745
144 4/8	24 5/8	25 7/8	15 5/8	5	7	Van Buren County	IA	Baree Weber	2005	745
144 4/8	22 4/8	23 5/8	17 7/8	6	7	Coles County	IL	Dan Milburn	2005	745
144 3/8	23 7/8	23 5/8	17 2/8	6	7	Rock County	NE	Dick Mauch	1963	751
144 3/8	23 6/8	23 6/8	16 4/8	4	5	Columbia County	WI	Ronald Bordson	1976	751
144 3/8	22 5/8	23 1/8	18 5/8	4	5	Darke County	OH	Ron Fansler	1994	751
144 3/8	22 4/8	23 0/8	17 3/8	5	5	McMullen County	TX	Bob Gilbert	1999	751
144 3/8	24 4/8	25 3/8	19 1/8	5	5	Ogle County	IL	John Bushnell	2003	751
144 2/8	25 7/8	26 2/8	23 0/8	4	4	Bucks County	PA	Robert Weaver	1923	756
144 2/8	21 5/8	21 4/8	17 2/8	5	5	Miner County	SD	William Hueners	1965	756
144 2/8	21 7/8	21 7/8	14 2/8	5	5	Brown County	SD	Harold Larson	1966	756
144 2/8	20 6/8	20 7/8	15 4/8	5	5	Juneau County	WI	Gordon Stittleburg	1966	756
144 2/8	24 2/8	24 1/8	20 0/8	4	5	Crawford County	IL	John Hale	1992	756
144 2/8	23 5/8	21 0/8	18 2/8	5	6	Ashland County	WI	Mark Francis Ellias	1994	756
144 2/8	24 7/8	25 1/8	20 6/8	4	4	Bucks County	PA	Tom Hooven	1997	756
144 2/8	21 4/8	23 0/8	13 6/8	5	5	St. Louis County	MO	Terry Comely	1998	756
144 1/8	21 3/8	21 2/8	17 2/8	5	7	Marshall County	KS	Jack Thornton	1965	764
144 1/8	25 0/8	24 0/8	17 7/8	5	4	Union County	IL	Pat Mitchell	1974	764
144 1/8	21 1/8	21 0/8	17 1/8	5	5	Maverick County	TX	Dean Oatman	1982	764
144 1/8	23 1/8	22 7/8	19 4/8	5	4	Des Moines County	IA	Ray Waschkat	1985	764
144 1/8	22 6/8	22 4/8	17 1/8	5	5	Woodbury County	IA	Ritch A. Stolpe	1985	764
144 1/8	25 4/8	26 3/8	17 4/8	6	5	Scotland County	MO	Jim Johnson	1991	764
144 1/8	22 7/8	19 2/8	21 5/8	5	5	Marquette County	WI	Dennis P. Gohlke	1992	764
144 1/8	23 1/8	23 5/8	16 4/8	6	5	Dubuque County	IA	Jim Boxleiter	1996	764
144 1/8	23 0/8	23 5/8	16 1/8	5	6	Collingsworth County	TX	Richard Childs	2010	764
144 0/8	21 4/8	22 0/8	16 6/8	5	5	Adams County	IL	Gerald Morton	1963	773
144 0/8	22 4/8	23 0/8	18 6/8	5	5	Fayette County	IA	Kenneth Durnin	1971	773
144 0/8	24 4/8	23 0/8	17 4/8	5	5	Warren County	IA	Jack Rush	1998	773
144 0/8	21 2/8	21 4/8	15 6/8	5	5	Gage County	NE	Larry Lottman	2003	773
143 7/8	21 4/8	22 4/8	18 3/8	5	5	Iroquois County	IL	Jerry Putnam	1987	777
143 7/8	25 1/8	25 7/8	17 7/8	4	5	Perry County	IL	Ron Pyron	1997	777
143 7/8	25 3/8	25 6/8	22 2/8	5	5	Carroll County	IL	James Carlson	2004	777
143 6/8	22 3/8	21 0/8	18 6/8	6	10	Ionia County	MI	Bob Jones	1966	780
143 6/8	23 7/8	23 4/8	19 0/8	4	5	Iron County	WI	Lee C. Dix	1968	780
143 6/8	22 6/8	23 2/8	19 6/8	6	5	Waseca County	MN	Larry W. Born	1975	780
143 6/8	23 0/8	21 7/8	15 6/8	5	5	Floyd County	VA	William M. Davis	1999	780
143 6/8	24 0/8	24 0/8	17 4/8	4	4	Suffolk County	NY	Hugh P. Miles	1999	780
143 6/8	24 2/8	25 3/8	19 0/8	4	4	Davis County	IA	Bruce Keller	2008	780
143 6/8	21 1/8	21 0/8	15 1/8	6	5	Hocking County	OH	Scott Flowers	2011	780
143 6/8	24 2/8	24 4/8	17 6/8	6	6	Waupaca County	WI	James Schoenike	2012	780
143 5/8	23 5/8	23 5/8	17 3/8	5	6	Neosho County	KS	Hugh B. Woolard	1976	788
143 5/8	25 0/8	24 7/8	17 5/8	5	4	Miami County	OH	Philip C. Gudorf	1982	788
143 5/8	21 0/8	21 1/8	19 6/8	6	6	Monroe County	OH	Darby A. Bender, Jr.	1991	788
143 5/8	23 2/8	23 6/8	19 1/8	5	5	Greenwood County	KS	George Kaplan	2000	788
143 5/8	20 7/8	20 6/8	17 1/8	7	7	Pittsburg County	OK	Tony Robison	2002	788
143 5/8	19 5/8	20 4/8	15 3/8	5	5	Switzerland County	IN	Joe Bacon	2005	788
143 4/8	23 7/8	23 0/8	20 0/8	5	7	Traverse County	MN	Roland L. Hausmann	1958	794
143 4/8	22 5/8	22 7/8	21 4/8	6	6	Ellis County	KS	Lee Couture	1969	794
143 4/8	24 3/8	25 0/8	17 2/8	5	5	Miami County	KS	Fred Supulver	1969	794
143 4/8	23 0/8	22 4/8	16 6/8	5	6	Sheboygan County	WI	Gary Mueller	1971	794
143 4/8	24 6/8	24 4/8	18 2/8	5	5	Henderson County	IL	Steve Fausel	1986	794
143 4/8	24 5/8	24 5/8	18 6/8	4	5	Missoula County	MT	James R. Clapham	1993	794
143 4/8	24 2/8	24 0/8	18 4/8	5	5	Winneshiek County	IA	Randy E. Doyle	1994	794
143 4/8	25 0/8	26 4/8	17 2/8	6	5	Highland County	OH	Roger Dale Burton	1996	794
143 4/8	25 2/8	20 7/8	17 2/8	5	4	Jackson County	MO	Gregory Robinson	2003	794
143 4/8	23 0/8	23 1/8	15 0/8	5	5	Fulton County	IL	Steve Miller	2009	794
143 3/8	25 0/8	23 6/8	16 3/8	6	5	Otter Tail County	MN	J. P. Maurins	1956	804
143 3/8	22 2/8	22 0/8	17 2/8	7	7	Morrison County	MN	Glen Marklowitz	1972	804
143 3/8	24 6/8	25 2/8	20 2/8	6	5	Jackson County	OH	Gordon Gibbs	1990	804

Minimum Score 125 *Continued*

Score	Length of Main Beam R	Length of Main Beam L	Inside Spread	Number of Points R	Number of Points L	Area	State/ Province	Hunter's Name	Date	Rank
143 3/8	21 4/8	22 0/8	19 3/8	5	5	Eau Claire County	WI	Scott Guthrie	1999	804
143 3/8	21 3/8	21 6/8	18 1/8	4	4	Black Hawk County	IA	Michael Judas	2000	804
143 3/8	22 6/8	23 0/8	17 7/8	5	5	Chickasaw County	IA	Gene Hall	2005	804
143 3/8	22 1/8	21 0/8	13 5/8	5	5	Crawford County	IL	Bob Snell	2008	804
143 2/8	25 0/8	24 0/8	16 6/8	5	6	Phillips County	AR	Stanley Zellner	1964	811
143 2/8	25 0/8	23 7/8	19 4/8	5	5	Geauga County	OH	Rudy Grecar	1971	811
143 2/8	23 2/8	22 4/8	18 7/8	5	6	Winona County	MN	Bill Clink	1984	811
143 2/8	26 6/8	25 0/8	18 6/8	5	6	Spokane County	WA	Paul Fisher	1988	811
143 2/8	22 5/8	22 4/8	18 0/8	5	4	Goodhue County	MN	Michael Schmidt	1990	811
143 2/8	23 6/8	23 7/8	17 4/8	4	6	Lake County	IL	Carl H. Spaeth	1992	811
143 2/8	22 4/8	22 4/8	16 4/8	4	5	Jefferson County	WI	Paul Alane	1996	811
143 2/8	24 5/8	24 4/8	19 6/8	4	5	Jennings County	IN	Brad Reinholt	1999	811
143 2/8	26 6/8	23 0/8	17 6/8	4	4	Kenosha County	WI	Carl H. Spaeth	2003	811
143 2/8	22 7/8	23 2/8	18 6/8	4	5	Chickasaw County	IA	Randy Platz	2006	811
143 1/8	23 0/8	23 2/8	17 1/8	5	4	Graham County	KS	Russell Hull	1965	821
143 1/8	22 5/8	22 4/8	19 5/8	5	5	Missoula County	MT	Greg Munther	1983	821
143 1/8	24 1/8	23 2/8	19 2/8	6	5	Cherry County	NE	Gary Galloway	1987	821
143 1/8	23 6/8	25 6/8	18 2/8	4	6	Vermilion County	IL	Donald Sollars	1996	821
143 1/8	22 2/8	23 1/8	16 3/8	5	5	La Salle County	IL	Bill Weygand	1999	821
143 1/8	24 6/8	24 4/8	18 1/8	4	4	McPherson County	KS	Jim Willems	2010	821
143 0/8	23 5/8	22 0/8	19 4/8	5	5	Wright County	IA	Ronald Gorden	1958	827
143 0/8	25 1/8	25 6/8	18 4/8	6	7	Aitkin County	MN	Ervin A. Buck	1961	827
143 0/8	24 0/8	24 5/8	20 5/8	5	5	Pope County	IL	Bob E. Sims	1964	827
143 0/8	25 4/8	21 0/8	14 6/8	5	6	Ottawa County	KS	Scotty Baugh	1967	827
143 0/8	22 7/8	21 7/8	16 2/8	5	5	Edwards County	KS	Gerald L. Schaller	1972	827
143 0/8	23 5/8	24 6/8	17 4/8	5	5	Alfalfa County	OK	David W. Dowell	1986	827
143 0/8	22 7/8	23 4/8	18 4/8	4	5	Bayfield County	WI	Wayne Zirn	1988	827
143 0/8	26 0/8	25 2/8	26 1/8	4	3	Edmonton	ALB	Warren Witherspoon	1993	827
143 0/8	21 4/8	21 3/8	13 4/8	6	5	Putnam County	GA	George Ralph Harper	1999	827
143 0/8	22 1/8	21 0/8	17 6/8	6	6	Winnebago County	IL	Thomas T. King	2002	827
142 7/8	20 4/8	20 6/8	16 3/8	5	5	Brown County	SD	Wayne Miller	1971	837
142 7/8	21 3/8	21 7/8	16 1/8	6	5	Becker County	MN	Kurt Lepping	1972	837
142 7/8	21 1/8	22 4/8	16 1/8	5	5	Jackson County	WI	Clark Gallup	1972	837
142 7/8	24 1/8	24 2/8	20 1/8	4	5	Douglas County	NE	Walter Ruff, Jr.	1973	837
142 7/8	22 7/8	23 0/8	17 1/8	5	6	Redwood County	MN	Kenneth A. Gilb	1980	837
142 7/8	23 1/8	24 3/8	17 3/8	5	6	Hardin County	OH	Mark Preston	1989	837
142 7/8	24 3/8	24 3/8	18 2/8	7	6	Green County	WI	Steve J. Gobeli	1992	837
142 7/8	26 2/8	24 7/8	22 3/8	4	4	Nemaha County	KS	Tom Lierz	1996	837
142 7/8	22 4/8	21 5/8	21 1/8	6	6	Mitchell County	IA	Michael D. Rehnelt	1996	837
142 7/8	21 3/8	20 6/8	16 1/8	5	5	Hardin County	IA	Kevin Heitland	1999	837
142 7/8	20 5/8	19 7/8	16 7/8	5	5	Aurora County	SD	Ron Adamson	2010	837
142 6/8	23 5/8	23 0/8	19 5/8	4	5	Green Lake County	WI	Mark Novitske	1969	848
142 6/8	25 3/8	25 4/8	16 6/8	4	4	Darke County	OH	Wayne Goubeaux	1974	848
142 6/8	22 3/8	23 1/8	17 6/8	5	4	Clay County	KS	Larry Reed	1984	848
142 6/8	22 0/8	22 1/8	17 2/8	5	5	Hamilton County	KS	Scott Showalter	1985	848
142 6/8	21 4/8	21 3/8	17 4/8	5	5	Franklin County	KS	Steven M. Hale	1994	848
142 6/8	27 1/8	26 3/8	23 5/8	5	5	Vinton County	OH	John Oberschlake	1996	848
142 6/8	24 2/8	22 4/8	18 4/8	4	4	Jefferson County	NE	Larry G. Lottman	2000	848
142 6/8	22 4/8	23 0/8	19 2/8	5	4	Jefferson County	KS	Michael K. Dunnaway	2005	848
142 5/8	22 6/8	22 5/8	20 3/8	5	4	Berkshire County	MA	Richard Scorzafava	1985	856
142 5/8	23 5/8	24 5/8	18 0/8	6	5	Mission Creek	BC	Colin L. Fazan	1991	856
142 5/8	24 4/8	24 1/8	16 1/8	5	5	New Castle County	DE	Jeff Flanagan	1993	856
142 5/8	23 3/8	23 3/8	18 5/8	5	5	Dearborn County	IN	Earl Baumann	2001	856
142 5/8	27 2/8	27 4/8	19 1/8	5	7	Pike County	OH	Robert "Matt" Fout	2011	856
142 4/8	22 5/8	22 4/8	19 0/8	5	5	Newton County	IN	Jim Manes	1963	861
142 4/8	23 3/8	23 4/8	18 4/8	4	5	Winnebago County	IA	Duane Peterson	1966	861
142 4/8	22 0/8	21 1/8	14 6/8	6	6	Dunn County	WI	John J. Logan	1970	861
142 4/8	24 2/8	24 3/8	20 6/8	5	5	Robertson County	KY	Glen Arnold	1981	861
142 4/8	25 4/8	25 6/8	17 5/8	5	5	Gallia County	OH	Alan Runyon	1988	861
142 4/8	22 5/8	23 1/8	17 0/8	4	4	Crawford County	IL	Jim Sexton	1989	861
142 4/8	21 0/8	21 1/8	16 6/8	5	5	McHenry County	IL	Richard Tudor	1992	861
142 4/8	22 3/8	20 5/8	16 7/8	5	6	Spink County	SD	Ron Miller	2004	861

Score	Length of R Main Beam	Length of L Main Beam	Inside Spread	Number of R Points	Number of L Points	Area	State/ Province	Hunter's Name	Date	Rank
142 4/8	20 1/8	21 5/8	16 6/8	7	6	Saskatoon Mtn.	ALB	Brent Watson	2007	861
142 4/8	22 7/8	21 7/8	16 2/8	6	7	Powder River County	MT	Jeffrey A. Noble	2009	861
142 4/8	25 4/8	24 7/8	18 2/8	8	7	Traill County	ND	Paul Speral	2012	861
142 3/8	24 4/8	25 1/8	18 3/8	5	5	Knox County	IL	Brad Wunder	1990	872
142 3/8	23 6/8	22 5/8	17 5/8	4	4	St. Croix County	WI	Bruce R. Guck	1992	872
142 3/8	22 1/8	22 0/8	16 7/8	6	5	Lincoln County	NE	Dave Hinton	1993	872
142 3/8	21 3/8	22 1/8	14 7/8	6	5	Madison County	MT	Jim Kennedy	1998	872
142 3/8	22 6/8	22 3/8	15 5/8	4	5	Kingman County	KS	Jimmy Montgomery	2001	872
142 3/8	24 4/8	23 4/8	20 3/8	5	5	Montgomery County	MD	Lothar Weber	2002	872
142 2/8	23 3/8	24 0/8	20 0/8	4	5	McLean County	ND	Robert Loftin	1959	878
142 2/8	23 2/8	22 6/8	17 6/8	5	6	Grant County	WI	Bob Woods	1962	878
142 2/8	21 1/8	22 2/8	18 1/8	6	4	Morrison County	MN	Dale Nieters	1971	878
142 2/8	27 3/8	28 5/8	19 6/8	4	5	Harrison County	OH	Joe Cola	1976	878
142 2/8	24 1/8	24 1/8	17 0/8	6	5	Le Sueur County	MN	Gene Solyntjes	1976	878
142 2/8	22 4/8	23 6/8	19 0/8	7	5	Jones County	IA	Donald Bohlken	1978	878
142 2/8	21 7/8	21 0/8	17 0/8	6	6	Dodge County	MN	Myles Keller	1980	878
142 2/8	20 7/8	20 7/8	15 4/8	5	5	Miller County	MO	John Patterson	1989	878
142 2/8	23 4/8	23 4/8	18 5/8	5	6	Worcester County	MA	Russell Gray	1994	878
142 2/8	23 3/8	24 0/8	18 4/8	4	5	Calhoun County	MI	George I. Swan, Jr.	2004	878
142 2/8	26 2/8	25 2/8	18 2/8	5	4	Knox County	IL	Don Owen	2007	878
142 1/8	23 1/8	22 7/8	18 6/8	5	5	Huron County	OH	Thomas Sheldon	1956	889
142 1/8	23 6/8	23 1/8	15 3/8	6	6	Morgan County	GA	Jerry Wall	1966	889
142 1/8	25 1/8	25 2/8	20 3/8	6	6	Faribault County	MN	Timothy Anderson	1970	889
142 1/8	24 2/8	25 7/8	22 1/8	5	4	Pope County	AR	Danny L. Mathis	1971	889
142 1/8	24 2/8	24 0/8	18 3/8	6	5	Buffalo County	WI	Myles Keller	1973	889
142 1/8	22 1/8	23 0/8	17 5/8	7	5	Saline County	MO	Jerry Underwood	1989	889
142 1/8	19 3/8	19 6/8	18 7/8	5	5	Weld County	CO	Dave Culter	1990	889
142 1/8	22 3/8	21 7/8	18 3/8	5	5	Winneshiek County	IA	Lonnie Tiedt	1990	889
142 1/8	22 3/8	24 1/8	18 5/8	5	6	Worth County	IA	Mark Johnson	1994	889
142 1/8	22 7/8	23 2/8	19 6/8	6	6	Putnam County	MO	Randy Cooling	1998	889
142 0/8	22 6/8	23 2/8	16 0/8	5	5	Fergus County	MT	Mike Sweeney	1986	899
141 7/8	24 2/8	24 5/8	23 3/8	4	6	Jackson County	IA	Thomas L. Berkley	1959	900
141 7/8	23 0/8	22 0/8	20 1/8	5	4	Pottawattamie County	IA	Gary A. Green	1968	900
141 7/8	21 0/8	19 6/8	15 7/8	5	5	Lincoln County	SD	Kai R. Anderson	1972	900
141 7/8	21 5/8	21 5/8	23 0/8	4	4	Fremont County	IA	Curtis L. Athen	1995	900
141 7/8	22 0/8	22 3/8	14 1/8	6	7	Pittsburg County	OK	J. Michael Goodnight	1999	900
141 7/8	22 6/8	23 5/8	18 1/8	4	5	Henry County	IN	Daniel E. Waples	2001	900
141 7/8	24 0/8	24 4/8	18 3/8	4	4	Somerset County	MD	Michael J. Hinman	2005	900
141 7/8	23 0/8	22 7/8	15 7/8	6	6	Traill County	ND	Steven W. Larson	2005	900
141 7/8	23 3/8	23 4/8	14 3/8	5	5	Wyandot County	OH	Monte D. Saam	2008	900
141 6/8	24 3/8	24 2/8	20 0/8	4	4	Morrison County	MN	Rodney W. Olson	1958	909
141 6/8	21 6/8	22 3/8	15 2/8	5	5	Eau Claire County	WI	Gordy Robinson	1962	909
141 6/8	23 6/8	23 7/8	17 4/8	5	4	Dodge County	MN	Cy Champa	1969	909
141 6/8	24 7/8	24 2/8	20 2/8	4	4	Todd County	MN	Ted Pilgrim	1981	909
141 6/8	22 3/8	22 3/8	15 4/8	5	5	Caldwell County	MO	Michael C. Burr	1989	909
141 6/8	24 7/8	24 7/8	19 6/8	4	4	Chase County	KS	Ron Smith	1992	909
141 6/8	21 7/8	21 6/8	17 4/8	5	5	Jackson County	WI	Dan R. Merritt	1995	909
141 6/8	22 2/8	23 1/8	19 2/8	4	4	Lake County	IL	Carl Spaeth	2000	909
141 6/8	22 4/8	22 4/8	18 0/8	5	5	Sauk County	WI	Brian Taggart	2002	909
141 6/8	24 2/8	22 7/8	18 5/8	6	5	McMullen County	TX	Peter Swenson	2004	909
141 6/8	22 2/8	22 6/8	17 6/8	5	5	Madison County	MT	Jim Kennedy	2004	909
141 5/8	24 0/8	24 3/8	21 6/8	6	3	Allegan County	MI	Stan Skorch	1967	920
141 5/8	22 7/8	23 1/8	20 4/8	6	5	Winnebago County	IL	Dave Judy	1995	920
141 5/8	22 4/8	23 3/8	16 1/8	5	5	Hamilton County	OH	Josh Hamilton	1997	920
141 5/8	22 4/8	23 0/8	17 5/8	4	5	Richland County	MT	Charles S. Caley	2000	920
141 4/8	21 1/8	21 4/8	19 6/8	5	5	Hyde County	SD	Gordon Sampson	1974	924
141 4/8	21 3/8	21 0/8	15 0/8	5	6	McHenry County	IL	James Coley	1989	924
141 4/8	21 5/8	22 1/8	15 2/8	6	6	Madison County	MT	J. Dudley Ottley, Sr.	1995	924
141 4/8	22 2/8	22 4/8	17 2/8	5	5	Suffolk County	NY	Michael Fenezia	1999	924
141 3/8	23 3/8	22 4/8	17 1/8	7	6	Hughes County	SD	Ross Krull	1969	928
141 3/8	24 6/8	24 6/8	19 2/8	5	7	Putnam County	WV	James H. Myers	1983	928
141 3/8	26 1/8	26 1/8	18 5/8	5	5	Fairfield County	OH	Thomas Moore	1988	928

WHITETAIL DEER [Typical Antlers]

Minimum Score 125 — Continued

Score	Length of Main Beam R	Length of Main Beam L	Inside Spread	Number of Points R	Number of Points L	Area	State/ Province	Hunter's Name	Date	Rank
141 3/8	23 5/8	23 2/8	17 3/8	6	5	Vinton County	OH	Les Loranzan, Jr.	1994	928
141 3/8	21 7/8	19 3/8	17 5/8	5	5	Pottawattamie County	IA	Doug Clayton	1994	928
141 3/8	23 1/8	23 5/8	18 7/8	5	5	Lehigh County	PA	Dale F. Arner	1995	928
141 3/8	21 6/8	22 5/8	16 7/8	6	5	Pike County	IL	David W. Reese	1997	928
141 3/8	23 2/8	23 4/8	17 1/8	6	6	Sawyer County	WI	Jeffrey P. Tomesh	1997	928
141 3/8	24 3/8	24 1/8	18 3/8	4	4	Wyoming County	NY	Jim Heckathorn	2008	928
141 2/8	24 6/8	26 2/8	14 6/8	4	5	Gage County	NE	Gene Tupa	1998	937
141 2/8	22 5/8	23 2/8	18 0/8	5	5	Vernon County	WI	Jeffrey R. Fechner	2004	937
141 2/8	25 6/8	26 6/8	18 0/8	6	4	Pottawatomie County	OK	Monroe Walters	2011	937
141 1/8	26 2/8	27 6/8	19 0/8	4	5	Dodge County	NE	Gary Trost	1961	940
141 1/8	24 0/8	23 1/8	23 3/8	5	5	Pittsburg County	OK	Dave Jilge	1981	940
141 1/8	24 2/8	24 6/8	17 1/8	5	4	Tuscaloosa County	AL	Bobby Hemphill	1981	940
141 1/8	22 4/8	23 2/8	15 6/8	7	6	Texas County	OK	Max Crocker	1986	940
141 1/8	23 3/8	23 1/8	18 1/8	4	4	Harper County	KS	Bryan Naccarato	1996	940
141 1/8	22 0/8	21 7/8	14 5/8	5	5	Jackson County	WI	Matt Staff	1999	940
141 1/8	21 5/8	22 3/8	18 5/8	5	5	Schuyler County	IL	Tom Spence	2000	940
141 1/8	20 6/8	21 1/8	15 7/8	6	5	Scott County	IL	Steve Tice	2004	940
141 0/8	22 2/8	21 7/8	16 0/8	4	5	Hand County	SD	Robert Werdel	1963	948
141 0/8	21 5/8	22 0/8	16 0/8	4	5	Nobles County	MN	Rod McNab	1974	948
141 0/8	25 0/8	24 5/8	17 6/8	5	5	Waterloo	ONT	John Wyszynski	1986	948
141 0/8	23 0/8	23 0/8	17 2/8	4	4	Burleigh County	ND	Jim Domaskin	1989	948
141 0/8	23 2/8	22 7/8	17 0/8	5	5	Guthrie County	IA	John D. Hambleton	1990	948
141 0/8	22 2/8	22 4/8	15 2/8	5	5	Franklin County	IA	Arlynn Ahrens	1990	948
141 0/8	23 0/8	24 1/8	17 0/8	5	6	Weld County	CO	Chuck Brewer	1990	948
141 0/8	25 7/8	25 1/8	17 6/8	4	4	Camden County	GA	John Bragg	1992	948
141 0/8	21 0/8	20 6/8	15 2/8	6	5	Dodge County	MN	Terry Krahn	1994	948
141 0/8	21 7/8	22 3/8	20 0/8	4	4	Lyon County	IA	Michael Judas	2000	948
141 0/8	24 1/8	24 6/8	17 7/8	4	5	Vermilion County	IL	Jeff Parkerson	2001	948
141 0/8	20 2/8	21 1/8	16 7/8	5	6	Porter County	IN	Ellis E. Ponton	2002	948
141 0/8	25 4/8	25 5/8	16 7/8	6	7	Christian County	IL	Jim Davis	2005	948
141 0/8	25 6/8	25 3/8	18 4/8	5	4	Steuben County	IN	Gregory J. Raatz	2006	948
140 7/8	27 1/8	26 4/8	21 7/8	4	6	Dodge County	NE	Ivan T. Ross	1960	962
140 7/8	21 7/8	21 1/8	16 7/8	5	6	Hughes County	SD	Gerald Snyder	1962	962
140 7/8	24 5/8	24 5/8	17 1/8	4	4	Ripley County	IN	Melvin M. Weddell	1968	962
140 7/8	21 7/8	20 4/8	17 5/8	5	5	Macoupin County	IL	Richard E. Carter	1986	962
140 7/8	23 1/8	20 7/8	19 1/8	5	5	Waupaca County	WI	Dan Bauman	1989	962
140 7/8	23 0/8	22 2/8	15 5/8	4	5	Maverick County	TX	F. H. Becker III	1990	962
140 7/8	23 5/8	24 2/8	22 6/8	7	6	Hot Springs County	WY	Larry Dickerson	1993	962
140 7/8	22 3/8	22 7/8	17 1/8	5	6	Tompkins County	NY	John N. Wilson	1994	962
140 7/8	22 4/8	21 3/8	17 7/8	5	5	Pike County	IL	Carl G. Muth, Jr.	2000	962
140 7/8	25 0/8	24 5/8	20 3/8	8	5	La Salle County	IL	George Ploch	2000	962
140 7/8	20 4/8	21 3/8	17 6/8	6	5	Clark County	MO	Tim Bradley	2003	962
140 6/8	24 2/8	24 6/8	19 5/8	5	6	Blue Earth County	MN	Earl D. Kopischke	1967	973
140 6/8	21 4/8	22 1/8	18 0/8	7	5	Weld County	CO	Roger Bechler	1974	973
140 6/8	22 4/8	22 4/8	16 6/8	5	5	Lincoln County	MN	Starlyn Mulder	1983	973
140 6/8	21 7/8	20 4/8	17 6/8	4	4	Las Animas County	CO	Patrick E. Powell	1989	973
140 6/8	24 2/8	24 5/8	16 1/8	5	7	Champaign County	IL	Richard Walden	1993	973
140 6/8	19 7/8	21 1/8	17 2/8	5	5	Furnas County	NE	Rick A. Collins	1997	973
140 6/8	21 2/8	21 6/8	18 0/8	6	5	Sussex County	DE	Charlie Sewell	2000	973
140 6/8	25 1/8	24 4/8	17 3/8	7	9	Chautauqua County	NY	Jeff Sample	2005	973
140 6/8	23 5/8	22 3/8	16 3/8	7	5	Logan County	OH	Leland Yoakam	2011	973
140 5/8	20 7/8	21 0/8	14 2/8	5	5	Dodge County	WI	Donald D. Voss	1960	982
140 5/8	25 0/8	26 0/8	19 4/8	6	5	Edgar County	IL	Richard Griffin	1973	982
140 5/8	23 6/8	24 0/8	18 1/8	5	5	Buchanan County	IA	Larry Chesmore	1989	982
140 5/8	20 1/8	20 2/8	16 5/8	4	4	Fauquier County	VA	Marvin Breeden	1992	982
140 5/8	22 0/8	22 3/8	15 1/8	7	5	Park County	WY	Marion O. DeBusk	1997	982
140 5/8	23 1/8	22 4/8	16 3/8	7	6	Jefferson County	OH	Guy Agin	1997	982
140 5/8	22 6/8	22 2/8	18 6/8	5	6	Belmont County	OH	Greg Chopko	2001	982
140 5/8	21 6/8	21 7/8	16 3/8	6	5	Atascosa County	TX	Dale R. Sanford	2002	982
140 5/8	23 3/8	24 7/8	16 6/8	5	6	Jackson County	OK	Lawrence K. Wilks	2003	982
140 5/8	22 6/8	22 6/8	17 4/8	7	5	Tuscarawas County	OH	Gary J. Czatt	2004	982
140 4/8	24 5/8	22 7/8	18 2/8	7	6	Guthrie County	IA	Dale Kromrie	1964	992

Score	Length of R Main Beam	L	Inside Spread	Number of R Points	L	Area	State/ Province	Hunter's Name	Date	Rank
140 4/8	21 1/8	21 1/8	15 5/8	5	5	Dickinson County	MI	Bernard Schmidt	1964	992
140 4/8	22 4/8	22 4/8	17 2/8	6	6	Pendleton County	KY	Gerald Bezold	1971	992
140 4/8	24 2/8	24 4/8	17 0/8	5	6	Waupaca County	WI	Allen D. Blum	1973	992
140 4/8	23 0/8	22 1/8	16 0/8	5	5	Allamakee County	IA	Todd Zeuske	1991	992
140 4/8	22 5/8	22 5/8	15 4/8	5	5	Clinton County	MO	David Eads	1993	992
140 4/8	23 5/8	22 5/8	20 7/8	5	5	Boone County	IA	Eric D. Souhrada	1995	992
140 4/8	22 6/8	21 7/8	18 4/8	5	5	Des Moines County	IA	Michael C. Dysh	1995	992
140 4/8	20 5/8	20 7/8	15 6/8	5	5	Sanders County	MT	Greg L. Munther	1996	992
140 4/8	24 0/8	23 4/8	20 2/8	4	4	Hunterdon County	NJ	Ronald S. Newman	1998	992
140 4/8	22 4/8	22 3/8	18 2/8	6	7	Osage County	OK	John K. DeLong	2008	992
140 3/8	23 5/8	23 6/8	16 4/8	6	7	Guthrie County	IA	Leland Purviance	1964	1003
140 3/8	27 0/8	26 1/8	19 5/8	5	5	Pulaski County	VA	Harold M. Peters	1965	1003
140 3/8	22 6/8	23 5/8	19 0/8	5	4	Morrison County	MN	Gary D. Wells	1972	1003
140 3/8	23 7/8	23 6/8	19 4/8	7	4	Goodhue County	MN	Dwight Dankers	1975	1003
140 3/8	23 4/8	24 0/8	16 4/8	4	5	La Salle County	TX	Jerald Carter	1993	1003
140 3/8	24 1/8	23 6/8	16 5/8	6	5	Miami County	KS	Randall Dilliner	1998	1003
140 3/8	24 4/8	25 2/8	21 5/8	7	7	Cochrane	ALB	Bry Loyd	1998	1003
140 3/8	22 6/8	22 6/8	16 3/8	5	5	Reagan County	TX	Ronnie Ray	1999	1003
140 2/8	20 5/8	21 5/8	25 0/8	6	6	Harrison County	IA	Larry Vaughn	1961	1011
140 2/8	23 7/8	23 1/8	18 4/8	4	4	Redwood County	MN	Irvin Plotz	1967	1011
140 2/8	25 1/8	25 2/8	16 2/8	8	9	Big Stone County	MN	Peter Behlen	1972	1011
140 2/8	19 0/8	20 1/8	17 2/8	5	5	Valley County	MT	Cory Kelm	1995	1011
140 2/8	22 1/8	21 4/8	17 4/8	5	5	Pike County	IL	Darren Doran	1995	1011
140 2/8	23 3/8	23 0/8	16 4/8	4	5	Keya Paha County	NE	Ronald Rockwell	2002	1011
140 2/8	24 7/8	25 1/8	18 2/8	5	5	Grant County	WI	Len K. Glass	2003	1011
140 2/8	20 0/8	20 3/8	18 6/8	5	5	Hancock County	OH	Craig E. Spieker	2008	1011
140 2/8	20 4/8	20 3/8	15 4/8	5	5	Comanche County	KS	Ronald Rockwell	2008	1011
140 1/8	24 5/8	24 2/8	18 6/8	5	5	Freeborn County	MN	Kent Dugstad	1973	1020
140 1/8	20 5/8	21 3/8	16 7/8	5	5	Jennings County	IN	Fred Hodson	1973	1020
140 1/8	22 1/8	21 2/8	16 1/8	5	4	Hitchcock County	NE	Roger Stewart	1976	1020
140 1/8	23 4/8	23 6/8	18 3/8	4	4	Coweta County	GA	Doug Miller	1979	1020
140 1/8	23 6/8	24 0/8	18 7/8	6	6	Duval County	TX	Ed Strayhorn	1991	1020
140 1/8	21 2/8	21 0/8	17 3/8	5	6	Becker County	MN	Arnold F. Ostgarden	1992	1020
140 1/8	23 0/8	22 4/8	17 7/8	4	4	Ravalli County	MT	Garret Decker	1996	1020
140 1/8	22 6/8	21 0/8	15 1/8	5	5	Webster County	IA	Dallas Keiswetter	2002	1020
140 1/8	22 5/8	23 2/8	17 3/8	4	4	Meeker County	MN	Steve Schultz	2007	1020
140 0/8	25 3/8	24 7/8	20 0/8	4	4	Houghton County	MI	Dr. V. R. Graber	1957	1029
140 0/8	22 6/8	22 3/8	17 4/8	5	5	Macoupin County	IL	Terry Jenkins	1979	1029
140 0/8	21 3/8	21 7/8	14 4/8	5	5	Hill County	MT	Steve Gorr	1984	1029
140 0/8	22 0/8	22 7/8	17 2/8	5	5	Highland County	OH	Martin H. Bullock	1989	1029
140 0/8	22 7/8	23 5/8	18 6/8	4	4	Todd County	MN	Steven R. Maack	1990	1029
140 0/8	23 7/8	23 4/8	17 0/8	4	5	Will County	IL	Richard L. Andre	1991	1029
140 0/8	23 4/8	22 6/8	16 6/8	5	6	Sullivan County	IN	John Hale	1992	1029
140 0/8	23 4/8	23 7/8	18 6/8	4	4	Marinette County	WI	Mark W. Leipold	1993	1029
140 0/8	25 3/8	24 5/8	19 0/8	6	4	Chickasaw County	IA	Randy Sinwell	2004	1029
140 0/8	23 5/8	23 5/8	16 6/8	4	4	Barron County	WI	Donald K. Isaacson	2005	1029
140 0/8	23 3/8	20 3/8	18 1/8	5	6	Fulton County	IL	Carl "Chopper" Muth, Jr.	2005	1029
140 0/8	22 6/8	22 4/8	16 1/8	6	4	Fillmore County	MN	Glenn E. Hisey	2006	1029
140 0/8	26 2/8	22 6/8	17 3/8	6	4	Granville County	NC	Larry B. Fischer III	2007	1029
140 0/8	23 0/8	23 7/8	19 6/8	5	4	Dinwiddie County	VA	Gary Wayne Adams	2008	1029
140 0/8	23 0/8	23 0/8	17 0/8	5	5	Winneshiek County	IA	Randy E. Doyle	2008	1029
139 7/8	22 6/8	23 3/8	16 3/8	5	5	Martin County	IN	Warren Trinkle	1968	1044
139 7/8	25 6/8	25 3/8	16 7/8	5	7	Athens County	OH	Matt Ross	1999	1044
139 7/8	22 6/8	22 4/8	18 5/8	4	4	Clinton County	OH	Greg Keeton	2000	1044
139 7/8	24 0/8	24 4/8	18 1/8	5	4	Juneau County	WI	Dave Puhl	2007	1044
139 6/8	22 5/8	23 2/8	20 6/8	5	5	Barton County	KS	Charles Batman	1968	1048
139 6/8	23 2/8	23 5/8	19 0/8	4	4	Wabasha County	MN	Myles Keller	1983	1048
139 6/8	22 6/8	23 3/8	17 6/8	5	5	Dane County	WI	Donald K. Isaacson	1988	1048
139 6/8	23 5/8	23 5/8	16 6/8	5	5	Preston County	WV	Lawrence P. Neidert	2007	1048
139 6/8	25 0/8	25 7/8	22 7/8	6	6	Dubuque County	IA	Terry Freiburger	2007	1048
139 6/8	23 1/8	22 2/8	15 6/8	5	5	Gallatin County	KY	Abraham Hall	2010	1048
139 6/8	22 6/8	22 2/8	18 2/8	5	5	Christian County	MO	Scott Hesterly	2011	1048

WHITETAIL DEER [Typical Antlers]

Minimum Score 125 *Continued*

Score	Length of Main Beam R	Length of Main Beam L	Inside Spread	Number of Points R	Number of Points L	Area	State/ Province	Hunter's Name	Date	Rank
139 5/8	22 0/8	22 3/8	18 6/8	6	6	Johnson County	IN	Howard VanSweringer	1961	1055
139 5/8	24 3/8	24 7/8	18 1/8	4	4	Lyon County	MN	Bruce Grow	1971	1055
139 5/8	24 4/8	24 6/8	19 1/8	5	4	Custer County	NE	Paul Ekberg	1974	1055
139 5/8	22 2/8	23 3/8	18 5/8	5	5	Custer County	MT	Dale Drilling	1980	1055
139 5/8	23 1/8	23 1/8	16 0/8	5	7	Waupaca County	WI	Terryl Newton	1993	1055
139 4/8	23 5/8	23 0/8	20 0/8	5	4	Hart County	KY	Robert L. Galloway	1967	1060
139 4/8	20 1/8	19 3/8	17 4/8	5	5	Clinton County	MI	Richard H. Wilt	1968	1060
139 4/8	22 4/8	23 4/8	19 0/8	5	5	Fillmore County	MN	Orvis J. Dahl	1971	1060
139 4/8	21 5/8	22 7/8	14 6/8	6	5	Christian County	KY	Bruce Pyle	1992	1060
139 4/8	24 2/8	23 1/8	17 2/8	5	4	Esterhazy	SAS	Murry Junek	1997	1060
139 4/8	21 1/8	22 0/8	17 4/8	4	4	Johnson County	NE	Russell Topp	1998	1060
139 4/8	21 3/8	21 5/8	17 4/8	5	5	Mountrail County	ND	Sjon Zunich	2006	1060
139 3/8	24 0/8	27 0/8	18 2/8	5	5	Fulton County	IL	Raymond Rumler	1968	1067
139 3/8	21 1/8	21 6/8	17 1/8	5	5	Outagamie County	WI	Thomas L. Haber	1975	1067
139 3/8	20 3/8	21 5/8	15 7/8	6	5	McKenzie County	ND	John Richardson	2011	1067
139 2/8	21 6/8	22 6/8	16 4/8	6	6	Shelby County	IL	David McSchooler	1968	1070
139 2/8	22 1/8	23 2/8	17 2/8	5	5	Buffalo County	WI	Leonard Anglewitz	1972	1070
139 2/8	24 2/8	23 4/8	17 6/8	4	4	McPherson County	KS	Glenn Waggoner	1977	1070
139 2/8	23 3/8	22 7/8	19 0/8	6	4	Saline County	KS	Hal Morris	1986	1070
139 2/8	21 4/8	22 6/8	17 2/8	5	5	Atascosa County	TX	Joe Braun	1989	1070
139 2/8	22 1/8	22 4/8	18 5/8	7	5	Iroquois County	IL	Robert G. Tammen	1991	1070
139 2/8	21 7/8	21 2/8	19 2/8	6	7	Houston County	MN	Scott Forsyth	1992	1070
139 2/8	21 5/8	21 3/8	16 4/8	5	6	Cumberland County	IL	Michael D. Kutch	1993	1070
139 2/8	21 7/8	22 7/8	16 0/8	4	4	Vernon County	WI	Jeffrey C. Schroeder	1997	1070
139 2/8	23 5/8	22 4/8	22 0/8	4	4	Edmonton	ALB	Rick Duggan	1998	1070
139 2/8	19 6/8	19 6/8	16 4/8	5	5	Columbia County	WI	Jason R. Wesbrock	2002	1070
139 2/8	23 2/8	23 7/8	19 1/8	6	5	Nemaha County	KS	Dennis Harper	2009	1070
139 1/8	24 0/8	24 0/8	20 7/8	4	4	Pottawattamie County	IA	James R. Kirlin	1968	1082
139 1/8	21 0/8	20 5/8	17 7/8	6	5	Calumet County	WI	Leo A. Broeckel	1969	1082
139 1/8	22 4/8	22 2/8	19 3/8	5	5	Ogle County	IL	Earl B. Thomas, Jr.	1984	1082
139 1/8	25 1/8	25 1/8	20 5/8	4	5	Des Moines County	IA	Duane R. Mabry	1994	1082
139 1/8	20 1/8	21 2/8	17 7/8	5	5	Sawyer County	WI	Norm Hasenfang	1997	1082
139 1/8	23 3/8	23 5/8	16 3/8	4	5	Edgar County	IL	Lowell Leturgez	2000	1082
139 1/8	21 1/8	21 1/8	15 1/8	4	5	Gallatin County	MT	Bob Morton	2005	1082
139 1/8	21 0/8	20 4/8	15 0/8	5	6	Toombs County	GA	Michael Manning	2005	1082
139 1/8	20 3/8	20 1/8	17 3/8	7	7	Maverick County	TX	H. Mike Palmer	2007	1082
139 0/8	23 2/8	23 5/8	18 5/8	5	4	Warren County	IA	Grant A. Poindexter	1964	1091
139 0/8	21 6/8	20 4/8	19 0/8	7	5	Butler County	KS	John Holzrechtes	1967	1091
139 0/8	21 1/8	20 6/8	17 5/8	6	5	Lyon County	MN	Harold G. Halfmann	1968	1091
139 0/8	25 0/8	25 2/8	19 2/8	4	4	Mower County	MN	Ronald Grothe	1971	1091
139 0/8	20 4/8	20 1/8	16 0/8	5	5	Fillmore County	MN	Richard Fryar	1972	1091
139 0/8	21 3/8	21 0/8	14 2/8	5	5	Phillips County	MT	Bill Beede	1980	1091
139 0/8	22 4/8	23 6/8	16 4/8	5	5	Renfrew	ONT	Glen McCutcheon	1989	1091
139 0/8	21 4/8	21 3/8	15 4/8	5	5	Wilson County	KS	Sterling Bruce Collins	1990	1091
139 0/8	21 3/8	21 2/8	18 0/8	7	5	Grundy County	IL	Mike O'Connor	1994	1091
139 0/8	23 1/8	22 6/8	17 6/8	6	5	Meeker County	MN	Jason Dudley Schultz	1996	1091
139 0/8	20 7/8	20 4/8	19 5/8	6	5	Spink County	SD	Wayne Miller	2000	1091
139 0/8	23 0/8	22 4/8	18 4/8	7	5	Olmsted County	MN	Randy E. Doyle	2002	1091
139 0/8	22 5/8	22 6/8	18 6/8	5	5	Adams County	WI	Tracy L. Hamman	2011	1091
138 7/8	22 5/8	21 4/8	16 1/8	5	6	Sullivan County	IN	Mike Burch	1976	1104
138 7/8	24 4/8	23 0/8	18 6/8	6	7	Cedar County	IA	Tom Foley	1978	1104
138 7/8	24 0/8	25 0/8	17 3/8	6	5	Woodson County	KS	Jerry Ramshaw	1989	1104
138 7/8	23 5/8	23 3/8	23 1/8	4	4	Hillsdale County	MI	John Stadler	1989	1104
138 7/8	23 4/8	23 6/8	19 1/8	6	6	Charles County	MD	William D. Wyland	1992	1104
138 7/8	23 2/8	23 1/8	17 0/8	5	5	Lafayette County	WI	Daniel F. Palzkill	1993	1104
138 7/8	19 6/8	20 4/8	17 7/8	5	5	Jackson County	WI	Matt Jessie	1994	1104
138 7/8	20 5/8	20 3/8	16 7/8	5	5	Richland County	WI	Michael Kieler	1997	1104
138 7/8	23 6/8	23 7/8	20 6/8	5	7	Kingman County	KS	Mark B. Steffen	2009	1104
138 7/8	18 2/8	18 0/8	16 5/8	6	7	Hancock County	OH	Bradley D. Clayton	2011	1104
138 6/8	25 0/8	23 7/8	20 4/8	4	4	Phillips County	AR	Everett Foley	1968	1114
138 6/8	23 5/8	23 4/8	18 6/8	4	4	St. Joseph County	MI	Jack R. Menges	1979	1114
138 6/8	23 4/8	23 4/8	18 4/8	5	4	Chase County	KS	Jerry D. Keller	1982	1114

Score	Length of R Main Beam	L	Inside Spread	Number of R Points	L	Area	State/ Province	Hunter's Name	Date	Rank
138 6/8	23 0/8	23 2/8	18 4/8	4	5	Jasper County	IL	Elmer R. Luce, Jr.	1990	1114
138 6/8	23 6/8	24 1/8	15 7/8	4	5	St. Marys County	MD	Pete Ropshaw	1997	1114
138 6/8	23 0/8	23 4/8	16 2/8	5	5	Hardin County	KY	Ricky Lowe	1997	1114
138 6/8	24 4/8	24 0/8	17 5/8	5	5	Greene County	OH	Darryl Woliver	1999	1114
138 6/8	23 4/8	24 1/8	15 0/8	4	4	Champaign County	OH	Thomas Flohre	2004	1114
138 6/8	21 7/8	22 7/8	14 2/8	4	5	Sumner County	KS	Ted Benvin	2004	1114
138 6/8	26 2/8	26 0/8	19 7/8	7	8	Mahoning County	OH	Gary Eugene Fansler	2007	1114
138 6/8	23 1/8	23 0/8	16 2/8	4	4	Vermilion	ALB	Tim Pardely	2007	1114
138 5/8	24 6/8	23 3/8	17 3/8	5	5	Vilas County	WI	B. C. Roemer	1963	1125
138 5/8	21 0/8	21 3/8	16 7/8	5	5	Martin County	IN	Tom Vieke	1967	1125
138 5/8	24 0/8	24 3/8	17 3/8	4	6	Logan County	IL	Lee Miller	1971	1125
138 5/8	22 5/8	22 5/8	16 1/8	5	5	Cherry County	NE	Lloyd C. Smith	1991	1125
138 5/8	24 2/8	23 2/8	20 7/8	5	5	Gallatin County	MT	Bob Morton	2000	1125
138 5/8	22 2/8	21 7/8	17 3/8	5	5	Clinton County	OH	Kenyon W. Woods	2001	1125
138 5/8	23 0/8	21 4/8	16 0/8	5	7	Dodge County	WI	Taylor M. Young	2008	1125
138 5/8	24 4/8	24 1/8	19 1/8	7	5	Pepin County	WI	Jerre Lerum	2009	1125
138 4/8	24 4/8	25 4/8	18 1/8	5	7	Emmet County	IA	Dr. Jerald T. Waite	1966	1133
138 4/8	24 6/8	23 5/8	16 6/8	4	4	Pike County	OH	Todd Frazier	1988	1133
138 4/8	21 2/8	20 6/8	17 4/8	6	5	Lee County	IA	Robert J. Lewis	1998	1133
138 4/8	22 6/8	23 2/8	17 4/8	5	5	Van Buren County	IA	Mike Bentler	2000	1133
138 4/8	24 4/8	25 0/8	20 2/8	5	5	Mercer County	PA	Nick Sernik	2007	1133
138 3/8	20 1/8	20 2/8	18 3/8	6	6	Marshall County	SD	Robert Peterson	1962	1138
138 3/8	22 4/8	20 3/8	16 7/8	6	5	Adams County	IL	David DeMoss	1967	1138
138 3/8	22 0/8	21 3/8	16 5/8	5	5	Chase County	KS	Larry Krom	1971	1138
138 3/8	22 1/8	22 7/8	16 7/8	4	4	O'Brien County	IA	Chuck Pemble	1973	1138
138 3/8	24 0/8	24 0/8	16 7/8	5	5	Bremer County	IA	Tom Bluhm	1988	1138
138 3/8	23 4/8	24 1/8	18 5/8	4	4	Jefferson County	OH	Jim "Reabo" Rea	1993	1138
138 3/8	23 4/8	23 3/8	16 5/8	6	5	Bingham County	ID	Doug Burkman	1998	1138
138 3/8	25 0/8	22 4/8	17 7/8	5	4	Lincoln County	WI	Michael Beiler	1999	1138
138 3/8	23 2/8	24 2/8	17 3/8	4	4	Hardin County	OH	Kyle Joseph	2001	1138
138 3/8	23 5/8	23 2/8	16 7/8	5	5	Issaquena County	MS	Mac Mabry	2010	1138
138 2/8	20 6/8	20 6/8	17 0/8	5	5	Kent County	MD	Paul W. Broadhurst	1962	1148
138 2/8	20 4/8	20 2/8	13 6/8	5	5	Adams County	IL	Russ Griffin	1970	1148
138 2/8	23 3/8	21 3/8	18 4/8	5	5	Ripley County	IN	Pat Wolf	1971	1148
138 2/8	23 5/8	22 1/8	20 2/8	5	6	Crawford County	WI	Jim Ferebee	1976	1148
138 2/8	25 7/8	26 6/8	19 6/8	5	4	Howard County	MD	Dave Wilson	1989	1148
138 2/8	22 3/8	22 6/8	17 3/8	6	5	Meeker County	MN	Larry M. Looman	1992	1148
138 2/8	20 7/8	21 3/8	16 4/8	5	6	McDonough County	IL	Burt Volker	1995	1148
138 2/8	21 5/8	22 6/8	18 3/8	6	7	Irion County	TX	Brett McCown	1997	1148
138 2/8	23 5/8	24 3/8	18 2/8	4	5	La Salle County	TX	Ricardo Longoria	1997	1148
138 2/8	21 6/8	23 0/8	19 0/8	5	5	Allegheny County	PA	Daniel K. Kuehn	2003	1148
138 2/8	24 4/8	22 3/8	20 2/8	5	6	Fulton County	PA	William C. Dovey, Jr.	2009	1148
138 1/8	25 4/8	24 0/8	17 2/8	4	5	Cottonwood County	MN	Rodney Ella	1971	1159
138 1/8	23 1/8	22 5/8	19 7/8	5	5	Pocahontas County	WV	Jim Manley II	1971	1159
138 1/8	22 5/8	24 6/8	20 1/8	6	7	Goodhue County	MN	Victor LoPresto	1979	1159
138 1/8	22 3/8	22 3/8	17 5/8	8	5	Jackson County	WI	Todd D. Reichert	1988	1159
138 1/8	22 0/8	20 7/8	15 5/8	8	6	Nelson County	ND	Douglas Magnus	1989	1159
138 1/8	24 4/8	24 4/8	18 7/8	5	5	Morrison County	MN	Gary Sigafoos	1993	1159
138 1/8	24 2/8	23 0/8	20 3/8	4	5	Spink County	SD	R. Craig Oberle	1993	1159
138 1/8	21 0/8	21 1/8	16 3/8	5	5	Madison County	MT	Jim Kennedy	2010	1159
138 1/8	23 6/8	23 5/8	17 0/8	4	6	Sherburne County	MN	Richard A. Iano	2011	1159
138 0/8	24 1/8	23 7/8	19 1/8	8	6	Frederick County	MD	Donald R. Shipley	1962	1168
138 0/8	22 2/8	22 3/8	18 4/8	4	4	Dane County	WI	Ernest Kalar	1966	1168
138 0/8	21 3/8	21 7/8	19 4/8	5	6	Washburn County	WI	John C. Gehlen	1971	1168
138 0/8	23 0/8	23 1/8	21 6/8	5	4	Sioux County	NE	Dan Coffee	1988	1168
138 0/8	20 0/8	24 5/8	17 4/8	4	6	Racine County	WI	Tom T. King	1992	1168
138 0/8	20 5/8	20 0/8	16 7/8	5	6	Coleman County	TX	Danny Mitcham	2006	1168
138 0/8	24 5/8	24 6/8	19 1/8	5	6	Rice County	KS	Shane Collier	2006	1168
138 0/8	21 0/8	21 2/8	15 4/8	6	5	Jo Daviess County	IL	Mike Feldermann	2007	1168
137 7/8	23 3/8	25 4/8	19 7/8	4	4	Hardin County	KY	Ancil Fields	1959	1176
137 7/8	22 3/8	22 7/8	21 1/8	6	8	Williams County	ND	Ray Hoveskeland	1959	1176
137 7/8	21 0/8	21 2/8	17 1/8	4	4	Platte County	NE	Lee Rupp	1968	1176

WHITETAIL DEER [Typical Antlers]

Minimum Score 125 *Continued*

Score	Length of Main Beam R	Length of Main Beam L	Inside Spread	Number of Points R	Number of Points L	Area	State/ Province	Hunter's Name	Date	Rank
137 7/8	23 0/8	23 4/8	17 5/8	5	5	Chippewa County	MN	Layton Albrecht	1978	1176
137 7/8	21 6/8	22 2/8	19 0/8	5	4	Kay County	OK	Guy L. LeMonnier, Jr.	1979	1176
137 7/8	23 1/8	24 0/8	15 5/8	4	4	Pike County	OH	Larry Cornett	1991	1176
137 7/8	23 3/8	24 7/8	15 2/8	4	5	Gage County	NE	Russell Topp	1995	1176
137 7/8	22 0/8	21 3/8	20 4/8	5	6	Suffolk County	NY	Michael Fenezia	1996	1176
137 7/8	21 5/8	21 5/8	17 5/8	5	6	Jo Daviess County	IL	Mike Felderman	2006	1176
137 7/8	21 5/8	21 2/8	14 5/8	5	4	Otoe County	NE	Dale D. Kopf	2007	1176
137 7/8	21 4/8	22 5/8	17 7/8	4	4	Linn County	KS	Morris Williams	2011	1176
137 6/8	24 5/8	23 7/8	19 2/8	4	5	Adams County	IL	William D. Force	1970	1187
137 6/8	22 1/8	20 1/8	18 1/8	6	5	Watonwan County	MN	Thomas E. Isley	1972	1187
137 6/8	21 2/8	21 7/8	19 6/8	5	5	Cowley County	KS	Michael R. Bowlin	1983	1187
137 6/8	23 5/8	23 5/8	19 6/8	5	4	Dundy County	NE	Randy Tylke	1986	1187
137 6/8	23 0/8	21 7/8	13 5/8	6	5	Marion County	IL	Bob Fabick	1993	1187
137 6/8	21 4/8	21 6/8	18 2/8	6	6	Cook County	IL	Jack Haviland	2002	1187
137 6/8	25 2/8	24 0/8	17 2/8	4	4	Wabasha County	MN	Bill Piar	2012	1187
137 5/8	20 4/8	19 7/8	15 5/8	5	5	Clark County	SD	Delmar Tobey	1959	1194
137 5/8	26 4/8	26 4/8	14 7/8	4	4	Rock County	NE	William Tutt	1963	1194
137 5/8	20 4/8	20 4/8	17 1/8	4	4	Custer County	MT	Dale Drilling	1976	1194
137 5/8	23 3/8	23 6/8	16 5/8	5	5	Crow Wing County	MN	Robert Woods	1979	1194
137 5/8	24 6/8	23 5/8	15 6/8	6	4	Washington County	IL	Paul Graves, Jr.	1988	1194
137 5/8	22 2/8	21 2/8	15 7/8	5	5	La Salle County	IL	Dirk Foltynewicz	1990	1194
137 5/8	22 1/8	21 3/8	19 3/8	5	5	Peoria County	IL	David Emken	1991	1194
137 5/8	23 1/8	23 4/8	15 3/8	5	5	Fulton County	IL	Patrick A. Cebuhar	1992	1194
137 5/8	23 6/8	22 6/8	14 4/8	5	6	Grant County	WI	Robert J. Lewis	1996	1194
137 5/8	22 6/8	20 5/8	15 5/8	5	5	McKenzie County	ND	Scotty A. Rehak	1996	1194
137 5/8	23 5/8	23 5/8	16 3/8	4	5	Wayne County	IA	Randy Cooling	2000	1194
137 5/8	21 2/8	20 0/8	16 3/8	5	5	Stark County	IL	George Bozdech	2003	1194
137 5/8	24 3/8	24 3/8	19 1/8	5	4	Will County	IL	Jack R. Haviland	2005	1194
137 4/8	23 4/8	23 6/8	18 2/8	5	4	Rice County	MN	Robert W. Berg	1965	1207
137 4/8	23 6/8	22 0/8	21 4/8	5	5	Spink County	SD	Ray McIntyre	1967	1207
137 4/8	22 5/8	22 1/8	17 2/8	6	6	Swan River	MAN	Terry Garmon	1994	1207
137 4/8	26 4/8	25 2/8	18 6/8	5	4	Preble County	OH	Les Loranzan, Jr.	1995	1207
137 4/8	23 4/8	23 3/8	17 6/8	4	5	Morrow County	OH	Vic Tumeo	1998	1207
137 4/8	23 3/8	22 2/8	16 2/8	5	6	Quitman County	MS	Russell E. Wright	2002	1207
137 4/8	19 6/8	20 0/8	13 6/8	6	8	Kenedy County	TX	Kip M. Melancon	2004	1207
137 4/8	24 1/8	23 7/8	13 0/8	5	5	Hartford County	CT	Richard J. Langer	2006	1207
137 4/8	22 7/8	23 6/8	17 5/8	7	4	Montgomery County	PA	David M. Stinley	2006	1207
137 3/8	22 0/8	21 4/8	19 7/8	5	4	Comanche County	OK	Edward J. Baumlin, Jr.	1960	1216
137 3/8	20 6/8	20 2/8	12 6/8	5	10	Kootenai County	ID	John Ruthuen	1961	1216
137 3/8	23 5/8	23 3/8	16 6/8	4	5	Allegan County	MI	Clayton Foster	1964	1216
137 3/8	25 1/8	24 4/8	20 7/8	4	4	Trempealeau County	WI	Randall J. Van Vleet	1966	1216
137 3/8	25 6/8	23 1/8	15 7/8	4	4	Madison County	IN	Lee Middleton	1966	1216
137 3/8	20 2/8	19 7/8	16 5/8	5	5	Stone County	MO	Charles A. Myers	1974	1216
137 3/8	25 2/8	26 4/8	18 2/8	8	7	Tuscarawas County	OH	Daniel A. Mozena	1991	1216
137 3/8	23 1/8	23 3/8	16 1/8	5	5	Washtenaw County	MI	James A. West	1992	1216
137 3/8	23 5/8	23 4/8	18 1/8	6	5	Vinton County	OH	Dave Couch	1997	1216
137 3/8	24 1/8	24 3/8	16 7/8	4	4	McHenry County	IL	Mark Gullickson	2001	1216
137 3/8	23 2/8	23 4/8	17 7/8	4	4	Calhoun County	MI	George I. Swan, Jr.	2002	1216
137 3/8	22 1/8	22 3/8	18 1/8	5	4	Olmsted County	MN	Daniel "Boone" Bell	2008	1216
137 2/8	23 2/8	24 7/8	18 5/8	8	5	Phelps County	NE	Dick Cepel	1959	1228
137 2/8	21 1/8	20 0/8	17 4/8	6	5	Brown County	SD	Richard Felch	1966	1228
137 2/8	23 4/8	23 5/8	18 2/8	6	6	Jefferson County	IN	George E Yazel	1988	1228
137 2/8	21 3/8	21 6/8	19 4/8	5	6	Huntingdon County	PA	Carl Zimmerman	1989	1228
137 2/8	22 6/8	22 2/8	18 0/8	6	5	Wright County	IA	Jeff Nelson	1990	1228
137 2/8	23 5/8	23 2/8	15 2/8	5	5	Delaware County	OH	Jimmie J. Dixon	1993	1228
137 2/8	21 4/8	21 0/8	14 4/8	5	5	Maverick County	TX	F. H. Becker III	1994	1228
137 2/8	24 2/8	24 0/8	16 2/8	4	4	Schuyler County	IL	Rob R. Hankins	2002	1228
137 2/8	24 2/8	24 3/8	16 2/8	5	4	Bayfield County	WI	Larry Fischer	2004	1228
137 2/8	23 2/8	23 6/8	18 6/8	5	6	Hancock County	IL	Bradley J. Swanson	2005	1228
137 2/8	20 5/8	20 4/8	14 1/8	7	7	Barber County	KS	Keith A. Dana	2007	1228
137 2/8	22 7/8	21 4/8	17 1/8	9	5	Kiowa County	KS	Rick Duggan	2009	1228
137 2/8	24 5/8	23 2/8	17 0/8	5	5	Switzerland County	IN	Abraham Hall	2010	1228

WHITETAIL DEER [Typical Antlers]

Minimum Score 125 *Continued*

Score	Length of R Main Beam	L	Inside Spread	Number of R Points	L	Area	State/ Province	Hunter's Name	Date	Rank
137 1/8	26 5/8	24 7/8	18 5/8	4	4	Bartholomew County	IN	Jimmy Middleton	1963	1241
137 1/8	22 4/8	21 4/8	19 7/8	5	5	Rock County	MN	Orville Hamm	1976	1241
137 1/8	23 6/8	23 3/8	18 5/8	5	4	Clayton County	IA	Eugene Arndt	1991	1241
137 1/8	20 3/8	20 2/8	19 1/8	5	5	Irion County	TX	William Jay Wilson	1992	1241
137 1/8	21 5/8	22 7/8	14 5/8	5	5	Washington County	OH	William G. Mason	1996	1241
137 1/8	25 0/8	26 0/8	17 2/8	6	5	Knox County	OH	Gary Sparks	1996	1241
137 1/8	26 4/8	25 3/8	18 5/8	5	5	Dubuque County	IA	John Duggan	1997	1241
137 1/8	22 1/8	22 0/8	17 3/8	4	5	Big Horn County	WY	Dave Moss	2008	1241
137 1/8	22 0/8	23 1/8	18 0/8	5	4	Isle of Wight County	VA	Curt Lytle	2011	1241
137 0/8	26 0/8	24 0/8	19 6/8	5	4	Hughes County	SD	Gerald Snyder	1963	1250
137 0/8	21 4/8	22 3/8	18 2/8	4	7	Black Hawk County	IA	Bob Wood	1971	1250
137 0/8	22 1/8	22 3/8	17 0/8	5	4	Custer County	MT	Jim Walters	1978	1250
137 0/8	22 6/8	21 4/8	19 4/8	5	4	Medina County	TX	Mike Palmer	1988	1250
137 0/8	23 1/8	23 4/8	18 5/8	5	4	Greene County	IN	Quince Hale	1989	1250
137 0/8	27 4/8	25 7/8	21 0/8	6	4	Lake County	IL	Steven Hysell	1990	1250
137 0/8	22 1/8	21 4/8	15 6/8	4	4	Concordia Parish	LA	Hank Kizer	1996	1250
137 0/8	21 5/8	22 0/8	25 7/8	6	5	Pittsburg County	OK	John Thomas Haynes	2002	1250
137 0/8	23 4/8	23 6/8	17 2/8	5	4	Carroll County	IL	James Carlson	2003	1250
137 0/8	22 5/8	23 6/8	16 0/8	4	4	Barber County	KS	Casmir Domurat	2005	1250
137 0/8	21 2/8	21 1/8	16 0/8	5	5	McKenzie County	ND	Dan Brockman	2007	1250
136 7/8	20 2/8	19 1/8	15 2/8	8	7	Ashland County	WI	Jim McGarvey	1965	1261
136 7/8	23 5/8	23 6/8	18 1/8	4	5	Jefferson County	IN	James Coldiron	1972	1261
136 7/8	21 5/8	23 5/8	16 7/8	6	5	Washtenaw County	MI	Alfred A. Reich	1984	1261
136 7/8	23 5/8	24 4/8	18 3/8	5	5	Trempealeau County	WI	Todd Nelson	1991	1261
136 7/8	19 3/8	18 6/8	16 7/8	6	6	La Salle County	TX	Dale R. Sanford	1992	1261
136 7/8	24 4/8	24 4/8	17 1/8	4	5	Tompkins County	NY	Stanley Ink	1999	1261
136 7/8	21 7/8	21 6/8	17 5/8	5	5	Parker County	TX	Steve Ray Dollar	2001	1261
136 7/8	21 7/8	21 6/8	15 3/8	5	5	Mountrail County	ND	Craig Richardson	2002	1261
136 7/8	22 2/8	21 7/8	16 1/8	4	4	Will County	IL	Robert Fisher	2004	1261
136 7/8	20 3/8	21 5/8	18 1/8	4	4	Pike County	IL	Brian D. Scarnegie	2005	1261
136 6/8	23 2/8	22 4/8	15 4/8	4	5	Ogle County	IL	Edwin Fitzgerald	1963	1271
136 6/8	22 0/8	20 6/8	16 5/8	5	6	Hamlin County	SD	James Larson	1963	1271
136 6/8	23 5/8	24 2/8	17 4/8	4	4	Mower County	MN	Robert Meyer	1965	1271
136 6/8	21 4/8	20 6/8	18 2/8	5	6	Watonwan County	MN	Gary G. Miest	1969	1271
136 6/8	22 1/8	20 6/8	17 4/8	5	5	Lake County	SD	Dennis DeBoer	1974	1271
136 6/8	19 6/8	19 6/8	15 4/8	5	4	Uvalde County	TX	Wyatt Birkner	1992	1271
136 6/8	24 3/8	25 2/8	18 2/8	5	6	Pittsburg County	OK	Stan Jones	1999	1271
136 6/8	20 4/8	21 4/8	22 6/8	5	6	Mason County	WV	John Wiltz	2000	1271
136 6/8	22 4/8	23 4/8	17 3/8	5	6	Maverick County	TX	Rex Dacus	2002	1271
136 6/8	22 0/8	21 0/8	12 4/8	6	5	Osage County	KS	Daniel K. Kuehn	2002	1271
136 6/8	22 4/8	22 4/8	18 6/8	4	4	Prairie River	SAS	Ken H. Taylor	2003	1271
136 6/8	20 4/8	20 4/8	16 6/8	5	5	Maverick County	TX	Fred Eichler	2003	1271
136 6/8	23 7/8	23 7/8	19 4/8	5	5	Fairfield County	CT	Bill Terry, Sr.	2006	1271
136 6/8	21 7/8	21 5/8	14 3/8	6	6	Van Buren County	IA	Baree Weber	2009	1271
136 6/8	21 2/8	21 4/8	17 2/8	4	5	Van Buren County	IA	Dennis Kelly	2010	1271
136 5/8	24 6/8	24 6/8	18 5/8	4	4	Braxton County	WV	John M. Friend	1965	1286
136 5/8	23 4/8	22 5/8	16 5/8	4	4	Marathon County	WI	Leroy Kazmierczak	1966	1286
136 5/8	22 0/8	22 2/8	18 1/8	5	5	Montgomery County	PA	Glenn Kuklick	1983	1286
136 5/8	21 4/8	21 3/8	16 7/8	5	7	La Salle County	IL	Kevin R. Mallie	1987	1286
136 5/8	22 6/8	23 2/8	15 7/8	5	5	Knox County	IL	David Emken	1994	1286
136 5/8	24 3/8	24 6/8	19 5/8	5	5	Anoka County	MN	William J. Courteau	1995	1286
136 5/8	22 6/8	22 5/8	15 7/8	5	5	Kent County	MD	William R. Fleegle	2000	1286
136 5/8	21 1/8	21 5/8	16 1/8	5	5	Lincoln County	KY	Stephen Burton	2009	1286
136 4/8	25 3/8	24 4/8	20 4/8	4	4	Roberts County	SD	Byron Siegel	1963	1294
136 4/8	24 6/8	23 6/8	21 6/8	4	4	Jo Daviess County	IL	Todd Muehleip	1975	1294
136 4/8	21 7/8	22 1/8	17 2/8	5	5	Coweta County	GA	Bobby Edwards	1981	1294
136 4/8	22 5/8	22 4/8	16 4/8	5	5	Wyoming County	NY	Ray Minnick	1985	1294
136 4/8	21 6/8	22 0/8	16 4/8	5	5	Henry County	IA	Marvin Foster	1992	1294
136 4/8	26 3/8	24 7/8	18 6/8	4	5	Lake County	IL	Carl Spaeth	1994	1294
136 4/8	21 2/8	21 3/8	18 0/8	6	6	Brown County	SD	Mike Sahli	2000	1294
136 3/8	23 5/8	24 2/8	17 6/8	4	5	Pope County	IL	Murray Schuchardt	1973	1301
136 3/8	23 6/8	23 6/8	18 5/8	5	7	Pope County	MN	Ernie Janish	1976	1301

Score	Length of R Main Beam	Length of L Main Beam	Inside Spread	Number of R Points	Number of L Points	Area	State/ Province	Hunter's Name	Date	Rank
136 3/8	23 4/8	23 3/8	15 3/8	5	5	Linn County	KS	Loren J. Sayers	1989	1301
136 3/8	22 4/8	16 5/8	18 4/8	6	6	Prairie River	SAS	Ken H. Taylor	1998	1301
136 3/8	23 3/8	22 4/8	17 4/8	4	5	Van Buren County	IA	Russell J. Miller	2001	1301
136 2/8	21 2/8	21 2/8	17 0/8	5	5	Sheridan County	ND	Robert Conklin	1967	1306
136 2/8	22 0/8	22 4/8	17 4/8	5	4	Taylor County	WI	Roger Williams	1968	1306
136 2/8	21 4/8	21 1/8	16 3/8	7	6	Goodhue County	MN	Terry Krahn	1987	1306
136 2/8	22 7/8	21 3/8	18 6/8	5	5	Oneida County	WI	Russell Ostermann	1989	1306
136 2/8	23 2/8	24 0/8	18 0/8	4	4	Baltimore County	MD	Bruce D. Hoover	1989	1306
136 2/8	22 2/8	21 0/8	15 1/8	7	7	Williams County	ND	Ron Borgeson	1997	1306
136 2/8	21 6/8	17 5/8	19 0/8	4	5	Muskegon County	MI	Keven Stafford	1998	1306
136 2/8	21 0/8	20 5/8	17 0/8	6	5	Beltrami County	MN	Robert Glick	1998	1306
136 2/8	24 0/8	23 4/8	18 2/8	5	4	Fayette County	IA	Mike Judas	2000	1306
136 2/8	21 2/8	21 1/8	14 6/8	7	8	Meeker County	MN	Jason Schultz	2010	1306
136 1/8	21 5/8	21 2/8	18 0/8	5	4	Spink County	SD	Louis Smith	1963	1316
136 1/8	23 3/8	23 4/8	19 7/8	6	6	Litchfield County	CT	Eugene Clini, Jr.	1970	1316
136 1/8	21 4/8	22 1/8	17 3/8	4	4	Fillmore County	MN	Glenn Hisey	1983	1316
136 1/8	20 1/8	20 1/8	18 5/8	5	5	Lincoln County	MT	Michael F. Shepard	1985	1316
136 1/8	20 4/8	21 4/8	16 3/8	5	5	La Salle County	TX	Dennis Faulkenberry	1991	1316
136 1/8	23 1/8	22 3/8	18 1/8	7	4	Dubuque County	IA	Kim Olberding	1999	1316
136 1/8	24 1/8	23 5/8	17 5/8	5	4	Holt County	MO	Juston W. Carr	2002	1316
136 1/8	25 4/8	25 0/8	19 1/8	4	4	Barber County	KS	Lew Webb	2004	1316
136 1/8	21 3/8	22 5/8	17 7/8	5	5	Buffalo County	WI	Glenn Klomaten	2006	1316
136 1/8	20 6/8	20 2/8	14 0/8	7	7	Madison County	MT	Jim Kennedy	2011	1316
136 0/8	22 7/8	23 0/8	17 1/8	5	6	Aitkin County	MN	Ervin A. Buck	1956	1326
136 0/8	19 6/8	19 5/8	19 4/8	4	4	Jackson County	IL	Henry Mika	1962	1326
136 0/8	23 2/8	23 4/8	17 2/8	5	5	Brookings County	SD	Douglas Tschetter	1964	1326
136 0/8	21 3/8	21 5/8	16 4/8	5	5	Juneau County	WI	Terry Taft	1981	1326
136 0/8	22 0/8	23 0/8	19 0/8	5	5	Jackson County	MI	William D. Burgess	1985	1326
136 0/8	24 5/8	24 7/8	14 4/8	4	4	Racine County	WI	Steve Holterman	1991	1326
136 0/8	23 5/8	22 7/8	18 4/8	5	4	Macon County	GA	Gary Edward Campbell	1996	1326
136 0/8	22 6/8	22 3/8	20 0/8	6	7	Madison County	MT	Jim Kennedy	1997	1326
136 0/8	20 1/8	19 0/8	15 4/8	5	5	Bexar County	TX	Deno Poulos	1998	1326
136 0/8	23 0/8	23 1/8	18 2/8	5	5	Brown County	NE	Joel Klammer	2001	1326
136 0/8	22 5/8	22 0/8	16 0/8	5	4	Kay County	OK	James C. Burnett	2004	1326
136 0/8	22 6/8	22 5/8	15 4/8	6	5	Webb County	TX	James Bailey	2004	1326
136 0/8	22 7/8	23 3/8	17 2/8	6	5	Grant County	WI	Gary Maciolek	2005	1326
136 0/8	25 0/8	24 1/8	18 4/8	5	5	Hartford County	CT	Richard J. Langer	2008	1326
136 0/8	22 0/8	22 2/8	17 4/8	5	5	Langlade County	WI	Kory Kais	2008	1326
135 7/8	24 6/8	21 6/8	20 1/8	4	5	Westchester County	NY	Colin M. Pierson	1967	1341
135 7/8	23 6/8	22 2/8	16 7/8	6	8	Adams County	IL	Joe Johnson	1967	1341
135 7/8	23 2/8	23 2/8	18 7/8	5	4	Burke County	GA	O. Jack Barrett, Jr.	1986	1341
135 7/8	21 7/8	22 0/8	20 1/8	6	7	Marshall County	OK	Joel A. Trammell	1989	1341
135 7/8	24 6/8	24 6/8	18 1/8	8	6	Clayton County	IA	Scott Butters	1997	1341
135 7/8	24 5/8	25 3/8	18 2/8	6	5	McLean County	IL	Tony Stolfa	2000	1341
135 7/8	20 0/8	20 7/8	15 3/8	5	5	Lawrence County	SD	Joseph Hallenbeck, Jr.	2005	1341
135 7/8	22 2/8	24 2/8	19 3/8	4	5	Calhoun County	MI	George I. Swan, Jr.	2006	1341
135 7/8	22 4/8	21 6/8	16 1/8	5	6	Davis County	IA	Russ Kay	2007	1341
135 6/8	24 4/8	22 7/8	19 2/8	5	5	Winnebago County	IL	Edward Fuller	1963	1350
135 6/8	22 6/8	23 2/8	20 6/8	4	4	Shawano County	WI	Darryl Erdman	1968	1350
135 6/8	25 5/8	24 4/8	19 7/8	6	4	Knox County	IL	Fred E. Miller	1968	1350
135 6/8	22 3/8	22 5/8	15 2/8	6	7	Dodge County	MN	Brad Blanchard	1973	1350
135 6/8	25 7/8	24 4/8	21 2/8	5	8	La Salle County	TX	Danny J. Carter	1992	1350
135 6/8	22 4/8	22 3/8	15 2/8	5	6	Autauga County	AL	Kenny Diaz	1994	1350
135 6/8	26 0/8	23 7/8	19 0/8	4	5	Fremont County	IA	Dan Spoelman	1997	1350
135 6/8	23 4/8	23 5/8	18 4/8	4	5	Morgan County	CO	Matt Archuleta	2002	1350
135 6/8	21 7/8	22 0/8	16 4/8	5	6	Sebastian County	AR	George Talkington	2007	1350
135 6/8	22 5/8	23 1/8	16 4/8	7	8	Steuben County	IN	Chris Spaw	2009	1350
135 6/8	23 6/8	23 2/8	15 2/8	4	4	Winona County	MN	Johnny F. Micheel	2009	1350
135 5/8	21 4/8	20 0/8	16 1/8	5	5	Sheboygan County	WI	Earl Uhl	1960	1361
135 5/8	22 7/8	21 6/8	21 1/8	4	4	Union County	KY	Randy Joe Duncan	1965	1361
135 5/8	22 1/8	22 1/8	19 3/8	5	5	Ringgold County	IA	William K. Seitz	1973	1361
135 5/8	22 4/8	23 2/8	14 0/8	4	7	Anderson County	KS	Steve Spangler	1983	1361

Score	Length of R Main Beam	L	Inside Spread	Number of R Points	L	Area	State/ Province	Hunter's Name	Date	Rank
135 5/8	22 2/8	22 2/8	15 7/8	4	5	Texas County	OK	Max Crocker	1986	1361
135 5/8	19 5/8	19 0/8	13 4/8	6	6	Pope County	IL	John F. Perso	1987	1361
135 5/8	24 3/8	24 1/8	17 1/8	4	4	Guernsey County	OH	Thomas J. Hentrick	1989	1361
135 5/8	23 1/8	23 0/8	18 5/8	4	4	Coshocton County	OH	Gene Mathias	1991	1361
135 5/8	23 6/8	23 2/8	19 4/8	5	6	Shawnee County	KS	Michael Troth	1995	1361
135 5/8	22 4/8	23 3/8	17 1/8	4	4	Big Horn County	WY	Dave Moss	1998	1361
135 5/8	24 4/8	22 6/8	18 1/8	5	5	Harrison County	IA	Bill Pearson	2001	1361
135 5/8	20 4/8	21 1/8	14 6/8	6	6	Madison County	MT	Jim Kennedy	2002	1361
135 4/8	21 0/8	23 0/8	18 6/8	6	5	Mitchell County	IA	Arthur Cepeda	1962	1373
135 4/8	23 4/8	23 4/8	18 6/8	4	4	Dodge County	WI	Daniel J. Rozek	1966	1373
135 4/8	21 6/8	21 4/8	19 4/8	5	5	Mower County	MN	Arthur McKenzie	1972	1373
135 4/8	23 0/8	23 6/8	18 0/8	5	5	Lyon County	KY	Kenneth McKay	1973	1373
135 4/8	20 5/8	20 5/8	13 2/8	5	5	Dickinson County	MI	Dave Bath	1973	1373
135 4/8	21 3/8	23 0/8	18 0/8	5	5	Ripley County	IN	G. Fred Asbell	1973	1373
135 4/8	21 7/8	21 6/8	18 0/8	5	5	Mower County	MN	Clark Gallup	1975	1373
135 4/8	22 7/8	22 6/8	19 0/8	6	5	Pittsburg County	OK	Bill Nelson	1977	1373
135 4/8	23 7/8	21 4/8	19 4/8	5	5	Logan County	CO	Pat Norris	1995	1373
135 4/8	21 3/8	21 2/8	15 7/8	6	5	Jefferson County	IN	Tom Bean	1996	1373
135 4/8	23 7/8	23 0/8	18 1/8	7	4	Fulton County	IL	Steve Miller	1997	1373
135 4/8	22 5/8	21 2/8	18 4/8	7	5	Hamilton County	OH	Greg L. Deatsman	2001	1373
135 4/8	24 2/8	23 3/8	17 2/8	5	5	Buffalo County	WI	Jeffrey R. Fechner	2002	1373
135 4/8	24 6/8	24 4/8	20 2/8	3	3	Brown County	NE	Joel A. Klammer	2004	1373
135 3/8	22 0/8	22 2/8	17 1/8	5	5	Columbia County	WI	Chester Sroka	1936	1387
135 3/8	23 2/8	23 0/8	21 0/8	6	6	Sully County	SD	R. L. Marso	1965	1387
135 3/8	23 7/8	23 0/8	17 5/8	4	4	Iowa County	WI	Paul Klingelhoets	1978	1387
135 3/8	22 3/8	21 5/8	15 5/8	5	5	Clark County	IN	Robert W. Thompson	1979	1387
135 3/8	24 1/8	24 1/8	19 7/8	4	4	Warren County	NJ	Bill L. Raub	1982	1387
135 3/8	22 5/8	22 5/8	18 3/8	4	4	Jones County	IA	William Janssen	1986	1387
135 3/8	22 2/8	22 5/8	18 5/8	5	5	Webster County	KY	E. Bert Combs	1991	1387
135 3/8	20 3/8	21 1/8	15 1/8	6	5	Parker County	TX	John L. Chapman	1991	1387
135 3/8	19 0/8	19 6/8	17 1/8	5	5	Washington County	WI	John Dohmen	1993	1387
135 3/8	24 2/8	23 5/8	16 3/8	4	5	Cole County	MO	John Schroer	1995	1387
135 3/8	26 1/8	25 2/8	16 7/8	6	4	Bingham County	ID	Doug Burkman	2001	1387
135 3/8	22 3/8	23 4/8	20 1/8	4	4	Madison County	MT	Jim Kennedy	2002	1387
135 3/8	24 7/8	25 2/8	18 5/8	4	4	Brown County	NE	Keith Karr	2004	1387
135 3/8	22 1/8	22 4/8	13 3/8	4	4	Dubuque County	IA	Tom Casey	2004	1387
135 3/8	22 1/8	22 3/8	15 1/8	4	4	Lyon County	IA	Mike Judas	2005	1387
135 3/8	25 0/8	24 6/8	16 5/8	6	6	Montgomery County	MD	Frank H. Wilmot	2006	1387
135 3/8	22 1/8	21 4/8	16 3/8	4	4	Norton County	KS	Nathan Andersohn	2009	1387
135 3/8	20 7/8	20 6/8	15 7/8	5	5	Williams County	ND	Milan Liesener	2010	1387
135 2/8	25 0/8	24 0/8	18 2/8	5	4	Lucas County	IA	Everett Parsons	1964	1405
135 2/8	21 3/8	22 7/8	16 0/8	5	8	Brown County	SD	Jack Eagleson	1966	1405
135 2/8	23 0/8	23 2/8	20 2/8	4	4	Cedar County	IA	Fred Wesselink	1971	1405
135 2/8	29 0/8	29 7/8	21 5/8	6	8	Grundy County	IL	Wesley Holm	1975	1405
135 2/8	22 0/8	21 6/8	16 0/8	5	5	Green Lake County	WI	Arwin Moldenhauer	1984	1405
135 2/8	25 5/8	24 5/8	16 7/8	6	4	Polk County	MN	Warren Nelson	1984	1405
135 2/8	25 1/8	25 1/8	17 5/8	5	6	Texas County	OK	Max Crocker	1987	1405
135 2/8	20 0/8	19 7/8	14 0/8	5	5	Mercer County	NJ	Glenn R. Kuklick	1990	1405
135 2/8	23 0/8	22 6/8	19 6/8	4	4	Prince Georges County	MD	Douglas S. Greenfield	1994	1405
135 2/8	21 5/8	22 0/8	14 6/8	5	6	Pittsburg County	OK	Larry Gibson	1995	1405
135 2/8	22 4/8	22 5/8	16 6/8	4	4	Claiborne County	MS	Lance Stroud	1998	1405
135 2/8	23 4/8	23 3/8	18 4/8	4	4	Buffalo County	WI	Eric Farr	2000	1405
135 2/8	21 1/8	23 4/8	17 0/8	5	5	Maverick County	TX	H. Mike Palmer	2004	1405
135 2/8	22 0/8	21 6/8	19 0/8	5	5	White County	IL	Walt Guthrie	2004	1405
135 2/8	21 4/8	21 2/8	16 6/8	7	7	Gallatin County	MT	Bob Morton	2006	1405
135 2/8	21 4/8	20 4/8	16 4/8	5	5	Van Buren County	IA	Brian Klein, Jr.	2006	1405
135 2/8	23 5/8	24 0/8	19 5/8	4	5	Carroll County	IL	Jim Carlson	2008	1405
135 2/8	20 4/8	20 3/8	16 4/8	5	5	McKenzie County	ND	John Richardson	2010	1405
135 2/8	23 6/8	23 4/8	19 5/8	5	5	Carroll County	IL	Jim Carlson	2011	1405
135 1/8	21 0/8	20 4/8	17 3/8	5	5	Westchester County	NY	Francis E. Hill	1958	1424
135 1/8	21 6/8	21 3/8	19 7/8	6	5	Brookings County	SD	Ray Buckley	1961	1424
135 1/8	21 5/8	21 5/8	17 7/8	4	4	Clark County	SD	Jack D. Chesmore	1971	1424

WHITETAIL DEER [Typical Antlers]

Minimum Score 125 *Continued*

Score	Length of Main Beam R	Length of Main Beam L	Inside Spread	Number of Points R	Number of Points L	Area	State/ Province	Hunter's Name	Date	Rank
135 1/8	21 5/8	21 6/8	18 0/8	5	6	Edmonson County	KY	Steve England	1974	1424
135 1/8	23 4/8	22 7/8	18 3/8	5	4	Hamilton County	OH	Jerome Buschle, Jr.	1983	1424
135 1/8	19 4/8	20 1/8	16 1/8	5	4	McLean County	IL	Robert P. Ryburn	1984	1424
135 1/8	22 3/8	21 4/8	16 5/8	5	5	Missoula County	MT	Monty Moravec	1985	1424
135 1/8	23 6/8	24 2/8	17 3/8	5	4	Juneau County	WI	Bennie L. Voigt	1988	1424
135 1/8	23 6/8	24 4/8	18 4/8	7	6	Pike County	IL	Richard Dewey	1990	1424
135 1/8	23 5/8	24 4/8	18 1/8	4	4	Gogebic County	MI	Eugene A. Pribek	1991	1424
135 1/8	21 0/8	22 0/8	15 3/8	5	5	Randolph County	WV	Daniel R. Gillenwater	1992	1424
135 1/8	22 2/8	23 4/8	21 6/8	4	6	Calhoun County	MI	George I. Swan, Jr.	1996	1424
135 1/8	22 2/8	22 4/8	15 1/8	4	6	Monroe County	NY	Terry Wilson	1999	1424
135 1/8	24 4/8	24 5/8	17 0/8	5	4	Jackson County	MI	Richard J. Galicki	1999	1424
135 1/8	22 7/8	24 1/8	13 4/8	11	7	Harris County	GA	Grant R. Woods	2001	1424
135 1/8	23 0/8	21 3/8	16 0/8	6	6	Cherokee County	KS	Doug Walden	2003	1424
135 1/8	24 4/8	24 0/8	16 7/8	4	4	Filmore County	NE	Jerry Franck	2005	1424
135 1/8	21 0/8	21 3/8	15 1/8	5	5	Nelson County	KY	Eric Johnson	2011	1424
135 0/8	22 0/8	20 6/8	17 0/8	4	4	Grant County	MN	Stanley D. Miles	1963	1442
135 0/8	22 3/8	22 1/8	19 4/8	5	5	Linn County	IA	David Padget	1985	1442
135 0/8	24 7/8	25 2/8	18 0/8	4	4	Douglas County	NE	Norman Armstrong	1988	1442
135 0/8	24 2/8	23 2/8	20 2/8	4	4	Champaign County	IL	Charles Flora	1990	1442
135 0/8	22 0/8	22 0/8	18 2/8	5	5	Meeker County	MN	Chuck Schultz	1992	1442
135 0/8	21 7/8	21 7/8	16 6/8	5	4	Ravalli County	MT	Ned Coorough	1993	1442
135 0/8	22 2/8	23 2/8	16 2/8	5	5	Grenada County	MS	Mac Moorman	1998	1442
135 0/8	19 7/8	20 4/8	15 0/8	5	5	Dawson County	MT	Vern Lindquist	2005	1442
135 0/8	22 3/8	22 5/8	15 2/8	5	4	Hinds County	MS	David Mosal	2011	1442
135 0/8	21 3/8	21 5/8	19 2/8	5	5	Warren County	IA	Peter George	2011	1442
134 7/8	20 1/8	21 4/8	16 3/8	5	5	Spokane County	WA	Harold Bratlie	1960	1452
134 7/8	20 6/8	20 6/8	16 7/8	4	4	Lawrence County	SD	Oliver Lewis	1960	1452
134 7/8	20 4/8	21 0/8	17 2/8	6	6	Cherokee County	OK	Addison Harrison	1965	1452
134 7/8	20 7/8	21 6/8	17 7/8	5	5	Morrison County	MN	Thomas Ackerman	1971	1452
134 7/8	24 5/8	24 1/8	21 3/8	7	6	Washington County	OH	Lyle W. Townson, Sr.	1974	1452
134 7/8	22 6/8	23 5/8	17 3/8	6	5	Dawson County	NE	Randy Wilson	1978	1452
134 7/8	24 4/8	23 7/8	15 7/8	4	4	Woolwrich	ONT	Peter Hartley	1990	1452
134 7/8	22 2/8	23 1/8	15 5/8	4	4	Washington County	MN	Wayne A. Edgerton	1993	1452
134 7/8	22 4/8	22 2/8	17 7/8	5	4	Pittsburg County	OK	Rodney Alexander	1997	1452
134 7/8	25 3/8	25 4/8	16 7/8	4	4	Webb County	TX	Roger Wintle	1997	1452
134 7/8	21 5/8	21 7/8	16 4/8	5	5	Pike County	IL	John Snyder	1999	1452
134 7/8	21 2/8	19 2/8	14 7/8	5	5	Decatur County	IA	Dale Karch	2000	1452
134 7/8	21 6/8	21 5/8	19 7/8	5	5	Jo Daviess County	IL	Paul K. Thompson, Sr.	2002	1452
134 7/8	23 1/8	22 6/8	16 1/8	4	5	Meigs County	OH	Josh Foster	2003	1452
134 7/8	23 2/8	23 3/8	19 1/8	4	4	Douglas County	KS	Mike Dunnaway	2003	1452
134 7/8	20 2/8	21 7/8	18 5/8	5	5	Hancock County	IL	Bradley J. Swanson	2004	1452
134 7/8	21 2/8	21 2/8	14 5/8	5	5	Van Buren County	IA	Baree Weber	2008	1452
134 7/8	25 1/8	26 2/8	23 7/8	5	6	Sherburne County	MN	Richard A. Iano	2009	1452
134 6/8	25 2/8	24 7/8	15 6/8	4	4	Newton County	IN	Charles Oliver, Sr.	1967	1470
134 6/8	20 0/8	20 0/8	16 5/8	5	6	Hughes County	SD	Dean Gretschmann	1970	1470
134 6/8	27 1/8	25 1/8	21 0/8	3	4	Spotsylvania County	VA	Michael Richards	1988	1470
134 6/8	21 6/8	21 5/8	16 2/8	5	5	La Salle County	IL	Micheal Underhill	1991	1470
134 6/8	21 7/8	23 0/8	17 2/8	6	4	McLean County	IL	David Grizzle	1991	1470
134 6/8	23 1/8	22 6/8	17 2/8	5	5	Lincoln County	NE	Dave Hinton	1992	1470
134 6/8	20 2/8	21 2/8	16 6/8	4	4	Clay County	NE	George F. Hoffmeister, Jr.	1992	1470
134 6/8	22 3/8	24 3/8	16 6/8	4	4	Christian County	MO	Roger Satterfield	1996	1470
134 6/8	21 5/8	23 1/8	18 0/8	5	5	Monroe County	NY	Daniel Scorza	1998	1470
134 6/8	21 4/8	21 7/8	18 6/8	5	5	Arapahoe County	CO	Louis Phillippe	1998	1470
134 6/8	24 4/8	24 7/8	20 0/8	4	4	McMullen County	TX	Bob Gilbert	1999	1470
134 6/8	21 4/8	21 1/8	17 0/8	5	5	Christian County	KY	Brian Bliese	2002	1470
134 6/8	23 4/8	22 5/8	16 2/8	5	4	Charles Mix County	SD	Dwight D. Pochant	2003	1470
134 6/8	24 1/8	25 0/8	17 6/8	4	5	Allen County	OH	Lee Yoakam	2004	1470
134 6/8	22 6/8	22 2/8	17 3/8	7	6	Antelope County	NE	Dan Schrage	2009	1470
134 5/8	20 6/8	22 4/8	18 7/8	7	6	Floyd County	IA	James K. Harris	1972	1485
134 5/8	22 2/8	22 0/8	16 2/8	7	6	Randolph County	WV	Robert E. Nace	1979	1485
134 5/8	24 5/8	24 7/8	18 3/8	4	5	Dodge County	WI	Ken Bialoszynski	1980	1485
134 5/8	22 1/8	23 1/8	15 3/8	5	5	Knox County	IL	David Emken	1988	1485

WHITETAIL DEER [Typical Antlers]

Minimum Score 125 *Continued*

Score	Length of R Main Beam	L	Inside Spread	Number of R Points	L	Area	State/ Province	Hunter's Name	Date	Rank
134 5/8	22 7/8	21 7/8	17 5/8	5	4	Morrison County	MN	Scott R. Kiekow	1992	1485
134 5/8	24 5/8	23 6/8	18 3/8	5	4	Lake County	IL	Carl H. Spaeth	1996	1485
134 5/8	22 0/8	22 0/8	20 1/8	5	5	Morgan County	CO	Pat Norris	1998	1485
134 5/8	23 2/8	24 1/8	17 7/8	5	5	Shelby County	IN	Joe Havlin	2004	1485
134 5/8	21 2/8	20 6/8	14 5/8	5	7	Potter County	SD	Randy Mink	2007	1485
134 4/8	23 5/8	21 1/8	17 2/8	4	4	Faribault County	MN	Sherwood F. Krosch	1969	1494
134 4/8	19 0/8	20 4/8	20 5/8	5	5	Morrison County	MN	Harold A. Walsh	1971	1494
134 4/8	19 7/8	19 5/8	17 6/8	5	5	Ripley County	IN	Robert Pitt	1974	1494
134 4/8	22 0/8	22 3/8	14 2/8	5	5	Buffalo County	WI	Douglas Henderson	1986	1494
134 4/8	23 1/8	23 3/8	20 4/8	4	4	Marion County	MO	Kurt Ebers	1989	1494
134 4/8	22 6/8	22 3/8	13 6/8	5	5	Jefferson County	OH	James C. Riggle	1990	1494
134 4/8	24 3/8	23 5/8	18 2/8	4	4	Licking County	OH	Ron Spence	1990	1494
134 4/8	20 6/8	20 7/8	14 4/8	6	6	Hitchcock County	NE	William Elfland	1990	1494
134 4/8	22 6/8	21 7/8	17 3/8	5	7	Madison County	MT	James Nygaard	1996	1494
134 4/8	21 6/8	20 3/8	17 2/8	5	5	Chautauqua County	NY	James M. Hoffman	1998	1494
134 4/8	22 0/8	21 6/8	18 2/8	5	5	Osage County	OK	Rick Gilley	2000	1494
134 4/8	23 4/8	22 6/8	18 2/8	5	4	Vinton County	OH	Martin E. Cain	2002	1494
134 4/8	22 5/8	23 1/8	19 0/8	5	5	Prairie River	SAS	Ken H. Taylor	2005	1494
134 3/8	23 4/8	22 1/8	17 1/8	4	5	Arkansas County	AR	Louis Rush	1959	1507
134 3/8	20 5/8	20 5/8	17 3/8	5	5	Rice County	MN	Gary Roemhildt	1965	1507
134 3/8	21 6/8	21 3/8	15 4/8	4	5	Cecil County	MD	Bernard Langhorne	1972	1507
134 3/8	20 5/8	20 4/8	17 1/8	5	5	Powell County	MT	Paul E. Tadlock	1972	1507
134 3/8	22 7/8	22 0/8	17 6/8	5	4	Pike County	IL	Dan Wombles	1981	1507
134 3/8	23 4/8	22 4/8	18 5/8	4	4	Kingman County	KS	Ken Marsh	1984	1507
134 3/8	23 4/8	23 6/8	19 1/8	5	4	Forest County	WI	Randy Schultz	1997	1507
134 3/8	22 6/8	21 7/8	18 3/8	6	6	Woodbury County	IA	Curtis P. Pedersen	1998	1507
134 3/8	19 3/8	19 2/8	15 1/8	5	6	Page County	IA	Bob Athen	1999	1507
134 3/8	21 1/8	22 5/8	16 5/8	6	5	Ogle County	IL	Marc A. Brumbly	2000	1507
134 2/8	21 0/8	21 2/8	16 4/8	5	5	Benton County	IA	Larry Walker	1961	1517
134 2/8	22 2/8	21 4/8	16 0/8	5	5	Perry County	IN	Jack E. Hungate	1963	1517
134 2/8	21 4/8	22 7/8	18 0/8	4	4	Barton County	KS	Dr. Nicholas J. Gray	1973	1517
134 2/8	20 4/8	21 4/8	17 4/8	5	5	Muskingum County	OH	Allen R. Smith	1980	1517
134 2/8	24 1/8	22 4/8	16 2/8	4	5	Will County	IL	Hugh M. MacCracken	1986	1517
134 2/8	21 5/8	21 1/8	16 6/8	5	5	Waukesha County	WI	Frank Turck, Jr.	1987	1517
134 2/8	24 1/8	23 6/8	16 4/8	4	6	Genesee County	MI	Clifford R. Neville, Sr.	1991	1517
134 2/8	22 5/8	22 5/8	16 4/8	4	4	Morton County	ND	Todd Schaedler	1994	1517
134 2/8	21 7/8	23 6/8	14 1/8	5	6	Pittsburg County	OK	Alan L. Rundell	1999	1517
134 2/8	23 5/8	22 7/8	15 2/8	4	4	Madison County	VA	Travis S. Lillard	2001	1517
134 2/8	20 3/8	20 4/8	17 0/8	5	5	Medina County	TX	Roger Wintle	2001	1517
134 1/8	23 2/8	23 3/8	20 1/8	5	5	Brown County	IN	Jason Thompson	1966	1528
134 1/8	21 2/8	22 6/8	18 2/8	7	5	Vanderburgh County	IN	William H. Davis	1967	1528
134 1/8	21 1/8	19 4/8	17 3/8	5	5	Genesee County	MI	Richard J. Parkhurst	1982	1528
134 1/8	21 3/8	21 2/8	16 7/8	5	5	Buffalo County	WI	Gary Aron	1996	1528
134 1/8	25 4/8	26 0/8	18 6/8	6	6	Licking County	OH	Jeff Varney	2000	1528
134 0/8	22 3/8	21 7/8	15 2/8	5	5	Ravalli County	MT	Gary Lee Hall	1990	1533
134 0/8	22 7/8	22 7/8	17 5/8	6	7	Frederick County	MD	Harry T. Lackey, Jr.	1990	1533
134 0/8	23 5/8	23 0/8	17 2/8	5	4	Dimmit County	TX	Hugh A. Fitzsimons III	1992	1533
134 0/8	21 6/8	21 2/8	18 4/8	4	4	Pottawattamie County	IA	Ron Rockwell	1994	1533
134 0/8	22 7/8	22 3/8	15 6/8	5	5	Walworth County	SD	Dave D. Knecht	1995	1533
134 0/8	22 6/8	20 3/8	16 1/8	7	6	Richland County	MT	R. Tim Reed	1995	1533
134 0/8	23 2/8	23 1/8	17 2/8	6	5	Webster County	IA	David Rial	1996	1533
134 0/8	20 7/8	20 3/8	19 2/8	5	5	Nemaha County	NE	Danny E. Williams	1998	1533
134 0/8	20 3/8	19 0/8	18 2/8	6	5	Jefferson County	WI	Paul Alane	1999	1533
134 0/8	25 3/8	24 5/8	21 6/8	5	6	Winneshiek County	IA	Randy E. Doyle	2002	1533
134 0/8	22 1/8	22 3/8	18 6/8	8	5	Jackson County	MI	John A. Hay	2004	1533
134 0/8	22 1/8	21 7/8	14 6/8	5	5	Irion County	TX	Andrew Stanco	2005	1533
134 0/8	22 6/8	22 3/8	16 4/8	4	4	Fayette County	OH	Whitlow Wyatt	2005	1533
134 0/8	22 3/8	23 1/8	14 4/8	4	5	Van Buren County	IA	Jim Reints	2009	1533
133 7/8	22 7/8	23 0/8	20 5/8	6	7	Dodge County	MN	Myles Keller	1976	1547
133 7/8	21 6/8	21 4/8	14 5/8	5	5	Sumner County	KS	Albert Whaley	1978	1547
133 7/8	23 1/8	23 3/8	16 1/8	5	4	Knox County	IL	Dave Emken	1982	1547
133 7/8	23 1/8	23 4/8	19 1/8	4	4	Livingston County	NY	David H. Kosowski	1989	1547

WHITETAIL DEER [Typical Antlers]

Minimum Score 125 *Continued*

Score	Length of Main Beam R	Length of Main Beam L	Inside Spread	Number of Points R	Number of Points L	Area	State/ Province	Hunter's Name	Date	Rank
133 7/8	23 1/8	23 7/8	15 1/8	4	5	Du Page County	IL	Richard C. Klimes	1993	1547
133 7/8	22 1/8	23 1/8	20 7/8	5	5	Winnebago County	IL	Dave Judy	1997	1547
133 7/8	25 3/8	25 3/8	23 3/8	4	4	Clinton County	IL	David Baum	2001	1547
133 7/8	22 6/8	21 6/8	16 5/8	5	6	Wyandotte County	KS	Jim Shanks	2002	1547
133 7/8	21 4/8	21 0/8	17 1/8	5	5	Hancock County	OH	Kenny Shoemaker	2003	1547
133 7/8	21 2/8	20 3/8	15 5/8	5	6	Tuscarawas County	OH	Steve Archer	2004	1547
133 7/8	24 5/8	24 3/8	16 6/8	6	6	Franklin County	IL	Darin DeNeal	2005	1547
133 7/8	18 4/8	18 7/8	14 5/8	5	5	Gallatin County	MT	Bob Morton	2009	1547
133 7/8	20 7/8	20 4/8	16 1/8	5	5	Rice County	KS	John L. Gardner	2010	1547
133 7/8	20 6/8	21 0/8	15 5/8	5	5	Perry County	PA	Michael J. Splain	2011	1547
133 6/8	22 6/8	24 7/8	17 2/8	5	4	Tama County	IA	Chad Sivertsen	1971	1561
133 6/8	24 2/8	25 2/8	19 7/8	4	6	McPherson County	KS	Mike Chambers	1975	1561
133 6/8	22 5/8	21 5/8	17 4/8	4	4	St. Marys County	MD	Marvin T. Breeden	1979	1561
133 6/8	21 7/8	20 4/8	19 2/8	5	5	La Porte County	IN	Ted J. Saliwanchik	1991	1561
133 6/8	24 2/8	23 4/8	18 7/8	5	4	Clayton County	IA	Terry Krahn	1992	1561
133 6/8	23 7/8	24 0/8	18 6/8	4	4	Monroe County	NY	Dominic Chirico	1997	1561
133 6/8	23 2/8	23 3/8	17 2/8	4	5	Marion County	KS	Tony Cassity	1999	1561
133 6/8	22 3/8	22 5/8	17 2/8	5	5	Will County	IL	John Noonan	2000	1561
133 6/8	23 2/8	23 4/8	19 2/8	5	5	Licking County	OH	Stephen B. Cooksey	2002	1561
133 6/8	22 1/8	23 3/8	17 3/8	5	6	Jackson County	MI	Russell P. Blair	2003	1561
133 6/8	20 6/8	21 0/8	15 2/8	8	5	Kenedy County	TX	Bob Gilbert	2005	1561
133 6/8	21 1/8	21 2/8	16 2/8	6	6	Meeker County	MN	Jason Schultz	2011	1561
133 5/8	23 3/8	23 2/8	17 1/8	4	4	Tripp County	SD	Larry Diehm	1965	1573
133 5/8	19 6/8	19 4/8	16 5/8	5	5	Brookings County	SD	Rodney Foster	1967	1573
133 5/8	26 6/8	25 7/8	19 3/8	3	4	Blue Earth County	MN	Dean Como	1973	1573
133 5/8	22 6/8	22 7/8	15 3/8	5	5	Murray County	MN	David D. Swanson	1979	1573
133 5/8	21 6/8	21 7/8	17 2/8	5	6	Washington County	MN	Leonard Ellingson	1979	1573
133 5/8	25 6/8	25 3/8	18 0/8	6	6	Beltrami County	MN	Ellsworth C. Underwood	1983	1573
133 5/8	24 0/8	26 0/8	21 7/8	3	5	Stearns County	MN	Steve Nelson	1991	1573
133 5/8	20 4/8	21 1/8	16 3/8	5	5	Valley County	MT	Jim Miller	1993	1573
133 5/8	21 5/8	21 5/8	21 6/8	5	5	Buffalo County	WI	Bob A. Brunkow	1994	1573
133 5/8	22 2/8	22 2/8	19 5/8	5	5	Clay County	AR	Phillip Vanderpool	1994	1573
133 5/8	22 1/8	21 3/8	19 3/8	5	6	Hardin County	OH	Kyle J. Joseph	1995	1573
133 5/8	21 5/8	22 1/8	16 7/8	4	6	Bexar County	TX	Deno Poulos	1997	1573
133 5/8	24 6/8	24 6/8	17 3/8	4	4	Hampden County	MA	Robbie Ober	1999	1573
133 5/8	22 3/8	23 3/8	17 4/8	4	6	Jackson County	MI	Eric Tooley	2001	1573
133 5/8	21 2/8	21 0/8	15 2/8	5	6	Okmulgee County	OK	Wayne Willis	2002	1573
133 5/8	20 4/8	20 7/8	14 1/8	6	6	Marquette County	WI	Jeffrey G. Frazer	2003	1573
133 5/8	24 7/8	24 3/8	18 3/8	4	4	Muskingum County	OH	Robert S. Hageter	2005	1573
133 5/8	23 2/8	23 3/8	19 3/8	5	4	Mountrail County	ND	Craig Richardson	2007	1573
133 4/8	22 2/8	22 0/8	17 2/8	5	5	Burleigh County	ND	Lyle F. Fischer	1956	1591
133 4/8	23 5/8	23 1/8	19 5/8	4	4	Washington County	NE	John Johnson	1958	1591
133 4/8	22 7/8	23 2/8	16 5/8	5	4	Hocking County	OH	Ted Schultz	1974	1591
133 4/8	19 5/8	19 3/8	15 0/8	5	5	Sargent County	ND	Frank Pfeifer	1975	1591
133 4/8	22 2/8	22 2/8	17 0/8	6	5	Dawson County	NE	Randy Ray Wilson	1988	1591
133 4/8	22 4/8	22 3/8	21 0/8	5	5	Bradford County	PA	Gary Dewey	1988	1591
133 4/8	21 5/8	21 7/8	18 0/8	5	4	Jefferson County	ID	Douglas S. Kauer	1996	1591
133 4/8	22 3/8	22 2/8	17 2/8	4	5	McDowell County	WV	Ray Burks	1998	1591
133 4/8	21 4/8	20 1/8	14 0/8	8	8	Craig County	OK	Mary Parish	2001	1591
133 4/8	22 1/8	21 2/8	18 6/8	5	5	Dodge County	WI	David R. Spellman	2005	1591
133 4/8	22 4/8	23 4/8	14 1/8	6	5	Warren County	IA	Edward Hyde	2006	1591
133 4/8	19 6/8	20 0/8	16 5/8	6	4	Calhoun County	MI	George Swan, Jr.	2008	1591
133 3/8	23 7/8	24 2/8	16 7/8	4	5	Green Lake County	WI	Frank Parker	1942	1603
133 3/8	21 0/8	21 2/8	19 6/8	5	4	Watonwan County	MN	Issac Davis	1962	1603
133 3/8	24 7/8	24 1/8	23 1/8	6	6	Morrison County	MN	Frank Salisbury	1963	1603
133 3/8	22 6/8	24 0/8	18 0/8	5	5	Dickinson County	IA	Keith F. Ellis	1973	1603
133 3/8	21 0/8	21 3/8	17 1/8	6	6	Juneau County	WI	James M. Carriveau	1977	1603
133 3/8	25 0/8	23 2/8	19 1/8	5	5	Jackson County	OH	Robert E. Thomas, Jr.	1986	1603
133 3/8	21 0/8	19 2/8	14 1/8	5	4	Haskell County	OK	Jerry C. Sturdy	1993	1603
133 3/8	22 4/8	22 3/8	15 3/8	5	6	Prairie County	AR	Chris Patton	1994	1603
133 3/8	24 1/8	24 2/8	14 3/8	4	4	Alachua County	FL	Craig C. Carpenter	1995	1603
133 3/8	23 7/8	23 7/8	21 7/8	6	5	Dodge County	WI	Mike Young	2001	1603

WHITETAIL DEER [Typical Antlers]

Minimum Score 125 *Continued*

Score	Length of R Main Beam	L	Inside Spread	Number of R Points	L	Area	State/ Province	Hunter's Name	Date	Rank
133 3/8	22 1/8	22 6/8	14 1/8	6	5	Greene County	IN	Dave Johnson	2001	1603
133 3/8	22 7/8	23 6/8	16 7/8	4	4	Ravalli County	MT	Marlon Clapham	2011	1603
133 2/8	18 4/8	18 6/8	14 7/8	5	6	Mitchell County	IA	Omar A. Toye	1958	1615
133 2/8	23 6/8	22 5/8	18 2/8	8	8	Olmsted County	MN	Roger E. Enderson	1961	1615
133 2/8	24 2/8	23 6/8	18 5/8	4	4	Pittsburg County	OK	Fred Parkison	1968	1615
133 2/8	21 1/8	20 7/8	18 2/8	5	5	Wabasha County	MN	Myles Keller	1977	1615
133 2/8	24 6/8	23 3/8	19 3/8	8	6	Fulton County	IL	Patrick Cebuhar	1988	1615
133 2/8	21 4/8	20 1/8	16 2/8	5	5	Pike County	IL	Chris Johnson	1990	1615
133 2/8	23 5/8	22 2/8	15 6/8	4	4	St. Marys County	MD	Keith Morgan	1992	1615
133 2/8	22 1/8	21 6/8	17 4/8	4	4	Hand County	SD	Ed Werdel	1994	1615
133 2/8	22 7/8	23 3/8	20 2/8	5	5	Sarpy County	NE	Keith Southworth	1995	1615
133 2/8	20 3/8	21 0/8	15 0/8	5	5	Gage County	NE	Brant Washburn	1996	1615
133 2/8	24 0/8	23 5/8	17 2/8	5	5	Franklin County	MO	Jim Bogert	1999	1615
133 2/8	22 6/8	23 0/8	17 2/8	4	5	Price County	WI	Kenneth Busalacchi	2000	1615
133 2/8	21 0/8	22 1/8	18 0/8	5	5	Concho County	TX	Bill Young	2007	1615
133 2/8	25 0/8	23 0/8	22 0/8	5	5	Delaware County	OH	Reno F. Carifa	2012	1615
133 1/8	22 2/8	21 5/8	16 1/8	4	4	Roberts County	SD	Roland L. Hausmann	1964	1629
133 1/8	24 5/8	23 7/8	20 1/8	4	4	Perry County	IL	Ron Lay	1974	1629
133 1/8	20 5/8	20 5/8	17 2/8	6	6	Cowley County	KS	Dr. Phil L. Bradley	1975	1629
133 1/8	26 0/8	26 3/8	19 6/8	5	5	Litchfield County	CT	Phillip M. Demetri	1976	1629
133 1/8	19 6/8	19 6/8	16 2/8	6	5	Eau Claire County	WI	David Nelson	1984	1629
133 1/8	23 2/8	23 0/8	18 3/8	5	5	Pottawatomie County	KS	Michael Schirer	1988	1629
133 1/8	22 3/8	21 7/8	19 1/8	5	4	Wellington County	ONT	R. Brian Oates	1989	1629
133 1/8	21 1/8	21 7/8	16 7/8	5	5	Clinton County	MI	Dan Bertalan	1990	1629
133 1/8	22 4/8	22 5/8	16 3/8	4	4	Van Buren County	IA	John "Rosey" Roseland	1995	1629
133 1/8	24 7/8	24 4/8	17 7/8	4	5	Delaware County	IN	Jeff S. Bell	2004	1629
133 1/8	22 1/8	22 0/8	18 3/8	6	5	Clark County	SD	Mark Turner	2005	1629
133 1/8	20 2/8	22 2/8	14 1/8	6	5	Laclede County	MO	Jim Atkinson	2010	1629
133 0/8	22 2/8	21 3/8	16 4/8	4	4	Luce County	MI	Rondell Bisbee	1953	1641
133 0/8	24 0/8	24 1/8	17 6/8	4	4	Dunn County	WI	George Woodington	1969	1641
133 0/8	19 0/8	19 3/8	16 6/8	5	5	Kingsbury County	SD	Arnold Aulner	1970	1641
133 0/8	20 7/8	21 0/8	17 4/8	5	5	Litchfield County	CT	Dan Ferrara, Jr.	1976	1641
133 0/8	23 5/8	22 1/8	17 4/8	5	6	Macon County	IL	Frank B. Graham	1977	1641
133 0/8	27 2/8	28 3/8	16 1/8	6	6	Lawrence County	IL	Hugh Sexton	1986	1641
133 0/8	24 4/8	24 4/8	17 4/8	5	4	Pittsburg County	OK	K. E. Pennington	1992	1641
133 0/8	22 1/8	22 5/8	17 4/8	5	5	Dakota County	MN	Lee S. Waltman	1996	1641
133 0/8	20 5/8	20 2/8	17 0/8	5	5	Missoula County	MT	Eric Kress	2004	1641
133 0/8	25 4/8	25 4/8	19 0/8	4	5	Webb County	TX	Bob Gilbert	2005	1641
133 0/8	21 6/8	22 5/8	15 6/8	4	5	Ripley County	IN	Jeff Cumberworth	2007	1641
133 0/8	21 6/8	22 2/8	15 6/8	5	6	Montgomery County	PA	Ken Barnett	2007	1641
133 0/8	19 7/8	20 2/8	17 2/8	5	5	Westmoreland County	PA	Daniel K. Kuehn	2010	1641
132 7/8	21 0/8	21 0/8	15 3/8	5	7	Jefferson County	IN	Jerome Sexton	1962	1654
132 7/8	20 0/8	18 6/8	15 5/8	5	6	Wilson County	KS	Warren Townsend	1970	1654
132 7/8	23 0/8	23 3/8	18 6/8	6	7	Sauk County	WI	Richard J. Osgood	1987	1654
132 7/8	23 3/8	22 7/8	18 7/8	4	4	Latah County	ID	Roger A. Rea	1990	1654
132 7/8	20 3/8	20 5/8	14 3/8	4	4	Jackson County	IA	Nick Fondell	1991	1654
132 7/8	20 3/8	20 5/8	14 5/8	6	5	La Salle County	TX	Roger Wintle	1992	1654
132 7/8	25 6/8	24 6/8	18 5/8	5	4	Dodge County	WI	Mike Young	1995	1654
132 7/8	21 5/8	20 6/8	14 5/8	5	5	Butler County	MO	Steve Rueck	1995	1654
132 7/8	22 2/8	20 7/8	18 2/8	5	6	Dundy County	NE	Ronald King	2007	1654
132 7/8	19 6/8	19 1/8	16 1/8	5	5	Bennett County	SD	Justin Smokov	2009	1654
132 6/8	23 0/8	23 4/8	20 0/8	4	4	Summit County	OH	Dana C. Feather	1960	1664
132 6/8	22 5/8	21 7/8	17 0/8	5	6	Waseca County	MN	Robert Barrie	1968	1664
132 6/8	19 7/8	22 7/8	17 4/8	6	6	Union County	IL	Fred W. Achilles	1978	1664
132 6/8	21 6/8	21 1/8	17 0/8	4	5	Grundy County	IL	Wesley Holm	1980	1664
132 6/8	21 3/8	21 1/8	17 4/8	4	4	Clinton County	IA	Kent Hoffmann	1982	1664
132 6/8	23 0/8	22 1/8	16 6/8	4	4	La Salle County	TX	Dennis Faulkenberry	1992	1664
132 6/8	20 7/8	20 7/8	15 6/8	5	5	Mower County	MN	John Mullen	1993	1664
132 6/8	20 2/8	20 5/8	16 4/8	5	5	Christian Valley	BC	Ray Quaedvlieg	1993	1664
132 6/8	22 0/8	22 3/8	20 0/8	6	7	Madison County	MT	Jim Kennedy	1996	1664
132 6/8	20 5/8	21 5/8	23 0/8	4	5	Vilas County	WI	Thomas M. Bablitch	2001	1664
132 5/8	23 3/8	24 4/8	21 5/8	4	6	Tazewell County	IL	Don Lounsberry	1964	1674

Minimum Score 125 *Continued*

Score	Length of Main Beam R	Length of Main Beam L	Inside Spread	Number of Points R	Number of Points L	Area	State/ Province	Hunter's Name	Date	Rank
132 5/8	24 2/8	23 0/8	19 2/8	6	6	Juneau County	WI	Michael Sigler	1978	1674
132 5/8	21 5/8	21 3/8	19 1/8	4	4	Clinton County	IA	Gary G. Olson	1987	1674
132 5/8	21 2/8	19 5/8	15 0/8	5	6	Gallatin County	MT	Bob Morton	1991	1674
132 5/8	19 5/8	21 5/8	17 4/8	5	6	Washburn County	WI	Jeff Tomesh	1991	1674
132 5/8	22 1/8	21 1/8	15 7/8	4	4	Comanche County	KS	Ronald Rockwell	2001	1674
132 5/8	21 0/8	21 5/8	16 1/8	5	5	Wabasha County	MN	Keith Ramthun	2007	1674
132 5/8	22 1/8	22 1/8	19 6/8	5	6	Oklahoma County	OK	Ken Jose	2007	1674
132 5/8	21 7/8	23 0/8	17 3/8	4	4	Mountrail County	ND	John Richardson	2009	1674
132 5/8	23 4/8	23 6/8	18 4/8	4	5	Las Animas County	CO	Fred Eichler	2009	1674
132 5/8	21 2/8	20 4/8	16 7/8	5	5	Ingham County	MI	Jim Kinsey	2010	1674
132 4/8	23 4/8	23 0/8	16 3/8	6	5	Jackson County	WI	Howard Knockel	1957	1685
132 4/8	25 4/8	26 0/8	21 0/8	4	4	Valley County	MT	Clare F. Mates	1961	1685
132 4/8	21 6/8	21 6/8	16 0/8	5	5	Cowley County	KS	William L. Walker	1965	1685
132 4/8	21 5/8	21 7/8	15 6/8	4	4	Iowa County	IA	Mel Berstler	1965	1685
132 4/8	24 5/8	24 6/8	17 3/8	7	5	Macon County	IL	Wayne Orrell	1976	1685
132 4/8	19 7/8	21 5/8	16 0/8	5	5	Lawrence County	SD	Micheal D. Hudson	1983	1685
132 4/8	18 5/8	18 7/8	15 0/8	5	5	Morton County	ND	Todd Schaedler	1988	1685
132 4/8	23 6/8	24 5/8	15 4/8	5	5	Lincoln County	NE	Steve Stumbo	1989	1685
132 4/8	24 4/8	23 3/8	15 3/8	7	5	Fulton County	IL	Rick Dilts	1990	1685
132 4/8	22 2/8	23 4/8	18 4/8	5	5	Fillmore County	MN	Steve W. Utley	1991	1685
132 4/8	20 7/8	22 0/8	17 0/8	5	5	Sheboygan County	WI	James Overbeck	1997	1685
132 4/8	22 0/8	19 0/8	15 4/8	5	7	McHenry County	IL	Joe Florent	2002	1685
132 4/8	19 5/8	20 4/8	16 4/8	5	6	Black Hawk County	IA	Mike Judas	2003	1685
132 3/8	20 5/8	20 6/8	14 3/8	4	4	Buffalo County	NE	Al Dawson	1961	1698
132 3/8	21 2/8	21 0/8	14 5/8	4	4	St. Joseph County	IN	Harry Ramsbey, Jr.	1963	1698
132 3/8	22 0/8	21 2/8	17 4/8	4	4	Morrison County	MN	Jay J. Jost	1968	1698
132 3/8	21 6/8	22 6/8	18 7/8	5	6	Spink County	SD	Roger Michels	1968	1698
132 3/8	22 6/8	21 7/8	15 3/8	4	4	Shawano County	WI	Gene M. Waite	1971	1698
132 3/8	22 6/8	22 2/8	17 3/8	4	4	Jefferson County	ID	Douglas Kauer	1994	1698
132 3/8	22 6/8	22 6/8	22 5/8	4	4	Bayfield County	WI	Rick L. Brinker	1998	1698
132 3/8	22 6/8	23 0/8	17 1/8	4	4	Schuyler County	IL	Sean S. Walp	2003	1698
132 3/8	19 4/8	20 5/8	16 5/8	5	5	McMullen County	TX	Jerome Bischoffberger	2007	1698
132 3/8	22 5/8	22 6/8	17 7/8	6	7	Kewaunee County	WI	Mark Nelis	2007	1698
132 3/8	20 6/8	21 1/8	17 7/8	5	5	Carroll County	MS	Trent Trussell	2008	1698
132 2/8	22 1/8	21 7/8	16 6/8	5	4	Waupaca County	WI	Al Wiltzius	1949	1709
132 2/8	22 0/8	22 4/8	19 2/8	4	4	Bullitt County	KY	Dell Pack	1962	1709
132 2/8	21 7/8	23 2/8	16 6/8	5	8	Cherokee County	IA	Darrell Magnussen	1962	1709
132 2/8	21 4/8	23 5/8	20 4/8	4	5	Roberts County	SD	Clayton Forrette	1967	1709
132 2/8	24 6/8	25 2/8	23 4/8	3	3	Stoddard County	MO	Ted Denkins	1971	1709
132 2/8	21 3/8	21 7/8	16 4/8	4	4	Westmoreland County	PA	Robert C. Kirschner	1971	1709
132 2/8	20 0/8	20 2/8	15 4/8	5	5	Ravalli County	MT	Dale E. Dye	1996	1709
132 2/8	22 0/8	22 2/8	14 4/8	4	4	Jim Hogg County	TX	Bob Gilbert	1997	1709
132 2/8	21 0/8	20 5/8	16 2/8	5	5	Delaware County	IA	Kelly Salow	2000	1709
132 2/8	22 6/8	22 5/8	15 4/8	6	5	Pike County	OH	Thomas E. McKelvey, Jr.	2001	1709
132 2/8	22 7/8	17 4/8	17 6/8	5	5	Wright County	MN	Allan E. Kasper	2005	1709
132 2/8	22 0/8	21 5/8	19 0/8	4	5	Monroe County	IA	Scott Hargrove	2008	1709
132 2/8	23 3/8	23 6/8	19 0/8	6	4	Carroll County	IL	Jim Carlson	2010	1709
132 2/8	22 1/8	21 5/8	16 4/8	4	4	Brown County	OH	Tom Weaver	2011	1709
132 1/8	18 1/8	19 7/8	18 3/8	5	5	Buffalo County	NE	Bill Orsborn	1962	1723
132 1/8	20 3/8	20 0/8	17 1/8	5	5	Rice County	KS	Robert Lagree	1967	1723
132 1/8	22 4/8	24 1/8	16 3/8	4	5	Jasper County	IA	Paul Casper	1972	1723
132 1/8	20 1/8	20 7/8	14 3/8	5	5	Buffalo County	WI	David J. Gard	1987	1723
132 1/8	23 0/8	22 5/8	18 0/8	5	7	Coshocton County	OH	Gene Mathias	1994	1723
132 1/8	21 5/8	21 2/8	16 5/8	4	4	Thomas County	NE	Gary M. Hornberger	1997	1723
132 1/8	23 4/8	23 2/8	15 1/8	5	5	Webb County	TX	Troy Hardesty	2000	1723
132 1/8	23 3/8	23 7/8	17 6/8	5	6	Charles County	MD	Phil Harris	2000	1723
132 1/8	21 0/8	21 0/8	14 3/8	6	5	McKenzie County	ND	Tim Novak	2005	1723
132 1/8	25 0/8	24 7/8	18 2/8	7	6	Sheridan County	KS	Jerry Bowen	2007	1723
132 1/8	22 5/8	19 7/8	18 3/8	5	5	La Salle County	IL	Richard L. Andre	2011	1723
132 0/8	21 4/8	21 2/8	18 2/8	4	4	Juneau County	WI	Arthur Witz	1966	1734
132 0/8	22 5/8	22 2/8	18 6/8	5	5	Iowa County	IA	Larry King	1967	1734
132 0/8	21 6/8	22 4/8	17 0/8	4	4	Harlan County	NE	Edward H. Backes	1969	1734

Score	Length of R Main Beam	L	Inside Spread	Number of R Points	L	Area	State/ Province	Hunter's Name	Date	Rank
132 0/8	20 0/8	20 5/8	14 6/8	4	4	Pittsburg County	OK	Fred Parkison	1969	1734
132 0/8	21 4/8	21 1/8	16 4/8	6	5	Dunn County	WI	John R. Bilderback	1970	1734
132 0/8	23 4/8	23 5/8	16 6/8	5	4	Wabasha County	MN	Myles Keller	1978	1734
132 0/8	20 6/8	21 3/8	18 5/8	5	5	Cambridge	ONT	Fred Law	1979	1734
132 0/8	21 6/8	21 5/8	16 1/8	6	5	Coweta County	GA	C. M. Edwards, Jr.	1982	1734
132 0/8	23 4/8	23 4/8	18 2/8	4	4	Guernsey County	OH	Dick Bayer	1987	1734
132 0/8	21 2/8	21 5/8	16 4/8	5	5	Fayette County	IL	Richard L. Perry	1989	1734
132 0/8	22 6/8	23 4/8	19 2/8	4	4	McHenry County	IL	Richard Tudor	1991	1734
132 0/8	20 6/8	20 5/8	17 6/8	5	5	Meeker County	MN	Chuck Schultz	1991	1734
132 0/8	19 5/8	19 5/8	16 0/8	5	5	Valley County	MT	Dennis Sturgis, Jr.	1992	1734
132 0/8	21 3/8	21 5/8	16 2/8	5	5	Adams County	WI	Howard L. Edwards	1994	1734
132 0/8	21 4/8	21 5/8	18 7/8	5	4	Edgar County	IL	Brent Webster	1995	1734
132 0/8	23 0/8	22 4/8	15 2/8	4	4	Meriwether County	GA	David Fagundes	1999	1734
132 0/8	23 0/8	23 0/8	17 6/8	4	4	Pike County	IL	Ronnie Cannon	1999	1734
132 0/8	22 4/8	22 4/8	19 4/8	4	4	Decatur County	GA	Charles S. Caley	1999	1734
132 0/8	20 0/8	20 2/8	16 6/8	7	6	Fulton County	IN	William D. Knarr	2001	1734
132 0/8	22 7/8	20 7/8	17 4/8	6	5	McHenry County	IL	Chad G. Lindholm	2002	1734
132 0/8	22 0/8	21 3/8	21 6/8	5	5	Randolph County	IN	Michael E. Coning	2004	1734
132 0/8	24 4/8	24 4/8	19 6/8	4	4	Jewell County	KS	Nathan Andersohn	2005	1734
131 7/8	22 0/8	22 7/8	22 2/8	6	4	Goodhue County	MN	James Jarvis	1964	1756
131 7/8	21 1/8	22 1/8	17 5/8	4	4	Brown County	SD	Barry Smith	1967	1756
131 7/8	22 0/8	23 1/8	15 1/8	4	5	Trempealeau County	WI	Clark Gallup	1967	1756
131 7/8	20 3/8	20 6/8	14 6/8	6	6	Adams County	IL	David E. DeMoss	1970	1756
131 7/8	22 2/8	21 1/8	20 5/8	4	5	Rock County	WI	Richard W. Pieterek	1972	1756
131 7/8	24 0/8	24 3/8	20 4/8	6	7	Harrison County	KY	Kendall Techau	1987	1756
131 7/8	23 2/8	22 2/8	20 1/8	5	4	Worcester County	MD	Mike Hill	1990	1756
131 7/8	20 5/8	20 4/8	15 5/8	5	5	Decatur County	IA	Ronald E. Tennant	1991	1756
131 7/8	21 5/8	21 3/8	19 4/8	6	7	Marion County	OH	Craig Ciola	1992	1756
131 7/8	24 0/8	22 0/8	16 1/8	5	5	Calhoun County	MI	Michael L. Newhouse	1995	1756
131 7/8	23 4/8	24 4/8	18 3/8	5	4	Stanton County	KS	Troy Cunningham	1999	1756
131 7/8	21 1/8	21 6/8	16 1/8	4	4	Miami County	OH	Mark K. Swallow	1999	1756
131 7/8	23 2/8	22 2/8	17 0/8	5	7	Vermilion County	IL	Robert Isom	2000	1756
131 7/8	23 6/8	23 6/8	17 2/8	4	7	Kent County	MD	Michael Travis	2005	1756
131 7/8	21 1/8	21 6/8	18 3/8	6	6	Douglas County	WI	Scott Soden	2005	1756
131 7/8	21 4/8	21 0/8	14 3/8	5	6	Montgomery County	TN	Keith Vaughn	2007	1756
131 7/8	22 4/8	22 3/8	17 5/8	4	4	Boone County	IN	Byron Padgett	2010	1756
131 6/8	19 0/8	18 6/8	13 6/8	6	6	Norton County	KS	Harold Fisher	1966	1773
131 6/8	22 3/8	22 1/8	17 2/8	4	5	Tuscarawas County	OH	Del Karnuth	1976	1773
131 6/8	20 6/8	21 7/8	19 4/8	5	5	Champaign County	IL	Robert E. Mabry	1978	1773
131 6/8	22 2/8	21 0/8	18 0/8	5	6	Kalamazoo County	MI	Richard Hettinga	1978	1773
131 6/8	21 5/8	22 3/8	16 4/8	4	4	St. Croix County	WI	Mark Crandall	1987	1773
131 6/8	23 1/8	23 3/8	17 0/8	4	4	Floyd County	IA	Joel Gray	1990	1773
131 6/8	19 7/8	19 5/8	13 4/8	5	5	Chickasaw County	IA	Dave Kerkove	1993	1773
131 6/8	21 2/8	22 4/8	13 6/8	4	5	Clinton County	MO	David Eads	1994	1773
131 6/8	22 4/8	22 0/8	16 6/8	4	4	Idaho County	ID	Wade L. Carstens	1997	1773
131 6/8	23 4/8	24 0/8	16 5/8	5	5	Marion County	IA	Donald C. Clark	1997	1773
131 6/8	23 3/8	23 0/8	20 6/8	4	4	Coshocton County	OH	Lee Cooper	2000	1773
131 6/8	20 6/8	20 6/8	19 0/8	5	5	Maverick County	TX	H. Mike Palmer	2002	1773
131 6/8	21 4/8	21 3/8	17 3/8	5	6	Young County	TX	Brad Hawpe	2005	1773
131 5/8	21 4/8	22 1/8	18 5/8	4	4	Forest County	WI	Max Wisnefske	1950	1786
131 5/8	19 7/8	19 7/8	12 3/8	5	5	Jefferson County	IN	Ted Taylor	1959	1786
131 5/8	23 4/8	22 5/8	14 1/8	4	5	Otter Tail County	MN	Gordon Swenson	1964	1786
131 5/8	25 1/8	25 2/8	16 7/8	4	4	Chatham County	NC	Jimmy Womble	1986	1786
131 5/8	23 3/8	23 3/8	19 5/8	5	6	Pepin County	WI	Michael Schmidt	1988	1786
131 5/8	21 6/8	22 4/8	15 5/8	6	6	Ft. Bend County	TX	Pat Byrne	1989	1786
131 5/8	21 3/8	21 6/8	18 1/8	4	4	Forest County	WI	Flint Gilbert	1991	1786
131 5/8	22 3/8	24 5/8	17 1/8	5	5	Calhoun County	IL	John Ciszczon	1991	1786
131 5/8	21 5/8	21 7/8	17 1/8	5	4	Lawrence County	IN	Arter Thompson II	1992	1786
131 5/8	22 5/8	22 4/8	23 0/8	4	4	Tuscarawas County	OH	Steve Archer	2001	1786
131 5/8	22 1/8	24 0/8	15 4/8	5	7	Adams County	OH	Ronald E. Perrine, Jr.	2002	1786
131 5/8	23 3/8	24 6/8	17 3/8	4	4	Fulton County	IL	Steve Miller	2004	1786
131 5/8	22 2/8	22 7/8	16 3/8	8	7	Hocking County	OH	Jerry Haynes	2006	1786

WHITETAIL DEER [Typical Antlers]

Minimum Score 125 *Continued*

Score	Length of R Main Beam	L	Inside Spread	Number of R Points	L	Area	State/ Province	Hunter's Name	Date	Rank
131 5/8	18 4/8	20 2/8	13 1/8	5	5	Logan County	OH	Van Roger Williams	2006	1786
131 5/8	21 5/8	20 3/8	15 5/8	5	5	Tuscola County	MI	Mark Jae Timko	2008	1786
131 5/8	23 1/8	23 0/8	14 3/8	5	5	Lincoln County	WV	Glen McClung III	2011	1786
131 4/8	20 4/8	20 1/8	17 2/8	5	5	Benton County	IA	Wayne Keefer	1956	1802
131 4/8	21 2/8	21 7/8	16 2/8	5	5	Johnson County	IN	James E. Thompson	1962	1802
131 4/8	20 2/8	22 1/8	14 2/8	5	5	Price County	WI	Frank W. Taylor	1964	1802
131 4/8	21 3/8	21 2/8	16 0/8	7	5	Roberts County	SD	Martin Carlson	1966	1802
131 4/8	23 1/8	21 5/8	17 4/8	5	4	Calvert County	MD	William L. Neal	1970	1802
131 4/8	19 2/8	19 2/8	15 2/8	5	5	Sheboygan County	WI	Donald P. Feidmann	1974	1802
131 4/8	23 3/8	23 4/8	16 4/8	4	4	Clinton County	IA	Gary Olson	1975	1802
131 4/8	24 2/8	22 6/8	16 6/8	4	4	Shelby County	OH	Wayne L. Goubeaux	1981	1802
131 4/8	21 3/8	20 6/8	18 2/8	5	5	Lake County	IL	Carl Spaeth	1988	1802
131 4/8	22 2/8	23 1/8	17 4/8	4	5	Lapeer County	MI	Larry Rae Faught	1989	1802
131 4/8	22 4/8	22 0/8	16 6/8	5	6	Logan County	CO	Terry Weimer	1990	1802
131 4/8	23 5/8	24 4/8	16 2/8	6	7	Cass County	MI	Patrick G. Dowling	1993	1802
131 4/8	24 4/8	21 3/8	18 1/8	5	7	Van Buren County	IA	Elmer R. Luce, Jr.	1993	1802
131 4/8	24 4/8	23 2/8	17 7/8	5	5	Franklin County	IN	Will McQueen	1993	1802
131 4/8	23 6/8	24 3/8	18 4/8	4	4	Montgomery County	MD	Frank H. Wilmot	1998	1802
131 4/8	20 5/8	20 5/8	17 6/8	6	5	Franklin County	IA	Arlynn Ahrens	1999	1802
131 4/8	20 5/8	21 3/8	16 4/8	5	5	Polk County	WI	Richard Montana	2002	1802
131 4/8	21 1/8	22 7/8	18 0/8	5	6	Brown County	IL	Woody R. Smith	2002	1802
131 4/8	18 6/8	18 0/8	17 0/8	5	5	Linn County	MO	Darvin B. McClellan	2002	1802
131 4/8	22 0/8	21 2/8	19 6/8	5	5	Jackson County	OH	Jim Ridge	2003	1802
131 3/8	22 2/8	22 6/8	18 7/8	5	5	Beltrami County	MN	Charles R. Bowman	1961	1822
131 3/8	21 7/8	21 6/8	17 7/8	4	4	Waushara County	WI	Reginald Vergin	1967	1822
131 3/8	21 2/8	20 7/8	18 1/8	5	5	Racine County	WI	Steven Holterman	1986	1822
131 3/8	22 7/8	22 5/8	21 1/8	4	4	Winneshiek County	IA	John Mullen	1989	1822
131 3/8	22 6/8	23 3/8	14 7/8	5	6	Hocking County	OH	Jim Bowman	1989	1822
131 3/8	22 2/8	22 4/8	17 7/8	4	4	McHenry County	IL	Ray Eisbrener	1994	1822
131 3/8	23 3/8	22 4/8	16 5/8	5	4	Jackson County	OH	Bob "Smokey" Lotts	2002	1822
131 3/8	23 0/8	23 7/8	18 4/8	5	5	Carroll County	IL	Jeff Carlson	2008	1822
131 2/8	19 2/8	19 4/8	17 0/8	5	5	Shawnee County	KS	Guy C. Michael	1968	1830
131 2/8	20 3/8	19 4/8	14 6/8	5	5	Ripley County	IN	Robert Pitt	1969	1830
131 2/8	24 5/8	22 5/8	16 0/8	4	5	Waushara County	WI	Randy Marks	1975	1830
131 2/8	20 5/8	22 0/8	17 4/8	5	5	Scott County	KY	Jerry Peavler	1975	1830
131 2/8	19 2/8	20 0/8	15 2/8	5	6	Dade County	MO	John Olsen	1989	1830
131 2/8	19 4/8	19 7/8	14 4/8	6	5	Crawford County	IL	John Hale	1994	1830
131 2/8	24 2/8	24 2/8	16 6/8	4	4	Rockland County	NY	Jim Paglianti	1996	1830
131 2/8	22 4/8	22 5/8	19 2/8	8	6	Dodge County	WI	Brian R. Lackas	1998	1830
131 2/8	20 7/8	21 0/8	16 4/8	5	4	Clearwater	ALB	Terry Brew	2000	1830
131 2/8	24 2/8	23 7/8	16 4/8	5	6	Van Buren County	IA	Bernard "Skeet" Batdorf	2003	1830
131 2/8	22 1/8	22 3/8	19 2/8	7	6	Ripley County	IN	Bryan Miller	2008	1830
131 1/8	22 2/8	22 1/8	20 1/8	4	5	Lafayette County	WI	Bob Wand	1971	1841
131 1/8	23 4/8	22 6/8	19 6/8	7	5	Franklin County	IA	Craig Eckhardt	1990	1841
131 1/8	21 4/8	22 2/8	16 5/8	5	5	Richland County	MT	Robert Granato	1994	1841
131 1/8	22 0/8	21 2/8	16 1/8	5	5	Ravalli County	MT	Ned Coorough	1994	1841
131 1/8	24 6/8	24 1/8	16 5/8	6	4	Dubuque County	IA	John Duggan	1994	1841
131 1/8	19 5/8	19 5/8	15 5/8	5	5	Fergus County	MT	Jason W. Birdwell	1997	1841
131 1/8	20 7/8	21 2/8	20 5/8	4	4	Duval County	TX	F. H. Becker	1998	1841
131 1/8	22 6/8	22 2/8	17 5/8	4	4	Jefferson County	OH	James C. Riggle	2001	1841
131 1/8	21 7/8	20 6/8	18 5/8	4	5	Woodbury County	IA	Mike Frankele	2003	1841
131 0/8	21 3/8	21 5/8	16 6/8	5	5	Flathead County	MT	Jack Whitney	1959	1850
131 0/8	23 5/8	23 5/8	17 0/8	4	5	Scott County	IA	Ron Anderson	1966	1850
131 0/8	21 7/8	22 0/8	14 6/8	4	4	Louisa County	IA	Larry King	1968	1850
131 0/8	22 0/8	23 5/8	16 0/8	6	6	Union County	OH	Bill Steele	1974	1850
131 0/8	19 5/8	20 3/8	18 0/8	5	5	Sheboygan County	WI	Steve Rortvedt	1974	1850
131 0/8	23 7/8	26 1/8	12 2/8	6	6	Morrison County	MN	John Strait	1979	1850
131 0/8	23 6/8	23 1/8	18 2/8	5	4	Butler County	PA	John E. Fry	1979	1850
131 0/8	20 7/8	20 3/8	15 4/8	4	5	Watonwan County	MN	George Nasman	1980	1850
131 0/8	23 1/8	23 5/8	17 0/8	5	5	Chase County	KS	Jerry Keller	1985	1850
131 0/8	24 1/8	23 7/8	17 2/8	5	6	Ross County	OH	Tony L. Wheaton	1988	1850
131 0/8	22 5/8	23 7/8	17 6/8	4	5	Chautauqua County	KS	Nels Hoadley	1989	1850

WHITETAIL DEER [Typical Antlers]

Minimum Score 125 *Continued*

Score	Length of R Main Beam	L	Inside Spread	Number of R Points	L	Area	State/ Province	Hunter's Name	Date	Rank
131 0/8	20 6/8	20 3/8	14 4/8	6	7	Pittsburg County	OK	Christopher Wolf	1990	1850
131 0/8	22 0/8	22 3/8	17 1/8	5	5	Meade County	SD	Ray Glover	1990	1850
131 0/8	22 7/8	21 6/8	17 4/8	4	4	Atchison County	MO	Patrick L. Athen	1992	1850
131 0/8	24 3/8	23 4/8	17 0/8	5	6	Delaware County	OH	Brent J. Smith	1995	1850
131 0/8	22 4/8	22 0/8	20 4/8	5	5	Hidalgo	MEX	Alvaro Palazuelos	1997	1850
131 0/8	23 1/8	22 5/8	15 2/8	5	7	Johnson County	NE	Max Gramann	1999	1850
131 0/8	21 3/8	20 7/8	17 0/8	6	6	Trempealeau County	WI	Bob Kostecki	1999	1850
131 0/8	18 3/8	19 1/8	16 4/8	5	5	Ector County	TX	Jerry Gray	2001	1850
131 0/8	22 5/8	22 2/8	17 2/8	4	5	Tuscarawas County	OH	Steve Archer	2005	1850
131 0/8	18 2/8	19 1/8	13 0/8	6	6	Keya Paha County	NE	Terry Marcukaitis	2006	1850
131 0/8	23 7/8	24 6/8	17 6/8	6	6	Anderson County	KS	Bob Morton	2007	1850
130 7/8	22 3/8	23 0/8	14 1/8	4	5	Sarpy County	NE	Ronald Beranek	1966	1872
130 7/8	21 6/8	23 1/8	15 4/8	6	5	Trempealeau County	WI	David Cater	1968	1872
130 7/8	23 2/8	24 1/8	15 6/8	4	5	Clayton County	IA	Bryan Sears	1969	1872
130 7/8	21 4/8	21 4/8	17 1/8	6	5	Trempealeau County	WI	Myles Keller	1975	1872
130 7/8	23 6/8	24 4/8	16 3/8	5	4	Sangamon County	IL	Keith Stigleman	1979	1872
130 7/8	23 2/8	25 2/8	19 7/8	5	5	Licking County	OH	Thomas Hughes	1981	1872
130 7/8	19 3/8	20 1/8	14 5/8	5	5	Ohio County	WV	Gene Petri	1990	1872
130 7/8	20 3/8	20 1/8	19 3/8	4	5	Logan County	CO	Frank Piacentino	1992	1872
130 7/8	25 4/8	26 0/8	14 3/8	6	4	Lincoln County	ONT	Darren W. Wilson	1994	1872
130 7/8	23 4/8	23 4/8	15 4/8	4	5	Hocking County	OH	Dave Hanson	1995	1872
130 7/8	19 4/8	20 0/8	15 7/8	4	4	Yuma County	CO	Rocky Chisholm	1996	1872
130 7/8	20 5/8	19 1/8	16 1/8	5	5	Calhoun County	MI	Richard L. Grosteffon, Jr.	1997	1872
130 7/8	21 4/8	22 0/8	18 1/8	4	4	Hughes County	OK	Robert Ingram	1998	1872
130 7/8	23 2/8	22 5/8	18 0/8	6	5	Franklin County	NY	Mike Savage	1998	1872
130 7/8	22 5/8	21 7/8	15 7/8	5	5	La Salle County	TX	Richard Kinas	1998	1872
130 7/8	21 2/8	22 2/8	18 6/8	7	5	Gallatin County	MT	Bob Morton	1999	1872
130 7/8	21 6/8	23 0/8	14 5/8	4	4	Butler County	MO	Steve Rueck	1999	1872
130 7/8	21 7/8	21 6/8	16 0/8	5	6	Madison County	MT	Jim Kennedy	2000	1872
130 7/8	24 1/8	24 3/8	17 1/8	4	4	Lawrence County	IN	Allen Clark	2000	1872
130 7/8	22 7/8	23 3/8	16 7/8	5	6	Dallas County	MO	Kevin Pinckney	2003	1872
130 7/8	22 5/8	23 1/8	16 5/8	5	5	Venango County	PA	Thomas A. McLaughlin	2005	1872
130 7/8	23 3/8	24 1/8	15 1/8	5	4	Keya Paha County	NE	Sid Morrison	2007	1872
130 6/8	23 2/8	21 1/8	14 2/8	5	4	Dewey County	SD	Bill Dunn	1969	1894
130 6/8	24 0/8	23 4/8	19 0/8	6	5	Burleigh County	ND	Todd Hecker	1986	1894
130 6/8	22 7/8	22 0/8	18 1/8	5	6	Bent County	CO	J. Keith Chastain	1987	1894
130 6/8	22 0/8	22 0/8	18 0/8	5	4	Missoula County	MT	Monty Moravec	1991	1894
130 6/8	23 3/8	22 7/8	18 0/8	5	5	Clark County	SD	Scott Lindgren	1992	1894
130 6/8	22 2/8	22 7/8	16 6/8	6	5	Vernon County	MO	Matthew A. Vesci	1993	1894
130 6/8	22 2/8	22 0/8	16 4/8	5	5	Jasper County	IL	Willie Runnels	1994	1894
130 6/8	20 4/8	21 0/8	13 0/8	6	6	Pike County	IL	Dan Janes	1995	1894
130 6/8	21 6/8	20 6/8	18 6/8	4	4	Boone County	IL	Dwight C. Gunderson	1998	1894
130 6/8	22 4/8	22 4/8	18 4/8	5	4	Prairie River	SAS	Ken H. Taylor	2002	1894
130 6/8	18 5/8	18 6/8	15 4/8	5	5	Yellowstone County	MT	Austin Rogers	2006	1894
130 5/8	23 4/8	21 4/8	18 1/8	4	5	Roberts County	SD	Franic W. Sherer, Jr.	1965	1905
130 5/8	20 7/8	21 4/8	14 5/8	5	6	Marquette County	WI	Tom Murphy	1974	1905
130 5/8	20 0/8	21 2/8	13 3/8	6	5	Pike County	IL	Richard Dewey	1987	1905
130 5/8	21 0/8	20 1/8	14 5/8	4	5	Tama County	IA	Monte Read	1989	1905
130 5/8	24 5/8	24 7/8	22 1/8	6	5	Fayette County	IA	Michael Judas	2001	1905
130 5/8	23 5/8	23 0/8	16 3/8	3	5	Loon Lake	SAS	Todd Andersen	2004	1905
130 5/8	21 7/8	22 0/8	15 5/8	4	4	Mountrail County	ND	Craig Richardson	2007	1905
130 5/8	25 2/8	23 4/8	16 5/8	6	5	Jo Daviess County	IL	Jeffrey R. Fechner	2007	1905
130 4/8	21 2/8	21 3/8	17 0/8	5	5	Allegan County	MI	Jack Yaeger	1942	1913
130 4/8	21 0/8	20 4/8	18 2/8	5	5	Putnam County	WV	Dan Lloyd	1966	1913
130 4/8	22 2/8	22 6/8	18 4/8	4	4	Westchester County	NY	John Frevele	1988	1913
130 4/8	21 0/8	21 2/8	16 2/8	4	4	Hocking County	OH	Martin R. Lohn	1988	1913
130 4/8	22 4/8	23 5/8	19 4/8	5	5	Washington County	OH	Steven R. Davis	1989	1913
130 4/8	22 0/8	21 6/8	19 2/8	6	7	Missoula County	MT	Jim Bradford	1991	1913
130 4/8	24 5/8	24 3/8	16 4/8	4	4	Coahoma County	MS	Charles "Buck" Neely	1992	1913
130 4/8	24 0/8	22 0/8	19 1/8	5	5	Gallatin County	MT	Bob Morton	1996	1913
130 4/8	21 5/8	21 2/8	19 7/8	6	6	Chester County	PA	Rick L. Wildermuth	1999	1913
130 4/8	22 2/8	22 4/8	17 6/8	5	4	Union County	SD	James V. Barwick	2000	1913

WHITETAIL DEER [Typical Antlers]

Minimum Score 125 *Continued*

Score	Length of Main Beam R	Length of Main Beam L	Inside Spread	Number of Points R	Number of Points L	Area	State/ Province	Hunter's Name	Date	Rank
130 4/8	21 6/8	21 3/8	13 6/8	4	4	Franklin County	IA	Arlynn Ahrens	2001	1913
130 4/8	20 4/8	21 0/8	13 2/8	6	6	Kiowa County	KS	Shane Collier	2005	1913
130 4/8	23 3/8	22 6/8	17 4/8	5	5	Bolivar County	MS	Leland Speakes, Jr.	2006	1913
130 3/8	21 2/8	20 6/8	14 1/8	5	5	Jefferson Prov. Grnds.	IN	Larry R. Smith	1966	1926
130 3/8	20 7/8	21 7/8	17 4/8	6	6	Belmont County	OH	Larry Michael	1991	1926
130 3/8	24 1/8	24 0/8	19 7/8	4	4	Knox County	OH	John W. Hunt	1993	1926
130 3/8	25 6/8	25 2/8	20 3/8	4	4	Covington County	MS	Jim W. Rutland	1994	1926
130 3/8	23 6/8	23 4/8	16 5/8	5	4	Weld County	CO	Nathan L. Andersohn	1995	1926
130 3/8	20 4/8	20 3/8	14 7/8	6	6	Nemaha County	NE	Danny E. Williams	1995	1926
130 3/8	21 6/8	19 4/8	17 1/8	5	5	Lake County	IL	Gary P. Koehler, DVM	1998	1926
130 3/8	22 0/8	21 0/8	19 1/8	5	5	Greene County	OH	John D. Sorrell	2001	1926
130 3/8	20 6/8	21 1/8	17 3/8	4	4	Charles Mix County	SD	Rick Parish	2001	1926
130 3/8	21 7/8	21 3/8	16 0/8	6	8	Des Moines County	IA	Steve J. Mann	2003	1926
130 3/8	22 0/8	22 2/8	18 1/8	4	5	Buffalo County	WI	Dan Brockman	2003	1926
130 3/8	20 7/8	20 1/8	17 1/8	5	5	Mountrail County	ND	James Domaskin	2008	1926
130 3/8	22 6/8	22 7/8	17 6/8	4	5	La Grange County	IN	Ronald Vaughn	2009	1926
130 2/8	20 3/8	20 6/8	16 2/8	4	4	Yankton County	SD	Jack Begley	1959	1939
130 2/8	26 7/8	26 3/8	17 3/8	8	6	Blue Earth County	MN	Ron Herz	1966	1939
130 2/8	22 1/8	23 1/8	20 2/8	4	5	Cheyenne County	NE	Marvin Clyncke	1983	1939
130 2/8	22 5/8	23 2/8	16 0/8	5	5	Licking County	OH	Glenn Dale Anderson	1986	1939
130 2/8	23 0/8	23 2/8	19 4/8	4	5	Fayette County	IA	Hugh Wright	1988	1939
130 2/8	21 1/8	21 4/8	17 2/8	4	4	Lake County	IL	Carl Spaeth	1988	1939
130 2/8	21 3/8	19 5/8	14 4/8	5	5	Todd County	KY	Larry Ross	1992	1939
130 2/8	23 1/8	22 6/8	15 2/8	5	5	Meeker County	MN	Brent Swanson	1993	1939
130 2/8	23 7/8	23 0/8	19 2/8	4	4	Frederick County	MD	Edward L. Reilly	1998	1939
130 2/8	23 7/8	24 1/8	19 0/8	4	4	Monroe County	IA	Larry Faught	2000	1939
130 2/8	20 5/8	21 1/8	16 4/8	5	5	Switzerland County	IN	Joe Bacon	2001	1939
130 2/8	23 3/8	24 5/8	17 3/8	5	6	Champaign County	OH	Thomas Flohre	2002	1939
130 2/8	22 6/8	22 4/8	18 0/8	5	5	Comanche County	KS	Pat Lefemine	2003	1939
130 2/8	22 5/8	22 4/8	20 4/8	4	4	Carroll County	IL	Jim Carlson	2006	1939
130 2/8	23 2/8	22 1/8	18 4/8	4	4	Maverick County	TX	Darren P. Cooper	2008	1939
130 1/8	24 4/8	25 7/8	19 7/8	4	3	Waupaca County	WI	John Schoenike	1952	1954
130 1/8	22 5/8	21 6/8	16 7/8	5	5	Oneida County	WI	Fred Felbab	1964	1954
130 1/8	22 6/8	24 4/8	17 4/8	6	5	Morrison County	MN	Stephen L. Marklund	1964	1954
130 1/8	22 6/8	24 0/8	18 2/8	6	5	Crow Wing County	MN	Robert Woods	1969	1954
130 1/8	24 0/8	23 4/8	18 3/8	4	5	Lyon County	MN	Bruce Grow	1972	1954
130 1/8	21 4/8	21 4/8	20 1/8	4	4	Garden County	NE	Larry Pierce	1976	1954
130 1/8	23 5/8	23 3/8	16 2/8	4	5	McMullen County	TX	W. Walker Lowry, Jr.	1989	1954
130 1/8	22 6/8	22 3/8	17 5/8	4	4	Kenedy County	TX	Dennis Dunn	1996	1954
130 1/8	17 0/8	17 7/8	15 7/8	5	5	Buffalo County	SD	Jerri Ann Haak	2009	1954
130 0/8	23 4/8	23 5/8	19 0/8	4	4	Wicomico County	MD	Donald J. Brown	1956	1963
130 0/8	21 0/8	19 3/8	17 4/8	5	4	Roberts County	SD	Byron Siegel	1962	1963
130 0/8	22 2/8	22 5/8	18 6/8	4	5	Berrien County	MI	Leon L. Williams	1965	1963
130 0/8	20 2/8	20 6/8	15 0/8	5	5	Bucks County	PA	Don Fitting	1967	1963
130 0/8	21 3/8	21 4/8	19 0/8	5	5	Vanderburgh County	IN	Scott E. Webster	1973	1963
130 0/8	21 7/8	21 2/8	16 1/8	6	6	Polk County	WI	Paul Petersen	1981	1963
130 0/8	21 0/8	22 0/8	18 2/8	6	6	Van Buren County	IA	Tom Casey	1992	1963
130 0/8	19 7/8	20 2/8	15 6/8	5	4	Pike County	IL	Richard Dewey	1995	1963
130 0/8	24 3/8	22 7/8	20 0/8	5	6	Sawyer County	WI	Patrick Venzke	1996	1963
130 0/8	23 4/8	22 6/8	16 4/8	4	5	Frederick County	MD	Edward L. Reilly	1997	1963
130 0/8	22 4/8	22 3/8	14 4/8	6	4	Wilson County	KS	Troy A. Graziadei	1997	1963
130 0/8	20 6/8	20 4/8	17 4/8	4	4	Weld County	CO	Timothy Bradley	1997	1963
130 0/8	19 2/8	19 5/8	17 0/8	4	4	Ward County	ND	John R. Plesuk	2002	1963
130 0/8	19 4/8	19 4/8	15 6/8	4	4	Dodge County	MN	Terry Krahn	2004	1963
130 0/8	19 2/8	20 4/8	16 0/8	5	5	Harrison County	TX	Richard Kinas	2006	1963
130 0/8	21 2/8	20 7/8	16 7/8	5	6	Charles Mix County	SD	Robert Struck	2006	1963
130 0/8	22 6/8	22 2/8	18 3/8	5	5	Norton County	KS	Joseph Bradley	2012	1963
129 7/8	23 6/8	21 6/8	18 0/8	5	5	Winnebago County	IL	Leo M. Ruefer, Jr.	1958	1980
129 7/8	21 7/8	22 0/8	16 5/8	4	4	Monroe County	WI	Jeff Skrade	1962	1980
129 7/8	23 3/8	23 2/8	16 5/8	5	6	Allegan County	MI	Clayton Foster	1964	1980
129 7/8	23 0/8	21 7/8	15 1/8	5	5	Jefferson County	KS	Delmar Tucking, Jr.	1966	1980
129 7/8	22 3/8	22 6/8	20 7/8	4	4	Ashland County	WI	William Sutton	1968	1980

WHITETAIL DEER [Typical Antlers]

Minimum Score 125 *Continued*

Score	Length of R Main Beam	L	Inside Spread	Number of R Points	L	Area	State/ Province	Hunter's Name	Date	Rank
129 7/8	19 7/8	19 3/8	20 2/8	5	5	Vilas County	WI	Art Heinze	1970	1980
129 7/8	19 2/8	19 5/8	15 3/8	5	5	Columbia County	WI	Jay Rosendick	1970	1980
129 7/8	22 4/8	21 7/8	17 1/8	4	4	Ogle County	IL	Henry E. Zimmerman	1970	1980
129 7/8	21 5/8	22 2/8	17 1/8	5	5	Sauk County	WI	Clair E. Keylock	1977	1980
129 7/8	19 7/8	20 4/8	16 5/8	5	5	Adair County	IA	Wallace R. Waddell	1980	1980
129 7/8	22 4/8	23 5/8	20 1/8	5	5	Franklin County	MA	John D. O'Brien, Jr.	1989	1980
129 7/8	19 7/8	20 0/8	14 5/8	5	5	Williams County	ND	Milan R. Liesener	1994	1980
129 7/8	24 0/8	23 0/8	19 0/8	4	6	Will County	IL	Mike O'Connor	1997	1980
129 7/8	21 6/8	21 5/8	16 1/8	4	5	Hillsdale County	MI	John T. Glassburn	1998	1980
129 7/8	21 5/8	20 7/8	18 1/8	5	5	Chickasaw County	IA	David J. Kerkove	2001	1980
129 7/8	23 2/8	23 1/8	17 7/8	4	4	Marquette County	WI	Timothy J. Cardinal	2010	1980
129 6/8	19 5/8	20 7/8	16 2/8	5	6	Adams County	IL	Mel Powell	1964	1996
129 6/8	20 2/8	18 0/8	17 2/8	5	5	Johnson County	WY	Jim Bartz	1969	1996
129 6/8	22 5/8	22 2/8	20 2/8	7	5	Morrison County	MN	Robert E. Nordstrom	1973	1996
129 6/8	20 3/8	20 4/8	15 4/8	5	5	Reagan County	TX	Ronald H. Ray	1992	1996
129 6/8	21 3/8	22 5/8	15 0/8	5	5	Lapeer County	MI	Thomas J. Baker	1995	1996
129 6/8	22 6/8	21 6/8	16 0/8	4	4	Sauk County	WI	Terry Woirol	1996	1996
129 6/8	21 7/8	21 4/8	17 0/8	6	5	Christian Valley	BC	Roy Gerrath	1996	1996
129 6/8	21 0/8	23 4/8	19 6/8	4	4	Calloway County	KY	Alan Ray Jones	1997	1996
129 6/8	20 7/8	20 3/8	15 0/8	5	5	Richardson County	NE	John Duggan	1999	1996
129 6/8	22 0/8	22 7/8	16 6/8	4	4	Fillmore County	MN	Richard Dean	2006	1996
129 6/8	21 7/8	22 3/8	15 6/8	5	4	Davis County	IA	Marc Rogers	2008	1996
129 5/8	23 2/8	24 0/8	17 6/8	4	4	Taylor County	WV	Jimmie R. Auvil	1959	2007
129 5/8	22 5/8	22 1/8	20 1/8	5	5	Cowley County	KS	Charles O'Daniel	1966	2007
129 5/8	22 3/8	22 4/8	17 5/8	4	5	Trigg County	KY	Ronald M. Goddard	1969	2007
129 5/8	21 0/8	20 7/8	14 3/8	5	5	Washington County	IA	Kevin Langan	1991	2007
129 5/8	22 3/8	21 6/8	15 3/8	4	5	Weld County	CO	Eric Stern	1997	2007
129 5/8	21 5/8	20 7/8	14 5/8	5	5	Fayette County	IA	Mike Judas	1998	2007
129 5/8	20 6/8	20 5/8	17 1/8	4	4	Waukesha County	WI	Jeffrey J. Gricar	1999	2007
129 5/8	23 1/8	23 2/8	18 1/8	5	5	Jasper County	IA	John Ponsetto	2000	2007
129 5/8	25 6/8	25 1/8	18 2/8	5	4	Putnam County	IN	Chris Gaskill	2003	2007
129 4/8	21 5/8	20 2/8	15 4/8	5	5	Sarpy County	NE	Russ Calloway	1962	2016
129 4/8	22 4/8	21 7/8	17 6/8	5	5	Monroe County	GA	Robert S. Carey	1973	2016
129 4/8	19 1/8	19 5/8	15 2/8	4	4	Pike County	IL	Lloyd Bateman	1985	2016
129 4/8	20 0/8	20 1/8	17 0/8	5	5	Dodge County	MN	Jay Deones	1986	2016
129 4/8	22 7/8	23 0/8	17 6/8	4	4	Oneida County	WI	Tim Johnson	1990	2016
129 4/8	25 4/8	24 4/8	17 2/8	6	4	Union County	IL	Tom Michael	1992	2016
129 4/8	20 7/8	21 2/8	14 6/8	5	4	Hillsdale County	MI	Gregory J. Raatz	1994	2016
129 4/8	22 0/8	24 6/8	17 6/8	6	7	Jackson County	OH	Larry J. Waugh	1994	2016
129 4/8	21 1/8	20 3/8	20 0/8	4	4	Weld County	CO	Alan Katchur	1994	2016
129 4/8	21 1/8	24 0/8	18 0/8	4	5	Decatur County	KS	Scott Hargrove	1997	2016
129 4/8	20 1/8	20 3/8	16 5/8	6	5	Williams County	ND	Richard Liesener	1999	2016
129 4/8	21 6/8	21 4/8	20 7/8	4	6	Buffalo County	SD	Randy Knippling	2010	2016
129 3/8	23 4/8	22 1/8	20 3/8	4	5	Bayfield County	WI	Kenneth R. Sweeny	1965	2028
129 3/8	20 6/8	20 7/8	20 2/8	7	9	White County	IN	Mark Mohler	1983	2028
129 3/8	23 6/8	22 1/8	18 5/8	4	4	Baltimore County	MD	Bruce D. Hoover	1987	2028
129 3/8	22 4/8	23 1/8	15 7/8	4	5	Lonoke County	AR	Keith Westbrook	1998	2028
129 3/8	24 4/8	22 7/8	24 7/8	5	6	Trego County	KS	Grant R. Woods	2000	2028
129 3/8	20 4/8	20 6/8	15 1/8	5	5	Hancock County	IL	Bradley J. Swanson	2007	2028
129 3/8	22 2/8	24 3/8	16 3/8	4	4	Comanche County	KS	Ron Rockwell	2007	2028
129 2/8	19 2/8	20 1/8	16 5/8	7	6	Columbia County	WA	Al Farrell	1956	2035
129 2/8	21 7/8	21 4/8	16 2/8	5	5	Kendall County	IL	Fred W. Achilles	1979	2035
129 2/8	22 3/8	23 4/8	16 6/8	4	4	Aitkin County	MN	Daniel L. Edelman	1982	2035
129 2/8	20 3/8	20 4/8	17 4/8	5	5	Canmore	ALB	J. C. Mackid	1982	2035
129 2/8	22 0/8	22 5/8	16 2/8	5	5	Muskingum County	OH	Dr. Jim Emerson	1982	2035
129 2/8	21 6/8	22 1/8	16 6/8	4	4	Des Moines County	IA	Ray Waschkat	1984	2035
129 2/8	20 5/8	21 7/8	16 0/8	4	4	Anderson County	KS	Jerry Howarter	1985	2035
129 2/8	21 4/8	21 3/8	21 0/8	4	5	Calhoun County	MI	Douglas L. Rial	1989	2035
129 2/8	21 3/8	21 6/8	15 4/8	4	4	Morrow County	OH	Vic Tumeo	1993	2035
129 2/8	22 1/8	22 4/8	17 6/8	4	5	Clark County	SD	Scott Lindgren	1995	2035
129 2/8	22 0/8	21 6/8	20 1/8	4	8	Benton County	IN	Richard D. Ellsworth, DDS	1995	2035
129 2/8	21 3/8	22 0/8	17 6/8	5	6	Boone County	MO	David Westmoreland	1997	2035

Minimum Score 125 *Continued*

Score	Length of Main Beam R	Length of Main Beam L	Inside Spread	Number of Points R	Number of Points L	Area	State/ Province	Hunter's Name	Date	Rank
129 2/8	23 1/8	23 1/8	16 6/8	6	5	Chickasaw County	IA	William H. Peterson	1998	2035
129 2/8	20 3/8	20 4/8	15 4/8	5	5	Douglas County	WI	Paul H. Funk	2001	2035
129 2/8	21 7/8	22 6/8	17 6/8	3	4	Pike County	IL	Daniel G. Wombles	2004	2035
129 2/8	20 1/8	20 3/8	14 6/8	5	6	Merrick County	NE	Shawn Church	2006	2035
129 2/8	22 1/8	21 6/8	17 3/8	5	6	Pittsburg County	OK	Robin L. Miller	2007	2035
129 1/8	20 6/8	21 6/8	16 5/8	5	5	Burnett County	WI	James G. Hurrle	1974	2052
129 1/8	22 1/8	22 4/8	17 5/8	6	5	Clark County	MO	Dennis Fish	1986	2052
129 1/8	20 1/8	20 2/8	16 2/8	6	7	Ohio County	WV	Gene Petri	1989	2052
129 1/8	21 2/8	20 3/8	15 7/8	4	4	Cass County	NE	Steve Rueck	1991	2052
129 1/8	24 0/8	24 2/8	16 5/8	5	4	Jackson County	OH	Edsel D. Duty	1991	2052
129 1/8	23 5/8	24 0/8	16 0/8	5	4	Knox County	IL	Dave Emken	1993	2052
129 1/8	23 0/8	23 2/8	18 4/8	6	5	Scotland County	MO	Jim Ferry	1994	2052
129 1/8	23 2/8	23 0/8	16 5/8	4	4	Pottawattamie County	IA	Daniel F. Krettek	1996	2052
129 1/8	21 1/8	22 0/8	13 5/8	6	5	Jefferson County	MO	Steve Hohensee	1996	2052
129 1/8	22 4/8	19 2/8	16 5/8	5	4	McKenzie County	ND	Craig Richardson	1997	2052
129 1/8	21 1/8	21 1/8	17 1/8	4	5	Hidalgo	MEX	Roberto Garza	1997	2052
129 1/8	21 3/8	21 2/8	16 7/8	4	4	Hitchcock County	NE	Jewell Leadford	1999	2052
129 0/8	22 6/8	22 5/8	17 5/8	5	5	Adams County	IL	Roger W. Seehafer	1964	2064
129 0/8	21 4/8	21 3/8	17 4/8	5	5	Monmouth County	NJ	Joseph N. Lazar	1967	2064
129 0/8	23 4/8	24 0/8	19 6/8	5	4	Shawano County	WI	Merton Giessel	1970	2064
129 0/8	19 6/8	20 6/8	19 0/8	4	4	Oneida County	WI	Gary Bohlman	1970	2064
129 0/8	22 1/8	22 1/8	16 2/8	5	4	Cass County	ND	Ellery Kundert	1985	2064
129 0/8	21 4/8	21 4/8	16 2/8	4	5	Pike County	MO	Richard Dewey	1985	2064
129 0/8	20 4/8	20 0/8	13 6/8	5	7	Todd County	MN	William H. Peterson	1993	2064
129 0/8	21 4/8	22 4/8	18 2/8	4	5	Knox County	IL	Larry Hesterly	1995	2064
129 0/8	22 2/8	22 3/8	18 0/8	5	4	Brown County	SD	Ron Miller	1995	2064
129 0/8	22 0/8	21 5/8	18 6/8	4	4	Hunterdon County	NJ	John S. Hoinowski	2003	2064
129 0/8	19 3/8	19 1/8	15 2/8	5	5	Kaufman County	TX	David M. Upchurch	2008	2064
129 0/8	23 5/8	22 0/8	15 6/8	4	4	Dubois County	IN	Scott J. Fromme	2009	2064
129 0/8	21 5/8	22 1/8	20 0/8	4	4	Breathitt County	KY	James G. Mullins	2010	2064
128 7/8	19 1/8	20 6/8	13 7/8	5	5	Martin County	IN	J. Steve Albertson	1969	2077
128 7/8	22 4/8	21 1/8	16 3/8	5	5	Grant County	WI	Kevin Freymiller	1975	2077
128 7/8	21 0/8	20 1/8	18 1/8	5	4	Clark County	WI	Clarence J. Biddle	1977	2077
128 7/8	23 7/8	23 1/8	16 0/8	5	5	Pike County	IL	Frank Dolbeare	1989	2077
128 7/8	21 2/8	21 2/8	15 5/8	6	6	Humboldt County	IA	Paul Detrick	1994	2077
128 7/8	19 4/8	19 4/8	15 3/8	5	5	Morgan County	CO	Larry Alma	1998	2077
128 7/8	22 0/8	21 5/8	16 5/8	5	5	Powell County	MT	John "Rosey" Roseland	2000	2077
128 7/8	22 3/8	23 4/8	20 1/8	6	5	Venango County	PA	Brian K. Perry	2003	2077
128 7/8	22 2/8	20 7/8	15 6/8	5	5	Jefferson County	AL	Greg Sweeney	2004	2077
128 7/8	22 4/8	22 6/8	16 1/8	7	7	Montcalm County	MI	Todd Andersen	2004	2077
128 7/8	22 0/8	20 7/8	14 4/8	6	7	Williams County	ND	Richard Liesener	2011	2077
128 6/8	23 0/8	25 2/8	16 4/8	4	4	Buffalo County	NE	Bill Orsborn	1963	2088
128 6/8	21 1/8	20 2/8	18 6/8	4	4	Johnson County	IA	Claire Doyle	1971	2088
128 6/8	20 6/8	21 2/8	17 0/8	5	5	Brown County	SD	Jerome J. Lingor	1978	2088
128 6/8	25 5/8	25 6/8	18 1/8	5	6	Blount County	TN	David Dotson	1981	2088
128 6/8	20 7/8	19 6/8	16 2/8	5	5	Winona County	MN	John R. Micheel	1986	2088
128 6/8	20 5/8	20 4/8	16 1/8	6	4	Stony Plain	ALB	Stan Chiras	1986	2088
128 6/8	22 2/8	22 4/8	18 4/8	5	4	Hubbard County	MN	Pascal Perrin	1987	2088
128 6/8	20 5/8	21 6/8	14 4/8	4	4	Iroquois County	IL	Warren Cary	1989	2088
128 6/8	24 2/8	24 2/8	19 0/8	4	4	Hampshire County	MA	John Higgins	1989	2088
128 6/8	21 0/8	20 3/8	16 0/8	5	5	Hayes County	NE	William E. Wuerthele	1991	2088
128 6/8	21 3/8	20 7/8	15 6/8	5	5	Wyandot County	OH	Michael H. Thorne	1999	2088
128 6/8	21 0/8	20 4/8	15 6/8	5	4	Elk County	KS	Matt Rosensweet	2000	2088
128 6/8	21 3/8	21 6/8	16 0/8	4	4	McHenry County	IL	Joe A. Florent	2004	2088
128 6/8	22 0/8	21 1/8	14 2/8	5	5	Pittsburg County	OK	Jeremy Dean	2007	2088
128 5/8	20 6/8	20 2/8	14 7/8	5	5	Mille Lacs County	MN	Milton J. Mattson	1952	2102
128 5/8	14 7/8	26 2/8	20 5/8	5	5	Ford County	KS	Rod Lies	1966	2102
128 5/8	18 1/8	18 5/8	15 3/8	5	5	Harford County	MD	Joseph Egner	1969	2102
128 5/8	20 4/8	20 4/8	18 0/8	6	6	Lawrence County	SD	Ronald Hazledine	1969	2102
128 5/8	21 6/8	22 0/8	17 5/8	4	4	Mercer County	IL	Kenneth E. Yeater	1971	2102
128 5/8	23 7/8	22 0/8	17 7/8	5	5	Van Buren County	IA	John "Rosey" Roseland	1990	2102
128 5/8	21 6/8	22 0/8	16 0/8	5	6	Hocking County	OH	Tony Casagrande	1994	2102

Score	Length of R Main Beam	L	Inside Spread	Number of R Points	L	Area	State/ Province	Hunter's Name	Date	Rank
128 5/8	21 3/8	21 0/8	17 5/8	4	4	St. Charles County	MO	Scott Cowen	1995	2102
128 5/8	20 4/8	20 3/8	14 1/8	5	5	Maverick County	TX	Bill R. Gantt	1995	2102
128 5/8	22 6/8	24 0/8	15 7/8	6	5	Suffolk County	NY	Michael Fenezia	1999	2102
128 5/8	22 2/8	21 6/8	17 3/8	4	4	Dubuque County	IA	John Duggan	1999	2102
128 5/8	24 6/8	23 0/8	18 1/8	4	4	Licking County	OH	Bill Jones	2000	2102
128 5/8	20 0/8	20 2/8	17 6/8	6	6	Scotts Bluff County	NE	Thane Anderson	2001	2102
128 5/8	23 1/8	22 2/8	17 5/8	4	4	Champaign County	OH	Neil Yoder	2001	2102
128 5/8	20 4/8	20 3/8	14 1/8	5	5	Schuyler County	IL	Tom Spence	2001	2102
128 5/8	20 3/8	20 2/8	14 5/8	5	5	Prowers County	CO	Rick Hooley	2004	2102
128 5/8	21 2/8	21 7/8	18 7/8	5	4	La Grange County	IN	Ronald Vaughn	2008	2102
128 5/8	24 2/8	24 5/8	21 3/8	6	6	Dodge County	WI	Michael Young	2010	2102
128 4/8	25 0/8	23 7/8	17 0/8	4	4	Stone County	AR	Alfred Hirt	1959	2120
128 4/8	20 4/8	21 0/8	17 7/8	7	5	Clay County	NE	Rollan Johnson	1961	2120
128 4/8	21 1/8	21 2/8	15 0/8	5	4	Marion County	IA	Thomas Tucker	1968	2120
128 4/8	22 3/8	23 2/8	17 5/8	5	6	Wright County	MN	Elwood Rokala	1971	2120
128 4/8	21 6/8	22 3/8	16 6/8	5	5	Jones County	IA	Tom Postel	1971	2120
128 4/8	22 6/8	23 0/8	23 7/8	4	4	Jasper County	SC	John V. Orr	1973	2120
128 4/8	19 5/8	20 3/8	15 4/8	5	5	Ozaukee County	WI	Ronald Mayer	1975	2120
128 4/8	22 0/8	20 4/8	18 2/8	4	4	Rosebud County	MT	Bob Brill	1976	2120
128 4/8	23 0/8	22 5/8	19 6/8	4	4	Delaware County	PA	Mark Gentry	1984	2120
128 4/8	19 0/8	19 1/8	13 4/8	5	5	Eddy County	ND	Tim Finley	1987	2120
128 4/8	21 0/8	20 5/8	14 4/8	6	8	Dimmit County	TX	Michael V. Stewart	1988	2120
128 4/8	22 6/8	22 2/8	18 4/8	4	4	Wayne County	OH	Keith R. Dotterer	1990	2120
128 4/8	20 4/8	20 5/8	18 6/8	4	4	Davis County	IA	Larry D. Jones	1993	2120
128 4/8	24 6/8	23 5/8	14 6/8	4	4	Baltimore County	MD	Shane A. Fitzgerald	1994	2120
128 4/8	23 0/8	22 7/8	17 4/8	5	4	Anson County	NC	William M. Stroupe	1994	2120
128 4/8	21 1/8	20 3/8	15 0/8	5	5	Edmunds County	SD	Dave Knecht	1997	2120
128 4/8	21 1/8	20 1/8	18 5/8	5	4	La Salle County	TX	Ronnie Gressett	1998	2120
128 4/8	19 1/8	18 2/8	15 0/8	5	5	Fergus County	MT	Joe Frazier	1999	2120
128 4/8	22 0/8	20 4/8	16 6/8	6	5	Spink County	SD	Ron Miller	2001	2120
128 4/8	24 6/8	24 0/8	15 5/8	4	5	Mercer County	OH	W. Stevens, Jr.	2003	2120
128 4/8	21 6/8	21 5/8	19 3/8	6	5	Independence County	AR	Heath Sanders	2004	2120
128 4/8	21 1/8	21 0/8	17 0/8	5	5	Calhoun County	MI	George Swan, Jr.	2005	2120
128 4/8	24 3/8	24 2/8	17 4/8	6	4	Osborne County	KS	Bob Snell	2009	2120
128 3/8	23 3/8	23 0/8	20 3/8	4	5	Williamson County	IL	Don Walker	1959	2143
128 3/8	19 1/8	19 6/8	15 2/8	6	4	Douglas County	NE	Cecil Smith	1962	2143
128 3/8	23 4/8	23 2/8	17 3/8	6	7	Poncil	MT	Bob Samson	1962	2143
128 3/8	20 6/8	20 5/8	14 1/8	5	4	Morrison County	MN	Ronald Thole	1964	2143
128 3/8	19 6/8	19 7/8	16 7/8	6	5	Columbia County	WI	David D. Luetkens	1965	2143
128 3/8	22 1/8	20 0/8	20 7/8	4	4	Brown County	SD	T. Michael Dunn	1966	2143
128 3/8	21 0/8	22 1/8	18 5/8	5	5	Bartholomew County	IN	Harold Frye	1966	2143
128 3/8	22 0/8	21 6/8	14 5/8	5	4	Allegany County	NY	Joseph Famiglietti	1972	2143
128 3/8	23 4/8	23 3/8	19 5/8	4	4	Union County	IL	Randy Edmonds	1974	2143
128 3/8	21 4/8	21 2/8	16 5/8	5	5	Cameron County	TX	Jerry Spencer	1980	2143
128 3/8	22 2/8	22 6/8	14 7/8	6	6	Clay County	SD	Marlowe Rames	1983	2143
128 3/8	21 1/8	20 5/8	15 3/8	6	6	Anoka County	MN	John Cardinal	1984	2143
128 3/8	20 6/8	20 7/8	17 0/8	6	4	Yuma County	CO	Bill Grammer	1986	2143
128 3/8	21 0/8	21 4/8	14 7/8	6	5	Tom Green County	TX	Mike Hegefeld	1989	2143
128 3/8	20 3/8	20 2/8	15 4/8	6	6	Clark County	IL	Ronald E. Karst	1992	2143
128 3/8	22 4/8	23 0/8	19 5/8	4	4	Dawson County	MT	Mark Ascheman	1993	2143
128 3/8	20 4/8	20 3/8	16 3/8	5	5	Yuma County	CO	Scott Armitage	1996	2143
128 3/8	20 3/8	20 3/8	16 7/8	4	4	Van Buren County	IA	Michael P. Kissel	2000	2143
128 3/8	23 5/8	22 6/8	17 7/8	4	4	Richland County	OH	John "Rosey" Roseland	2008	2143
128 2/8	22 1/8	21 1/8	15 6/8	5	5	Franklin County	KS	Gary Hunsicker	1966	2162
128 2/8	20 5/8	21 3/8	14 4/8	5	5	Waupaca County	WI	Dennis Arndt	1967	2162
128 2/8	22 2/8	22 3/8	17 0/8	4	4	Cherokee County	OK	Eddie Goss	1973	2162
128 2/8	20 5/8	21 4/8	17 0/8	5	5	Buffalo County	NE	Lynn Bombeck	1974	2162
128 2/8	23 1/8	22 7/8	20 3/8	4	5	Jefferson County	OH	Michael W. Brown	1976	2162
128 2/8	19 0/8	19 1/8	14 4/8	5	5	Travis County	TX	Russell Schmidt	1982	2162
128 2/8	22 5/8	22 3/8	17 4/8	3	5	Brown County	KS	Chuck McNally	1983	2162
128 2/8	23 2/8	23 3/8	20 4/8	4	5	Missoula County	MT	Greg L. Munther	1987	2162
128 2/8	21 2/8	21 6/8	15 0/8	5	6	Haskell County	OK	Jerry C. Sturdy	1987	2162

WHITETAIL DEER [Typical Antlers]

Minimum Score 125 *Continued*

Score	Length of Main Beam R	Length of Main Beam L	Inside Spread	Number of Points R	Number of Points L	Area	State/ Province	Hunter's Name	Date	Rank
128 2/8	24 1/8	23 3/8	16 6/8	4	4	Monroe County	NY	Wayne Meritt	1989	2162
128 2/8	20 7/8	20 4/8	16 4/8	5	5	Pike County	IL	Joel M. Riotto	1996	2162
128 2/8	19 0/8	18 0/8	13 0/8	5	5	Medina County	TX	Kelly C. Parrino	1997	2162
128 2/8	21 5/8	21 2/8	17 2/8	5	4	Marion County	KS	Tony Cassity	2001	2162
128 2/8	23 1/8	24 0/8	18 0/8	4	5	Vinton County	OH	Clark Miller	2005	2162
128 2/8	21 4/8	22 3/8	17 0/8	5	5	Allen County	OH	Lee Yoakam	2010	2162
128 2/8	19 7/8	20 3/8	14 6/8	5	5	McKenzie County	ND	Richard Liesener	2012	2162
128 1/8	20 1/8	21 0/8	17 5/8	5	5	Dorchester County	MD	Powell D. Cook	1964	2178
128 1/8	21 1/8	21 5/8	16 3/8	4	4	Calhoun County	MI	Clarence Bowers, Jr.	1982	2178
128 1/8	22 6/8	23 3/8	17 1/8	5	4	Brown County	SD	Ron Rockwell	1984	2178
128 1/8	23 0/8	23 5/8	16 1/8	4	5	Price County	WI	Chuck Dicka	1985	2178
128 1/8	20 4/8	20 4/8	15 4/8	6	5	Fergus County	MT	John "Rosey" Roseland	1988	2178
128 1/8	24 6/8	24 7/8	20 1/8	3	4	Hampden County	MA	Dick Scorzafava	1989	2178
128 1/8	19 1/8	19 2/8	15 3/8	5	7	Buffalo County	WI	Jeff Moore	1992	2178
128 1/8	22 4/8	23 1/8	17 7/8	6	5	Christopher Lake	SAS	Les King	1993	2178
128 1/8	20 3/8	20 4/8	13 1/8	6	5	Van Buren County	IA	Jeffrey R. Fechner	2001	2178
128 1/8	22 1/8	22 5/8	15 2/8	4	5	Dundy County	NE	Bryan Holley	2002	2178
128 1/8	23 3/8	23 1/8	16 3/8	6	6	Steuben County	NY	Tom Giambrone	2003	2178
128 1/8	21 1/8	21 0/8	15 1/8	4	4	Cedar County	MO	Kendall Gire	2004	2178
128 1/8	20 3/8	20 0/8	15 7/8	4	5	Ogle County	IL	Alan D. Schuler	2004	2178
128 1/8	20 3/8	20 5/8	15 3/8	4	4	Big Horn County	WY	David Moss	2005	2178
128 1/8	21 4/8	22 0/8	16 5/8	4	5	Throckmorton County	TX	Ron Fief	2010	2178
128 0/8	18 7/8	18 1/8	17 4/8	5	5	McLean County	ND	Bennie R. Maytum	1960	2193
128 0/8	22 7/8	23 0/8	19 0/8	6	8	Sibley County	MN	Darwin Grack	1972	2193
128 0/8	23 0/8	22 6/8	16 6/8	4	4	Westchester County	NY	Theodore J. Korkidas	1978	2193
128 0/8	21 1/8	21 6/8	18 0/8	6	8	Allegheny County	PA	Albert Polovich, Jr.	1981	2193
128 0/8	22 3/8	22 6/8	21 5/8	5	4	Orleans County	NY	Warren L. Lewis	1988	2193
128 0/8	22 3/8	22 3/8	18 6/8	4	5	Wayne County	OH	Tom P. Dotterer	1990	2193
128 0/8	19 5/8	18 4/8	17 4/8	5	5	Ravalli County	MT	Eric Treutel	1996	2193
128 0/8	19 6/8	20 0/8	14 1/8	6	6	Audrain County	MO	Tim Lawrence	1999	2193
128 0/8	19 0/8	18 3/8	17 0/8	5	6	Woodbury County	IA	Mike Frankele	2000	2193
127 7/8	21 4/8	20 0/8	18 1/8	4	4	Monroe County	IN	Jerry L. Swafford	1968	2202
127 7/8	21 5/8	21 2/8	15 5/8	6	6	Henderson County	IL	Randy Moore	1971	2202
127 7/8	19 2/8	20 3/8	15 7/8	6	5	Jones County	IA	Larry Stewart	1972	2202
127 7/8	23 3/8	23 3/8	19 1/8	4	5	Morgan County	CO	Mark Hansen	1993	2202
127 7/8	21 6/8	20 5/8	17 3/8	4	4	Shelby County	MO	Rusty Hammond	2007	2202
127 7/8	22 3/8	21 3/8	16 7/8	5	4	Webb County	TX	Bob Gilbert	2008	2202
127 7/8	21 6/8	23 0/8	17 5/8	5	5	Jefferson County	OH	Jeremy Sensibaugh	2008	2202
127 6/8	21 3/8	22 2/8	16 2/8	4	4	Rice County	MN	David Knutson	1969	2209
127 6/8	24 4/8	23 6/8	18 3/8	6	5	Muskingum County	OH	Charles F. Fineran	1974	2209
127 6/8	20 0/8	19 3/8	17 0/8	4	4	Tucker County	WV	Larry A. Williams	1975	2209
127 6/8	22 2/8	21 5/8	17 2/8	4	4	Phillips County	KS	Michael Sohm	1990	2209
127 6/8	20 4/8	21 4/8	16 4/8	4	4	Calhoun County	MI	Greg H. St. John	1992	2209
127 6/8	22 7/8	20 2/8	16 0/8	4	5	Valley County	MT	Nickie E. Roth	1993	2209
127 6/8	22 2/8	22 6/8	19 4/8	5	5	La Salle County	TX	H. Mike Palmer	1993	2209
127 6/8	19 5/8	19 4/8	14 6/8	5	5	Nuevo Laredo	MEX	Roy Marlow	1993	2209
127 6/8	22 4/8	22 3/8	17 2/8	5	4	Charles County	MD	Bradley F. Herbert	1996	2209
127 6/8	21 0/8	19 5/8	14 3/8	5	6	Yuma County	CO	Ronald L. King	1996	2209
127 6/8	23 6/8	21 7/8	18 2/8	4	4	Hartford County	CT	Richard J. Langer	2004	2209
127 6/8	22 4/8	21 6/8	18 0/8	4	4	Woodford County	IL	David Fultz	2005	2209
127 6/8	21 0/8	21 7/8	15 2/8	5	5	Nuevo Leon	MEX	Fred Eichler	2008	2209
127 6/8	19 4/8	18 6/8	16 0/8	5	5	Kent County	MI	Steve C. Vanhoose	2011	2209
127 5/8	21 4/8	22 1/8	17 3/8	5	5	Monroe County	MO	C. R. Jackson	1962	2223
127 5/8	22 3/8	22 3/8	16 5/8	6	5	Vilas County	WI	Carl R. Strauss	1968	2223
127 5/8	21 2/8	21 0/8	18 3/8	5	5	Warren County	NJ	Jerry W. Kauffman	1972	2223
127 5/8	22 0/8	22 6/8	14 7/8	4	4	Winnebago County	IA	Kenneth R. Coe, Jr.	1973	2223
127 5/8	24 4/8	24 5/8	19 0/8	4	5	Wayne County	OH	Ronald Stine	1975	2223
127 5/8	21 0/8	21 3/8	16 5/8	4	4	Dickson County	TN	Andy Jackson	1976	2223
127 5/8	20 2/8	22 1/8	14 5/8	5	4	Crawford County	IL	Dennis Sturgis, Jr.	1991	2223
127 5/8	19 4/8	19 4/8	18 2/8	6	7	Des Moines County	IA	Michael C. Dysh	1996	2223
127 5/8	20 7/8	20 4/8	17 1/8	4	4	Ravalli County	MT	Wayne Capp	1996	2223
127 5/8	22 5/8	22 5/8	16 1/8	5	6	Clinton County	IL	David Baum	2000	2223

Score	Length of R Main Beam	L	Inside Spread	Number of R Points	L	Area	State/ Province	Hunter's Name	Date	Rank
127 5/8	21 1/8	21 0/8	17 7/8	5	4	Young County	TX	Bart Brackeen	2000	2223
127 5/8	23 2/8	22 4/8	18 5/8	8	5	Osborne County	KS	Casmir Domurat	2003	2223
127 5/8	20 4/8	19 4/8	18 1/8	5	6	Maverick County	TX	H. Mike Palmer	2005	2223
127 5/8	21 3/8	22 4/8	16 5/8	5	6	Weld County	CO	Frank Piacentino	2006	2223
127 5/8	20 5/8	20 6/8	16 7/8	5	5	McKenzie County	ND	Milan Liesener	2007	2223
127 5/8	22 5/8	23 3/8	18 5/8	4	5	Mountrail County	ND	James Domaskin	2007	2223
127 5/8	19 0/8	20 2/8	19 3/8	5	5	Montgomery County	AR	Eugene Smith, Jr.	2009	2223
127 4/8	20 5/8	20 5/8	16 2/8	5	5	Boone County	MO	Roger Soukup	1978	2240
127 4/8	20 1/8	19 3/8	16 4/8	5	6	Cooper County	MO	Joe Stenger	1988	2240
127 4/8	19 1/8	19 0/8	15 2/8	5	5	Hardin County	IA	William Stonebraker	1990	2240
127 4/8	20 6/8	20 6/8	15 3/8	5	6	Ravalli County	MT	Ned Coorough	1991	2240
127 4/8	23 4/8	22 7/8	17 0/8	6	6	Waupaca County	WI	Daniel W. Schultz	1993	2240
127 4/8	21 5/8	22 2/8	14 4/8	5	5	La Salle County	TX	Roger Wintle	1993	2240
127 4/8	21 5/8	19 3/8	17 6/8	5	6	Hillsdale County	MI	Thomas R. Johnson	1994	2240
127 4/8	22 4/8	23 1/8	18 5/8	4	5	Will County	IL	Tom Spence	1994	2240
127 4/8	23 3/8	23 3/8	20 1/8	4	5	Bayfield County	WI	Brian L. Gerndt	1994	2240
127 4/8	22 6/8	23 3/8	16 2/8	4	4	Dane County	WI	Mike Hughes	1997	2240
127 4/8	19 5/8	21 5/8	16 6/8	5	5	Gallatin County	MT	Brian F. Koelzer	1998	2240
127 4/8	22 4/8	23 7/8	17 0/8	4	4	Fairfield County	CT	Bill Terry, Sr.	2003	2240
127 4/8	22 5/8	22 3/8	18 6/8	5	4	Montgomery County	IA	William Cmiel	2004	2240
127 4/8	22 6/8	23 0/8	20 0/8	4	4	Baltimore County	MD	Earle F. Yearsley	2006	2240
127 4/8	21 1/8	21 0/8	16 2/8	5	5	Haskell County	OK	Tim Burnett	2007	2240
127 4/8	23 4/8	22 7/8	19 0/8	4	4	McMullen County	TX	Steven M. Shedd	2008	2240
127 4/8	21 3/8	20 0/8	16 6/8	5	5	Dunn County	ND	James W. Casto, Jr.	2009	2240
127 3/8	22 0/8	22 1/8	17 3/8	4	4	Lee County	IA	Lewallen Foster	1972	2257
127 3/8	22 5/8	20 1/8	17 2/8	5	7	Andrew County	MO	Dan A. McLaughlin	1991	2257
127 3/8	21 1/8	20 5/8	14 3/8	4	5	Medina County	TX	Kip M. Melancon	1995	2257
127 3/8	22 0/8	22 2/8	18 3/8	4	5	Pottawatomie County	KS	Michael Schirer	1996	2257
127 3/8	23 3/8	24 6/8	16 1/8	6	5	Shelby County	IL	Mark K. Rothrock	1997	2257
127 3/8	20 1/8	20 5/8	15 5/8	8	6	Van Buren County	IA	Russ Miller	2000	2257
127 3/8	23 3/8	23 4/8	19 1/8	6	6	Floyd County	IA	Gene A. Hall	2000	2257
127 3/8	22 1/8	24 0/8	15 5/8	5	4	Montgomery County	TN	Jim Coode	2003	2257
127 2/8	22 2/8	23 1/8	16 2/8	5	5	Madison County	AR	Johnny Darris	1964	2265
127 2/8	23 5/8	23 7/8	16 4/8	4	3	Lawrence County	IL	Steven R. Tice	1967	2265
127 2/8	19 2/8	19 0/8	15 2/8	5	5	Blue Earth County	MN	Stanley R. Defries	1973	2265
127 2/8	21 7/8	21 2/8	15 6/8	5	4	Jefferson County	IN	Robert Pitt	1978	2265
127 2/8	20 3/8	22 2/8	16 7/8	7	6	Cerro Gordo County	IA	Vern Schnoebelen	1988	2265
127 2/8	18 6/8	18 7/8	17 2/8	6	6	Will County	IL	Tom S. Spence	1989	2265
127 2/8	23 2/8	22 5/8	14 4/8	4	4	Highland County	OH	Martin Bullock	1991	2265
127 2/8	23 1/8	22 5/8	16 5/8	4	6	Van Buren County	IA	John (Rosey) Roseland	1992	2265
127 2/8	20 4/8	21 0/8	13 0/8	6	6	Lincoln County	NE	Ron Rockwell	1994	2265
127 2/8	19 5/8	19 7/8	14 2/8	4	4	Valley County	MT	Todd W. Henck	1994	2265
127 2/8	22 5/8	23 4/8	19 2/8	4	6	Gallia County	OH	Randy Epling	1995	2265
127 2/8	20 1/8	20 4/8	17 4/8	6	5	Will County	IL	Kenneth A. Genovese	1995	2265
127 2/8	22 7/8	24 3/8	18 5/8	5	6	Chautauqua County	KS	Christopher Fechner	1999	2265
127 2/8	24 4/8	24 0/8	19 4/8	4	4	McKenzie County	ND	David Tofte	2001	2265
127 2/8	21 6/8	22 1/8	15 6/8	4	4	St. Louis County	MN	Jeff Pederson	2001	2265
127 2/8	22 2/8	22 2/8	18 6/8	5	5	Page County	IA	Bob Athen	2003	2265
127 2/8	21 3/8	21 4/8	17 1/8	4	5	Decatur County	IA	Brad Blanchard	2005	2265
127 2/8	20 0/8	19 5/8	13 6/8	5	5	McKenzie County	ND	Ginger Brockman	2008	2265
127 2/8	24 7/8	23 7/8	16 4/8	4	5	Shelby County	OH	Craig Ambos	2009	2265
127 2/8	21 3/8	21 4/8	20 0/8	4	4	Marquette County	WI	Jeffrey Frazer	2009	2265
127 1/8	19 5/8	20 5/8	15 1/8	4	4	Oneida County	WI	Fred Felbab	1961	2285
127 1/8	21 0/8	20 4/8	18 3/8	6	5	Marion County	IA	Donald C. Clark	1970	2285
127 1/8	18 0/8	20 5/8	15 5/8	5	5	Des Moines County	IA	Roy Veach	1970	2285
127 1/8	23 0/8	23 0/8	16 7/8	4	4	Union County	IA	Richard Siddens	1971	2285
127 1/8	23 0/8	22 5/8	18 4/8	6	4	Yuma County	CO	Tony Seahorn	1986	2285
127 1/8	20 6/8	20 6/8	17 5/8	4	4	Franklin County	MA	Richard Scorzafava	1987	2285
127 1/8	21 7/8	21 7/8	15 3/8	4	5	Jasper County	TX	Kevin W. Walker, Sr.	1990	2285
127 1/8	20 4/8	21 4/8	13 4/8	5	5	Knox County	IL	Larry Hesterly	1992	2285
127 1/8	24 6/8	25 7/8	19 0/8	5	5	Buckingham County	VA	Steven W. Blosser	1992	2285
127 1/8	23 2/8	23 1/8	19 1/8	4	5	Shelby County	KY	Steven D. Williams	1993	2285

Minimum Score 125 *Continued*

Score	Length of Main Beam R	Length of Main Beam L	Inside Spread	Number of Points R	Number of Points L	Area	State/ Province	Hunter's Name	Date	Rank
127 1/8	24 0/8	23 3/8	17 5/8	4	5	Licking County	OH	Stephen B. Cooksey	1999	2285
127 1/8	21 7/8	20 6/8	16 7/8	4	4	Weld County	CO	Wayne Depperschmidt	2000	2285
127 1/8	22 0/8	21 1/8	16 7/8	4	4	Montcalm County	MI	Todd Andersen	2003	2285
127 1/8	23 2/8	23 2/8	18 3/8	6	6	Concho County	TX	Steve Fegley	2005	2285
127 0/8	20 0/8	20 7/8	18 2/8	5	5	Harris County	TX	Daniel Barnes	1980	2299
127 0/8	22 1/8	21 6/8	17 6/8	5	5	Atascosa County	TX	Mike Palmer	1989	2299
127 0/8	21 1/8	20 6/8	13 2/8	5	5	Dodge County	WI	Mike Young	1992	2299
127 0/8	21 4/8	21 5/8	18 0/8	4	4	Motley County	TX	David Lynn Wigley	1993	2299
127 0/8	18 4/8	18 6/8	15 2/8	6	6	Harris County	TX	Stan Gouger, Jr.	1995	2299
127 0/8	21 6/8	21 4/8	16 7/8	5	6	Cowley County	KS	Brian K. Smith	1997	2299
127 0/8	21 4/8	21 3/8	14 0/8	5	4	Valley County	MT	Richard J. Langer	1999	2299
127 0/8	21 1/8	19 4/8	15 1/8	7	8	Raleigh County	WV	Bobby Lane	2000	2299
127 0/8	19 6/8	19 6/8	15 0/8	4	4	McKenzie County	ND	Jody Smith	2009	2299
126 7/8	22 5/8	23 0/8	15 5/8	5	5	Winnebago County	WI	Patrick O'Brien	1970	2308
126 7/8	22 5/8	22 6/8	17 4/8	5	4	Dade County	MO	Paul Watson	1970	2308
126 7/8	21 6/8	21 4/8	17 3/8	5	5	Racine County	WI	Patrick Blaskowski	1986	2308
126 7/8	20 6/8	20 2/8	15 7/8	5	5	Lapeer County	MI	John E. Wencley	1991	2308
126 7/8	21 6/8	21 1/8	12 5/8	5	6	Logan County	CO	Jim Grover	1997	2308
126 7/8	23 2/8	24 0/8	23 1/8	4	5	Lee County	IL	Charles W. Rehor	2000	2308
126 7/8	23 3/8	21 7/8	18 6/8	5	4	Coles County	IL	Gary F. Saveley	2001	2308
126 7/8	22 4/8	22 6/8	17 3/8	4	4	Todd County	MN	William H. Peterson	2005	2308
126 7/8	24 1/8	25 4/8	18 5/8	4	4	Lemhi County	ID	Benjamin Fahnholz	2005	2308
126 6/8	22 5/8	22 0/8	17 2/8	6	4	Winneshiek County	IA	Nick Bowlus	1952	2317
126 6/8	20 0/8	19 5/8	13 4/8	5	6	Pine County	MN	Marvin W. Brown	1968	2317
126 6/8	18 1/8	18 2/8	15 6/8	5	5	Todd County	MN	Paul Jenc	1973	2317
126 6/8	21 0/8	21 0/8	16 2/8	5	4	Woodford County	IL	Rickie D. Snell	1976	2317
126 6/8	23 0/8	22 5/8	18 4/8	4	4	Grant County	WV	Carl Muth	1977	2317
126 6/8	23 7/8	23 6/8	15 7/8	5	6	Williamson County	IL	Wallace Mick	1986	2317
126 6/8	21 1/8	21 4/8	17 2/8	5	6	Chase County	KS	Larry Walker, Jr.	1991	2317
126 6/8	20 2/8	21 2/8	22 6/8	4	4	Hunterdon County	NJ	Bill Berner	1994	2317
126 6/8	20 3/8	20 6/8	18 3/8	5	4	Sullivan County	IN	John Hale	1994	2317
126 6/8	23 4/8	23 4/8	16 4/8	4	4	Mercer County	MO	Brenda Hudson	1995	2317
126 6/8	24 1/8	23 1/8	18 0/8	4	4	Hidalgo	MEX	Patricio Sada	1995	2317
126 6/8	24 0/8	22 5/8	17 6/8	5	4	Lake County	IL	Carl Spaeth	1996	2317
126 6/8	22 2/8	22 2/8	16 2/8	4	4	Richland County	WI	Chris Duane Cina	1996	2317
126 6/8	21 1/8	20 0/8	18 2/8	4	4	Waupaca County	WI	David Schafhauser	1997	2317
126 6/8	22 4/8	23 1/8	17 6/8	4	4	Livingston County	NY	Melvin Mensinger	2000	2317
126 6/8	22 4/8	22 7/8	17 4/8	4	5	Prairie River	SAS	Ken H. Taylor	2007	2317
126 5/8	22 4/8	22 0/8	21 5/8	4	4	Pine County	MN	Buck Doran	1945	2333
126 5/8	22 5/8	23 0/8	19 0/8	5	5	Roberts County	SD	Roland L. Hausmann	1961	2333
126 5/8	21 4/8	21 6/8	17 2/8	7	7	Jackson County	MI	Gary A. Dawson	1980	2333
126 5/8	18 2/8	18 2/8	12 7/8	6	5	Johnson County	WY	Jerry N. Blossom	1983	2333
126 5/8	21 2/8	20 7/8	19 5/8	5	4	Roberts County	SD	Paul Sand	1986	2333
126 5/8	20 5/8	20 5/8	15 5/8	7	6	Ferry County	WA	Jerry Solie	1994	2333
126 5/8	22 3/8	23 6/8	22 1/8	4	5	Franklin County	OH	Steve Byerly	1995	2333
126 5/8	22 2/8	22 4/8	17 7/8	5	4	Ravalli County	MT	Wayne Capp	1996	2333
126 5/8	23 0/8	23 0/8	20 1/8	4	4	Baltimore County	MD	Daniel E. Kauffman	1999	2333
126 5/8	21 2/8	22 0/8	20 3/8	4	4	Olmsted County	MN	Jerry V. Finley	2004	2333
126 5/8	21 3/8	21 4/8	17 3/8	4	5	Kenedy County	TX	Bob Gilbert	2008	2333
126 5/8	21 1/8	20 5/8	16 4/8	9	8	Madison County	MT	Jim Kennedy	2008	2333
126 4/8	20 4/8	21 2/8	18 6/8	5	5	Warren County	PA	Fred Massa	1967	2345
126 4/8	21 2/8	21 3/8	15 2/8	4	4	Chester County	PA	Joseph R. Yannelli	1967	2345
126 4/8	22 1/8	22 0/8	18 2/8	5	4	Cayuga County	NY	John Pardee	1971	2345
126 4/8	22 4/8	21 6/8	17 4/8	4	4	Waupaca County	WI	Arlin Kersten, Jr.	1971	2345
126 4/8	19 7/8	19 6/8	16 4/8	5	5	Owen County	IN	G. Fred Asbell	1971	2345
126 4/8	21 2/8	21 3/8	12 4/8	4	4	Bayfield County	WI	Del Zwiefelhofer	1972	2345
126 4/8	24 6/8	25 0/8	17 0/8	5	4	Portage County	OH	Burt Thompson, Jr.	1976	2345
126 4/8	20 4/8	19 6/8	13 2/8	5	5	Olmsted County	MN	Tom Lofgren	1983	2345
126 4/8	21 6/8	21 0/8	17 6/8	5	6	Muskingum County	OH	Allen Randal Smith	1985	2345
126 4/8	21 4/8	21 7/8	16 6/8	6	4	Saginaw County	MI	Dave L. Tanney	1988	2345
126 4/8	20 5/8	21 0/8	18 2/8	5	5	Harris County	TX	Daniel K. Barnes	1991	2345
126 4/8	21 5/8	22 4/8	18 0/8	6	4	Morgan County	CO	Fred Espinoza	1992	2345

Score	Length of R Main Beam	L	Inside Spread	Number of R Points	L	Area	State/ Province	Hunter's Name	Date	Rank
126 4/8	23 3/8	23 6/8	17 0/8	4	4	Cass County	MI	James E. Akey	1994	2345
126 4/8	20 3/8	19 7/8	17 0/8	6	5	Sanders County	MT	Greg L. Munther	1995	2345
126 4/8	21 5/8	21 4/8	13 6/8	5	5	Jefferson County	ID	Doug Kauer	1997	2345
126 4/8	22 3/8	22 3/8	17 0/8	4	5	Ramsey County	MN	Dave Vomela	2002	2345
126 4/8	22 0/8	21 7/8	16 4/8	4	4	Armstrong County	PA	Allen Randal Smith	2006	2345
126 4/8	22 1/8	22 3/8	15 6/8	5	4	Jackson County	OH	Jim Ridge	2007	2345
126 3/8	24 2/8	23 0/8	20 6/8	4	6	Morris County	KS	Darrell Haeker	1963	2363
126 3/8	22 6/8	21 3/8	15 5/8	4	4	Lac qui Parle County	MN	Elmer Hubbs	1965	2363
126 3/8	21 5/8	20 2/8	17 0/8	5	6	Gallatin County	KY	Robert L. Hegge, Jr.	1973	2363
126 3/8	21 5/8	20 7/8	15 5/8	4	4	Plymouth County	IA	Guy H. Hempey	1974	2363
126 3/8	22 3/8	19 4/8	17 7/8	5	4	Champaign County	OH	Thomas J. Flohre	1991	2363
126 3/8	20 3/8	20 1/8	14 1/8	5	5	Boone County	MO	Ralph W. Olinger	1993	2363
126 3/8	19 7/8	20 7/8	16 7/8	5	4	Winneshiek County	IA	Tim Everding	1993	2363
126 3/8	21 1/8	20 3/8	15 7/8	5	5	Winona County	MN	Gary Bauer	1997	2363
126 3/8	22 0/8	22 7/8	17 5/8	4	5	Pope County	IL	Terry Staebel	1998	2363
126 3/8	22 0/8	23 3/8	14 3/8	4	4	Prowers County	CO	Barry J. Smith	2000	2363
126 3/8	21 3/8	22 3/8	16 1/8	4	4	Spink County	SD	Barry J. Smith	2001	2363
126 3/8	20 7/8	20 2/8	16 7/8	4	4	Dundy County	NE	Ronald L. King	2005	2363
126 3/8	20 7/8	20 7/8	18 5/8	4	4	Prowers County	CO	Garrett Chavez	2011	2363
126 3/8	18 6/8	18 6/8	17 0/8	7	8	Wapiti River	ALB	Dennis Dunn	2012	2363
126 2/8	22 7/8	22 0/8	17 2/8	5	5	Lee County	IL	Edmund R. Braun	1959	2377
126 2/8	22 0/8	20 6/8	17 6/8	4	4	Johnson County	IA	Paul F. Spicer	1963	2377
126 2/8	20 0/8	19 3/8	16 0/8	4	4	Harding County	SD	Gerald Swayze	1965	2377
126 2/8	20 6/8	20 2/8	15 6/8	5	5	Adams County	IL	Adrian K. Smith	1967	2377
126 2/8	21 4/8	21 4/8	16 2/8	4	4	Ford County	KS	Bob Stephenson	1967	2377
126 2/8	21 3/8	21 4/8	15 4/8	4	4	Morrison County	MN	Frank Hogan	1967	2377
126 2/8	23 7/8	23 5/8	17 3/8	5	4	Marathon County	WI	Dan Niehaus	1968	2377
126 2/8	20 1/8	20 7/8	14 4/8	5	5	Platte County	NE	Keith Bruhn	1968	2377
126 2/8	24 1/8	23 7/8	14 6/8	4	4	Winneshiek County	IA	Marc Headington	1971	2377
126 2/8	22 5/8	21 6/8	17 2/8	6	5	Livingston County	MO	Myles Keller	1980	2377
126 2/8	19 2/8	20 5/8	17 6/8	6	5	Clay County	MN	Steve Steinhoff	1983	2377
126 2/8	20 3/8	19 6/8	14 4/8	5	5	McKenzie County	ND	Tim Finley	1985	2377
126 2/8	20 2/8	20 0/8	15 2/8	5	5	Meeker County	MN	Brent Swanson	1986	2377
126 2/8	23 6/8	23 3/8	16 5/8	4	6	Iroquois County	IL	Bill Price	1987	2377
126 2/8	21 5/8	21 7/8	15 0/8	5	4	Miami County	KS	Randy Stoughton	1993	2377
126 2/8	19 2/8	18 5/8	16 0/8	5	5	Olmsted County	MN	Mark R. Batterson	1993	2377
126 2/8	22 1/8	21 2/8	15 4/8	4	6	Screven County	GA	Thad Beckum	1997	2377
126 2/8	28 6/8	22 6/8	15 2/8	4	4	Allegheny County	PA	S. Andrew Wright	2002	2377
126 2/8	23 3/8	21 0/8	16 6/8	5	7	Wyandot County	OH	Aaron Leightey	2007	2377
126 2/8	19 7/8	19 4/8	14 5/8	5	6	Cherokee County	IA	Alan Bud Martin	2009	2377
126 2/8	21 4/8	24 1/8	19 1/8	5	3	Fairfield County	CT	Bill Terry, Sr.	2011	2377
126 1/8	24 1/8	24 3/8	17 1/8	4	5	White County	GA	James Turner	1967	2398
126 1/8	25 2/8	24 5/8	18 5/8	4	4	York County	PA	Gregory E. Smith	1969	2398
126 1/8	22 1/8	21 7/8	16 5/8	5	4	Monmouth County	NJ	Charles C. Lasala	1970	2398
126 1/8	20 4/8	18 6/8	18 1/8	4	4	Pope County	IL	Murray Schuhart	1972	2398
126 1/8	22 3/8	21 7/8	16 1/8	4	4	Rock County	WI	Gregg C. Sutherland	1992	2398
126 1/8	22 4/8	22 1/8	14 3/8	5	5	Rusk County	WI	Brian Tessmann	2005	2398
126 1/8	22 6/8	21 4/8	20 5/8	5	5	Sullivan County	PA	Daniel L. Garman	2008	2398
126 0/8	20 3/8	20 3/8	14 0/8	4	4	Cherry County	NE	Dean Bergman	1975	2405
126 0/8	18 4/8	17 5/8	13 6/8	5	5	Delaware County	NY	Alan Beyer	1976	2405
126 0/8	20 4/8	21 6/8	17 0/8	4	4	Dunn County	WI	Richard Paul	1977	2405
126 0/8	20 7/8	21 3/8	16 6/8	4	6	Brookings County	SD	Michael Kjellsen	1982	2405
126 0/8	20 1/8	20 2/8	14 4/8	5	6	Des Moines County	IA	Duane R. Mabry	1984	2405
126 0/8	24 0/8	24 4/8	16 2/8	6	4	Van Buren County	IA	Don Kieler	1986	2405
126 0/8	22 1/8	21 6/8	17 6/8	4	4	Hardin County	IL	Thomas R. Grabowski	1991	2405
126 0/8	20 2/8	20 7/8	17 7/8	5	6	Vermilion County	IL	Jim Richards	1993	2405
126 0/8	21 7/8	19 3/8	14 6/8	5	5	Vinton County	OH	Clark Miller	1994	2405
126 0/8	21 5/8	20 7/8	18 0/8	4	4	Montgomery County	PA	Paul Warren	1995	2405
126 0/8	20 1/8	21 2/8	18 2/8	4	4	Davis County	IA	Gary Wissmueller	1996	2405
126 0/8	23 1/8	23 3/8	20 5/8	5	6	Brown County	OH	Kendall Techau	1997	2405
126 0/8	20 4/8	22 1/8	18 2/8	5	5	Lawrence County	TN	Michael Manning	1998	2405
126 0/8	23 6/8	23 1/8	19 0/8	4	4	Davis County	IA	Mark Wagner	1998	2405

WHITETAIL DEER [Typical Antlers]

Continued

Score	Length of Main Beam R	Length of Main Beam L	Inside Spread	Number of Points R	Number of Points L	Area	State/ Province	Hunter's Name	Date	Rank
126 0/8	22 4/8	23 6/8	17 4/8	5	4	Grant County	WI	Wayne Willkomm	1999	2405
126 0/8	21 4/8	22 0/8	17 2/8	4	6	Gage County	NE	Larry L. Maschmeier	2003	2405
126 0/8	20 7/8	21 3/8	19 6/8	4	7	Fergus County	MT	John "Rosey" Roseland	2007	2405
126 0/8	22 6/8	22 4/8	16 6/8	4	4	Concho County	TX	David R. Toney	2007	2405
126 0/8	22 2/8	21 6/8	19 0/8	3	4	Black Hawk County	IA	Mike Judas	2007	2405
126 0/8	20 5/8	20 4/8	16 0/8	6	5	Maverick County	TX	H. Mike Palmer	2008	2405
126 0/8	20 6/8	21 6/8	15 6/8	4	4	Flathead County	MT	Chuck Williams	2008	2405
125 7/8	23 6/8	25 7/8	21 7/8	7	5	Westchester County	NY	Steve Solomon	1969	2426
125 7/8	20 0/8	20 0/8	14 5/8	5	5	Oconto County	WI	Chuck Matyska	1972	2426
125 7/8	22 7/8	22 5/8	15 3/8	4	4	Pope County	AR	William E. Campbell	1974	2426
125 7/8	22 4/8	23 0/8	15 1/8	4	4	Fayette County	IL	Edward G. Myers	1975	2426
125 7/8	20 6/8	22 7/8	15 5/8	5	5	Butler County	PA	John Schmiedlin	1976	2426
125 7/8	20 4/8	18 4/8	15 5/8	5	5	Johnson County	WY	Jimmy Womble	1986	2426
125 7/8	21 2/8	22 2/8	16 7/8	4	4	Jefferson County	WI	Edward L. Anfang	1994	2426
125 7/8	20 0/8	21 0/8	17 7/8	4	4	Dodge County	MN	Brad Blanchard	1995	2426
125 7/8	18 1/8	18 5/8	16 6/8	6	5	Madison County	MT	J. Dudley Ottley	1997	2426
125 7/8	25 3/8	24 4/8	17 3/8	4	4	Wayne County	NY	John N. Bacon	1998	2426
125 7/8	21 7/8	21 4/8	18 3/8	4	4	Bingham County	ID	Doug Burkman	1999	2426
125 7/8	20 0/8	20 2/8	13 3/8	4	5	Otoe County	NE	Rod Aden	2000	2426
125 7/8	22 0/8	20 6/8	11 7/8	5	5	Owen County	KY	Alan Morris	2000	2426
125 7/8	20 5/8	20 7/8	19 1/8	5	4	Ogle County	IL	Marc A. Brumbly	2000	2426
125 7/8	20 7/8	21 2/8	18 1/8	5	4	Highland County	OH	Dan Sowders	2004	2426
125 7/8	19 3/8	19 2/8	17 7/8	5	5	Keya Paha County	NE	Terry Marcukaitis	2005	2426
125 7/8	20 4/8	19 3/8	14 5/8	5	5	Fillmore County	MN	Scott Pierce	2007	2426
125 6/8	23 0/8	23 4/8	19 2/8	4	4	Bath County	VA	W. C. Bedall, Jr.	1958	2443
125 6/8	19 5/8	19 3/8	16 2/8	5	5	Wood County	WI	Dennis Palmer	1967	2443
125 6/8	21 4/8	21 5/8	18 6/8	4	5	Martin County	MN	Robert Barnett	1970	2443
125 6/8	20 0/8	21 6/8	20 1/8	6	4	Palo Alto County	IA	Earl J. Gustafson	1971	2443
125 6/8	25 2/8	25 3/8	21 0/8	4	4	Westchester County	NY	Carl Ranieri	1986	2443
125 6/8	19 7/8	19 0/8	16 0/8	5	4	Washington County	MN	Wayne Edgerton	1991	2443
125 6/8	20 1/8	20 3/8	14 6/8	5	5	Cabell County	WV	Raymond Kingery	1993	2443
125 6/8	21 4/8	21 5/8	17 6/8	5	5	Carroll County	OH	Randy S. Collins	1994	2443
125 6/8	22 4/8	21 3/8	15 2/8	5	5	Logan County	CO	Wayne Depperschmidt	1995	2443
125 6/8	22 3/8	19 6/8	14 4/8	5	5	Webb County	TX	Troy Hardesty	1996	2443
125 6/8	21 0/8	21 7/8	17 2/8	5	5	Traill County	ND	Richard L. Aamold	2001	2443
125 6/8	22 2/8	21 3/8	16 0/8	4	5	Meigs County	OH	Mike Parrish	2002	2443
125 6/8	22 0/8	22 0/8	17 4/8	4	5	Shelby County	OH	Craig Ambos	2003	2443
125 6/8	21 0/8	20 3/8	14 2/8	5	5	Licking County	OH	Willard Jarvis	2003	2443
125 6/8	21 3/8	21 4/8	17 0/8	6	8	Pittsburg County	OK	Don Bales	2008	2443
125 6/8	19 1/8	19 1/8	13 5/8	5	6	Williams County	ND	Mark Hughes	2009	2443
125 5/8	21 4/8	21 3/8	18 3/8	4	4	Brown County	SD	Leo J. Weber	1965	2459
125 5/8	21 5/8	21 4/8	17 5/8	4	4	Lyman County	SD	Dennis Lien	1966	2459
125 5/8	19 2/8	19 3/8	16 3/8	5	5	Sauk County	WI	Andre J. Jestafie	1968	2459
125 5/8	22 1/8	25 0/8	20 1/8	5	6	Meade County	KS	Kent Davis	1971	2459
125 5/8	21 1/8	21 5/8	17 5/8	5	4	Frontier County	NE	Steven P. Salmieri	1987	2459
125 5/8	22 3/8	21 6/8	19 4/8	6	4	Faribault County	MN	Bob Prange	1988	2459
125 5/8	21 2/8	21 5/8	19 1/8	5	5	Jefferson County	OH	William Gary Rusinovich	1989	2459
125 5/8	21 3/8	22 2/8	17 5/8	5	4	Wayne County	MO	Don Reynolds	1989	2459
125 5/8	24 0/8	23 5/8	17 2/8	6	6	Sanilac County	MI	James C. Whitehead	1993	2459
125 5/8	20 3/8	19 7/8	13 1/8	5	5	Monroe County	WI	Dave G. Carr	1994	2459
125 5/8	24 1/8	21 7/8	19 0/8	5	5	Woodbury County	IA	Blake Miller	1998	2459
125 5/8	21 6/8	22 6/8	15 3/8	4	4	Halifax County	VA	Timothy Miles Hamlet	1999	2459
125 5/8	19 6/8	20 7/8	18 1/8	6	5	Ionia County	MI	Kurt R. Buche	2003	2459
125 5/8	21 1/8	21 7/8	16 5/8	4	5	Fountain County	IN	Tyler Ratcliff	2005	2459
125 5/8	22 0/8	22 3/8	17 2/8	6	5	Olmsted County	MN	Walter P. Krapohl	2005	2459
125 5/8	21 7/8	20 1/8	21 5/8	4	5	Hancock County	IL	Bradley J. Swanson	2006	2459
125 5/8	20 6/8	21 0/8	16 5/8	5	5	Webb County	TX	Bob Gilbert	2007	2459
125 4/8	24 7/8	21 6/8	17 2/8	4	4	Marion County	IA	Thomas Tucker	1962	2476
125 4/8	20 4/8	20 7/8	19 0/8	4	4	Clark County	MO	Allen L. Courtney	1971	2476
125 4/8	20 2/8	21 3/8	16 6/8	5	5	Otter Tail County	MN	Terry Tamke	1971	2476
125 4/8	21 7/8	21 7/8	19 4/8	5	4	Warren County	IL	Bob Rawlings	1985	2476
125 4/8	19 5/8	19 6/8	13 7/8	4	6	Cedar County	NE	Kevin Schmidt	1989	2476

WHITETAIL DEER [Typical Antlers]

Minimum Score 125 *Continued*

Score	Length of R Main Beam	L	Inside Spread	Number of R Points	L	Area	State/ Province	Hunter's Name	Date	Rank
125 4/8	21 4/8	23 0/8	15 2/8	5	4	Lake County	IL	Lynn Anderson	1989	2476
125 4/8	21 5/8	21 0/8	18 4/8	4	5	Hardin County	IA	William J. Stonebraker	1991	2476
125 4/8	19 6/8	19 7/8	15 2/8	5	6	Otsego County	MI	Jerome H. Lubbers	1992	2476
125 4/8	21 4/8	21 2/8	16 4/8	6	4	Brown County	IL	Jeff Darr	1993	2476
125 4/8	20 3/8	20 5/8	14 6/8	5	5	Davis County	IA	Mark P. Wagner	1995	2476
125 4/8	23 1/8	23 1/8	18 4/8	5	4	Waupaca County	WI	Joel Dobberpuhl	1997	2476
125 4/8	19 1/8	18 7/8	14 6/8	7	6	Harris County	TX	Gordon Trout	1998	2476
125 4/8	22 2/8	21 2/8	16 4/8	4	4	Clinton County	MO	Keith Westbrook	2000	2476
125 4/8	20 7/8	20 4/8	15 6/8	4	5	Dawson County	MT	Vern Lindquist	2003	2476
125 4/8	18 6/8	18 7/8	16 2/8	4	4	Renville County	ND	Irvin Prough	2003	2476
125 4/8	21 3/8	21 6/8	15 2/8	6	5	Clark County	SD	Mark Turner	2004	2476
125 4/8	22 2/8	20 5/8	20 5/8	7	6	Cass County	IN	Timothy B. Hatfield	2006	2476
125 4/8	20 6/8	22 0/8	19 4/8	6	6	Logan County	KY	Shane Collier	2006	2476
125 4/8	20 1/8	19 5/8	16 4/8	4	4	Johnston County	OK	Gary Higgins	2007	2476
125 3/8	20 5/8	21 4/8	16 5/8	5	4	Stephenson County	IL	Vaughn Zimmerman	1969	2495
125 3/8	24 0/8	24 6/8	17 3/8	4	5	Wabasha County	MN	David Mohler	1971	2495
125 3/8	20 7/8	20 3/8	15 3/8	4	4	Newport News County	VA	Dr. Glenn A. Parker	1972	2495
125 3/8	21 7/8	22 6/8	16 3/8	4	5	Warren County	IA	Michael Woolman	1974	2495
125 3/8	21 4/8	20 6/8	15 7/8	5	5	Cheboygan County	MI	Steven Schrauben	1984	2495
125 3/8	21 1/8	20 3/8	15 3/8	5	5	Waukesha County	WI	Daniel Buchta	1987	2495
125 3/8	21 1/8	22 3/8	18 1/8	5	6	Barry County	MO	David Counts	1995	2495
125 3/8	21 1/8	21 4/8	16 7/8	4	4	Clermont County	OH	Will Mullis	1997	2495
125 3/8	18 6/8	18 4/8	12 5/8	4	4	Plymouth County	IA	John C. Willer	1998	2495
125 3/8	21 4/8	21 2/8	19 1/8	7	4	Marshall County	IN	Larry L. Herr	1999	2495
125 3/8	23 1/8	23 1/8	19 3/8	5	6	Cass County	IL	Mark Livingston	2004	2495
125 3/8	19 0/8	20 1/8	15 1/8	5	5	Dubuque County	IA	John Duggan	2004	2495
125 3/8	21 3/8	20 6/8	16 7/8	5	5	Edwards County	IL	Eli Landingham	2008	2495
125 3/8	22 0/8	21 0/8	20 1/8	6	4	Jo Daviess County	IL	Matthew C. Rehor	2011	2495
125 3/8	19 5/8	19 4/8	17 1/8	7	6	Pittsburg County	OK	Charles "Chuck" Vann	2011	2495
125 2/8	19 3/8	20 0/8	15 0/8	4	4	Missaukee County	MI	Eugene D. Garris, DDS	1966	2510
125 2/8	26 5/8	25 2/8	22 2/8	4	3	Fulton County	IL	Arnold C. Hegele	1967	2510
125 2/8	20 6/8	20 7/8	15 6/8	4	4	Sussex County	NJ	Tom Barber	1973	2510
125 2/8	22 6/8	22 2/8	17 2/8	4	3	Washington County	IA	Danny B. Jirsa	1974	2510
125 2/8	21 6/8	21 5/8	17 2/8	5	5	Jo Daviess County	IL	Bob Wolff	1977	2510
125 2/8	24 1/8	23 4/8	17 4/8	4	4	Blue Earth County	MN	Ray Smothers	1985	2510
125 2/8	24 0/8	23 0/8	17 0/8	4	5	Harford County	MD	Warren J. Barth	1986	2510
125 2/8	21 1/8	20 7/8	16 6/8	4	4	Litchfield County	CT	John Swiklas	1987	2510
125 2/8	21 3/8	21 1/8	19 2/8	4	4	Will County	IL	Robert L. Kamenjarin	1990	2510
125 2/8	22 1/8	21 7/8	14 2/8	5	6	Atascosa County	TX	Mike Palmer	1990	2510
125 2/8	22 1/8	21 4/8	16 0/8	4	4	McMullen County	TX	Bill Crome, Jr.	1993	2510
125 2/8	19 6/8	19 0/8	16 2/8	6	6	Brown County	NE	Joel Klammer	1995	2510
125 2/8	19 4/8	19 1/8	15 6/8	5	5	Dodge County	WI	John R. Puetz	1996	2510
125 2/8	20 6/8	21 1/8	16 6/8	5	4	Decatur County	IA	Dave Sabrowsky	1997	2510
125 2/8	21 6/8	21 4/8	15 4/8	5	5	Ashtabula County	OH	W. John Gaul, Jr.	1999	2510
125 2/8	21 0/8	21 0/8	15 0/8	5	5	Henry County	IA	Jeffrey A. Japsen	2000	2510
125 2/8	21 6/8	21 1/8	16 0/8	5	6	Dubuque County	IA	Michael Casey	2002	2510
125 2/8	20 2/8	20 6/8	16 7/8	5	4	Carroll County	IL	James Carlson	2003	2510
125 2/8	21 6/8	21 0/8	16 4/8	4	4	Lucas County	IA	Kent Brigham	2004	2510
125 2/8	22 1/8	22 7/8	18 4/8	4	4	Brown County	OH	Harold E. Uhl	2005	2510
125 2/8	21 7/8	20 6/8	19 0/8	5	4	Pittsburg County	OK	James L. Fowler	2005	2510
125 2/8	21 5/8	21 5/8	15 6/8	4	4	Missoula County	MT	Marlon Clapham	2005	2510
125 1/8	19 1/8	19 7/8	14 7/8	5	5	Ozaukee County	WI	Don Schwerin	1966	2532
125 1/8	20 7/8	20 6/8	15 3/8	4	4	Graham County	KS	Russell Hull	1975	2532
125 1/8	20 4/8	20 5/8	14 7/8	5	4	Clinton County	MI	Dan Bertalan	1986	2532
125 1/8	21 0/8	20 7/8	16 1/8	5	4	Hunterdon County	NJ	Michael P. Wanat	1988	2532
125 1/8	21 2/8	21 1/8	14 1/8	4	4	Darke County	OH	Dave Shellhaas	1992	2532
125 1/8	19 3/8	19 5/8	18 1/8	5	5	Latah County	ID	Roger Stewart	1992	2532
125 1/8	18 2/8	18 1/8	14 7/8	5	5	Flathead County	MT	John Voelker	1994	2532
125 1/8	22 1/8	22 5/8	16 1/8	5	4	Bullock County	AL	Robert "Grub" Matthews	1994	2532
125 1/8	22 5/8	23 5/8	16 4/8	6	4	Muskingum County	OH	Dan Jennings	1995	2532
125 1/8	21 2/8	19 6/8	15 7/8	4	5	Pike County	IL	Stephen J. Micio	1997	2532
125 1/8	21 4/8	22 5/8	16 3/8	4	4	Gage County	NE	Russell Topp	1999	2532

WHITETAIL DEER [Typical Antlers]

Minimum Score 125 *Continued*

Score	Length of R Main Beam	L	Inside Spread	Number of R Points	L	Area	State/ Province	Hunter's Name	Date	Rank
125 1/8	19 5/8	20 1/8	14 3/8	5	5	Warren County	OH	Kevin Wilson	2001	2532
125 1/8	21 5/8	22 0/8	15 1/8	4	4	Woodford County	IL	John Wesley Zerby	2005	2532
125 1/8	19 7/8	19 7/8	16 7/8	4	4	Fergus County	MT	John "Rosey" Roseland	2008	2532
125 1/8	23 0/8	21 2/8	14 3/8	4	4	Cowley County	KS	Stephen C. Van Hoose	2008	2532
125 1/8	23 3/8	24 3/8	17 1/8	5	5	Shelby County	KY	Joseph C. Fehribach	2009	2532
125 1/8	20 6/8	19 4/8	16 5/8	6	6	Antelope County	NE	Martin Vetick	2009	2532
125 0/8	20 6/8	20 6/8	16 6/8	5	5	Sheboygan County	WI	Earl Uhl	1962	2549
125 0/8	22 4/8	23 0/8	15 4/8	5	5	Hughes County	SD	Robert A. Clough	1967	2549
125 0/8	20 5/8	20 5/8	17 2/8	4	4	Livingston County	NY	Robert A. Carone	1969	2549
125 0/8	22 0/8	22 0/8	17 4/8	6	6	Dodge County	WI	Charles E. Songstad	1970	2549
125 0/8	21 0/8	21 1/8	18 2/8	4	4	Pecos County	TX	Joe Christman	1989	2549
125 0/8	23 2/8	23 5/8	22 2/8	3	4	Ogle County	IL	Mark E. Guffey	1990	2549
125 0/8	21 0/8	20 5/8	17 6/8	4	4	Anoka County	MN	Blair Rawlings	1996	2549
125 0/8	20 6/8	21 1/8	16 6/8	5	5	Lyon County	IA	Michael Judas	1998	2549
125 0/8	22 2/8	22 3/8	18 2/8	4	4	Doniphan County	KS	David Hurd II	1998	2549
125 0/8	21 3/8	20 2/8	14 6/8	4	4	Meeker County	MN	David H. Boland	2000	2549
125 0/8	22 0/8	22 2/8	18 2/8	4	4	Montour County	PA	Dan Reinaker	2000	2549
125 0/8	19 6/8	20 1/8	14 2/8	5	5	Hubbard County	MN	Barry W. Babcock	2003	2549
125 0/8	22 1/8	22 6/8	17 6/8	5	5	Rock Island County	IL	Mike Rayburn	2003	2549
125 0/8	21 3/8	20 4/8	19 0/8	5	5	Milk River	ALB	Tom Taylor	2003	2549
125 0/8	20 6/8	20 6/8	18 2/8	5	4	Powell County	MT	John "Rosey" Roseland	2007	2549
125 0/8	24 0/8	23 1/8	14 6/8	5	4	Tuscarawas County	OH	Charles S. Caley	2008	2549

WHITETAIL DEER [Typical Velvet Antlers]

Minimum Score 125 *Odeocoileus virginianus* and certain related subspecies

Score	Length of R Main Beam	L	Inside Spread	Number of R Points	L	Area	State/ Province	Hunter's Name	Date	Rank
170 2/8	25 0/8	25 6/8	18 0/8	5	5	Switzerland County	IN	Rob Wiley	2000	*
153 5/8	24 6/8	24 5/8	17 1/8	5	5	Menominee County	WI	Al Crowe	1994	*
136 4/8	21 3/8	22 5/8	14 6/8	4	5	Sanders County	MT	Greg L. Munther	1990	*
136 4/8	21 2/8	21 3/8	16 2/8	5	5	McKenzie County	ND	Daniel F. Cote	2001	*
135 1/8	20 4/8	20 4/8	14 7/8	5	5	Valley County	MT	Kathy Terakedis	2000	*
134 5/8	21 2/8	22 0/8	16 7/8	5	5	Jewell County	KS	Nathan Andersohn	2003	*
130 4/8	20 3/8	21 0/8	17 4/8	4	4	Russell	MAN	Ron Klimack	2003	*
129 4/8	21 7/8	22 3/8	17 0/8	4	4	Latah County	ID	Ryan Malm	2000	*
127 4/8	21 4/8	21 1/8	16 2/8	7	6	Saskatoon Mtn.	ALB	Brent Watson	2006	*

Hunter: Jim Willems
Species: Typical Coues' Deer
Score: 95 5/8
Date: 1/18/11
Location: Sonora, Mexico

WHITETAIL DEER [Non-Typical Antlers]

Minimum Score 155 *Odeocoileus virginianus* and certain related subspecies

Score	Length of R Main Beam	L	Inside Spread	Number of R Points	L	Area	State/ Province	Hunter's Name	Date	Rank
279 7/8	27 7/8	28 1/8	21 3/8	21	18	Hall County	NE	Del Austin	1962	1
249 6/8	28 0/8	27 7/8	20 2/8	8	10	Greenwood County	KS	Clifford Pickell	1968	2
229 7/8	25 2/8	25 6/8	19 1/8	12	15	Yuma County	CO	David "Jake" Powell	1986	3
229 2/8	29 0/8	29 0/8	17 5/8	16	11	Crawford County	KS	Dennis D. Jameson	2003	4
222 7/8	21 6/8	22 6/8	26 1/8	9	10	Van Buren County	IA	Gene Wensel	2004	5
222 1/8	26 5/8	25 6/8	17 5/8	15	12	Hancock County	IA	J. M. Monson	1977	6
220 3/8	25 3/8	24 5/8	18 3/8	13	9	Riley County	KS	Melvin D. Padgett	1989	7
220 2/8	26 4/8	24 1/8	18 3/8	12	9	Linn County	KS	Randy A. Reeves	2011	8
218 1/8	26 2/8	25 6/8	17 3/8	10	9	Clay County	IA	Blaine R. Salzkorn	1970	9
212 0/8	25 4/8	27 1/8	17 2/8	9	8	Dodge County	WI	John "Jack" Hoey	1955	10
209 2/8	26 5/8	27 6/8	21 4/8	10	10	McPherson County	KS	Lonnie Ensminger	1968	11
206 7/8	30 5/8	29 3/8	21 6/8	8	9	McPherson County	KS	Dennis G. Bordner	1994	12
206 0/8	22 3/8	24 1/8	17 4/8	9	11	Saunders County	NE	Nordean E. Bade	1964	13
205 2/8	29 0/8	29 2/8	19 4/8	7	8	Washington County	IA	Ed Lash	1999	14
204 1/8	20 4/8	25 0/8	16 6/8	14	12	Chase County	KS	Stephen J. Stivaly	2003	15
203 4/8	26 6/8	27 6/8	17 0/8	7	13	Dodge County	MN	Lawrence Sowieja	1955	16
203 3/8	25 0/8	24 2/8	19 0/8	10	13	Marquette County	WI	Joseph E. Bell	1969	17
203 0/8	25 0/8	25 4/8	19 3/8	12	10	Geauga County	OH	Rudy Grecar	1969	18
202 2/8	23 3/8	25 6/8	19 6/8	12	9	Clay County	SD	Patrick Hudson	1969	19
200 5/8	23 4/8	24 1/8	18 4/8	11	8	Grayson County	TX	Forrest L. Robertson	1995	20
198 5/8	26 2/8	25 0/8	22 6/8	5	5	Douglas County	NE	Ivan Mascher	1961	21
198 3/8	24 3/8	24 3/8	20 7/8	10	10	Lyon County	MN	Edward Matthys	1966	22
198 1/8	27 0/8	26 3/8	20 2/8	10	7	Hocking County	OH	Hugh Cox	1964	23
197 4/8	25 0/8	23 7/8	22 2/8	7	7	Johnson County	IA	Dennis R. Ballard	1971	24
197 1/8	26 1/8	26 3/8	17 3/8	8	7	Jackson County	MO	Jim Martin	1984	25
197 0/8	26 1/8	26 4/8	21 5/8	8	9	Marshall County	IL	Larry Rowe	1975	26
196 5/8	25 7/8	26 4/8	16 4/8	7	8	Worcester County	MD	John W. Bromley III	2005	27
196 2/8	24 7/8	24 6/8	18 4/8	8	7	Louisa County	IA	Tony Thomas	1995	28
195 5/8	28 6/8	28 2/8	20 2/8	9	10	Martin County	MN	Ben Johnson	1973	29
195 5/8	24 1/8	24 2/8	19 0/8	9	7	Waushara County	WI	Randy Chamberlain	1984	29
195 0/8	26 6/8	25 0/8	16 3/8	9	13	Juneau County	WI	Maurice Sterba	1955	31
194 5/8	26 3/8	24 7/8	20 7/8	7	8	Guernsey County	OH	Dick Bayer	1985	32
194 1/8	23 3/8	24 2/8	17 7/8	8	7	Clayton County	IA	Brian Beck	2007	33
193 7/8	25 1/8	27 2/8	18 0/8	11	11	Blaine County	MT	Gene Wensel	1981	34
193 6/8	29 6/8	28 0/8	20 2/8	9	9	Lake County	IN	Walter Sobczak	1979	35
192 5/8	25 0/8	27 0/8	22 1/8	6	8	Republic County	KS	Don Dejmal	1983	36
192 4/8	21 4/8	20 5/8	15 4/8	11	10	Redwood County	MN	Mark A. Steinle	1973	37
191 6/8	23 2/8	20 7/8	16 5/8	6	15	Murray County	MN	Delbert Peck	1956	38
191 4/8	23 3/8	25 1/8	18 3/8	10	7	Tippecanoe County	IN	Jim Foster	1993	39
189 7/8	26 1/8	26 3/8	24 3/8	7	7	Logan County	OH	Larry Pooler	1989	40
189 6/8	24 4/8	23 0/8	16 3/8	9	7	Chisago County	MN	Reinhold L. Lind	1956	41
189 5/8	27 1/8	27 1/8	18 4/8	6	8	Vinton County	OH	Dean Thompson	2011	42
189 2/8	21 4/8	23 1/8	21 3/8	10	8	Clayton County	IA	Jim Monat	1981	43
189 1/8	21 5/8	24 2/8	17 3/8	13	7	Graham County	KS	Don Berry	1970	44
188 1/8	26 5/8	26 0/8	19 4/8	7	10	Barnes County	ND	William Cruff	1961	45
187 6/8	24 0/8	25 4/8	22 6/8	8	6	Clayton County	IA	Virgil High	2002	46
187 5/8	27 6/8	25 2/8	18 0/8	8	6	Monroe County	IA	Cecil Dicks	1961	47
187 4/8	26 2/8	26 0/8	18 0/8	8	7	Juneau County	WI	George "Butch" Sterbenz	1947	48
186 6/8	24 2/8	25 1/8	19 3/8	9	10	Traverse County	MN	Roland L. Hausmann	1964	49
186 2/8	23 1/8	22 5/8	18 6/8	9	9	Otter Tail County	MN	D. F. Vraspir	1959	50
186 1/8	24 0/8	23 7/8	16 2/8	11	10	Stevens County	WA	John Wantulok	1992	51
185 5/8	26 6/8	28 4/8	18 6/8	13	7	Hancock County	IL	Tracy Longenecker	2000	52
185 3/8	27 6/8	27 4/8	23 5/8	8	8	Allamakee County	IA	LeRoy B. Spiker	1968	53
185 3/8	27 3/8	26 7/8	20 7/8	6	6	Rice County	MN	Wayne Jahnke	1975	53
185 2/8	26 1/8	25 0/8	16 6/8	9	8	Sedgwick County	KS	Alfred Weaver	1965	55
185 2/8	24 0/8	23 6/8	19 6/8	11	8	Buffalo County	WI	Jacob Reich	2008	55
185 1/8	22 5/8	22 3/8	18 7/8	8	8	Kent County	MD	W. Ralph Fleegle	2001	57
184 7/8	26 4/8	25 2/8	18 1/8	9	7	Jackson County	IA	Eric C. Proshuto	2007	58
184 4/8	25 7/8	26 3/8	17 2/8	6	6	Monroe County	IA	Robert Smith	2007	59
184 1/8	25 1/8	26 3/8	21 3/8	7	6	St. Charles County	MO	Larry D. Stelzer	1962	60
183 3/8	24 3/8	25 2/8	19 0/8	11	10	Morrison County	MN	Ralph Hakel	1974	61
183 2/8	27 3/8	24 5/8	21 7/8	6	9	Holt County	NE	Lyle Ruff	1967	62

WHITETAIL DEER [Non-Typical Antlers]

Minimum Score 155

Score	Length of Main Beam R	Length of Main Beam L	Inside Spread	Number of Points R	Number of Points L	Area	State/ Province	Hunter's Name	Date	Rank
183 2/8	28 4/8	26 6/8	16 4/8	10	9	Christian County	MO	Roger J. Newell	1984	62
183 1/8	26 7/8	26 0/8	18 0/8	7	6	Grayson County	TX	Wayne Tarpley	1998	64
182 7/8	24 2/8	25 5/8	19 2/8	9	7	Arkansas County	AR	Tommy Horton	1971	65
182 7/8	24 3/8	25 2/8	18 0/8	11	5	Olmsted County	MN	Dan Matheson	1973	65
182 5/8	24 4/8	25 2/8	19 1/8	8	7	Labette County	KS	Jaret Drummond	2004	67
182 4/8	19 5/8	26 7/8	17 2/8	11	11	Will County	IL	Richard Heintz	1971	68
181 6/8	23 2/8	22 6/8	20 1/8	8	7	Adams County	IL	Festal McCarty	1967	69
181 3/8	28 0/8	28 0/8	20 5/8	11	6	Hardin County	IA	Howard Nelson	1963	70
181 2/8	22 1/8	24 5/8	19 0/8	12	8	Desha County	AR	John T. Greer	1962	71
181 1/8	24 1/8	24 1/8	17 6/8	9	6	Coles County	IL	Gerald L. Davis	1973	72
181 0/8	24 5/8	24 6/8	18 5/8	8	6	Grundy County	MO	Paul F. Duffner	2000	73
180 5/8	27 5/8	26 3/8	21 1/8	9	6	Linn County	IA	Craig Shepard	1980	74
180 2/8	23 1/8	24 4/8	19 6/8	7	7	McKenzie County	ND	Rodney Fretland	1998	75
180 1/8	23 6/8	24 7/8	19 6/8	8	7	Mercer County	MO	John D. Green	1996	76
179 6/8	24 4/8	24 2/8	21 0/8	7	8	Creek County	OK	Marion Lewis	1975	77
179 5/8	24 4/8	25 6/8	19 5/8	9	11	Fillmore County	MN	Wayne Pfremmer	1972	78
179 5/8	21 7/8	23 7/8	19 0/8	8	8	Edmonton	ALB	Tony Zambon	2002	78
179 2/8	23 7/8	24 5/8	18 3/8	8	8	Ross County	OH	Paul Leroy Tull	2008	80
179 1/8	24 1/8	23 4/8	17 2/8	6	7	St. Charles County	MO	Martin F. Meier	1999	81
178 6/8	21 2/8	25 3/8	20 5/8	9	8	Lincoln County	SD	H. L. Tuggle	1975	82
178 5/8	24 7/8	25 2/8	18 7/8	7	6	Washington County	OH	David Wallace	2004	83
178 3/8	23 3/8	22 7/8	19 6/8	8	8	Traverse County	MN	Roland L. Hausmann	1953	84
178 3/8	25 1/8	24 4/8	19 6/8	8	10	Mineral County	MT	Gene Wensel	1981	84
178 1/8	23 1/8	24 7/8	15 4/8	6	9	Pepin County	WI	Brennan Fayerweather	2007	86
176 6/8	28 4/8	28 2/8	19 4/8	8	9	Dodge County	WI	Erwin C. Koehler	1957	87
176 0/8	23 2/8	24 2/8	20 1/8	8	9	DeKalb County	IL	Kevin M. Grivetti	1995	88
175 7/8	27 4/8	28 1/8	23 3/8	8	7	Guernsey County	OH	Jack L. Milligan	1971	89
175 4/8	22 4/8	22 7/8	19 2/8	10	10	Woodbury County	IA	Everett Gothier	1962	90
175 3/8	24 5/8	23 7/8	20 1/8	8	7	Grant County	MN	Lee Offerdahl	1972	91
175 3/8	25 4/8	23 2/8	16 5/8	8	8	Hennepin County	MN	Robert R. Herman	1993	91
175 2/8	22 0/8	21 6/8	21 1/8	9	9	Jackson County	MI	Scott A. Weaver	1996	93
175 0/8	23 2/8	23 5/8	18 4/8	7	6	Pottawatomie County	KS	Tod Edwin Anthony	1993	94
174 7/8	22 5/8	22 5/8	20 0/8	6	7	Goliad County	TX	Mark Thompson	1995	95
174 7/8	22 6/8	22 2/8	17 1/8	6	8	Jefferson County	IN	Art Moystner	2003	95
174 4/8	22 1/8	22 3/8	16 4/8	8	8	Benson County	ND	Curtis A. Ehnert	1977	97
174 4/8	22 7/8	22 7/8	17 7/8	6	8	Marinette County	WI	Chuck Druckrey	2004	97
174 2/8	26 6/8	25 1/8	19 7/8	9	6	Clay County	IA	Darrell Magnussen	1962	99
173 7/8	24 7/8	24 5/8	19 5/8	6	7	La Crosse County	WI	Daniel Koenen	2000	100
173 6/8	25 4/8	25 2/8	22 0/8	8	9	Spink County	SD	Milton Haag	1959	101
173 6/8	25 3/8	24 5/8	17 6/8	7	7	Cass County	MI	James Akey	1995	101
173 4/8	24 7/8	25 5/8	21 6/8	6	6	Barton County	KS	Norman Kimber	1967	103
173 4/8	22 4/8	23 7/8	14 1/8	6	8	Trumbull County	OH	Peter Bradley	1969	103
173 3/8	26 5/8	27 3/8	18 7/8	6	9	Berrien County	MI	Steven P. Spenner	2009	105
173 2/8	20 4/8	21 0/8	18 5/8	8	8	Brown County	SD	Barry J. Smith	1995	106
173 2/8	24 0/8	23 4/8	17 7/8	6	6	Lyon County	IA	Michael Judas	1996	106
172 7/8	24 0/8	21 4/8	14 4/8	10	10	McIntosh County	OK	Clark Utley	1976	108
172 7/8	24 4/8	24 2/8	18 3/8	7	6	Dodge County	WI	John R. Puetz	2005	108
172 3/8	28 0/8	27 3/8	15 4/8	5	7	Cass County	MI	Patrick G. Dowling	1999	110
172 1/8	23 1/8	22 6/8	18 6/8	7	7	Hardin County	IA	Stan Caldwell	2002	111
171 6/8	22 3/8	21 5/8	20 5/8	8	9	Dubuque County	IA	Dick Theis	1975	112
171 0/8	24 0/8	24 4/8	18 1/8	7	7	Dodge County	WI	Dallas Johnson	1955	113
171 0/8	26 3/8	22 6/8	22 4/8	4	10	Lee County	IA	Gary Frost	1967	113
170 6/8	23 5/8	22 6/8	16 4/8	10	8	St. Charles County	MO	Jack A. Jones	1992	115
170 6/8	26 0/8	25 1/8	25 5/8	7	11	Henderson County	IL	Terry L. Cook	1997	115
170 5/8	23 7/8	23 2/8	20 3/8	5	5	Clay County	KS	Tom E. Bowman	1999	117
170 2/8	22 1/8	23 6/8	19 0/8	7	9	Warren County	IA	Grant A. Poindexter	1991	118
170 1/8	22 5/8	21 7/8	19 5/8	6	6	Comanche County	KS	Ken Beck	2001	119
169 2/8	23 4/8	23 7/8	18 5/8	6	8	Rice County	MN	Vernon J. Kleve	1972	120
169 2/8	23 3/8	23 3/8	17 2/8	7	7	Vermilion County	IL	Jeff Parkerson	2000	120
169 1/8	29 1/8	27 5/8	22 4/8	9	4	Coles County	IL	Jim Eveland	2008	122
168 7/8	22 4/8	21 6/8	19 4/8	7	8	Meeker County	MN	Ralph Hakel	1964	123
168 7/8	26 7/8	26 3/8	19 5/8	7	6	Dane County	WI	Craig Lunaas	1997	123

WHITETAIL DEER [Non-Typical Antlers]

Minimum Score 155 *Continued*

Score	Length of R Main Beam	L	Inside Spread	Number of R Points	L	Area	State/ Province	Hunter's Name	Date	Rank
168 4/8	26 3/8	27 5/8	17 6/8	7	6	Van Buren County	IA	Don Davidson	2002	125
167 7/8	25 0/8	25 2/8	20 4/8	7	7	St. Clair County	IL	Bryan J. Apostol	1995	126
167 6/8	24 3/8	24 3/8	19 7/8	9	7	Scott County	IA	Gordon Vrana	1967	127
167 6/8	21 7/8	22 2/8	17 2/8	8	6	Carroll County	IL	Carl Spaeth	1998	127
167 5/8	22 5/8	20 6/8	18 6/8	7	8	Dubuque County	IA	Joseph C. Hinderman	2006	129
167 4/8	23 6/8	24 2/8	18 1/8	6	6	Dane County	WI	Mike Hughes	2000	130
167 2/8	24 0/8	24 3/8	17 1/8	7	6	Sauk County	WI	Charles Davenport	1969	131
166 7/8	24 1/8	24 4/8	17 7/8	7	7	Worcester County	MA	Larry Berestka	2005	132
166 6/8	25 1/8	24 5/8	17 7/8	7	7	Kenedy County	TX	Steve Ray Dollar	1990	133
166 5/8	28 3/8	26 6/8	15 1/8	8	9	Shelby County	IA	Billy Custer	1968	134
166 3/8	20 6/8	23 6/8	15 2/8	9	6	Iron County	WI	Joseph Hahn	2001	135
165 7/8	24 2/8	22 2/8	17 3/8	6	6	Arkansas County	AR	Bruce Wiggins	1959	136
165 7/8	22 4/8	22 4/8	13 3/8	8	9	Webb County	TX	James D. Bailey	2007	136
165 4/8	23 5/8	22 2/8	16 2/8	5	8	Boundary County	ID	Gary Stueve	1991	138
165 4/8	26 4/8	25 7/8	22 3/8	5	6	Menard County	IL	Jim Elliott	2001	138
165 2/8	20 3/8	21 0/8	15 2/8	8	10	Lee County	IA	Gary Frost	1991	140
165 2/8	23 7/8	24 0/8	18 4/8	5	9	Pittsburg County	OK	Justin Ammons	2004	140
165 1/8	25 1/8	24 2/8	19 0/8	8	6	Vermilion County	IL	Jim Melton	1997	142
165 0/8	23 2/8	23 6/8	21 6/8	8	6	Fayette County	IA	Mike Judas	2007	143
164 6/8	18 0/8	20 4/8	16 1/8	9	9	Chippewa County	MN	Steven P. Ellingson	1975	144
164 5/8	23 5/8	25 3/8	17 4/8	11	4	Jewell County	KS	Dean Derby	2008	145
163 7/8	27 1/8	25 1/8	20 5/8	6	6	Dickinson County	IA	Eldon L. Kraninger	1969	146
163 7/8	27 1/8	28 7/8	24 5/8	5	7	Grundy County	IL	Ed Vitko, Jr.	1974	146
163 7/8	25 7/8	24 7/8	17 1/8	6	7	Owen County	IN	David L. Hash	2011	146
163 5/8	23 4/8	20 7/8	17 6/8	7	8	Pepin County	WI	Mike J. Breitung	1988	149
163 1/8	26 1/8	25 5/8	15 2/8	8	7	Richardson County	NE	John Duggan	2004	150
162 7/8	26 6/8	25 5/8	20 0/8	6	5	Weld County	CO	Frank Piacentino	2001	151
162 5/8	24 4/8	21 7/8	16 4/8	6	9	Coshocton County	OH	Bill Randles	1995	152
162 5/8	25 6/8	24 6/8	19 7/8	7	5	Coffey County	KS	Craig Giancola	2003	152
162 4/8	22 3/8	23 2/8	13 2/8	8	7	Douglas County	OR	Hugh C. Bennett	1958	154
162 0/8	23 4/8	24 6/8	16 6/8	8	8	Coshocton County	OH	Richard Morgan	1987	155
162 0/8	21 4/8	21 0/8	14 2/8	7	7	Williams County	ND	Richard Liesener	2002	155
161 7/8	24 7/8	25 5/8	22 0/8	7	9	Cascade County	MT	Kits Smith	1980	157
161 7/8	24 4/8	28 4/8	17 3/8	6	11	Warren County	IA	Bob R. Branchcomb	1988	157
161 6/8	23 1/8	23 5/8	18 2/8	8	6	Jo Daviess County	IL	Michael Feldermann	2010	159
161 3/8	23 1/8	22 4/8	19 4/8	6	9	Vinton County	OH	Martin E. Cain	2004	160
161 1/8	22 7/8	22 0/8	18 2/8	6	7	Calhoun County	IL	Herman W. Kovar, Sr.	2003	161
161 0/8	22 6/8	23 4/8	21 2/8	7	6	Douglas County	NE	Ivan Mascher	1969	162
161 0/8	20 5/8	22 6/8	18 0/8	7	6	Montgomery County	PA	Donald C. Marin	2004	162
160 6/8	25 4/8	23 4/8	17 0/8	8	8	Columbiana County	OH	David Tice	1987	164
160 5/8	19 6/8	21 6/8	17 0/8	11	9	Adams County	IL	Ray Gedaminski	1967	165
160 5/8	24 7/8	25 6/8	17 0/8	7	7	Athens County	OH	K. Mike Johnson	2003	165
160 4/8	24 2/8	24 3/8	17 6/8	7	6	Iroquois County	IL	Steven S. Caraker	2009	167
160 0/8	21 3/8	21 5/8	24 3/8	7	8	Knox County	OH	Antonio Mortera II	1999	168
159 7/8	24 7/8	24 4/8	22 3/8	6	6	Hunterdon County	NJ	Ernie Hofer	2000	169
159 4/8	24 7/8	21 0/8	16 2/8	7	10	Madison County	MT	Jim Kennedy	2006	170
159 1/8	21 5/8	21 6/8	13 3/8	8	10	Scott County	KY	Vic Morrison	1972	171
159 1/8	25 6/8	24 7/8	21 3/8	6	8	Rice County	KS	Chris Baldwin	2000	171
158 6/8	25 4/8	25 2/8	14 1/8	7	7	Callaway County	MO	Kevin Pinckney	2005	173
158 5/8	21 2/8	21 1/8	14 5/8	6	7	Sullivan County	IN	John P. Hale	1986	174
157 5/8	20 4/8	21 2/8	15 0/8	8	7	Pontotoc County	OK	Ed Rutherford	2004	175
157 4/8	22 7/8	23 3/8	14 6/8	7	5	Nemaha County	NE	Perry Oates	2003	176
156 6/8	17 5/8	22 1/8	18 3/8	10	6	Washington County	RI	Stephen Tyrrell	1998	177
156 4/8	24 3/8	25 0/8	20 1/8	7	8	Lyon County	MN	Harold G. Halfmann	1972	178
156 4/8	24 6/8	23 3/8	20 2/8	7	9	Winnebago County	IL	Jim Dorney	1975	178
156 4/8	23 6/8	23 5/8	18 6/8	7	5	Columbia County	WI	Jotham E. Pease	1998	178
156 2/8	23 2/8	20 0/8	16 4/8	8	5	Holt County	NE	Darrell Clyde	1963	181
156 1/8	24 4/8	27 7/8	23 3/8	8	9	Cottonwood County	MN	Joe Earl	1959	182
156 0/8	21 0/8	22 0/8	18 5/8	7	7	Lake County	SD	Lonnie Iverson	1987	183
155 6/8	21 6/8	21 7/8	16 5/8	6	7	Grundy County	IL	Wesley Holm	1988	184
155 5/8	25 3/8	24 1/8	18 6/8	6	5	Chickasaw County	IA	David J. Kerkove	1995	185
155 5/8	22 2/8	22 3/8	19 0/8	6	6	Waukesha County	WI	James A. Hitchcock	1997	185

WHITETAIL DEER [Non-Typical Antlers]

Continued

Score	Length of Main Beam R	Length of Main Beam L	Inside Spread	Number of Points R	Number of Points L	Area	State/ Province	Hunter's Name	Date	Rank
155 5/8	21 7/8	22 6/8	14 3/8	9	7	Pittsburg County	OK	Chris Victory	2003	185
155 4/8	25 1/8	25 1/8	16 3/8	9	9	Pottawatomie County	KS	Richard L. Ruetti	1970	188
155 3/8	21 2/8	21 3/8	17 1/8	6	8	Lyon County	MN	Daryl Miller	1968	189
155 0/8	25 1/8	24 3/8	16 0/8	5	7	Waukesha County	WI	Dan Bauman	1996	190

Hunter: Randy A. Reeves
Species: Non-typical Whitetail Deer
Score: 220 2/8
Date: 2/14/12
Location: Linn County, KS

AMERICAN ELK [Typical Antlers]

Minimum Score 260 — *Cervus elaphus nelsoni* and certain related subspecies

Score	Length of R Main Beam	L	Inside Spread	Number of R Points	L	Area	State/ Province	Hunter's Name	Date	Rank
396 2/8	58 2/8	57 6/8	39 0/8	6	6	Socorro County	NM	Bill Clark	2006	1
393 2/8	57 6/8	57 6/8	43 7/8	7	7	Coconino County	AZ	Chad J. Connor	1995	2
385 0/8	52 6/8	53 5/8	43 2/8	7	7	Catron County	NM	Larry A. Green	1997	3
376 6/8	47 2/8	50 3/8	39 2/8	6	7	Adams County	ID	Jack D. Sheppard	1966	4
376 1/8	48 1/8	47 7/8	52 6/8	6	7	Johnson County	WY	Tom Roush	1993	5
373 1/8	49 1/8	48 4/8	41 7/8	7	6	Albany County	WY	Wes Walton	1990	6
371 7/8	52 3/8	53 0/8	46 5/8	6	6	Mineral County	MT	Scott J. Stern	1983	7
370 3/8	56 0/8	56 2/8	33 6/8	7	6	Madison County	MT	Lee Poole	1992	8
370 0/8	52 0/8	51 1/8	40 0/8	6	6	Catron County	NM	Len Abernathy	1994	9
369 4/8	53 0/8	56 6/8	43 7/8	7	7	Sheridan County	WY	Mike Barrett	1998	10
369 2/8	51 0/8	55 4/8	48 4/8	6	6	Madison County	MT	Jeff Engler	1977	11
368 6/8	54 4/8	54 4/8	47 0/8	6	6	Coconino County	AZ	Cathy Grgas	2005	12
367 3/8	47 7/8	49 3/8	45 1/8	8	7	Larimer County	CO	H. Troxell/B. Alexander	1970	13
366 5/8	53 0/8	53 2/8	45 3/8	7	7	Shoshone County	ID	D. A. Johnson	1962	14
364 6/8	50 3/8	46 2/8	39 2/8	6	6	Grant County	OR	William C. Sanowski	1976	15
363 2/8	54 7/8	55 4/8	46 6/8	6	6	Park County	WY	Michael Till	2000	16
362 2/8	50 2/8	49 4/8	46 2/8	7	7	Jackson County	CO	Alfred H. O'Brien	1972	17
361 7/8	47 7/8	48 2/8	35 1/8	6	6	Park County	WY	John R. Buche	2001	18
360 6/8	51 1/8	49 7/8	37 0/8	6	6	Petroleum County	MT	Kerby A. Durbin	1987	19
359 2/8	51 0/8	52 5/8	42 0/8	6	9	Coconino County	AZ	Robert Cecrle	1972	20
359 2/8	58 2/8	56 7/8	42 0/8	6	6	Coconino County	AZ	Jim David	1974	20
359 2/8	56 3/8	53 7/8	40 0/8	7	6	Greenlee County	AZ	N. Richard McMullan	1992	20
358 6/8	51 7/8	55 0/8	47 6/8	6	6	Sheridan County	WY	Mike Barrett	1986	23
358 4/8	52 3/8	52 0/8	39 0/8	7	7	Cibola County	NM	Thomas R. Grabowski	1997	24
358 2/8	58 0/8	57 4/8	43 0/8	7	7	Coconino County	AZ	Joel D. Prickett	2012	25
357 7/8	54 6/8	57 1/8	45 5/8	6	6	Albany County	WY	Richard L. Andre	1990	26
357 0/8	54 2/8	56 0/8	41 0/8	6	6	Caribou County	ID	Howard E. Johnson	1968	27
357 0/8	56 5/8	56 7/8	46 6/8	6	6	Gila County	AZ	David L. Crockett	1973	27
357 0/8	53 5/8	53 6/8	43 2/8	6	7	Strawberry Lake	CO	Frank Fraser	1977	27
356 7/8	54 1/8	53 6/8	53 7/8	6	7	Park County	MT	Charles Alkire	1964	30
356 2/8	45 4/8	45 5/8	41 4/8	6	7	Jackson County	CO	Tom Kelley	2010	31
355 4/8	54 7/8	56 5/8	42 6/8	6	6	Beaverhead County	MT	Robert B. McKay	1973	32
355 1/8	50 2/8	50 4/8	40 1/8	7	6	Sheep Creek	ALB	Paul S. Unger	2005	33
355 0/8	50 2/8	52 1/8	52 1/8	7	7	Jackson County	CO	Jack K. Dodge	2008	34
354 3/8	56 0/8	55 3/8	39 1/8	6	7	Converse County	WY	Edward Coy	1972	35
353 3/8	55 1/8	55 2/8	41 1/8	6	7	Fergus County	MT	Mark Schwomeyer	2007	36
352 7/8	54 4/8	53 4/8	42 0/8	6	7	Teton County	MT	Jack Howard	1972	37
352 2/8	55 2/8	52 2/8	47 2/8	7	8	Powder River County	MT	Gary D. Sumwalt	2008	38
352 0/8	53 4/8	51 6/8	36 6/8	6	6	Converse County	WY	Larry Hardwig	2003	39
352 0/8	48 3/8	46 6/8	37 4/8	6	6	Catron County	NM	Jon Preston	2006	39
352 0/8	49 3/8	51 7/8	35 0/8	7	7	Beaver County	UT	Frederick S. Cliff	2011	39
351 0/8	51 7/8	51 6/8	38 2/8	6	6	Canmore	ALB	Gunter Lemke	1987	42
350 1/8	54 7/8	53 4/8	34 5/8	6	6	Navajo County	AZ	Don Parks, Jr.	1995	43
350 1/8	54 7/8	53 2/8	37 7/8	6	6	Navajo County	AZ	Steve Osminski	2005	43
349 6/8	57 4/8	56 4/8	40 4/8	6	7	Coconino County	AZ	Gary A. Linendoll	1991	45
349 3/8	52 2/8	51 7/8	36 7/8	6	6	Fergus County	MT	Charles R. Bowman	1966	46
349 3/8	46 2/8	46 4/8	40 5/8	6	6	Socorro County	NM	Jon Preston	2010	46
349 1/8	59 4/8	58 6/8	45 5/8	6	6	San Miguel County	NM	James L. Maves	1991	48
349 1/8	53 5/8	50 1/8	40 2/8	7	8	White Pine County	NV	Lew Webb	2001	48
348 3/8	50 0/8	51 7/8	34 4/8	7	7	Pierce County	WA	John Lyday	1984	50
348 2/8	50 0/8	50 4/8	41 6/8	6	7	Carbon County	UT	Rick G. Huempfner	1989	51
348 0/8	53 4/8	53 6/8	43 6/8	6	6	Sheridan County	WY	Mark S. Hutchins	1992	52
348 0/8	57 0/8	57 0/8	32 4/8	6	6	Coconino County	AZ	Todd Zeuske	2000	52
347 4/8	50 2/8	49 5/8	45 6/8	7	6	Fremont County	ID	Bob Baird	1961	54
346 7/8	53 4/8	54 6/8	42 5/8	7	6	Sheridan County	WY	Daniel K. Doke	1998	55
346 6/8	49 6/8	51 2/8	40 4/8	6	6	Catron County	NM	Gardner Rowell	2000	56
346 3/8	50 4/8	49 5/8	40 7/8	6	6	Garfield County	UT	Jerry Woodland	2008	57
345 6/8	51 5/8	54 2/8	36 4/8	6	6	Summit County	UT	Monty Dyke	1992	58
345 5/8	54 6/8	54 1/8	38 7/8	6	6	Catron County	NM	Elmer R. Luce, Jr.	1995	59
345 0/8	52 2/8	52 1/8	37 4/8	7	6	Big Horn County	WY	Randy Burtis	2008	60
345 0/8	49 7/8	47 3/8	36 2/8	6	6	Powder River County	MT	Mike Barrett	2009	60
344 6/8	53 4/8	52 7/8	34 0/8	7	7	King County	WA	Robin F. Buck	1989	62

AMERICAN ELK [Typical Antlers]

Minimum Score 260 *Continued*

Score	Length of R Main Beam	L	Inside Spread	Number of R Points	L	Area	State/ Province	Hunter's Name	Date	Rank
344 0/8	54 0/8	55 1/8	35 0/8	6	6	Columbia County	WA	Daniel Dengate	2010	63
343 7/8	52 0/8	51 2/8	41 5/8	6	6	Catron County	NM	Tim Friday	1995	64
343 4/8	51 4/8	52 1/8	38 4/8	6	7	Socorro County	NM	Richard Dewey	1986	65
343 3/8	49 1/8	49 7/8	40 7/8	6	6	Beaverhead County	MT	Bill Fuller	2005	66
342 7/8	54 3/8	54 3/8	42 2/8	8	8	Grand County	CO	G. Fred Asbell	1980	67
342 2/8	56 6/8	56 4/8	35 4/8	7	7	Catron County	NM	David H. Boland	1992	68
342 0/8	48 6/8	49 1/8	35 4/8	6	6	Mineral County	CO	Kenneth L. Ryan	1996	69
341 7/8	49 5/8	50 2/8	37 4/8	7	8	Fergus County	MT	Keith Meckling	1992	70
341 4/8	45 6/8	47 5/8	37 4/8	6	6	Las Animas County	CO	Douglas F. Murray	1974	71
340 7/8	54 7/8	54 5/8	35 7/8	7	7	Boise County	ID	George A. Foote	1962	72
340 7/8	50 7/8	49 4/8	42 3/8	6	6	Las Animas County	CO	Pat Powell	2000	72
340 6/8	56 6/8	59 5/8	33 6/8	6	6	Coconino County	AZ	Earnest E. Milton	1976	74
340 6/8	54 0/8	51 7/8	41 2/8	7	6	Petroleum County	MT	John 'Rosey' Roseland	1983	74
340 6/8	52 1/8	49 6/8	46 6/8	6	6	Maugher County	MT	Jim Kennedy	2008	74
340 2/8	53 3/8	54 2/8	40 4/8	6	6	Missoula County	MT	David M. Anderson	1978	77
340 2/8	55 4/8	57 1/8	38 2/8	6	6	Coconino County	AZ	Duane A. Wilson, Jr.	1994	77
339 7/8	52 6/8	53 3/8	38 1/8	6	6	Yavapai County	AZ	Gene Barcak	1991	79
339 5/8	48 6/8	54 7/8	45 1/8	7	6	Catron County	NM	Ray Hammond	1999	80
339 1/8	54 4/8	54 4/8	48 1/8	6	6	Navajo County	AZ	Larry D. Jones	1993	81
338 5/8	55 2/8	55 5/8	34 7/8	6	7	Idaho County	ID	Robert Maves	1998	82
338 4/8	49 2/8	49 6/8	29 0/8	6	6	Catron County	NM	Eugene Arndt	1993	83
338 3/8	55 5/8	54 0/8	36 7/8	6	6	Sheridan County	WY	Mike Barrett	1981	84
338 2/8	49 7/8	50 2/8	36 6/8	7	7	Natrona County	WY	Kevin Christopherson	1997	85
338 2/8	50 2/8	49 2/8	50 4/8	6	6	Kittitas County	WA	Bruce Hedlund	2002	85
338 2/8	51 1/8	49 6/8	41 0/8	7	6	Taos County	NM	Sean D. Kelly	2004	85
338 1/8	47 0/8	50 7/8	39 3/8	5	6	Lincoln County	WY	Gary L. Sims	1991	88
337 7/8	47 3/8	51 2/8	46 3/8	6	6	Sheridan County	WY	Mike Barrett	2001	89
337 4/8	52 3/8	54 5/8	41 4/8	7	6	Coconino County	AZ	Alan Blanchard	1972	90
337 3/8	53 0/8	49 2/8	42 1/8	6	5	Mesa County	CO	Mike Flores	1970	91
337 3/8	46 5/8	48 0/8	41 1/8	6	6	Carbon County	WY	Larry K. Hardwig	2011	91
337 2/8	48 2/8	49 4/8	41 6/8	6	8	Beaverhead County	MT	Theodore T. Ralls, Sr.	1990	93
336 6/8	53 5/8	53 3/8	39 4/8	6	6	Taos County	NM	Jeff Lampe	1992	94
335 4/8	47 4/8	48 6/8	49 0/8	6	6	Clearwater County	ID	Robert J. Kreisher	1958	95
335 4/8	52 0/8	52 6/8	45 4/8	7	7	Skamania County	WA	Ted Jaycox	1985	95
335 0/8	54 2/8	55 1/8	40 2/8	6	7	Madison County	MT	Scott L. Koelzer	1999	97
335 0/8	47 1/8	48 7/8	39 6/8	6	7	Moffat County	CO	Roger L. Grove	2010	97
334 1/8	51 0/8	51 7/8	45 1/8	7	6	Sanders County	MT	Greg Hertel	2009	99
334 1/8	52 5/8	51 6/8	38 1/8	7	7	Madison County	MT	Michael Feldermann	2010	99
333 7/8	45 4/8	46 6/8	35 1/8	6	6	Phillips County	MT	Brad Eide	2001	101
333 6/8	49 4/8	49 4/8	45 0/8	6	6	Baker County	OR	Russell B. Jones	1946	102
333 5/8	46 0/8	44 3/8	38 6/8	8	6	Socorro County	NM	Bob Bartoshesky	2007	103
333 4/8	53 5/8	53 4/8	43 0/8	6	6	Wallowa County	OR	Kirk Skovlin	2003	104
333 3/8	45 6/8	47 4/8	39 7/8	6	6	Catron County	NM	Gary A Littauer	1984	105
332 6/8	51 1/8	51 7/8	32 0/8	7	6	King County	WA	Curtis H. Fowler	1987	106
332 0/8	49 2/8	50 5/8	38 2/8	7	7	Caribou County	ID	Joseph M. Hulse	1967	107
331 4/8	54 0/8	53 5/8	39 4/8	6	6	Boulder County	CO	Ron Readmond	1991	108
331 4/8	51 1/8	50 4/8	37 4/8	6	6	Judith Basin County	MT	Kelly Norskog	1997	108
331 1/8	48 3/8	48 1/8	38 3/8	6	6	Sheridan County	WY	Mike Barrett	1984	110
331 1/8	52 4/8	52 3/8	35 5/8	6	6	Archuleta County	CO	Milton H. Sherman, Jr.	2008	110
330 5/8	52 4/8	54 4/8	35 7/8	6	6	Catron County	NM	Eugene Smith	1993	112
330 4/8	44 6/8	46 4/8	38 1/8	6	6	Latah County	ID	John K. Pell	1964	113
330 2/8	52 3/8	48 0/8	36 2/8	7	6	Sheridan County	WY	Randy Burtis	2010	114
330 0/8	52 5/8	52 6/8	41 2/8	6	6	Catron County	NM	Russell Arndt	1998	115
330 0/8	53 4/8	53 6/8	42 0/8	6	6	Coconino County	AZ	Michael A. Pfander	2005	115
329 7/8	51 4/8	48 6/8	39 3/8	6	6	Powell County	MT	Marlon J. Clapham	1986	117
329 6/8	49 1/8	49 0/8	37 4/8	6	6	Rio Blanco County	CO	Myles Keller	1974	118
329 2/8	52 3/8	52 4/8	44 4/8	6	6	Valley County	ID	B. C. Cunningham	1969	119
329 2/8	52 1/8	54 2/8	37 6/8	6	6	Meagher County	MT	Jim Kennedy	2009	119
329 1/8	47 7/8	49 4/8	40 1/8	6	6	Adams County	ID	Curt Lemmons	1965	121
329 1/8	51 5/8	53 7/8	44 3/8	7	7	Moffat County	CO	Glenn W. Pritchard	1972	121
329 0/8	56 0/8	54 3/8	40 2/8	8	6	Yakima County	WA	Jerry Harris	1963	123
328 3/8	55 4/8	51 0/8	44 7/8	6	6	White Pine County	NV	Patrick C. Allen	2011	124

Minimum Score 260 *Continued*

Score	Length of R Main Beam	L	Inside Spread	Number of R Points	L	Area	State/ Province	Hunter's Name	Date	Rank
328 2/8	52 0/8	52 6/8	37 2/8	6	6	Coconino County	AZ	Richard D. Tone	1989	125
328 0/8	54 1/8	51 3/8	40 0/8	6	7	Coconino County	AZ	James L. Hyde	1977	126
327 6/8	50 5/8	51 5/8	38 6/8	7	6	Gallatin County	MT	Scott L. Koelzer	1993	127
327 5/8	53 0/8	53 2/8	39 1/8	6	6	Lemhi County	ID	Mike Virgin	1996	128
327 3/8	47 4/8	47 0/8	34 3/8	7	7	Coconino County	AZ	Richard D. Tone	1991	129
327 2/8	51 1/8	51 0/8	37 4/8	6	6	Rio Arriba County	NM	Larson Panzy	2000	130
327 1/8	50 3/8	48 2/8	43 3/8	6	6	Park County	MT	Dale Alt	1968	131
327 0/8	50 6/8	51 7/8	40 4/8	6	6	Clearwater County	ID	C. Randall Byers	1976	132
326 7/8	48 0/8	49 0/8	42 7/8	6	6	Grand County	CO	Kevin E. Vecchiarelli	2010	133
326 1/8	44 6/8	46 6/8	35 5/8	6	6	Lemhi County	ID	Ray Torrey	1967	134
326 1/8	54 2/8	54 5/8	37 5/8	6	6	Grand County	CO	Paul Adams	1977	134
326 0/8	51 1/8	50 6/8	46 2/8	7	6	Boulder County	CO	Duke Prentup	1977	136
325 3/8	50 2/8	51 5/8	40 7/8	6	6	Coconino County	AZ	Gary D. Bills	1991	137
325 0/8	47 1/8	47 3/8	35 6/8	7	7	Silver Bow County	MT	Bob Gossack	1977	138
324 3/8	48 2/8	50 4/8	42 1/8	6	6	Socorro County	NM	Steve McCluskey	2005	139
324 1/8	48 0/8	49 2/8	40 1/8	6	6	Jackson County	CO	Tom G. Kelley	1998	140
323 7/8	53 6/8	51 3/8	39 3/8	7	7	Coconino County	AZ	Joseph O. Fogleman	1992	141
323 6/8	49 0/8	49 6/8	38 4/8	6	6	Natrona County	WY	Tom Bradach	2004	142
323 4/8	49 1/8	52 7/8	42 0/8	6	6	Mora County	NM	Robert Snell	2002	143
322 7/8	50 6/8	51 0/8	44 3/8	6	7	Phillips County	MT	Ray Hoveskeland	1967	144
322 7/8	48 5/8	46 5/8	36 5/8	6	6	Deer Lodge County	MT	Eddie McGreevey	1977	144
322 4/8	51 3/8	48 7/8	40 0/8	6	7	Lewis & Clark County	MT	Jerry Biresch	1970	146
322 4/8	49 2/8	49 4/8	34 4/8	6	6	Fremont County	ID	Brian Frickey	1999	146
322 3/8	51 4/8	49 5/8	38 5/8	6	6	Rio Arriba County	NM	Larson E. Panzy	1999	148
321 6/8	49 7/8	47 1/8	36 2/8	6	6	Meagher County	MT	Jim Kennedy	2010	149
321 2/8	46 7/8	46 2/8	35 6/8	6	6	Gallatin County	MT	Lee Poole	1998	150
320 5/8	48 5/8	47 5/8	38 7/8	7	7	Madison County	MT	Richard King	1993	151
320 4/8	54 5/8	53 4/8	36 0/8	6	6	Meagher County	MT	Jim Kennedy	2005	152
320 4/8	47 4/8	49 1/8	31 6/8	7	7	Fergus County	MT	Kelly Norskog	2009	152
320 3/8	46 1/8	46 6/8	39 3/8	6	6	Rio Grande County	CO	Bing Kemp	1966	154
319 7/8	50 5/8	50 4/8	43 1/8	6	6	Sandoval County	NM	James E. Montgomery	1992	155
319 4/8	57 0/8	56 7/8	44 2/8	6	6	Custer County	ID	Bryan Tilly	2003	156
319 2/8	55 0/8	55 0/8	35 6/8	6	6	Douglas County	CO	Rick Duggan	2010	157
319 0/8	49 7/8	49 3/8	44 6/8	7	6	Clearwater County	ID	Burgess Blevins	1993	158
318 7/8	43 3/8	43 4/8	36 5/8	7	7	Lincoln County	MT	Terry V. Crooks	1997	159
318 6/8	48 5/8	48 1/8	42 2/8	6	7	Larimer County	CO	Tony Seahorn	1972	160
318 5/8	44 4/8	44 0/8	34 6/8	6	6	Mineral County	CO	David Powell	1969	161
318 4/8	46 6/8	46 6/8	44 0/8	6	6	Sublette County	WY	John Gedroez	2003	162
318 3/8	51 4/8	49 7/8	33 3/8	6	6	Mineral County	MT	Farrell Cooper	1975	163
318 2/8	45 7/8	46 1/8	38 4/8	6	6	Fergus County	MT	Bob Allen	1992	164
318 1/8	47 5/8	48 0/8	36 4/8	6	7	Petroleum County	MT	Chris G. Sanford	1997	165
318 0/8	46 6/8	49 2/8	36 6/8	6	6	Rich County	UT	John Harding	2003	166
317 6/8	44 5/8	45 7/8	41 0/8	6	6	Riley County	KS	Trevor Haddix	2002	167
317 2/8	46 0/8	45 7/8	34 4/8	6	6	Carbon County	WY	Rhett Heflin	1999	168
317 1/8	47 6/8	48 0/8	36 7/8	6	6	Wallowa County	OR	Russ Hultberg	1996	169
317 0/8	45 4/8	44 1/8	37 0/8	6	6	Umatilla County	OR	Tom Nieradka	2004	170
316 6/8	47 6/8	48 5/8	35 2/8	6	6	Coconino County	AZ	Joseph O. Fogleman	1991	171
316 3/8	49 4/8	48 7/8	36 7/8	6	6	Coconino County	AZ	Troy A. Graziadei	1998	172
316 3/8	45 6/8	47 4/8	44 5/8	6	6	Moffat County	CO	Robert Lowers, Jr.	2008	172
316 2/8	48 2/8	48 2/8	34 6/8	7	7	Coconino County	AZ	William M. Lanese	1981	174
316 0/8	54 3/8	52 1/8	38 0/8	6	6	Sweet Grass County	MT	Marlin F. Dunlap	1983	175
316 0/8	45 0/8	45 3/8	33 6/8	6	6	Sheridan County	WY	Dave Moss	1999	175
315 7/8	51 7/8	51 0/8	34 5/8	6	6	Sheridan County	WY	Dan Doke	1990	177
315 6/8	49 1/8	50 5/8	41 4/8	7	8	Converse County	WY	Frank Moore	1989	178
315 6/8	40 0/8	47 7/8	42 4/8	6	6	Caribou County	ID	Coby Tigert	1990	178
315 6/8	46 6/8	48 4/8	32 4/8	6	6	Catron County	NM	Jorge Garcia-Segovia	1992	178
315 5/8	47 0/8	47 2/8	34 2/8	7	8	Mesa County	CO	Mark Livingston	2005	181
315 4/8	43 0/8	42 1/8	41 0/8	6	6	Fording River	BC	Walter Conibear	2006	182
315 3/8	51 3/8	51 4/8	39 3/8	6	6	Coconino County	AZ	Jim Wheeler	1997	183
315 2/8	51 0/8	50 3/8	34 6/8	6	6	Meagher County	MT	Jim Kennedy	2006	184
315 2/8	51 4/8	50 7/8	36 4/8	6	6	Meagher County	MT	Jim Kennedy	2007	184
315 1/8	47 1/8	48 2/8	37 3/8	6	6	Butte County	ID	Kirk W. Reese	1995	186

AMERICAN ELK [Typical Antlers]

Minimum Score 260 *Continued*

Score	Length of R Main Beam	L	Inside Spread	Number of R Points	L	Area	State/ Province	Hunter's Name	Date	Rank
315 0/8	54 3/8	54 5/8	43 4/8	7	6	Sandoval County	NM	Jim Willems	2010	187
314 7/8	43 4/8	44 1/8	37 1/8	6	6	Caribou County	ID	Brody Gibson	2002	188
314 6/8	43 4/8	42 7/8	37 2/8	6	6	Socorro County	NM	David C. Wade	2000	189
314 5/8	41 6/8	49 1/8	38 3/8	6	6	Catron County	NM	Dr. Charles W. Hendricks	1997	190
314 1/8	51 0/8	51 2/8	41 5/8	6	6	Gallatin County	MT	C. W. Smith	1974	191
314 0/8	54 3/8	54 4/8	37 0/8	6	5	Coconino County	AZ	James R. Dreves	1985	192
314 0/8	45 6/8	46 4/8	42 4/8	7	7	Big Horn County	WY	Mike Belcourt	1986	192
314 0/8	45 5/8	45 5/8	52 3/8	7	7	Gallatin County	MT	Rex Rogers	1995	192
313 5/8	48 6/8	47 1/8	41 3/8	6	6	Ravalli County	MT	Dan Wiediger	2005	195
313 3/8	45 6/8	44 1/8	32 1/8	6	6	Garfield County	MT	Ron Borgeson	1997	196
313 3/8	51 3/8	50 4/8	43 1/8	6	6	Rich County	UT	Rick Wilson	1999	196
313 2/8	52 0/8	50 6/8	40 6/8	6	6	Fergus County	MT	Mike Bentler	2001	198
313 0/8	50 1/8	50 5/8	40 0/8	6	6	Judith Basin County	MT	Kelly Norskog	1996	199
312 7/8	46 7/8	50 1/8	43 7/8	6	6	Shoshone County	ID	Harry Barker	1962	200
312 7/8	50 0/8	52 4/8	36 7/8	6	6	Moffat County	CO	Glenn W. Pritchard	1991	200
312 7/8	47 6/8	49 1/8	32 3/8	7	7	Catron County	NM	Robin Klemme	1997	200
312 5/8	47 0/8	47 3/8	39 5/8	6	6	Caribou County	ID	Dean Humphreys	1962	203
312 5/8	46 7/8	47 6/8	32 3/8	6	6	Park County	CO	William Elfland	1991	203
312 0/8	49 7/8	49 7/8	39 0/8	6	6	Rio Blanco County	CO	John Richardson	1990	205
312 0/8	41 7/8	43 1/8	34 4/8	6	6	Otero County	NM	Mark S. Hahn	1997	205
311 0/8	51 6/8	50 0/8	41 6/8	6	6	Teton County	WY	Fred Bear	1953	207
311 0/8	44 5/8	44 1/8	39 0/8	5	6	Colfax County	NM	Gene E. Ohmstede	1999	207
311 0/8	46 0/8	44 4/8	37 4/8	7	6	Fergus County	MT	Justin Deacon	1999	207
310 6/8	45 5/8	46 2/8	35 0/8	7	8	Fergus County	MT	Steven J. Nelson	1977	210
310 6/8	45 6/8	47 0/8	36 2/8	6	7	Ravalli County	MT	Alan Lear	1988	210
310 6/8	48 1/8	46 3/8	38 6/8	6	7	Park County	MT	Thomas A. Stock	2002	210
310 5/8	51 3/8	51 2/8	38 3/8	6	6	Little Belt Mtns.	MT	Jim Ekness	1978	213
310 4/8	49 0/8	49 0/8	43 6/8	6	6	La Plata County	CO	Tom Price	1967	214
310 4/8	43 2/8	43 6/8	38 2/8	6	6	Baker County	OR	Billy J. Cruise	1970	214
309 7/8	47 0/8	45 5/8	40 5/8	6	6	Rio Arriba County	NM	Scott Pope	2004	216
309 6/8	43 1/8	43 4/8	35 0/8	6	6	Routt County	CO	Steve Gorr	1971	217
309 5/8	51 0/8	50 6/8	39 7/8	6	6	Flathead County	MT	Paul Roney	1976	218
309 2/8	44 6/8	47 4/8	44 4/8	5	6	Fremont County	ID	Larry Bauer	1960	219
309 1/8	48 0/8	45 4/8	37 1/8	6	6	Fergus County	MT	Donny Roy	1996	220
309 0/8	43 3/8	42 6/8	39 6/8	7	7	Coconino County	AZ	Richard D. Tone	1998	221
308 7/8	54 5/8	53 1/8	47 7/8	6	5	Albany County	WY	Jeremy Artery	2004	222
308 5/8	42 0/8	42 5/8	37 5/8	6	6	Huerfano County	CO	Ken Beck	1998	223
308 4/8	45 1/8	47 4/8	40 4/8	6	6	Elmore County	ID	Matt March, Jr.	1987	224
308 4/8	47 6/8	46 6/8	37 7/8	7	7	Elmore County	ID	Anthony L. Mudd	1992	224
308 3/8	48 3/8	47 3/8	37 1/8	6	6	Lincoln County	MT	John C. McDivitt	1992	226
308 1/8	48 3/8	49 4/8	39 1/8	6	6	Coconino County	AZ	Richard D. Tone	1987	227
307 7/8	47 3/8	48 6/8	36 3/8	6	5	Baca County	CO	Max Crocker	1989	228
307 6/8	44 7/8	46 2/8	49 2/8	6	6	Madison County	MT	Everett W. Ayers, Sr.	1992	229
307 4/8	45 5/8	42 7/8	37 4/8	7	7	Huerfano County	CO	Wilbur F. Lay, Jr.	1974	230
307 3/8	42 4/8	43 3/8	51 7/8	7	7	Park County	CO	Marko Green	1989	231
307 0/8	45 1/8	45 0/8	41 6/8	6	6	Garfield County	UT	Jerry Woodland	2001	232
306 6/8	50 3/8	49 3/8	37 6/8	6	7	Pueblo County	CO	James L. Smith	1996	233
306 4/8	52 2/8	51 5/8	41 6/8	5	6	Albany County	WY	Jerry Bowen	2005	234
306 1/8	50 2/8	51 5/8	39 4/8	6	7	Phillips County	MT	Michael J. Domeika	1999	235
306 0/8	49 2/8	48 7/8	34 2/8	6	6	Catron County	NM	William C. Davis	1986	236
306 0/8	53 2/8	53 2/8	39 4/8	6	6	Ravalli County	MT	Dave L. Fretz	1987	236
305 5/8	43 3/8	48 3/8	37 5/8	8	6	Clark County	ID	Jason McConeghy	2002	238
305 5/8	45 2/8	42 5/8	37 2/8	6	7	Coconino County	AZ	Chris D. Krueger	2008	238
305 2/8	50 2/8	51 2/8	39 4/8	6	6	Larimer County	CO	Craig Nelson	1968	240
305 2/8	47 0/8	45 2/8	38 1/8	7	6	Lincoln County	NM	Henry Vega	1987	240
305 2/8	46 6/8	46 3/8	37 0/8	6	6	Beaverhead County	MT	Gene Loder	1993	240
305 1/8	46 7/8	48 6/8	45 5/8	6	6	Montrose County	CO	Darby Bender, Jr.	2003	243
305 1/8	48 3/8	48 1/8	37 3/8	6	6	Baker County	OR	Larry D. Jones	2011	243
305 0/8	44 4/8	45 0/8	47 3/8	5	6	Catron County	NM	Leonard Scarborough	1994	245
304 6/8	47 4/8	51 2/8	35 0/8	7	6	Stillwater County	MT	Charlie W. Green	2007	246
304 5/8	46 0/8	49 6/8	27 1/8	6	7	Park County	WY	Marion O. DeBusk	1992	247
304 4/8	45 7/8	45 7/8	35 0/8	6	6	Missoula County	MT	Byron Schurg	1989	248

AMERICAN ELK [Typical Antlers]

Minimum Score 260 *Continued*

Score	Length of R Main Beam	L	Inside Spread	Number of R Points	L	Area	State/ Province	Hunter's Name	Date	Rank
304 2/8	45 4/8	42 0/8	35 2/8	6	6	Mesa County	CO	Baree Weber	1995	249
304 2/8	47 4/8	50 4/8	37 0/8	6	6	Cibola County	NM	Mark Siedschlag	1998	249
304 2/8	50 1/8	49 4/8	34 2/8	6	6	Catron County	NM	Richard Van Valkenburg	1998	249
304 0/8	47 3/8	45 7/8	35 6/8	6	7	Beaverhead County	MT	Nate Pelkey	2010	252
303 6/8	43 4/8	42 2/8	42 1/8	6	7	Costilla County	CO	Jim Kennedy	2005	253
303 4/8	49 1/8	42 1/8	41 0/8	6	6	Grand County	CO	Jim Cleland	1977	254
303 3/8	44 0/8	43 0/8	42 7/8	5	6	Adams County	ID	Rick Mason	1984	255
303 0/8	49 0/8	47 3/8	36 2/8	6	6	Big Horn County	WY	Paul S. Warren	2005	256
302 7/8	44 6/8	45 6/8	38 5/8	6	6	Swan River	MT	Joe Lawrence	1969	257
302 4/8	47 7/8	47 2/8	38 5/8	6	6	Eagle County	CO	Roger Rothhaar	1974	258
302 4/8	46 5/8	43 5/8	33 4/8	6	6	Fergus County	MT	John "Rosey" Roseland	2003	258
302 3/8	47 6/8	47 1/8	37 3/8	7	6	Petroleum County	MT	Joe Sevart	1999	260
301 5/8	49 2/8	50 1/8	38 7/8	6	7	Ravalli County	MT	Ned Coorough	1992	261
301 4/8	46 5/8	47 3/8	34 6/8	6	6	Idaho County	ID	Jacob Deberg	2002	262
301 3/8	46 2/8	45 2/8	40 1/8	6	6	Clearwater County	ID	David Thorkildsen	1996	263
301 3/8	45 0/8	47 2/8	40 7/8	7	6	Converse County	WY	Ken Ball	2004	263
301 1/8	43 7/8	44 1/8	32 1/8	6	6	Clearwater County	ID	Jim Walters	1981	265
301 1/8	46 6/8	47 7/8	36 3/8	6	6	Larimer County	CO	Randy A. Reeves	1993	265
301 1/8	44 5/8	45 4/8	35 3/8	6	6	Big Horn County	WY	Randy Burtis	2009	265
301 0/8	47 0/8	43 7/8	37 7/8	7	7	Larimer County	CO	Brad Riley	1999	268
301 0/8	45 7/8	45 5/8	37 6/8	6	6	Pitkin County	CO	Patrick Hooker	2010	268
300 7/8	45 6/8	44 5/8	39 1/8	6	5	Teton County	MT	Bill Schenck	1979	270
300 7/8	46 6/8	47 2/8	42 1/8	6	6	Sandoval County	NM	Lloyd Wood	2010	270
300 6/8	42 1/8	43 7/8	36 4/8	6	6	Sheridan County	WY	Kerry Struckman	1992	272
300 5/8	44 3/8	44 5/8	44 6/8	6	6	Beaverhead County	MT	Monty Moravec	1989	273
300 5/8	43 4/8	42 5/8	30 5/8	6	6	Rosebud County	MT	Craig Compher, Sr.	2002	273
300 3/8	45 7/8	45 3/8	40 3/8	6	6	Madison County	MT	Lee Poole	1993	275
300 2/8	51 2/8	51 3/8	41 0/8	6	6	Idaho County	ID	Mike S. Ihnat	2004	276
300 0/8	44 2/8	54 2/8	37 6/8	6	6	Coconino County	AZ	Lew Webb	2003	277
299 7/8	51 0/8	54 0/8	41 0/8	6	7	Fergus County	MT	Charles R. Bowman	1966	278
299 7/8	46 2/8	47 3/8	48 4/8	7	6	Custer County	ID	Mike Zavadlov	2002	278
299 5/8	48 0/8	48 0/8	38 7/8	6	6	La Plata County	CO	Daniel E. Parkinson	2006	280
299 3/8	43 1/8	46 1/8	27 7/8	7	6	Catron County	NM	Glen R. Cousins	2001	281
299 1/8	45 2/8	47 6/8	34 1/8	6	7	Catron County	NM	Jules Pacheco	1990	282
299 1/8	50 1/8	47 4/8	35 4/8	7	6	Valley County	ID	Gerald B. Wilson	2012	282
299 0/8	45 5/8	46 4/8	37 2/8	6	6	Powell County	MT	Paul Brunner	1979	284
298 4/8	52 6/8	52 2/8	34 2/8	7	8	Converse County	WY	Frank N. Moore	1997	285
298 3/8	43 1/8	45 0/8	31 1/8	6	6	Phillips County	MT	Larry I. Nordlund	1993	286
298 3/8	47 7/8	50 4/8	39 4/8	7	6	Rio Arriba County	NM	Larson Panzy	2003	286
298 1/8	44 1/8	43 6/8	37 7/8	6	6	Gunnison County	CO	Andrew J. Logan	2009	288
297 7/8	44 5/8	46 7/8	31 1/8	7	8	Coconino County	AZ	Mark Vancas	1980	289
297 6/8	46 4/8	46 5/8	41 4/8	6	6	Boulder County	CO	Steve Gorr	1969	290
297 4/8	45 5/8	44 5/8	35 6/8	6	6	Fergus County	MT	Don Davidson	2003	291
297 3/8	48 3/8	47 5/8	37 3/8	6	7	Moffat County	CO	Glenn Pritchard	1990	292
297 3/8	43 1/8	46 1/8	52 0/8	6	6	Albany County	WY	Mark Nelson	2007	292
297 1/8	49 0/8	45 7/8	41 7/8	6	6	Bear Lake County	ID	Randy K. Guinn	1993	294
297 1/8	44 7/8	44 7/8	46 6/8	6	6	Beaverhead County	MT	Greg L. Munther	1996	294
296 7/8	45 3/8	46 0/8	37 1/8	6	6	Big Horn County	WY	Brad Jones	2011	296
296 5/8	46 0/8	45 1/8	33 5/8	6	6	Moffat County	CO	Jim Willems	1994	297
296 5/8	45 1/8	42 6/8	29 3/8	6	7	Catron County	NM	Ken Grosslight	1998	297
296 1/8	43 2/8	45 3/8	43 1/8	6	6	Sheridan County	WY	Mike Barrett	1982	299
296 0/8	44 2/8	45 0/8	39 4/8	6	6	Gallatin County	MT	Bob Savage	1968	300
295 7/8	42 5/8	40 5/8	38 2/8	6	7	Larimer County	CO	Ben Alexander	1972	301
295 6/8	42 6/8	42 6/8	41 6/8	6	6	Meagher County	MT	Jim Kennedy	2000	302
295 3/8	50 0/8	48 7/8	39 1/8	6	6	Phillips County	MT	Lynn Paris	1995	303
294 7/8	57 2/8	54 6/8	43 1/8	6	6	Catron County	NM	Jorge L. Coppen	2010	304
294 6/8	42 7/8	43 6/8	41 2/8	6	6	Gallatin County	MT	Vince Galli	1998	305
294 6/8	43 0/8	44 2/8	32 6/8	6	6	Park County	WY	Abraham Hall	2003	305
294 6/8	52 4/8	53 3/8	33 2/8	6	6	Catron County	NM	James C. Carlson	2010	305
294 2/8	46 0/8	46 4/8	31 6/8	6	6	Fremont County	ID	Drake R. Atwood	2010	308
294 0/8	44 7/8	44 5/8	31 4/8	6	6	Big Horn County	WY	Dave Moss	1988	309
293 7/8	46 3/8	48 0/8	40 5/8	6	6	Gunnison County	CO	Robert C. Goodman	1974	310

AMERICAN ELK [Typical Antlers]

Minimum Score 260 *Continued*

Score	Length of R Main Beam	L	Inside Spread	Number of R Points	L	Area	State/ Province	Hunter's Name	Date	Rank
293 7/8	45 5/8	43 3/8	33 5/8	6	6	Klamath County	OR	Greg Chakarun	2005	310
293 6/8	49 3/8	48 0/8	38 6/8	6	6	Sheridan County	WY	Gary T. Laya	1988	312
293 5/8	42 6/8	43 3/8	39 7/8	6	6	West	MT	Raymond Alt	1962	313
293 4/8	49 5/8	50 2/8	42 2/8	6	5	Sheridan County	WY	Mike Barrett	1980	314
293 4/8	48 5/8	49 0/8	38 2/8	6	6	Sheridan County	WY	Randy Burtis	2004	314
293 3/8	46 0/8	45 5/8	30 5/8	6	7	Lemhi County	ID	William C. Shuster	1986	316
292 7/8	41 4/8	41 3/8	38 5/8	6	6	Boulder County	CO	Ron Readmond	1990	317
292 7/8	45 5/8	44 1/8	36 1/8	6	6	Petroleum County	MT	Don Davidson	1992	317
292 5/8	44 0/8	44 6/8	38 1/8	6	6	Larimer County	CO	Todd K. Johnson	1993	319
292 5/8	42 2/8	45 7/8	42 0/8	7	7	Sanders County	MT	Glen A. Drexler	1998	319
292 4/8	48 0/8	50 4/8	33 4/8	6	6	Sanders County	MT	Doug Gunderson	1978	321
292 3/8	46 3/8	43 3/8	38 3/8	6	6	Coconino County	AZ	Gary R. Williams	1999	322
292 2/8	41 2/8	44 4/8	39 6/8	6	6	Idaho County	ID	Richard C. Nichols	1975	323
292 2/8	43 5/8	42 4/8	34 0/8	6	6	Sheridan County	WY	Mike Barrett	1996	323
292 1/8	45 5/8	45 6/8	38 5/8	6	6	Catron County	NM	Robin Johnson	2001	325
292 0/8	46 0/8	45 0/8	34 0/8	6	6	Catron County	NM	Brad Blanchard	1989	326
291 6/8	45 2/8	45 5/8	41 0/8	6	6	Summit County	CO	Howard Moser	1972	327
291 3/8	49 0/8	48 4/8	37 1/8	6	6	Sublette County	WY	John Gedroez	2004	328
291 0/8	48 4/8	49 4/8	40 4/8	6	6	Custer County	CO	Douglas R. Jones	1970	329
291 0/8	45 0/8	45 3/8	33 1/8	6	7	Nye County	NV	David Powning	2002	329
290 6/8	42 4/8	43 0/8	35 6/8	6	6	Park County	WY	William P. Mastrangel	1955	331
290 6/8	42 4/8	43 6/8	38 2/8	6	6	Madison County	MT	Randy Brown	1991	331
289 7/8	41 5/8	46 3/8	40 3/8	6	6	Mesa County	CO	George P. Sofronas	1986	333
289 6/8	45 4/8	45 6/8	42 6/8	6	6	Los Alamos County	NM	Bruce Barrie	1989	334
289 3/8	47 2/8	47 6/8	38 5/8	6	5	Sheridan County	WY	Daniel K. Doke	2008	335
289 1/8	42 5/8	43 4/8	35 7/8	6	6	Rio Arriba County	NM	Ryan Panzy	2002	336
289 1/8	43 7/8	43 7/8	37 5/8	6	6	Rio Blanco County	CO	Mark Klonicke	2010	336
288 6/8	50 3/8	49 6/8	31 2/8	6	6	Umatilla County	OR	Mike Billman	1998	338
288 2/8	48 1/8	47 4/8	38 6/8	6	6	Grant County	NM	Richard E. Hewett	2006	339
288 0/8	40 0/8	41 0/8	34 4/8	6	6	Fergus County	MT	Kelly Norskog	1993	340
287 7/8	43 1/8	46 1/8	35 3/8	6	6	Caribou County	ID	Tom Norris	1995	341
287 5/8	47 2/8	46 3/8	33 1/8	6	6	Clackamas County	OR	Tim Metzger	1999	342
287 4/8	45 2/8	44 0/8	39 0/8	6	6	Baker County	OR	Larry D. Jones	1987	343
287 4/8	42 0/8	41 0/8	37 2/8	6	6	Phillips County	MT	Kenneth E. Ruzicka	1988	343
287 4/8	42 0/8	42 4/8	36 2/8	6	6	Routt County	CO	Ronald Breitsprecher	2000	343
287 2/8	50 0/8	50 6/8	35 4/8	6	6	Grand County	CO	Robert Pitt	1978	346
286 6/8	44 1/8	44 0/8	35 6/8	6	6	Coconino County	AZ	Perry V. Dunn	1999	347
286 5/8	43 4/8	42 4/8	39 5/8	6	6	Routt County	CO	Floyd Montgomery	1994	348
286 4/8	41 4/8	43 6/8	39 4/8	6	6	Clackamas County	OR	Larry D. Jones	1983	349
286 4/8	40 5/8	40 5/8	35 6/8	6	6	Ravalli County	MT	Rod Osburn	1989	349
286 3/8	49 3/8	46 6/8	34 7/8	6	6	Clear Creek County	CO	Troy A. Cunningham	2004	351
286 3/8	46 3/8	45 7/8	35 7/8	6	6	Las Animas County	CO	Fred Eichler	2007	351
286 2/8	48 2/8	51 5/8	40 4/8	6	6	Rio Arriba County	NM	Eudane Vicenti	1993	353
286 2/8	48 2/8	48 6/8	43 2/8	6	6	Fergus County	MT	Don Davidson, Jr.	1994	353
286 0/8	45 6/8	44 6/8	45 2/8	6	6	Custer County	ID	Lynn M. Gregorash	2002	355
285 6/8	40 0/8	39 5/8	35 2/8	5	5	Morgan County	UT	Rick C. Wilson	1996	356
285 6/8	40 4/8	42 2/8	32 4/8	7	6	Fergus County	MT	Donny Roy	1998	356
285 5/8	47 3/8	48 2/8	45 3/8	6	6	Lemhi County	ID	Dennis DesJardins	1969	358
284 6/8	47 7/8	47 6/8	33 4/8	6	6	Gallatin County	MT	William Elfland	1998	359
284 6/8	42 6/8	43 0/8	35 0/8	6	6	Ravalli County	MT	Greg Munther	2000	359
284 4/8	47 6/8	50 0/8	36 6/8	6	6	Sheridan County	WY	Dave Moss	1995	361
284 4/8	40 6/8	41 6/8	32 2/8	6	6	Carbon County	WY	Dan Miner	2003	361
284 2/8	47 7/8	46 3/8	34 4/8	6	6	Apache County	AZ	Ronald King	1985	363
284 0/8	49 6/8	48 4/8	37 2/8	6	6	Colfax County	NM	John L. Chapman	1989	364
283 4/8	48 4/8	44 7/8	43 4/8	6	5	Eagle County	CO	Douglas W. Ator	2012	365
283 2/8	44 6/8	40 5/8	32 6/8	6	6	Clearwater County	ID	Jay Deones	1983	366
283 1/8	50 3/8	48 1/8	31 6/8	6	7	Bighorn Mtns.	WY	David Shoop	1979	367
283 1/8	43 6/8	42 7/8	35 7/8	6	6	Sheridan County	WY	Dale Stahl	1995	367
283 0/8	42 7/8	44 5/8	41 0/8	6	6	Baker County	OR	Don Rajnus	1962	369
283 0/8	45 1/8	47 5/8	43 2/8	6	6	Grand County	CO	G. Fred Asbell	1973	369
282 5/8	50 0/8	47 2/8	42 1/8	7	6	Gallatin County	MT	Scott L. Koelzer	1979	371
282 5/8	39 2/8	43 7/8	33 3/8	5	6	Valley County	ID	Dennis Gratton	1983	371

AMERICAN ELK [Typical Antlers]

Minimum Score 260 *Continued*

Score	Length of R Main Beam	L	Inside Spread	Number of R Points	L	Area	State/ Province	Hunter's Name	Date	Rank
282 5/8	42 6/8	43 0/8	42 7/8	6	6	Dolores County	CO	Mike Zion	1985	371
282 5/8	41 2/8	41 6/8	35 5/8	6	6	Mesa County	CO	Jeffrey Price	1989	371
282 5/8	47 0/8	43 4/8	24 3/8	6	6	Sheridan County	WY	Mike Barrett	1990	371
282 5/8	43 6/8	44 6/8	36 3/8	6	6	Moffat County	CO	Scott George	2000	371
282 5/8	40 6/8	40 7/8	41 3/8	6	6	Colfax County	NM	Robert S. Bartoshesky	2002	371
282 1/8	45 4/8	44 7/8	44 7/8	5	6	Wallowa County	OR	Dale F. Story	1967	378
282 1/8	43 3/8	44 2/8	32 3/8	6	6	Sheridan County	WY	Mike Barrett	1983	378
281 5/8	44 6/8	45 6/8	39 3/8	6	6	Phillips County	MT	Robert Monhollon	1986	380
281 5/8	45 4/8	45 5/8	33 5/8	6	6	Coconino County	AZ	Tom Vanasche	2003	380
281 1/8	42 4/8	44 1/8	35 6/8	6	6	Larimer County	CO	Gary Galloway	1984	382
280 7/8	44 6/8	45 2/8	31 7/8	6	6	Pierce County	WA	William H. Wakeley	1997	383
280 7/8	45 7/8	45 6/8	36 2/8	8	6	Wallowa County	OR	Tom Vanasche	2005	383
280 6/8	44 6/8	44 1/8	35 6/8	6	6	Yakima County	WA	Kevin Spencer	1989	385
280 3/8	41 7/8	40 0/8	36 5/8	6	6	Fergus County	MT	Mike Bentler	2004	386
280 0/8	45 4/8	45 3/8	33 6/8	6	6	Rio Blanco County	CO	Rolland M. Esterline	1969	387
280 0/8	40 6/8	40 1/8	31 0/8	6	6	Meagher County	MT	Jim Kennedy	2002	387
279 6/8	44 0/8	45 7/8	32 2/8	6	6	Catron County	NM	Dr. Charles W. Hendricks	1999	389
279 5/8	41 2/8	42 5/8	41 1/8	6	6	Sheridan County	WY	Mike Barrett	1987	390
279 4/8	41 2/8	39 0/8	33 0/8	6	6	Wallowa County	OR	James R. Brackenbury	1970	391
279 4/8	46 5/8	45 0/8	40 4/8	6	5	Idaho County	ID	Richard C. Nichols	1973	391
279 4/8	41 1/8	40 4/8	37 2/8	6	6	Summit County	CO	Bill Easterly	2005	391
279 3/8	45 6/8	45 1/8	39 1/8	6	6	Phillips County	MT	Dave Zimmer	1983	394
279 1/8	49 6/8	49 5/8	40 1/8	6	6	Summit County	CO	Brad Ham	1998	395
279 1/8	45 6/8	48 3/8	30 5/8	6	6	Las Animas County	CO	Peter Swenson	2002	395
279 0/8	45 4/8	44 0/8	40 6/8	6	6	Granite County	MT	Garret Decker	1991	397
279 0/8	42 6/8	43 6/8	37 4/8	6	6	Sheridan County	WY	Mike Barrett	1994	397
279 0/8	43 6/8	44 7/8	35 0/8	6	6	Meagher County	MT	Don Davidson, Jr.	2008	397
278 6/8	43 4/8	42 4/8	39 4/8	6	6	Meagher County	MT	Jim Kennedy	2004	400
278 3/8	42 2/8	40 4/8	36 5/8	6	6	La Plata County	CO	Alan J. Andrews	2000	401
278 1/8	44 3/8	41 3/8	38 3/8	6	6	Phillips County	MT	Ronald W. Vanden Brink	2002	402
278 0/8	44 1/8	43 4/8	42 0/8	6	6	Sweet Grass County	MT	Charlie W. Green	2011	403
277 7/8	46 3/8	45 2/8	37 7/8	6	6	Sheridan County	WY	Brian Mutch	2005	404
277 4/8	44 3/8	43 3/8	30 0/8	7	7	Caribou County	ID	Royce Brown	1986	405
277 4/8	48 5/8	48 6/8	34 2/8	6	5	Montrose County	CO	Troy R. Lamb	2004	405
277 2/8	48 0/8	48 6/8	36 4/8	6	6	Gallatin County	MT	Bob Morton	2003	407
277 2/8	46 6/8	45 7/8	32 6/8	6	6	Apache County	AZ	Daniel W. Peloso	2007	407
277 1/8	41 7/8	39 3/8	42 5/8	6	6	Clearwater County	ID	Jon Skinner	1988	409
277 0/8	43 0/8	43 0/8	31 6/8	6	6	Grant County	OR	Andy Day	1981	410
276 4/8	45 0/8	45 4/8	34 4/8	6	6	Jackson County	CO	Tom Kelley	1995	411
276 4/8	42 6/8	42 1/8	36 2/8	6	6	Gallatin County	MT	William M. Elfland	1999	411
276 0/8	40 4/8	41 1/8	32 4/8	6	6	Routt County	CO	Charles DeLong	1991	413
275 6/8	44 7/8	43 5/8	30 4/8	6	6	Idaho County	ID	Jerry Vega	1989	414
275 5/8	42 1/8	43 6/8	38 7/8	6	6	Ravalli County	MT	Rod Osburn	1980	415
275 5/8	45 4/8	44 4/8	34 7/8	6	6	Moffat County	CO	Glenn Pritchard	1988	415
275 5/8	46 2/8	45 1/8	35 1/8	6	6	Sheridan County	WY	Ron Niziolek	1990	415
275 5/8	45 0/8	47 7/8	39 1/8	6	6	Archuleta County	CO	William J. Langer	2005	415
274 6/8	39 7/8	45 2/8	35 2/8	7	6	Catron County	NM	Shawn C. Edwards, DVM	2010	419
274 4/8	40 3/8	40 2/8	33 0/8	6	6	Eagle County	CO	John Duggan	2004	420
274 3/8	41 3/8	42 0/8	31 1/8	6	6	Saguache County	CO	Jerry Woodland	1973	421
274 3/8	49 0/8	50 4/8	35 7/8	6	5	Rio Arriba County	NM	Jaymes R. Panzy	2003	421
274 1/8	42 2/8	43 7/8	38 0/8	7	6	Gallatin County	MT	David F. Gibson	1974	423
274 1/8	49 5/8	41 3/8	31 3/8	6	5	Dunn County	ND	Craig Richardson	1989	423
274 1/8	44 5/8	44 7/8	36 3/8	6	6	Custer County	CO	Mark Wuerthele	1995	423
274 0/8	43 3/8	43 4/8	31 2/8	6	6	Clearwater County	ID	Jay Deones	1984	426
273 7/8	46 2/8	47 2/8	37 3/8	6	6	Routt County	CO	Joe Wilhelm	1995	427
273 6/8	48 4/8	49 0/8	33 6/8	6	6	Lemhi County	ID	Greg Munther	1963	428
273 6/8	44 6/8	42 7/8	36 0/8	6	6	Coconino County	AZ	Daniel P. Spatchek	1994	428
273 5/8	42 5/8	42 1/8	35 3/8	6	6	Ravalli County	MT	Jerry Johnson	2003	430
273 4/8	41 1/8	42 6/8	39 0/8	6	6	Fremont County	WY	Ken Davis	1991	431
273 4/8	40 5/8	42 0/8	40 4/8	6	6	Skamania County	WA	Michael Culbertson	2003	431
273 0/8	48 1/8	46 0/8	34 4/8	6	6	Custer County	ID	Randy Guinn	1987	433
273 0/8	44 6/8	42 2/8	36 0/8	6	6	Carbon County	WY	Rodney C. Hill	1990	433

AMERICAN ELK [Typical Antlers]

Minimum Score 260 *Continued*

Score	Length of R Main Beam	Length of L Main Beam	Inside Spread	Number of R Points	Number of L Points	Area	State/ Province	Hunter's Name	Date	Rank
273 0/8	41 3/8	42 3/8	31 4/8	6	6	Gallatin County	MT	Jerry Karsky	1998	433
272 7/8	39 6/8	40 0/8	37 7/8	6	6	Clackamas County	OR	Ed Bensel	1959	436
272 5/8	44 4/8	44 7/8	35 1/8	6	6	Sublette County	WY	John B. Rice, Jr.	1990	437
272 5/8	39 7/8	38 5/8	37 7/8	5	6	Bonneville County	ID	Lendul David Tanney	1999	437
272 3/8	41 7/8	39 5/8	35 3/8	6	6	Skagit County	WA	Steve Gorr	1982	439
272 2/8	38 6/8	39 4/8	34 4/8	6	6	Idaho County	ID	Stanley D. Miles	1976	440
272 1/8	41 7/8	40 7/8	39 5/8	6	5	Rio Arriba County	NM	Larson Panzy	2002	441
272 0/8	43 7/8	44 4/8	38 0/8	6	6	Saguache County	CO	Buster Mize	1968	442
271 7/8	47 4/8	49 1/8	32 5/8	6	6	Chelan County	WA	Claude E. Gates	1973	443
271 6/8	46 4/8	46 5/8	33 6/8	6	6	Sheridan County	WY	Mike Barrett	1985	444
271 5/8	43 1/8	43 6/8	36 1/8	5	6	Park County	MT	Dennis Vance	1975	445
271 5/8	38 5/8	39 5/8	36 1/8	6	6	Larimer County	CO	Kenneth D. Allen	1981	445
271 4/8	45 2/8	46 3/8	36 0/8	6	6	Routt County	CO	John G. Whitten	2011	447
271 3/8	42 5/8	42 0/8	33 7/8	6	6	Catron County	NM	Jeffrey D. Eddy	2002	448
271 1/8	43 3/8	43 4/8	38 3/8	6	6	Lemhi County	ID	David A. Bronson	1991	449
271 1/8	38 0/8	42 5/8	43 4/8	7	6	Grant County	OR	Boyd Guthrie	1999	449
270 6/8	45 0/8	42 0/8	32 2/8	6	6	Hinsdale County	CO	Scott Hargrove	2000	451
270 5/8	45 3/8	45 3/8	38 1/8	6	6	Judith Basin County	MT	Jeff Jackson	1995	452
270 5/8	43 3/8	43 7/8	31 1/8	6	6	Catron County	NM	Rapheal Kukowski	1999	452
270 4/8	42 7/8	42 6/8	37 0/8	6	6	Blaine County	ID	Aaron J. Fortune	1996	454
270 3/8	43 5/8	42 6/8	34 7/8	6	6	Catron County	NM	Bill Clink	1989	455
270 3/8	47 4/8	46 0/8	30 1/8	6	6	Rio Arriba County	NM	C. Randall Byers	1994	455
270 1/8	41 0/8	42 6/8	36 5/8	6	6	Big Horn County	WY	Dave Moss	1992	457
270 0/8	48 6/8	47 5/8	32 4/8	6	6	Coconino County	AZ	Wade L. Carstens	1983	458
269 7/8	38 2/8	36 6/8	28 1/8	6	6	Bear Lake County	ID	Dennis Burdick	1969	459
269 7/8	36 2/8	39 4/8	33 3/8	5	6	Rio Blanco County	CO	Brad Murray	1990	459
269 7/8	41 4/8	39 7/8	40 1/8	6	6	Sheridan County	WY	Bruce Burr	2010	459
269 5/8	40 5/8	40 0/8	35 7/8	6	6	Jackson County	CO	Louis Phillippe	1999	462
269 1/8	41 4/8	42 2/8	37 5/8	6	6	Hinsdale County	CO	Arland N. Rininger	1994	463
269 0/8	35 5/8	35 5/8	33 2/8	6	6	Park County	CO	Jarold Allen Shriver	1988	464
269 0/8	41 6/8	45 0/8	33 2/8	6	6	Teton County	MT	Don Davidson	1990	464
268 3/8	42 0/8	45 6/8	43 3/8	6	6	Baker County	OR	Dwayne Powell	2000	466
268 2/8	43 7/8	45 0/8	34 6/8	6	6	Lemhi County	ID	Ben L. Fahnholz	2004	467
268 2/8	41 4/8	41 6/8	40 6/8	6	6	Meagher County	MT	Tanner Ozburn	2007	467
268 1/8	39 0/8	40 3/8	30 0/8	7	6	Lincoln County	WY	Troy Miller	1973	469
268 0/8	42 3/8	40 5/8	32 6/8	6	5	Baker County	OR	Les Thoreby	1963	470
268 0/8	43 5/8	45 7/8	38 0/8	6	6	Madison County	MT	Scott Lemieux	2007	470
267 5/8	43 3/8	43 5/8	37 1/8	6	6	Meagher County	MT	Warren Werp	2008	472
267 4/8	40 7/8	38 6/8	31 6/8	6	6	Routt County	CO	Thomas E. Garrett	1998	473
267 4/8	45 4/8	44 2/8	28 0/8	6	6	Cibola County	NM	Richard L. Andre	2008	473
267 1/8	39 4/8	40 4/8	38 3/8	6	6	Albany County	WY	Jerry Bowen	1988	475
266 6/8	41 0/8	40 4/8	32 6/8	6	6	Lemhi County	ID	Scott Spaeth	1978	476
266 6/8	42 5/8	42 7/8	29 0/8	6	6	Rio Blanco County	CO	Bill McMahan	1989	476
266 4/8	44 1/8	43 0/8	37 4/8	6	6	Lemhi County	ID	Frederick J. Gimbel	1993	478
265 4/8	49 4/8	45 5/8	31 6/8	6	6	Dolores County	CO	Scott Roberts	1975	479
265 4/8	40 6/8	40 3/8	34 0/8	6	6	Stillwater County	MT	Mike O'Connor	2011	479
265 1/8	41 2/8	39 2/8	29 1/8	6	7	Mineral County	CO	Rod Wintz	1968	481
265 1/8	46 3/8	45 5/8	36 1/8	6	6	Boise County	ID	Jeffrey Stevens	1993	481
264 7/8	45 0/8	45 3/8	36 3/8	5	6	Garfield County	MT	Richard Hjort	1993	483
264 6/8	38 5/8	37 4/8	32 6/8	6	6	Teton County	WY	Paul Birkholz	1966	484
264 4/8	44 6/8	45 7/8	34 4/8	7	7	Boise County	ID	Robert Hiller	1969	485
264 4/8	45 0/8	43 1/8	31 6/8	6	6	Sheridan County	WY	Daniel Doke	2007	485
264 3/8	40 7/8	41 0/8	38 1/8	6	6	Meagher County	MT	Donovan A. Khalar	1995	487
263 7/8	46 4/8	46 7/8	39 7/8	6	5	Boise County	ID	Jack Brennan	1964	488
263 7/8	41 3/8	43 2/8	40 6/8	7	6	Garfield County	CO	Andrew Bair	2003	488
263 4/8	40 6/8	39 6/8	34 6/8	6	6	Carbon County	WY	Tracy Villwok	2008	490
263 3/8	51 3/8	49 3/8	36 4/8	6	6	Rio Blanco County	CO	Brad Murray	1992	491
263 0/8	40 7/8	42 5/8	38 4/8	6	6	Caribou County	ID	Chad Jones	2003	492
262 6/8	38 4/8	43 0/8	37 3/8	8	6	Meagher County	MT	James C. Kennedy	1999	493
262 4/8	39 5/8	39 1/8	36 0/8	6	6	Sheridan County	WY	Mike Barrett	2009	494
262 2/8	45 0/8	45 3/8	32 6/8	5	5	Grand County	CO	G. Fred Asbell	1979	495
262 2/8	46 1/8	43 5/8	38 4/8	6	6	Saguache County	CO	Burton R. Thompson, Jr.	1995	495

AMERICAN ELK [Typical Antlers]

Minimum Score 260 *Continued*

Score	Length of R Main Beam	L	Inside Spread	Number of R Points	L	Area	State/ Province	Hunter's Name	Date	Rank
262 0/8	44 5/8	45 5/8	35 2/8	6	6	Moffat County	CO	Thomas E. Powell	2007	497
261 7/8	42 5/8	41 3/8	34 3/8	6	6	Montezuma County	CO	Gene Benavidez	1998	498
261 6/8	38 0/8	40 0/8	37 6/8	6	5	Taos County	NM	Dr. D. A. Henbest	1972	499
261 5/8	35 6/8	35 4/8	34 5/8	6	6	Fremont County	ID	Todd J. Frickey	1987	500
261 5/8	41 4/8	41 0/8	32 3/8	6	6	Sheridan County	WY	Dave Moss	2002	500
261 4/8	42 0/8	39 3/8	31 6/8	6	6	Madison County	MT	Robert J. Kleinhans	1992	502
261 2/8	46 2/8	44 3/8	37 6/8	6	6	Sheridan County	WY	Mike Barrett	1988	503
260 7/8	37 6/8	37 6/8	34 3/8	6	6	Grant County	OR	Colby Moulton	1986	504
260 6/8	38 0/8	37 5/8	34 2/8	6	6	Clearwater County	ID	Chuck Lynde	1988	505
260 4/8	40 6/8	44 4/8	32 4/8	6	6	Coconino County	AZ	Patrick Donald Dufek	1992	506
260 4/8	47 2/8	46 6/8	40 2/8	6	6	Lemhi County	ID	Benjamin C. Fahnholz	2006	506
260 3/8	37 5/8	40 0/8	37 7/8	6	6	Lewis & Clark County	MT	Jeff Jackson	1992	508
260 2/8	37 0/8	38 0/8	35 0/8	6	6	Rio Blanco County	CO	Charles R. Clark	1991	509

Hunter: Fred Eichler
Species: Stone's Sheep
Score: 144 4/8
Date: 8/6/07
Location: Two Peaks, Yukon

AMERICAN ELK [Non-Typical Antlers]

Minimum Score 300 *Continued*

Score	Length of R Main Beam	L	Inside Spread	Number of R Points	L	Area	State/ Province	Hunter's Name	Date	Rank
409 2/8	55 2/8	56 4/8	48 6/8	7	8	Coconino County	AZ	Robert Duane Hartwig	2004	1
409 0/8	54 7/8	53 5/8	40 4/8	8	7	Lincoln County	MT	Terry V. Crooks	1996	2
385 5/8	50 4/8	49 7/8	35 4/8	8	7	Petroleum County	MT	Gregory A. Herrin	1997	3
378 3/8	58 5/8	56 0/8	43 2/8	6	8	Coconino County	AZ	Scott Schuff	1998	4
372 3/8	53 4/8	49 6/8	37 2/8	8	9	Pueblo County	CO	Ron R. Minton	2009	5
365 5/8	51 0/8	48 6/8	38 6/8	6	7	Fergus County	MT	Donny Roy	1999	6
362 6/8	52 0/8	50 1/8	40 1/8	7	8	Teton County	WY	Frank J. Cina	1999	7
346 2/8	54 2/8	57 1/8	38 0/8	7	9	Catron County	NM	Jerry Leair	1999	8
345 7/8	47 2/8	47 5/8	50 4/8	8	9	Valley County	ID	Julian Salutregui	1998	9
342 6/8	49 7/8	49 0/8	35 0/8	6	7	Phillips County	MT	Lynn Paris	2000	10
339 4/8	49 1/8	48 1/8	34 3/8	6	7	Meagher County	MT	Brian Stremcha	2004	11

Hunter: Richard King
Species: Typical Coues' Deer
Score: 74 2/8
Date: 1/22/11
Location: Cochise County, AZ

ROOSEVELT'S [Olympic] ELK

Minimum Score 225 — *Cervus elaphus roosevelti*

Score	Length of R Main Beam	L	Inside Spread	Number of R Points	L	Area	State/ Province	Hunter's Name	Date	Rank
334 7/8	45 7/8	46 3/8	37 6/8	7	9	Campbell River	BC	Rick Duggan	2004	1
318 4/8	51 1/8	48 0/8	38 5/8	6	7	Vancouver Island	BC	Ed Kellow	1993	2
308 4/8	44 3/8	44 6/8	39 0/8	7	6	Douglas County	OR	Christopher Tipton	2007	3
301 5/8	44 3/8	48 2/8	37 2/8	6	7	Grays Harbor County	WA	Frank Stinchfield	2001	4
300 5/8	48 7/8	47 1/8	42 0/8	6	6	Curry County	OR	Allen Boice	1998	5
294 0/8	44 2/8	46 0/8	41 2/8	6	6	Jefferson County	WA	Sam Windle	1993	6
285 3/8	41 2/8	40 4/8	34 3/8	6	7	Douglas County	OR	Christopher Tipton	2002	7
282 0/8	45 0/8	43 6/8	42 0/8	6	6	Jefferson County	WA	Dave Robertson	1970	8
275 3/8	37 0/8	37 4/8	36 1/8	7	7	Tillamook County	OR	Alan Richardson	1980	9
274 3/8	47 6/8	47 5/8	36 0/8	8	5	Jefferson County	WA	Eric Olson	1985	10
272 7/8	42 6/8	43 2/8	36 2/8	7	6	Grays Harbor County	WA	Frank Stinchfield	2010	11
271 1/8	46 0/8	45 3/8	33 3/8	6	6	Afognak Island	AK	David Harper	1970	12
270 2/8	44 6/8	44 1/8	44 4/8	4	5	Grays Harbor County	WA	Bill Brown	1959	13
269 3/8	39 4/8	42 1/8	36 7/8	6	6	Douglas County	OR	Christopher Tipton	2010	14
268 7/8	40 5/8	41 1/8	34 1/8	5	7	Coos County	OR	Thomas E. Tipton	1986	15
268 1/8	38 0/8	39 2/8	37 3/8	6	6	Clatsop County	OR	Ed Beisley	1967	16
262 0/8	39 6/8	40 2/8	41 0/8	5	5	Coos County	OR	Tom Tipton	1991	17
257 2/8	43 3/8	43 0/8	38 7/8	5	6	Douglas County	OR	Larry D. Jones	2004	18
251 7/8	39 7/8	38 7/8	33 0/8	6	6	Coos County	OR	Dennis Dunn	2011	19
251 5/8	41 4/8	36 2/8	40 6/8	6	6	Jefferson County	WA	Sanford Windle	1972	20
247 6/8	34 4/8	36 6/8	37 1/8	5	5	Grays Harbor County	WA	Wayne McReynolds	1981	21
247 0/8	39 6/8	38 4/8	39 6/8	5	5	Olympic Peninsula	WA	Lloyd Beebe	1951	22
246 6/8	39 5/8	37 3/8	34 0/8	6	6	Jefferson County	WA	Eugene Wells	1960	23
246 5/8	37 6/8	40 7/8	33 2/8	8	7	Columbia County	OR	Keith Jones	2003	24
243 6/8	43 4/8	41 4/8	33 0/8	5	5	Clallam County	WA	Glenn St. Charles	1952	25
241 6/8	39 7/8	39 1/8	34 6/8	6	5	Clatsop County	OR	David Lawrence	1982	26
234 2/8	39 5/8	38 4/8	33 6/8	5	5	Clallam County	WA	Dean R. Swerin	1985	27
230 6/8	35 2/8	34 4/8	32 2/8	5	5	Clallam County	WA	Wayne Haag	1980	28
230 2/8	38 5/8	36 4/8	38 0/8	5	5	Coos County	OR	Thomas Tipton	1984	29
230 0/8	35 6/8	34 7/8	34 0/8	6	6	Yamhill County	OR	Doug Bain	1994	30
227 1/8	36 2/8	31 0/8	29 5/8	5	7	Clatsop County	OR	Randy Jennings	1980	31

Hunter: Bill Van Buskick
Species: Typical Coues' Deer
Score: 91 7/8
Date: 1/31/11
Location: Cochise County, AZ

TULE ELK

Minimum Score 225 *Cervus elaphus nannodes*

Score	Length of R Main Beam	L	Inside Spread	Number of R Points	L	Area	State/ Province	Hunter's Name	Date	Rank
303 0/8	44 2/8	43 7/8	45 5/8	7	7	Mendocino County	CA	Rick Duggan	2010	1

Hunter: Rick Duggan
Species: Tule Elk
Score: 303 0/8
Date: 8/17/10
Location: Mendocino County, CA

ALASKA-YUKON MOOSE

Minimum Score 170 *Alces alces gigas*

Score	Greatest Spread	Width of Palm R	Width of Palm L	Number of Points R	Number of Points L	Area	State/ Province	Hunter's Name	Date	Rank
229 1/8	70 1/8	15 0/8	14 7/8	11	11	Doghouse Creek	AK	Rick Schikora	2002	1
224 6/8	67 6/8	19 1/8	16 4/8	12	14	Koyukuk River	AK	Larry D. Jones	2000	2
224 2/8	67 4/8	19 0/8	14 5/8	10	12	Mosquito Flats	AK	William C. Shuster	1998	3
223 2/8	68 0/8	14 3/8	16 2/8	11	14	Innoko River	AK	Roger Stewart	1993	4
223 1/8	62 5/8	13 3/8	14 2/8	11	11	Sugarloaf Mtn.	AK	James C. Walters	1990	5
220 3/8	57 3/8	16 0/8	16 2/8	12	11	Little Delta River	AK	William Wright	1959	6
219 7/8	64 7/8	13 2/8	14 0/8	12	12	Kuskokwim River	AK	A. Domas III	2004	7
217 4/8	63 6/8	18 4/8	18 3/8	14	11	Grayling Creek	AK	Carl Spaeth	1995	8
217 3/8	64 1/8	17 0/8	15 0/8	14	15	Moose John River	AK	Mike Parsons	1991	9
215 7/8	63 3/8	13 2/8	12 1/8	15	14	Koyukuk River	AK	Rickie D. Snell	1990	10
215 7/8	63 1/8	16 2/8	17 1/8	12	13	Stony River	AK	Thomas R. Johnson	1997	10
215 0/8	56 2/8	14 0/8	13 2/8	13	17	Koyukuk River	AK	Patricio Sada	1998	12
213 0/8	55 6/8	14 0/8	14 3/8	13	13	Kobuk River	AK	Joe Ellsworth	1998	13
212 7/8	65 7/8	13 0/8	17 6/8	9	10	Mulchatna River	AK	Jay Massey	1973	14
212 1/8	58 3/8	15 0/8	14 6/8	16	15	Brooks Range	AK	Roger Stewart	1985	15
211 2/8	66 6/8	15 0/8	14 4/8	9	9	Anchorage	AK	Dr. Rex Hancock	1961	16
211 2/8	66 6/8	21 3/8	16 0/8	7	9	Meshik River	AK	Art Kragness	1970	16
211 1/8	55 1/8	16 2/8	13 0/8	11	10	Mac Millan River	YUK	David Emken	1994	18
210 6/8	64 6/8	15 3/8	12 6/8	13	11	Mystery Creek	AK	Kevin M. Krause	1997	19
210 1/8	64 5/8	14 1/8	13 6/8	8	9	Nabesna	AK	Bill Ellis	1965	20
209 6/8	69 0/8	13 0/8	14 4/8	9	9	Port Heiden	AK	Margaret Cooley	1966	21
208 6/8	64 2/8	11 4/8	12 6/8	11	14	Koyukuk River	AK	David Emken	1998	22
208 1/8	59 7/8	13 5/8	15 2/8	13	13	Brooks Range	AK	Ted Grover	1985	23
208 0/8	65 4/8	13 0/8	12 0/8	9	7	Alaska Peninsula	AK	Jack Niles	1970	24
207 5/8	59 1/8	14 3/8	14 2/8	14	15	Coleen River	AK	Lyle Willmarth	1990	25
207 2/8	60 2/8	12 1/8	16 6/8	10	10	Yenlo Mtns.	AK	John F. Sumrall	1979	26
206 2/8	50 6/8	16 5/8	16 2/8	13	17	Mac Millan River	YUK	T. Sanford Roberts	1990	27
206 0/8	57 6/8	13 2/8	16 7/8	15	10	Grayling Creek	AK	Karl R. Spaeth	1997	28
205 2/8	64 0/8	14 0/8	12 7/8	8	12	Alaska Peninsula	AK	Donald B. McIntosh	1969	29
204 1/8	57 3/8	14 1/8	15 5/8	15	10	Cooper Creek	AK	Joe W. Harrison	1993	30
203 7/8	55 7/8	14 4/8	13 3/8	14	13	Grayling Creek	AK	Carl Spaeth	2001	31
203 6/8	66 0/8	11 5/8	12 7/8	13	11	Port Heiden	AK	Jim Dougherty	1968	32
203 2/8	56 0/8	13 0/8	13 6/8	16	12	Hogatza River	AK	Scott Anspaugh	2012	33
203 1/8	59 1/8	18 3/8	17 7/8	10	11	Earn Lake	YUK	Kevin M. Fisher	1999	34
202 7/8	58 1/8	12 6/8	13 4/8	15	13	Earn Lake	YUK	Dr. R. D. Keeler	1986	35
202 4/8	49 6/8	14 2/8	14 4/8	10	12	Brooks Range	AK	Brent Chapman	1978	36
202 4/8	58 2/8	15 2/8	14 1/8	13	12	Chulitna River	AK	Rickie D. Snell	1982	36
202 0/8	63 1/8	14 2/8	13 5/8	10	8	Moose John River	AK	Kent Brigham	1985	38
201 7/8	62 1/8	17 0/8	20 1/8	8	12	Mac Millan River	YUK	Michael J. Windemuller	2001	39
201 6/8	59 0/8	11 7/8	13 0/8	13	12	Brooks Range	AK	Mike Rosetti	1985	40
200 7/8	58 5/8	12 5/8	15 0/8	13	13	Wrangell Mtns.	AK	Robert Warpack	1988	41
200 2/8	64 2/8	14 4/8	14 0/8	10	11	Dog Salmon River	AK	Robert C. Keadle	1972	42
199 7/8	50 1/8	14 5/8	13 6/8	13	13	Bonnet Plume Lake	YUK	Rick Stockburger	2000	43
199 3/8	56 1/8	11 2/8	11 2/8	9	9	Nelchina	AK	Henry Wichers	1962	44
198 7/8	63 7/8	11 1/8	13 0/8	8	10	Gulkana Basin	AK	Thomas L. A. Pucci	1970	45
198 6/8	56 2/8	12 4/8	14 0/8	12	14	Ogilvie River	YUK	Rick Duggan	2004	46
198 4/8	64 4/8	14 3/8	13 6/8	10	7	Koyukuk River	AK	Michael W. Carroll	1998	47
198 2/8	52 2/8	17 0/8	13 4/8	8	10	Nenana River	AK	Dr. Harley Scholz	1973	48
198 1/8	63 5/8	11 3/8	11 6/8	10	12	Rainy Pass	AK	Dr. Henry C. McDonald	1970	49
198 1/8	53 5/8	14 3/8	15 2/8	12	9	Grayling Creek	AK	James C. Carlson	1995	49
197 4/8	65 4/8	13 0/8	14 5/8	6	9	Ugashik River	AK	Robert Borland	1972	51
197 4/8	61 4/8	10 6/8	12 1/8	8	11	Ugashik Lake	AK	John Wallace	1974	51
197 4/8	59 2/8	11 4/8	15 2/8	12	12	Chulitna River	AK	Rickie D. Snell	1983	51
197 1/8	62 1/8	14 4/8	11 1/8	12	10	Susitna River	AK	Jake Sonnentag	1969	54
197 1/8	64 7/8	12 6/8	17 1/8	10	11	Mount Gillis	YUK	Rick Martin	1999	54
196 6/8	63 6/8	13 2/8	16 4/8	6	6	Alaska Peninsula	AK	Phillip Durr	1969	56
196 5/8	59 7/8	11 4/8	12 5/8	12	9	Cheeneetnuk River	AK	Dan J. Ryle, Jr.	2002	57
196 5/8	57 3/8	12 6/8	13 4/8	10	10	Wrangell Mtns.	AK	Stephen A. Kenesie	2009	57
196 4/8	63 0/8	11 1/8	11 1/8	10	10	Chilikadrotna River	AK	Patrick J. Lefemine	1991	59
196 4/8	55 0/8	13 1/8	14 1/8	10	11	Koyukuk River	AK	Mike Misch	1994	59
196 1/8	59 5/8	13 2/8	15 7/8	9	8	Selawik River	AK	Todd Bandemer	2003	61
195 6/8	56 6/8	11 4/8	11 6/8	14	14	Kuskokwim River	AK	Brian Brockette	1996	62

ALASKA-YUKON MOOSE

Minimum Score 170 *Continued*

Score	Greatest Spread	Width of R Palm	Width of L Palm	Number of Points R	Number of Points L	Area	State/ Province	Hunter's Name	Date	Rank
195 1/8	53 5/8	14 2/8	13 2/8	10	9	Artic Red River	NWT	Raymond E. Stongle	1995	63
194 4/8	64 0/8	17 5/8	15 3/8	11	7	Alaska Peninsula	AK	Jim Dougherty	1962	64
194 2/8	61 6/8	12 7/8	13 7/8	7	12	Koyukuk River	AK	Thomas J. Hentrick	1984	65
193 5/8	59 3/8	11 4/8	12 2/8	12	11	Dishna River	AK	Luther L.N. Trower, Jr.	1999	66
193 5/8	58 3/8	10 0/8	11 3/8	7	9	Iowa Ridge	AK	Fred P. Biasiello	2008	66
193 5/8	56 3/8	12 2/8	12 4/8	11	12	Hart River	YUK	Tom Vanasche	2011	66
193 3/8	53 7/8	11 6/8	11 6/8	10	10	Kenai Peninsula	AK	Robert LaFollette	1962	69
193 2/8	56 2/8	13 6/8	13 4/8	10	13	Can Creek	AK	Dale Dye	2000	70
193 0/8	53 2/8	12 2/8	12 4/8	10	12	Galena	AK	Greg Campbell	1996	71
192 7/8	56 3/8	13 7/8	12 7/8	9	10	Little Tok River	AK	Dennis L. Lattery	1977	72
192 5/8	55 1/8	10 5/8	12 5/8	11	13	Wrangell Mtns.	AK	Loren Willey	1973	73
192 3/8	55 7/8	11 6/8	12 0/8	11	11	Eklutna Lake	AK	Ron C. Harvey	1989	74
192 1/8	51 3/8	12 0/8	13 0/8	12	12	Koyukuk River	AK	Jeffery R. Barnett	1997	75
192 0/8	52 6/8	14 2/8	16 2/8	8	10	Clarence Lake	AK	John Schoenike	1966	76
191 6/8	54 6/8	12 3/8	12 2/8	9	12	Brooks Range	AK	Joseph Stanevich	1986	77
190 7/8	59 1/8	11 4/8	10 4/8	8	10	Innoko River	AK	Samuel D. Adams	1997	78
190 3/8	57 3/8	13 0/8	13 0/8	12	12	Teslin Lake	YUK	Paul Schafer	1977	79
190 3/8	62 3/8	13 6/8	15 1/8	11	12	Derby Creek	AK	Greg Kempf	1991	79
190 2/8	53 2/8	13 5/8	14 5/8	8	11	Moose John River	AK	Monty Moravec	1992	81
189 6/8	57 2/8	15 4/8	13 4/8	4	9	Iliamna	AK	Thomas R. Grabowski	1994	82
189 0/8	55 0/8	13 0/8	14 2/8	11	11	Latna River	AK	Ed Wentzler	1995	83
188 4/8	63 0/8	11 5/8	12 3/8	8	8	Kenai Peninsula	AK	Dale L. Lofstedt	1969	84
187 7/8	56 1/8	13 1/8	13 0/8	8	9	Koyukuk River	AK	Mike O'Connor	1996	85
187 3/8	56 7/8	13 0/8	14 6/8	9	11	Selawik River	AK	Bruce R. Walker	2003	86
187 3/8	63 1/8	12 2/8	11 0/8	6	7	Dick Creek	AK	Mark Wayne Smith	2010	86
187 1/8	60 7/8	11 6/8	10 4/8	11	11	Wood River	AK	David A. Van Dyke	2007	88
186 6/8	53 4/8	11 3/8	11 5/8	7	9	Flat River	NWT	Lew Webb	2004	89
186 3/8	52 5/8	13 5/8	14 0/8	10	11	Middle Fork	AK	Glen Williams	1969	90
186 2/8	56 2/8	11 2/8	12 4/8	8	8	Cook Inlet	AK	George Moerlein	1961	91
185 3/8	56 1/8	12 5/8	12 4/8	12	10	Ambler River	AK	Don Owen	2004	92
184 6/8	55 6/8	11 3/8	12 4/8	5	10	Cinder River	AK	Francis Hosch	1965	93
184 3/8	53 5/8	9 4/8	13 4/8	12	13	Palmer	AK	A. H. Stange, Jr.	1962	94
184 3/8	63 3/8	12 4/8	11 0/8	12	8	Mishik River	AK	George Wright	1969	94
183 6/8	50 4/8	12 5/8	12 1/8	8	8	Nabesna	AK	George Moerlein	1962	96
183 6/8	59 2/8	14 4/8	11 3/8	7	9	Minto Flats	AK	John A. Miller	1999	96
183 0/8	54 6/8	11 1/8	11 2/8	10	10	Alaska Peninsula	AK	Jerry Putnam	1973	98
182 7/8	52 1/8	12 3/8	11 7/8	8	10	Ivishak River	AK	Ed Strayhorn	1987	99
182 5/8	57 5/8	11 2/8	16 1/8	10	10	Little Delta River	AK	Keith R. Clemmons	1957	100
182 5/8	59 1/8	12 5/8	15 0/8	5	12	Koyukuk River	AK	Troy A. Graziadei	1993	100
182 5/8	56 7/8	13 1/8	13 2/8	7	10	Koyukuk River	AK	Jeff Barnett	1993	100
182 2/8	57 4/8	12 2/8	12 6/8	7	8	Nome	AK	Erv Plotz	1979	103
182 1/8	56 5/8	9 7/8	12 0/8	7	9	Toolik River	AK	Reggie Spiegelberg	1986	104
181 5/8	53 5/8	12 6/8	12 4/8	11	14	Koyukuk River	AK	Greg L. Munther	1995	105
181 4/8	57 2/8	14 3/8	16 6/8	7	9	Port Heiden	AK	Bill L. Carlos	1970	106
181 4/8	55 4/8	12 4/8	12 6/8	6	10	Fort Yukon	AK	Ron Rockwell	1985	106
181 1/8	57 7/8	9 7/8	10 0/8	10	11	Ugashik Lake	AK	Scott Showalter	1972	108
181 1/8	54 1/8	8 6/8	10 2/8	8	10	King Salmon	AK	Joe Fogleman	1982	108
180 4/8	57 6/8	12 7/8	11 3/8	10	8	Middle Fork	AK	Norm Goodwin	1969	110
180 4/8	52 0/8	15 1/8	15 3/8	8	10	Sheenjek River	AK	David W. Doran	1992	110
180 2/8	55 6/8	11 3/8	12 1/8	10	10	Koyukuk River	AK	Doug Whitehead	1999	112
180 1/8	52 5/8	11 5/8	14 1/8	11	13	King Salmon	AK	Chuck Maloney	1974	113
179 6/8	59 2/8	9 3/8	12 3/8	8	7	Susitna River	AK	Robert Pitt	1968	114
179 5/8	53 3/8	11 0/8	11 2/8	11	12	Berry Creek	AK	Larry Jones	1962	115
179 5/8	53 1/8	9 4/8	13 3/8	10	13	Kenai Peninsula	AK	George Moerlein	1969	115
179 3/8	56 3/8	9 7/8	9 6/8	8	9	Stony River	AK	Bob Ameen	1986	117
179 1/8	55 7/8	13 1/8	12 0/8	6	7	Fort Yukon	AK	Barry J. Smith	1985	118
178 4/8	53 2/8	12 7/8	11 0/8	11	13	Koyukuk River	AK	Mark Reyher	1997	119
177 3/8	58 7/8	12 1/8	12 3/8	8	9	Coldfoot	AK	Butch A. Todd	1990	120
177 2/8	55 0/8	13 7/8	12 1/8	7	12	Whitehorse	YUK	Scott Koelzer	1977	121
177 2/8	61 0/8	11 0/8	14 7/8	9	13	Noatak River	AK	Guy M. Leibenguth	1995	121
177 1/8	53 7/8	12 3/8	15 0/8	10	11	Fox Mtn.	YUK	Todd R. Zeuske	1995	123
177 1/8	52 1/8	11 4/8	11 6/8	11	10	Nowitna River	AK	David C. Manca	1998	123

ALASKA-YUKON MOOSE

Minimum Score 170 *Continued*

Score	Greatest Spread	Width of R Palm	L	Number R of Points	L	Area	State/ Province	Hunter's Name	Date	Rank
177 0/8	62 0/8	9 0/8	9 2/8	5	7	Port Heiden	AK	James R. Scott	1966	125
176 4/8	50 0/8	11 6/8	9 3/8	11	8	Jim River	AK	Patrick A. Campanella	1997	126
176 2/8	55 4/8	9 3/8	9 7/8	9	8	Eklutna Lake	AK	Thomas J. Rutz	1991	127
176 1/8	56 3/8	10 7/8	11 3/8	7	9	Innoko River	AK	John McCullough	1998	128
175 6/8	48 2/8	10 0/8	9 0/8	11	11	Anvil Mtns.	YUK	James V. Barwick	1994	129
175 4/8	60 4/8	11 6/8	10 0/8	7	8	Susitna River	AK	David A. Drover	1971	130
175 3/8	46 3/8	11 3/8	13 1/8	11	15	Kaktovik	AK	Grant Poindexter	1987	131
174 6/8	58 6/8	8 0/8	9 6/8	8	8	McCarty Creek	AK	Stanley J. Rogers, Jr.	1972	132
174 3/8	57 5/8	12 5/8	12 0/8	13	8	Tustumena Lake	AK	Gary Wall	1974	133
174 2/8	54 6/8	10 0/8	10 0/8	12	9	Delta River	AK	Dr. R. Congdon	1960	134
174 2/8	54 0/8	8 0/8	10 1/8	11	9	Lake Telequana	AK	Greg S. Fields	1992	134
173 5/8	54 3/8	12 3/8	11 4/8	8	9	Red Creek	YUK	Fred Eichler	2008	136
172 0/8	50 4/8	11 0/8	9 5/8	11	11	Black Stone River	YUK	Rick Duggan	2002	137
171 5/8	57 3/8	10 6/8	9 0/8	13	7	Hayes Creek	AK	Keith R. Clemmons	1962	138
171 4/8	49 4/8	10 2/8	10 6/8	12	9	Delta River	AK	Richard R. Cooper	1959	139
171 2/8	59 0/8	12 0/8	10 1/8	11	9	Big River Flats	AK	Lonnie Rumley	1988	140
171 2/8	58 6/8	10 2/8	13 2/8	6	9	Rohn Roadhouse	AK	Fred Eichler	1994	140
170 7/8	58 1/8	12 0/8	10 6/8	8	5	Port Heiden	AK	John E. Lawson	1970	142
170 4/8	52 6/8	12 1/8	11 4/8	10	9	Unit 13D	AK	Dayle Paulson	1969	143

Hunter: Jason Roe
Species: Shiras' Moose
Score: 154 3/8
Date: 2/14/12
Location: Grand County, CO

CANADA MOOSE

Minimum Score 135 — *Alces alces americana* and *Alces alces andersoni*

Score	Greatest Spread	Width of R Palm	Width of Palm L	Number R of Points	Number of Points L	Area	State/ Province	Hunter's Name	Date	Rank
217 2/8	56 0/8	14 4/8	14 6/8	12	11	Birch Mtn.	ALB	Fredrick J. Gimbel	1994	1
210 7/8	58 5/8	14 3/8	15 4/8	13	12	Churchill River	MAN	Duane Carroll	2003	2
201 4/8	55 2/8	12 4/8	11 4/8	12	12	Mt. Lady Laurier	BC	Peter Halbig	1968	3
200 5/8	59 1/8	12 6/8	13 1/8	8	12	Cindy Lake	BC	Brian Koelzer	2003	4
200 0/8	50 4/8	13 5/8	15 0/8	15	17	Disella Lake	BC	T. J. Conrads	2001	5
199 0/8	56 0/8	12 6/8	14 5/8	11	11	Redwillow River	BC	Steven W. Hiebert	1992	6
199 0/8	57 4/8	14 0/8	13 2/8	10	10	Teslin River	BC	Scott L. Koelzer	1996	6
197 0/8	58 6/8	16 4/8	13 5/8	9	7	Cold Fish Lake	BC	Steve Gorr	1975	8
196 2/8	59 5/8	14 4/8	13 7/8	10	13	Skeena	BC	Larry Garoutte	1972	9
196 2/8	53 6/8	12 3/8	15 4/8	13	12	Nacht Creek	BC	Scott L. Koelzer	1991	9
194 6/8	61 0/8	12 4/8	11 2/8	11	11	Lake Leroy	BC	Lee Poole	1998	11
191 6/8	53 6/8	12 2/8	12 0/8	8	8	Lake Discella	BC	Sid Morrison	1995	12
187 5/8	57 0/8	13 3/8	12 1/8	11	13	Turnagin River	BC	Glenn Hisey	1978	13
187 0/8	54 0/8	13 6/8	11 4/8	10	9	Ear Falls	ONT	Robert Scott Ellery	1995	14
185 3/8	54 3/8	11 4/8	11 3/8	13	11	Mossy River	SAS	Jerome J. Huseby	1966	15
183 4/8	60 6/8	10 4/8	11 5/8	10	9	Terminus Valley	BC	Paul P. Schafer	1975	16
183 2/8	54 0/8	14 4/8	13 2/8	10	8	Skinny Lake	BC	Brian Koelzer	2006	17
182 3/8	49 3/8	14 3/8	13 0/8	11	11	Bonnie Lake	BC	Dale Karch	2004	18
181 1/8	51 5/8	12 0/8	11 5/8	12	13	Halfway River	BC	Duane L. Scroggins	1977	19
180 2/8	49 0/8	12 6/8	10 0/8	12	11	Keily Creek	BC	Thomas M. Taylor	2001	20
179 6/8	59 2/8	9 3/8	11 7/8	9	12	Penobscot County	ME	Scott E. Phillips	1998	21
178 5/8	51 2/8	11 1/8	12 6/8	10	10	Kechika River	BC	Scott L. Koelzer	1976	22
177 2/8	52 2/8	10 7/8	9 0/8	10	8	Thrimble Lake	BC	Chester Schardt	1966	23
176 1/8	48 3/8	11 2/8	12 1/8	11	13	Hyland Lake	BC	Randall S. Smith	1998	24
176 1/8	49 3/8	12 5/8	13 2/8	12	8	Ignace	ONT	Thomas M. Stepanek	2001	24
176 0/8	55 2/8	13 6/8	14 2/8	9	14	Long Range Mtns.	NFL	Waldemar D. Maya	1965	26
175 2/8	65 4/8	10 4/8	12 3/8	5	7	Aroostook County	ME	Craig A. Warren	2009	27
175 0/8	52 0/8	10 1/8	10 7/8	10	9	Central	BC	Ronald Lauretti	1973	28
175 0/8	53 4/8	10 4/8	12 0/8	8	10	Kechika River	BC	Paul F. Schafer	1974	28
175 0/8	50 6/8	13 6/8	14 1/8	12	10	Hyland Lake	BC	Gene A. Welle	1997	28
172 2/8	61 4/8	10 2/8	9 5/8	9	8	Lake County	MN	Chuck Schultz	1987	31
171 6/8	48 4/8	10 4/8	12 2/8	12	11	South Branch	NFL	Paul Erdbrink	1966	32
171 5/8	44 1/8	15 5/8	13 1/8	13	11	Cape Anquille Mtns.	NFL	Terrance Estes	1966	33
170 4/8	44 4/8	10 7/8	12 4/8	13	12	Quibell	ONT	Fred Bear	1945	34
170 2/8	52 0/8	11 3/8	10 0/8	7	10	Tua Lake	BC	Jeff Koelzer	1991	35
170 0/8	55 2/8	13 0/8	12 3/8	8	7	Aroostook County	ME	Cory Mattson	1996	36
166 6/8	54 2/8	11 0/8	9 5/8	7	11	Dease Lake	BC	Don Davidson	1993	37
165 2/8	50 0/8	14 0/8	12 5/8	5	7	Cottonwood River	BC	Scott L. Koelzer	1995	38
164 5/8	55 7/8	8 3/8	10 4/8	9	10	Little Dease Creek	BC	Ben Weatherall	2010	39
164 4/8	48 4/8	9 4/8	9 4/8	10	9	Sikanni Chief River	BC	Joel M. Riotto	1998	40
164 1/8	49 3/8	12 5/8	11 2/8	12	10	Tweedsmuir Park	BC	Glenn St. Charles	1954	41
163 6/8	47 6/8	10 4/8	10 2/8	11	9	Pink Mtn.	BC	Gerald R. Dishion	1987	42
163 1/8	51 7/8	11 3/8	12 3/8	8	8	Cook County	MN	Kirk D. Grupa	1997	43
162 4/8	48 0/8	9 0/8	9 0/8	10	11	Kapuskasing	ONT	Ron Alguire	1963	44
162 4/8	50 0/8	8 2/8	8 0/8	10	13	Andrew Lake	ALB	Warren Witherspoon	1994	44
162 1/8	48 3/8	11 6/8	11 6/8	11	8	Cook County	MN	Terry Krahn	1994	46
161 4/8	50 0/8	6 4/8	9 5/8	7	7	Tatla Lake	BC	Bill Nickerson	1985	47
161 3/8	49 1/8	10 0/8	9 2/8	8	8	Dease Lake	BC	Gene Welle	1995	48
161 1/8	45 5/8	10 5/8	11 6/8	11	11	Lloyds River	NFL	Harold A. Hill	1966	49
160 7/8	47 1/8	10 6/8	10 4/8	10	6	Grande Cache	ALB	Tim Rice	2005	50
160 4/8	46 0/8	10 4/8	10 4/8	10	9	Muskwa River	BC	W. Jay Boynton III	1970	51
160 1/8	50 7/8	10 0/8	9 5/8	9	8	Toad River	BC	Jerry Leair	1986	52
159 4/8	48 0/8	11 2/8	9 6/8	9	10	Chic-Chocs	QUE	Ken H. Taylor	2005	53
158 7/8	50 7/8	12 1/8	11 7/8	13	11	Kirkland Lake	ONT	Luther Gordon	1963	54
158 6/8	52 4/8	8 5/8	9 6/8	7	11	Aroostook County	ME	Richard J. Langer	2001	55
158 0/8	49 0/8	8 0/8	8 5/8	7	8	Williams Lake	BC	Gary Swan	1968	56
157 3/8	47 3/8	11 6/8	9 2/8	11	9	St. Louis County	MN	Gus Maxfield	1987	57
157 0/8	47 2/8	11 0/8	9 5/8	9	8	Maple Creek	SAS	Howard Harder	2011	58
156 6/8	51 0/8	8 5/8	8 5/8	8	8	Blue Lake	BC	Rick Duggan	2005	59
156 5/8	46 5/8	10 2/8	12 0/8	8	8	Main Brook	NFL	Scott Langston	2000	60
156 4/8	48 2/8	10 6/8	9 0/8	8	11	Tsayta Lake	BC	Bob Duncan	1986	61
156 1/8	50 1/8	10 6/8	11 0/8	10	8	Red Lake	ONT	Kevin D. Springman	2002	62

CANADA MOOSE

Minimum Score 135 *Continued*

Score	Greatest Spread	Width of Palm R	Width of Palm L	Number of Points R	Number of Points L	Area	State/ Province	Hunter's Name	Date	Rank
155 3/8	34 2/8	8 0/8	10 0/8	8	10	Smithers	BC	Chris VanderHorst	1974	63
155 2/8	44 0/8	8 2/8	8 2/8	8	6	Algoma	ONT	Dan Bertalan	1991	64
155 2/8	46 0/8	8 6/8	9 6/8	9	8	Toad River	BC	Peter C. Swenson	2005	64
155 1/8	51 5/8	10 3/8	10 4/8	8	9	Algoma District	ONT	Carol Wert	1963	66
154 1/8	44 3/8	9 7/8	9 7/8	8	11	Thutade Lake	BC	Donald N. Lehman	1973	67
154 1/8	45 7/8	13 4/8	11 5/8	10	8	Pink Mtn.	BC	William Newman	2009	67
154 0/8	49 2/8	10 0/8	10 3/8	9	8	Jackfish Creek	ONT	Darryl Miller	1987	69
153 6/8	49 2/8	10 0/8	10 2/8	5	6	Main Brook	NFL	Don Rozema	2002	70
153 5/8	45 3/8	10 6/8	11 2/8	7	12	Lloyds River	NFL	Harold A. Hill	1964	71
152 7/8	42 7/8	9 1/8	9 7/8	10	10	Zone 1	NFL	Harold A. Hill	1961	72
152 7/8	47 1/8	8 4/8	10 1/8	8	8	Trout Lake	ONT	Mike Brees	1998	72
151 6/8	47 6/8	11 5/8	9 6/8	9	10		NFL	Bill Hirst	1960	74
151 6/8	44 0/8	10 6/8	9 4/8	9	9	Duti Lake	BC	Walter J. Sawicki	1972	74
151 5/8	53 3/8	8 7/8	9 6/8	8	6	Stikine River	BC	James A. Farnsworth	1973	76
150 7/8	43 7/8	8 1/8	8 2/8	8	9	Gander River	NFL	David Lamoreaux	1997	77
150 4/8	41 0/8	10 4/8	11 2/8	10	7	Red Lake	ONT	Joseph B. Burgess	1993	78
150 2/8	42 2/8	9 1/8	8 4/8	9	11	Dumbell Lake	ONT	Gregory D. Keeton	1998	79
149 6/8	46 2/8	8 5/8	9 3/8	7	7	Little Johnny Lake	BC	Larry D. Jones	1988	80
149 0/8	40 6/8	7 6/8	9 3/8	7	7	Monashee Mtn.	BC	Thomas M. Basch	2002	81
148 4/8	47 2/8	10 4/8	11 3/8	8	9	Penobscot County	ME	Brian E. Thomas	2011	82
148 3/8	42 1/8	9 2/8	9 2/8	10	10	South Branch	NFL	W. P. Hirst	1964	83
148 0/8	48 0/8	10 4/8	10 2/8	3	7	Coos County	NH	Stephen C. Van Hoose	1997	84
147 5/8	44 7/8	9 4/8	10 4/8	7	5	Willow Lake	BC	Bob Gilbert	2008	85
146 6/8	39 4/8	8 1/8	9 5/8	8	10	King George IV Lake	NFL	Bill Hirst	1966	86
146 6/8	50 6/8	8 6/8	8 0/8	8	7	Penobscot County	ME	Scott E. Phillips	1999	86
146 0/8	46 0/8	8 4/8	8 0/8	9	8	Tarnezell Lake	BC	Dr. Rex Hancock	1960	88
145 6/8	47 4/8	8 3/8	8 4/8	9	8	Princess Lake	NFL	John Iannuzzo	1967	89
145 4/8	45 4/8	9 1/8	8 3/8	9	8	Tatlatui Lake	BC	G. Fred Asbell	1975	90
145 4/8	45 2/8	7 5/8	9 4/8	9	11	Edmonton	ALB	H. Mike Palmer	2005	90
142 4/8	47 0/8	10 2/8	7 7/8	7	9	Myrnam	ALB	Orest Popil	1995	92
142 2/8	42 2/8	10 3/8	9 2/8	8	8	Lake Discella	BC	Ken Grosslight	1995	93
142 0/8	49 4/8	8 4/8	8 6/8	9	8	Zec St. Patrice	QUE	Henry P. Bouchard	1993	94
142 0/8	44 6/8	10 2/8	6 5/8	10	7	Bow Lake	ONT	Steven Hearne	2006	94
141 4/8	44 0/8	8 3/8	8 3/8	7	6	Keily Creek	BC	Tom Vanasche	1998	96
140 3/8	42 1/8	9 2/8	10 6/8	9	8	Stork Lake	ONT	Shawn Sand	2006	97
140 2/8	48 2/8	7 0/8	7 2/8	7	7	Princess Lake	NFL	John Musacchia	1966	98
138 0/8	34 4/8	11 2/8	10 5/8	8	10	St. George Lake	NFL	Bill Carlos	1968	99
137 7/8	41 5/8	8 1/8	10 6/8	7	8	Princess Lake	NFL	Ken Rapp	1966	100
137 1/8	44 7/8	8 0/8	8 4/8	9	9	Gowganda	ONT	James Kelley	1993	101
136 1/8	39 3/8	8 2/8	9 4/8	6	9	Buchans Plateau	NFL	Lesley Bartus	1998	102

CANADA MOOSE [Velvet Antlers]

Minimum Score 135 *Alces alces americana* and *Alces alces andersoni*

Score	Greatest Spread	Width of Palm R	Width of Palm L	Number of Points R	Number of Points L	Area	State/ Province	Hunter's Name	Date	Rank
142 2/8	44 0/8	7 5/8	7 4/8	6	4	St. Anthony	NFL	William G. Mason	2000	*
144 4/8	46 0/8	9 3/8	11 7/8	5	6	Rabbit River	BC	Marty McMillian	2002	*
168 6/8	52 2/8	13 0/8	13 5/8	6	8	Tetsa River	BC	Lew Webb	2002	*

SHIRAS' [Wyoming] MOOSE

Minimum Score 125 — *Alces alces shirasi*

Score	Greatest Spread	Width of Palm R	Width of Palm L	Number of Points R	Number of Points L	Area	State/ Province	Hunter's Name	Date	Rank
185 6/8	54 0/8	9 3/8	8 4/8	8	11	Sheridan County	WY	Richard E. Jones	1987	1
183 7/8	49 7/8	13 3/8	14 4/8	12	10	Johnson County	WY	Scott A. Wodahl	2002	2
177 1/8	43 1/8	15 6/8	13 3/8	13	14	Bingham County	ID	David Cederberg	1995	3
169 2/8	53 0/8	10 7/8	11 7/8	9	12	Park County	MT	Sam A. Terakedis	1995	4
165 6/8	42 2/8	11 2/8	10 6/8	11	12	Big Horn County	WY	Mike Coble	2010	5
165 4/8	47 4/8	10 4/8	10 4/8	9	8	Utah County	UT	Chad J. Hall	1997	6
164 1/8	55 3/8	7 2/8	9 0/8	9	9	Teton County	WY	Rick Parish	1996	7
159 6/8	43 0/8	15 4/8	12 0/8	6	7	Teton County	WY	Daniel B. White	1978	8
159 1/8	49 5/8	12 7/8	10 7/8	8	10	Gallatin County	MT	Keith Wheat	1960	9
158 0/8	49 0/8	10 6/8	10 2/8	7	9	Madison County	MT	Paul Anthony Pernak	2001	10
154 6/8	45 4/8	11 0/8	10 5/8	10	8	Sheridan County	WY	Jerry Bailey	1992	11
154 3/8	45 7/8	10 7/8	10 5/8	7	7	Grand County	CO	Jason Roe	2011	12
154 0/8	49 4/8	9 1/8	8 7/8	8	9	Flathead County	MT	C. P. Mendenhall	1960	13
154 0/8	45 2/8	9 2/8	9 2/8	10	11	Caribou County	ID	Brett Dee Hymas	1986	13
153 4/8	45 4/8	10 0/8	12 6/8	9	11	Deer Lodge County	MT	Terry L. Button	1986	15
150 6/8	39 4/8	9 4/8	11 2/8	10	11	Sublette County	WY	Bryan Radakovich	1990	16
149 1/8	49 7/8	9 6/8	9 4/8	7	7	Lincoln County	WY	Ken Hoehn	1994	17
147 1/8	46 3/8	10 1/8	9 6/8	10	9	Lincoln County	WY	Gary Gale	1988	18
146 3/8	45 1/8	9 1/8	9 1/8	8	7	Sheridan County	WY	Cecil Benner	1990	19
144 6/8	42 6/8	9 4/8	12 4/8	8	7	Grand County	CO	Doug J. Carlton	2010	20
143 1/8	44 1/8	8 6/8	8 1/8	4	6	Jackson County	CO	Kevin M. Ayers	2002	21
143 1/8	41 7/8	9 4/8	10 0/8	9	10	Lincoln County	WY	Jerry Bowen	2004	21
141 1/8	42 1/8	9 6/8	11 2/8	7	8	Johnson County	WY	Russell Schoonover	2005	23
140 2/8	42 2/8	10 5/8	10 6/8	9	6	Lincoln County	WY	Keith Dana	1977	24
139 3/8	42 1/8	8 4/8	7 6/8	5	7	Idaho County	ID	Ray Torrey	1968	25
138 2/8	42 0/8	9 6/8	7 6/8	9	7	Lincoln County	MT	Thomas A. DeShazer	1965	26
137 7/8	47 1/8	9 2/8	11 3/8	7	9	Sublette County	WY	R. H. Siegert	1969	27
137 7/8	36 3/8	10 3/8	9 2/8	9	8	Teton County	WY	Kevin Anderson	1998	27
137 7/8	40 3/8	8 6/8	9 7/8	7	10	Johnson County	WY	Tom Taylor	2003	27
136 7/8	42 1/8	6 5/8	9 6/8	8	10	Cache County	UT	Larry Cross	1986	30
136 0/8	43 2/8	8 6/8	8 2/8	7	7	Sublette County	WY	August S. Gray	1991	31
135 7/8	44 7/8	8 4/8	7 3/8	7	8	Park County	WY	Mike Yonker	1987	32
134 4/8	38 6/8	8 1/8	7 5/8	8	9	Lincoln County	MT	Don Davidson	1989	33
133 7/8	41 7/8	9 7/8	7 7/8	9	7	Sublette County	WY	Guy M. Leibenguth	2001	34
131 6/8	44 0/8	7 6/8	7 7/8	7	7	Lincoln County	WY	Kevin Jackson	1980	35
131 4/8	47 0/8	8 1/8	8 7/8	7	7	Gallatin County	MT	Scott L. Koelzer	1992	36
131 4/8	43 6/8	7 6/8	8 3/8	6	5	Big Horn County	WY	Kendel Cheatham	2008	36
131 1/8	44 5/8	6 4/8	8 1/8	6	9	Sheridan County	WY	Mike Barrett	2008	38
129 5/8	43 1/8	8 2/8	8 2/8	7	7	Grand County	CO	Rick Duggan	1999	39
128 0/8	38 4/8	11 2/8	11 6/8	8	7	Sublette County	WY	Jim Willems	2006	40
127 5/8	38 1/8	6 6/8	6 6/8	7	7	Jefferson County	ID	David Brizendine	2007	41
127 2/8	41 2/8	6 4/8	6 4/8	9	8	Missoula County	MT	Richard Schaub	1997	42
125 0/8	40 6/8	8 4/8	8 0/8	8	7	Lincoln County	WY	Al Haines	2002	43

SHIRAS' [Wyoming] MOOSE [Velvet Antlers]

Minimum Score 125 — *Alces alces shirasi*

Score	Greatest Spread	Width of Palm R	Width of Palm L	Number of Points R	Number of Points L	Area	State/ Province	Hunter's Name	Date	Rank
140 2/8	44 2/8	11 2/8	9 4/8	8	6	Park County	WY	Gregory R. Meinecke	2002	*

MUSKOX

Minimum Score 90 — *Ovibos moschatus* and certain related subspecies

Score	Length of R Horn	Length of L Horn	Width R of Boss	Width L of Boss	Greatest Spread	Area	State/ Province	Hunter's Name	Date	Rank
115 0/8	27 7/8	26 6/8	10 4/8	10 3/8	26 7/8	Kugluktuk	NUN	Patrick C. Allen	2005	1
114 6/8	27 0/8	27 0/8	10 3/8	10 3/8	25 7/8	Horton River	NWT	Phil Muller	2000	2
114 2/8	29 1/8	28 0/8	9 0/8	9 3/8	28 7/8	Ellice River	NUN	Steve Munier	1988	3
112 2/8	26 2/8	25 2/8	8 5/8	8 5/8	26 7/8	Seward Peninsula	AK	Mark Wayne Smith	2011	4
110 4/8	25 2/8	25 2/8	8 6/8	8 5/8	28 4/8	Nome	AK	Mark Wayne Smith	2001	5
110 2/8	27 2/8	26 0/8	8 5/8	8 5/8	28 0/8	Nuluk River	AK	Al Grierson	1998	6
109 4/8	26 1/8	26 0/8	9 4/8	9 5/8	25 5/8	Banks Island	NWT	Larry Hoff	1986	7
109 2/8	26 0/8	26 5/8	9 5/8	9 1/8	28 4/8	Caribou Creek	AK	Mark Wayne Smith	2009	8
109 0/8	27 3/8	27 2/8	9 4/8	9 2/8	26 3/8	McNaughton Lake	NUN	Rick Stockburger	2004	9
108 4/8	25 0/8	25 2/8	8 4/8	8 3/8	26 0/8	Kougarok River	AK	Mark Wayne Smith	2005	10
108 2/8	26 6/8	28 5/8	8 6/8	8 2/8	26 4/8	Nelson Island	AK	Dexter Lemon	1994	11
107 6/8	25 1/8	27 3/8	9 6/8	9 5/8	27 3/8	Banks Island	NWT	Rusty Neely	1987	12
107 6/8	25 1/8	25 3/8	8 3/8	8 2/8	26 0/8	Nuluk River	AK	Al Grierson	1997	12
107 6/8	26 7/8	27 1/8	8 4/8	8 4/8	25 7/8	Kougarok River	AK	Mark Wayne Smith	2007	12
107 4/8	25 0/8	25 1/8	8 1/8	8 1/8	26 3/8	Nunivak Island	AK	Timothy A. Gleason	1992	15
107 0/8	28 2/8	26 3/8	9 3/8	9 4/8	26 7/8	Banks Island	NWT	Dennis Kamstra	1987	16
106 6/8	25 5/8	26 2/8	8 5/8	9 0/8	27 6/8	Banks Island	NWT	Billy Ellis	1982	17
106 6/8	25 5/8	25 6/8	8 6/8	8 2/8	27 6/8	Nelson Island	AK	Dexter Lemon	1985	17
106 6/8	25 5/8	25 4/8	8 3/8	8 3/8	25 4/8	Nunivak Island	AK	Doug Baily	2005	17
104 6/8	24 6/8	24 1/8	8 2/8	8 4/8	27 3/8	Combo Lake	NWT	Nathan L. Andersohn	1997	20
104 6/8	25 4/8	27 4/8	8 6/8	8 1/8	24 2/8	Caribou Creek	AK	Mark Wayne Smith	2010	20
104 4/8	25 4/8	25 2/8	9 0/8	9 1/8	25 4/8	Banks Island	NWT	Ronald L. Sherer	1986	22
104 4/8	25 1/8	26 1/8	9 5/8	9 2/8	25 7/8	West Victoria Island	NWT	John Janelli	1987	22
104 4/8	25 6/8	25 0/8	7 7/8	8 0/8	29 0/8	Kougarok River	AK	Mark Wayne Smith	2004	22
104 4/8	25 2/8	26 4/8	8 6/8	8 7/8	29 0/8	Baldy Dome	AK	Mark Wayne Smith	2006	22
104 4/8	26 4/8	25 2/8	9 2/8	9 3/8	25 7/8	Seward Peninsula	AK	Steve Hohensee	2009	22
104 2/8	26 6/8	28 1/8	9 1/8	9 1/8	25 1/8	Cambridge Bay	NUN	Dave Dusthimer	2000	27
104 0/8	24 4/8	24 4/8	8 4/8	9 0/8	25 0/8	Banks Island	NWT	Mark F. Vancas	1998	28
103 6/8	26 0/8	26 4/8	8 1/8	8 1/8	27 3/8	Nunivak Island	AK	Carl E. Brent	1993	29
103 6/8	26 4/8	25 7/8	8 3/8	8 2/8	25 1/8	Nunivak Island	AK	Travis Peterson	2005	29
103 2/8	27 0/8	27 0/8	8 1/8	8 2/8	25 5/8	Paulatuk	NWT	Erv Plotz	1980	31
103 2/8	26 2/8	25 7/8	9 1/8	9 3/8	23 3/8	Holman	NWT	Bryce Olson	2009	31
103 0/8	25 6/8	25 3/8	9 0/8	9 1/8	25 2/8	Holman	NWT	James Matuszewski	2000	33
102 6/8	24 4/8	24 6/8	9 3/8	9 2/8	26 0/8	Banks Island	NWT	Frank C. Eifert	1987	34
102 4/8	25 1/8	25 5/8	9 3/8	9 3/8	24 7/8	Holman	NWT	Warren Witherspoon	1997	35
102 0/8	24 5/8	27 3/8	8 7/8	8 6/8	27 5/8	Nome	AK	Mark Wayne Smith	2000	36
101 6/8	23 7/8	24 0/8	8 1/8	8 0/8	26 2/8	Nunivak Island	AK	Rick D. Snell	1994	37
101 0/8	24 7/8	26 4/8	8 0/8	8 2/8	26 0/8	Nunivak Island	AK	Michael J. Lettis	1989	38
100 4/8	25 2/8	26 2/8	7 7/8	7 6/8	28 1/8	Nunivak Island	AK	Bruce J. Werba	1977	39
100 4/8	24 3/8	24 3/8	8 5/8	8 3/8	27 4/8	Nunivak Island	AK	Ron C. Harvey	1994	39
100 2/8	22 6/8	24 2/8	8 6/8	8 4/8	25 0/8	Cape Peel	NUN	Robert Snell	2005	41
99 6/8	24 6/8	27 0/8	8 7/8	8 5/8	25 4/8	Holman	NWT	Leonard Rinke	2005	42
99 4/8	24 6/8	22 7/8	9 3/8	9 0/8	25 1/8	Cambridge Bay	NUN	Ricardo Longoria	1998	43
99 4/8	23 7/8	24 3/8	9 4/8	9 2/8	24 0/8	Paulatuk	NWT	Martin E. Cain	2006	43
99 2/8	24 7/8	23 3/8	8 3/8	8 4/8	24 7/8	Paulatuk	NWT	Tom Foss	1997	45
98 6/8	24 2/8	25 3/8	8 6/8	8 4/8	24 1/8	Holman	NWT	Larry Streiff	1998	46
98 4/8	23 5/8	24 1/8	8 0/8	8 0/8	25 0/8	Holman	NWT	Duane Hicks	1997	47
97 4/8	24 7/8	24 2/8	7 0/8	7 1/8	25 0/8	Paulatuk	NWT	Richard Gerhart	2000	48
97 0/8	24 3/8	24 5/8	8 2/8	8 0/8	24 2/8	Cambridge Bay	NUN	Rick Wilson	2007	49
96 6/8	22 5/8	24 0/8	8 2/8	8 0/8	26 2/8	Nunivak Island	AK	Bob Hammond	1988	50
96 4/8	25 4/8	23 4/8	8 0/8	8 3/8	24 4/8	Holman	NWT	Bill Shuster	2003	51
95 6/8	23 4/8	23 4/8	6 6/8	6 5/8	26 4/8	Bella Creek	AK	Mark Wayne Smith	2008	52
95 2/8	24 6/8	24 1/8	8 5/8	9 2/8	24 1/8	Banks Island	NWT	Roger Wintle	2003	53
94 6/8	23 1/8	23 4/8	8 2/8	8 5/8	24 2/8	Victoria Island	NUN	Dale Karch	2005	54
94 4/8	24 5/8	23 6/8	6 6/8	6 6/8	26 7/8	Nunivak Island	AK	C. Vernon Humble	1976	55
94 4/8	21 0/8	22 7/8	9 2/8	9 2/8	25 1/8	Gjoa Haven	NUN	Gary Bogner	1991	55
94 4/8	23 2/8	24 7/8	8 4/8	8 4/8	26 4/8	Victoria Island	NWT	Thomas M. Basch	2003	55
94 2/8	23 2/8	24 3/8	8 6/8	8 3/8	24 3/8	Victoria Island	NWT	Rick Kroll	1993	58
94 2/8	24 4/8	23 6/8	8 1/8	7 6/8	24 2/8	Banks Island	NWT	Jay Deones	2001	58
93 4/8	23 2/8	23 2/8	8 4/8	8 2/8	26 6/8	Holman	NWT	Rick Duggan	2006	60

PRONGHORN

Minimum Score 67 — *Autilocapra americana americana* and related subspecies

Score	Length of R Horn	Length of L Horn	Circumference R of Base	Circumference L of Base	Inside Spread	Area	State/ Province	Hunter's Name	Date	Rank
88 6/8	15 6/8	15 7/8	7 2/8	7 0/8	14 2/8	Grant County	ND	Archie Malm	1958	1
86 2/8	14 5/8	15 2/8	7 0/8	7 0/8	9 0/8	Big Horn County	WY	Kendel D. Cheatham	2010	2
85 6/8	16 0/8	16 1/8	7 0/8	7 1/8	10 5/8	Natrona County	WY	Alfred Gemrich	2004	3
84 6/8	16 2/8	16 2/8	7 0/8	7 0/8	11 1/8	Converse County	WY	Charles DeLong	2001	4
84 6/8	16 1/8	16 0/8	6 6/8	6 6/8	10 6/8	Big Horn County	WY	Brett Tippets	2007	4
84 4/8	15 3/8	15 1/8	7 3/8	7 3/8	9 7/8	Owyhee County	ID	Randy "Rann" Haight	2005	6
84 2/8	16 4/8	16 5/8	7 2/8	7 1/8	13 1/8	Lake County	OR	Tom Nieradka	2012	7
84 0/8	16 2/8	15 5/8	6 7/8	6 7/8	11 6/8	Perkins County	SD	Spike Jorgensen	1964	8
83 4/8	15 2/8	15 2/8	6 7/8	6 7/8	9 3/8	Campbell County	WY	James N. Monat	1990	9
83 4/8	14 5/8	15 0/8	7 0/8	7 0/8	9 4/8	Sweetwater County	WY	Keith Dana	1994	9
83 4/8	15 1/8	15 0/8	7 7/8	7 6/8	11 7/8	Emery County	UT	Rick Stockburger	1997	9
82 4/8	15 1/8	16 0/8	6 5/8	6 3/8	6 3/8	McLean County	ND	Edward J. Weigel	1966	12
82 0/8	17 4/8	18 2/8	6 6/8	6 2/8	15 6/8	Guadalupe County	NM	C. E. Foster, Jr.	1961	13
82 0/8	16 1/8	16 5/8	6 3/8	6 3/8	8 7/8	Moffat County	CO	Marv Clyncke	1998	13
81 4/8	15 2/8	15 3/8	6 5/8	6 4/8	10 6/8	Lake County	OR	Jeff Eder	1990	15
81 2/8	15 5/8	15 3/8	7 0/8	7 0/8	11 2/8	Big Horn County	WY	Dave Moss	2000	16
81 2/8	15 5/8	15 7/8	6 4/8	6 4/8	11 0/8	Las Animas County	CO	Barry J. Smith	2009	16
80 6/8	15 5/8	15 5/8	6 3/8	6 2/8	11 6/8	Lincoln County	NM	Robert W. Davis	1959	18
80 6/8	16 0/8	16 1/8	6 4/8	6 4/8	11 6/8	Thomas County	NE	Andrew L. Glidden	1993	18
80 6/8	16 0/8	16 0/8	6 2/8	6 2/8	13 0/8	Harding County	SD	Jerry Notch	2008	18
80 4/8	15 5/8	15 6/8	6 0/8	5 7/8	11 5/8	Converse County	WY	William R. Horn, Jr.	1999	21
80 2/8	14 5/8	14 5/8	7 1/8	7 3/8	8 6/8	Natrona County	WY	Dr. J. A. Martin	1964	22
80 2/8	15 7/8	15 6/8	6 4/8	6 4/8	8 4/8	Yavapai County	AZ	Jim Machac	1988	22
80 0/8	16 6/8	16 5/8	6 1/8	6 1/8	13 6/8	Albany County	WY	Dave A. Current	1969	24
80 0/8	15 4/8	15 3/8	6 5/8	6 4/8	7 6/8	Weld County	CO	Rick Parish	1995	24
80 0/8	14 6/8	15 1/8	6 5/8	6 5/8	9 7/8	Albany County	WY	Jerry Bowen	2007	24
80 0/8	14 3/8	14 5/8	6 5/8	6 4/8	10 0/8	Campbell County	WY	Marty Auch	2010	24
79 6/8	15 4/8	14 6/8	7 1/8	7 3/8	9 3/8	Mellette County	SD	John Anderson	1970	28
79 6/8	15 4/8	15 3/8	6 4/8	6 4/8	11 0/8	Sweetwater County	WY	Keith Dana	1991	28
79 6/8	15 1/8	15 1/8	6 2/8	6 2/8	9 7/8	Weld County	CO	Tim Bradley	2006	28
79 6/8	14 5/8	15 0/8	7 2/8	7 1/8	9 5/8	Carter County	MT	Michael Moose	2008	28
79 4/8	14 3/8	14 2/8	6 3/8	6 2/8	11 0/8	Natrona County	WY	Alfred J. Gemrich	2001	32
79 4/8	14 5/8	14 5/8	6 3/8	6 4/8	13 7/8	Albany County	WY	Jerry Bowen	2009	32
79 0/8	14 1/8	14 4/8	6 3/8	6 2/8	11 5/8	Millard County	UT	Keith Dana	1988	34
79 0/8	15 1/8	15 1/8	6 7/8	6 7/8	11 3/8	Weld County	CO	Fred Eichler	1995	34
78 6/8	14 0/8	14 0/8	6 4/8	6 4/8	9 3/8	Sweetwater County	WY	Mark L. Preston	1992	36
78 6/8	14 5/8	14 6/8	6 3/8	6 3/8	10 2/8	Sweetwater County	WY	Mike Barrett	1992	36
78 4/8	16 0/8	15 7/8	6 0/8	6 2/8	11 4/8	Chaves County	NM	Dr. D. A. Henbest	1957	38
78 4/8	15 1/8	14 7/8	6 5/8	6 5/8	10 0/8	Mountrail County	ND	Bennie J. Burtts	1967	38
78 4/8	15 3/8	15 1/8	6 4/8	6 4/8	9 5/8	Petroleum County	MT	Clark Jenner	1990	38
78 4/8	13 7/8	13 7/8	6 7/8	7 0/8	10 1/8	Converse County	WY	Ronald D. Stoynoff	1992	38
78 4/8	15 0/8	15 0/8	5 4/8	5 4/8	9 6/8	Grant County	NM	Art Ramirez	1994	38
78 4/8	14 7/8	14 6/8	6 5/8	6 4/8	13 7/8	Big Horn County	WY	Dave Moss	2001	38
78 2/8	17 3/8	17 1/8	5 5/8	5 6/8	14 6/8	Guadalupe County	NM	James L. Henry	1961	44
78 2/8	13 6/8	13 7/8	6 5/8	6 4/8	8 7/8	Natrona County	WY	Jack Conner	1990	44
78 2/8	14 7/8	15 0/8	6 4/8	6 5/8	10 3/8	Rosebud County	MT	Neil Hamilton	2003	44
78 0/8	15 5/8	15 1/8	6 2/8	6 2/8	18 2/8	Sweetwater County	WY	Don Dvoroznak	1976	47
78 0/8	15 3/8	15 2/8	6 7/8	6 4/8	13 7/8	Moffat County	CO	Tony Seahorn	1980	47
78 0/8	15 4/8	15 3/8	6 1/8	6 1/8	12 3/8	Saguache County	CO	Dan Bertalan	1991	47
78 0/8	14 1/8	14 0/8	6 2/8	6 2/8	7 5/8	Powder River County	MT	Terry Krahn	2005	47
77 6/8	15 4/8	15 4/8	6 2/8	6 1/8	11 5/8	McKinley County	NM	Lee Burnett	1975	51
77 6/8	15 7/8	15 0/8	6 0/8	5 6/8	10 4/8	Lincoln County	CO	Steve Winkelman	1978	51
77 6/8	15 0/8	15 0/8	6 2/8	6 1/8	8 5/8	Fremont County	WY	Gary Laya	1991	51
77 6/8	14 0/8	14 0/8	6 5/8	6 3/8	9 0/8	Hot Springs County	WY	Larry Dickerson	1992	51
77 6/8	15 2/8	15 3/8	5 7/8	5 7/8	13 2/8	Medicine Hat	ALB	Paul Unger	1994	51
77 6/8	14 1/8	14 2/8	6 5/8	6 5/8	10 1/8	Harding County	SD	Henry Loving	1997	51
77 6/8	14 7/8	15 0/8	6 4/8	6 4/8	6 4/8	Fremont County	WY	Gary Laya	2006	51
77 4/8	13 7/8	13 7/8	7 1/8	7 0/8	10 6/8	Moffat County	CO	Michael Schirer	1995	58
77 4/8	14 3/8	14 2/8	6 6/8	6 4/8	9 3/8	Natrona County	WY	Michael Clevinger	2011	58
77 2/8	16 3/8	16 4/8	6 0/8	6 0/8	10 0/8	Lincoln County	NM	Charles L. Hughes	1960	60
77 2/8	14 4/8	14 3/8	6 0/8	6 0/8	8 7/8	Mesa County	CO	Bob Black	1983	60
77 2/8	14 0/8	14 0/8	6 0/8	6 0/8	14 6/8	Garfield County	MT	Paul Schafer	1984	60
77 2/8	15 0/8	15 2/8	6 1/8	6 1/8	11 1/8	Rosebud County	MT	Gene Welle	1988	60

Score	Length of R Horn	L	Circumference R of Base	L	Inside Spread	Area	State/ Province	Hunter's Name	Date	Rank
77 2/8	14 2/8	14 2/8	6 5/8	6 6/8	11 3/8	Sweetwater County	WY	Brad Hugh Jacobs	1992	60
77 2/8	14 3/8	14 3/8	6 2/8	6 1/8	9 6/8	Powder River County	MT	John Martinson	1993	60
77 2/8	15 2/8	15 2/8	6 5/8	6 4/8	10 2/8	Sweetwater County	WY	Mike Barrett	1994	60
77 2/8	15 6/8	16 2/8	6 1/8	6 0/8	8 0/8	Johnson County	WY	Craig Grant	2005	60
77 2/8	14 6/8	14 6/8	5 7/8	5 6/8	9 7/8	Las Animas County	CO	Glen Persinger	2010	60
77 0/8	15 0/8	14 7/8	6 2/8	6 2/8	9 3/8	Carbon County	WY	Rod Schmidt	1992	69
77 0/8	14 5/8	14 5/8	6 5/8	6 5/8	11 5/8	Converse County	WY	Ron Grant	1997	69
77 0/8	13 1/8	13 4/8	6 5/8	6 5/8	7 3/8	Weld County	CO	H. Mike Palmer	1999	69
77 0/8	14 4/8	14 1/8	6 4/8	6 4/8	12 0/8	Carbon County	WY	Martin E. Cain	2010	69
76 6/8	14 1/8	14 5/8	6 2/8	6 2/8	14 2/8	Fremont County	WY	Jim Puthoff	1969	73
76 6/8	15 4/8	14 3/8	7 0/8	6 7/8	10 2/8	Weld County	CO	Alan J. Douthit	1995	73
76 6/8	13 2/8	13 4/8	6 5/8	6 5/8	11 5/8	Powder River County	MT	Kerry A. White	2006	73
76 4/8	15 5/8	15 5/8	6 0/8	6 0/8	8 2/8	Beaverhead County	MT	Neil L. Jacobson	1990	76
76 4/8	15 0/8	15 1/8	6 4/8	6 3/8	11 7/8	Thomas County	NE	Andrew L. Glidden	1990	76
76 4/8	14 7/8	14 7/8	6 1/8	6 1/8	11 6/8	Fremont County	WY	Gary Laya	1992	76
76 4/8	14 3/8	14 0/8	5 7/8	5 7/8	10 4/8	Jackson County	CO	Bruce Ayers	1995	76
76 4/8	14 1/8	14 2/8	6 3/8	6 4/8	9 4/8	Big Horn County	WY	Mike Barrett	2004	76
76 4/8	15 2/8	15 1/8	6 0/8	6 0/8	10 4/8	Albany County	WY	Mike Yancey	2006	76
76 4/8	14 3/8	14 2/8	6 5/8	6 5/8	13 0/8	Gunnison County	CO	Dean Derby	2007	76
76 4/8	14 4/8	14 5/8	5 7/8	5 7/8	11 5/8	Gallatin County	MT	Bob Morton	2007	76
76 4/8	15 3/8	15 0/8	6 6/8	6 4/8	12 0/8	Weld County	CO	Frank Piacentino	2008	76
76 2/8	16 0/8	15 6/8	5 7/8	6 1/8	11 4/8	Converse County	WY	Jack M. Conner	1981	85
76 2/8	15 2/8	15 2/8	6 4/8	6 4/8	10 4/8	Hettinger County	ND	Bill Clink	1988	85
76 2/8	13 5/8	13 5/8	6 5/8	6 5/8	8 4/8	Converse County	WY	William L. Randles	1991	85
76 2/8	15 0/8	15 1/8	6 2/8	6 2/8	9 3/8	Saguache County	CO	Gary W. Bohochik	1992	85
76 2/8	14 6/8	14 5/8	5 6/8	5 6/8	10 3/8	Petroleum County	MT	Craig Richardson	1995	85
76 2/8	14 6/8	14 4/8	6 1/8	6 1/8	11 6/8	Las Animas County	CO	Fred Eichler	2004	85
76 0/8	15 0/8	14 7/8	6 0/8	6 0/8	14 6/8	Mountrail County	ND	Wayne A. Metcalf	1972	91
76 0/8	15 1/8	15 1/8	6 2/8	6 2/8	9 7/8	Natrona County	WY	Alfred J. Gemrich	2002	91
76 0/8	13 6/8	13 6/8	6 2/8	6 1/8	9 5/8	Converse County	WY	Gary Tarver	2010	91
75 6/8	14 4/8	14 6/8	6 1/8	6 1/8	10 0/8	Natrona County	WY	Jack M. Conner	1982	94
75 6/8	15 1/8	15 1/8	6 2/8	6 2/8	10 3/8	Converse County	WY	Lonny G. Herrick	1983	94
75 6/8	14 2/8	14 3/8	5 6/8	6 1/8	6 5/8	Carbon County	WY	Clarence E. Faber	1990	94
75 6/8	13 6/8	13 3/8	6 1/8	6 0/8	12 4/8	Sublette County	WY	Mike Lamade	1990	94
75 6/8	13 3/8	13 3/8	6 5/8	6 4/8	8 2/8	Sweetwater County	WY	Roger Wintle	1996	94
75 6/8	13 4/8	13 5/8	6 3/8	6 3/8	10 0/8	Converse County	WY	Lee Waltman	1999	94
75 6/8	14 0/8	13 7/8	6 3/8	6 3/8	10 6/8	Albany County	WY	Jim Willems	2009	94
75 6/8	14 4/8	14 6/8	5 7/8	5 6/8	11 0/8	Las Animas County	CO	Dale Charles	2011	94
75 4/8	12 1/8	12 2/8	5 7/8	6 0/8	8 2/8	Campbell County	WY	Dr. R. F. Helzerman	1960	102
75 4/8	17 2/8	16 4/8	6 0/8	5 6/8	15 0/8	Guadalupe County	NM	M. K. Vance	1962	102
75 4/8	14 3/8	14 2/8	6 0/8	6 2/8	12 6/8	McLean County	ND	Don Sorge	1970	102
75 4/8	14 7/8	14 7/8	6 0/8	6 0/8	12 0/8	Williams County	ND	Terry L. Halgrimson	1970	102
75 4/8	16 0/8	16 0/8	6 0/8	5 7/8	11 0/8	Moffat County	CO	John R. Morris II	1983	102
75 4/8	14 0/8	14 0/8	6 2/8	6 2/8	10 0/8	Powder River County	MT	Raleigh D. Buckmaster	1983	102
75 4/8	14 0/8	14 1/8	6 6/8	6 4/8	8 1/8	Moffat County	CO	Tracy L. Gulliksen	1987	102
75 4/8	15 3/8	15 1/8	6 7/8	6 7/8	7 4/8	Custer County	ID	Matt March	1990	102
75 4/8	13 6/8	13 6/8	6 5/8	6 3/8	9 6/8	Carbon County	WY	Larry Cross	1992	102
75 4/8	15 2/8	15 1/8	5 7/8	5 7/8	9 1/8	Albany County	WY	Jerry Bowen	2002	102
75 4/8	13 4/8	13 4/8	6 7/8	6 6/8	11 1/8	Las Animas County	CO	Andrew C. Bair	2006	102
75 4/8	14 4/8	14 3/8	6 0/8	6 0/8	9 2/8	Moffat County	CO	Marvin Clyncke	2008	102
75 4/8	12 7/8	13 0/8	6 6/8	6 6/8	11 5/8	Converse County	WY	David J. Lucas	2011	102
75 2/8	16 1/8	16 6/8	5 6/8	5 5/8	11 2/8	Klamath County	OR	Steve H. Bell	1973	115
75 2/8	14 4/8	14 5/8	6 1/8	6 0/8	12 2/8	Moffat County	CO	Ronald King	1990	115
75 2/8	15 4/8	16 1/8	5 7/8	5 5/8	9 1/8	Harney County	OR	Stanley Miles	1990	115
75 2/8	14 3/8	14 4/8	6 3/8	6 3/8	14 4/8	Moffat County	CO	Bud Boker	1991	115
75 2/8	15 6/8	15 4/8	6 0/8	6 0/8	9 7/8	Yavapai County	AZ	Don Parks, Jr.	1992	115
75 2/8	15 4/8	16 2/8	6 0/8	6 0/8	8 5/8	Sioux County	NE	John J. Schaffer	1993	115
75 2/8	13 6/8	13 2/8	6 3/8	6 3/8	11 4/8	Sweetwater County	WY	Vic R. Dana	1998	115
75 0/8	13 6/8	14 0/8	6 2/8	6 2/8	8 6/8	Sweetwater County	WY	Mike Ward	1976	122
75 0/8	14 4/8	14 4/8	5 5/8	5 5/8	10 4/8	Converse County	WY	William G. Mason	1986	122
75 0/8	14 3/8	14 3/8	6 1/8	6 1/8	10 2/8	Sweetwater County	WY	Donald Ace Morgan	1992	122
75 0/8	13 6/8	14 3/8	6 7/8	7 1/8	12 4/8	Moffat County	CO	Glenn W. Pritchard	1992	122
75 0/8	15 0/8	14 6/8	6 3/8	6 2/8	9 1/8	Yavapai County	AZ	Steve Rawlins	1993	122

PRONGHORN

Minimum Score 67 | *Continued*

Score	Length of R Horn	L	Circumference R of Base	L	Inside Spread	Area	State/ Province	Hunter's Name	Date	Rank
75 0/8	12 6/8	13 2/8	6 4/8	6 4/8	12 1/8	Graham County	AZ	John C. Rhodes	1994	122
75 0/8	12 4/8	12 4/8	6 5/8	6 5/8	8 6/8	Hot Springs County	WY	Micheal H. Eastman	1995	122
75 0/8	14 0/8	14 0/8	6 1/8	6 0/8	11 7/8	Big Horn County	WY	Dave Moss	1998	122
74 6/8	14 4/8	14 5/8	6 3/8	6 1/8	7 0/8	Converse County	WY	Norm Goodwin	1981	130
74 6/8	14 4/8	14 5/8	6 0/8	6 1/8	9 6/8	Converse County	WY	Steve Gorr	1982	130
74 6/8	15 2/8	14 7/8	5 7/8	5 7/8	7 4/8	Washoe County	NV	Charlie Powning	1992	130
74 6/8	14 6/8	14 5/8	5 7/8	5 7/8	11 1/8	Natrona County	WY	Elmer R. Luce, Jr.	1992	130
74 6/8	14 5/8	14 6/8	6 4/8	6 4/8	12 1/8	Moffat County	CO	Jerry Douthit	1997	130
74 6/8	13 6/8	13 7/8	5 7/8	5 6/8	8 1/8	Gunnison County	CO	Pat Reilly	2005	130
74 6/8	13 7/8	14 0/8	6 1/8	6 1/8	8 1/8	Washakie County	WY	David Gluyas	2008	130
74 4/8	14 6/8	14 4/8	6 2/8	6 2/8	12 3/8	Campbell County	WY	William P. Mastrangel	1957	137
74 4/8	15 3/8	15 3/8	6 3/8	6 2/8	11 2/8	Lincoln County	NM	Harvey May	1960	137
74 4/8	14 2/8	14 6/8	6 0/8	6 1/8	12 0/8	Ward County	ND	Bennie Burtts	1964	137
74 4/8	14 1/8	14 0/8	6 6/8	6 5/8	8 7/8	Wheeler County	NE	Lynn M. Briggs	1965	137
74 4/8	13 4/8	13 6/8	7 1/8	7 2/8	8 5/8	Sioux County	NE	Bill Carlos	1969	137
74 4/8	14 0/8	14 2/8	6 4/8	6 4/8	11 5/8	Lyman County	SD	Loran Hills	1970	137
74 4/8	14 2/8	14 2/8	6 4/8	6 4/8	10 0/8	Baca County	CO	Bill McEndree	1982	137
74 4/8	13 5/8	13 2/8	6 3/8	6 3/8	12 1/8	Moffat County	CO	Dale Drilling	1985	137
74 4/8	13 3/8	13 3/8	6 5/8	6 5/8	10 0/8	Sweetwater County	WY	Bill Clink	1986	137
74 4/8	13 3/8	13 3/8	6 4/8	6 4/8	7 4/8	Sweetwater County	WY	Glenn E. Hisey	1986	137
74 4/8	14 4/8	15 1/8	6 1/8	6 1/8	14 1/8	Washoe County	NV	Darrel Reed	1990	137
74 4/8	13 7/8	13 5/8	6 1/8	6 1/8	9 1/8	Moffat County	CO	Michael L. Cone	1995	137
74 4/8	13 7/8	14 0/8	6 2/8	6 1/8	9 7/8	Petroleum County	MT	Mark D. Hughes	1995	137
74 4/8	14 2/8	14 0/8	6 1/8	6 1/8	8 1/8	Natrona County	WY	Dale A. Storey	1995	137
74 4/8	13 2/8	14 2/8	5 7/8	6 0/8	9 0/8	Las Animas County	CO	Eldon J. Jandreau	2002	137
74 2/8	13 4/8	13 7/8	5 5/8	5 5/8	8 1/8	Sweet Grass County	MT	Charles Alkire	1964	152
74 2/8	14 5/8	14 6/8	6 3/8	6 2/8	10 3/8	Sioux County	NE	Wayne Scherbarth	1969	152
74 2/8	13 1/8	13 0/8	6 2/8	6 2/8	7 6/8	Sweetwater County	WY	Keith Dana	1983	152
74 2/8	14 4/8	14 7/8	6 0/8	6 0/8	9 3/8	Natrona County	WY	Gerry C. Stinski	1988	152
74 2/8	13 6/8	13 6/8	6 1/8	6 1/8	11 0/8	Sweetwater County	WY	Quince Hale	1989	152
74 2/8	14 1/8	13 6/8	6 3/8	6 3/8	11 6/8	Moffat County	CO	Cary Laman	1991	152
74 2/8	14 0/8	14 1/8	6 2/8	6 2/8	11 0/8	Moffat County	CO	Bruno Ammann	1991	152
74 2/8	14 2/8	14 2/8	5 5/8	5 5/8	9 5/8	Converse County	WY	Gene Mathias	1991	152
74 2/8	14 0/8	14 2/8	6 4/8	6 4/8	9 5/8	Moffat County	CO	Casey Oliver	1996	152
74 2/8	14 1/8	14 3/8	6 1/8	6 1/8	9 0/8	Moffat County	CO	Scott George	1996	152
74 2/8	13 5/8	13 1/8	6 7/8	6 6/8	9 4/8	Converse County	WY	Bob Gilbert	2004	152
74 2/8	14 7/8	14 7/8	5 7/8	5 7/8	14 3/8	Jackson County	CO	Tom Kelley	2009	152
74 2/8	14 0/8	14 0/8	6 0/8	6 0/8	12 1/8	Moffat County	CO	Todd Clyncke	2010	152
74 2/8	14 4/8	14 3/8	6 0/8	5 7/8	12 4/8	Albany County	WY	Jerry Bowen	2010	152
74 0/8	15 0/8	14 5/8	5 3/8	5 3/8	8 3/8	Moffat County	CO	Glenn Pritchard	1990	166
74 0/8	14 1/8	14 1/8	6 3/8	6 2/8	12 3/8	Sierra County	NM	Peter La Scala	1991	166
74 0/8	14 7/8	15 1/8	5 7/8	5 6/8	9 5/8	Sweetwater County	WY	Bill Clink	1993	166
74 0/8	13 4/8	13 4/8	6 3/8	6 4/8	9 6/8	Johnson County	WY	Gary G. Olson	1993	166
74 0/8	13 2/8	13 3/8	6 1/8	6 0/8	5 6/8	Moffat County	CO	E. Damon Handley	1996	166
74 0/8	13 3/8	13 2/8	6 1/8	6 0/8	9 7/8	Natrona County	WY	Alex Gemrich	1998	166
74 0/8	15 1/8	14 7/8	5 6/8	5 4/8	7 4/8	Garfield County	UT	Joseph T. Allman	2007	166
74 0/8	14 7/8	14 6/8	5 7/8	5 6/8	11 3/8	Albany County	WY	Jim Willems	2010	166
73 6/8	14 0/8	14 0/8	6 1/8	6 2/8	13 7/8	Williams County	ND	Robert Halseth	1967	174
73 6/8	14 2/8	14 0/8	6 2/8	6 2/8	10 4/8	Sweetwater County	WY	Dan Winder	1973	174
73 6/8	14 2/8	14 1/8	6 5/8	6 4/8	8 5/8	Converse County	WY	Don Schram	1982	174
73 6/8	12 7/8	12 7/8	7 0/8	6 7/8	10 7/8	Natrona County	WY	Pat McAteer	1984	174
73 6/8	14 2/8	14 2/8	5 7/8	5 6/8	11 6/8	Moffat County	CO	Ron Rockwell	1988	174
73 6/8	14 3/8	14 4/8	6 4/8	6 4/8	8 0/8	Natrona County	WY	Ron Niziolek	1990	174
73 6/8	13 4/8	13 4/8	6 1/8	6 1/8	9 0/8	Sheridan County	WY	Gerhard Eimer	1991	174
73 6/8	13 7/8	13 7/8	6 1/8	6 1/8	12 2/8	Weld County	CO	Jim Tatro	1995	174
73 6/8	13 6/8	14 1/8	6 3/8	6 3/8	9 3/8	Big Horn County	WY	H. Mike Palmer	1999	174
73 6/8	14 1/8	14 0/8	5 7/8	5 6/8	10 0/8	Elko County	NV	Jesse Buckner	2004	174
73 6/8	13 7/8	13 7/8	6 3/8	6 3/8	9 1/8	Albany County	WY	Jerry Bowen	2004	174
73 6/8	14 5/8	14 6/8	5 6/8	5 5/8	11 0/8	Niobrara County	WY	Mike Kucera	2007	174
73 6/8	13 7/8	13 7/8	6 4/8	6 5/8	11 7/8	Converse County	WY	Michael Clevinger	2008	174
73 6/8	14 0/8	14 1/8	6 1/8	6 2/8	11 0/8	Las Animas County	CO	Mandie Hill	2008	174
73 4/8	15 4/8	15 1/8	5 6/8	5 7/8	8 3/8	Moffat County	CO	Henry Wichers	1957	188
73 4/8	16 0/8	16 4/8	5 6/8	5 6/8	15 6/8	Guadalupe County	NM	Jack McCaw	1961	188

Score	Length of R Horn	L	Circumference R of Base	L	Inside Spread	Area	State/ Province	Hunter's Name	Date	Rank
73 4/8	15 7/8	16 0/8	5 7/8	5 6/8	10 5/8	Coconino County	AZ	Charles Meriwether	1968	188
73 4/8	13 2/8	13 2/8	6 7/8	6 6/8	10 0/8	Moffat County	CO	Curtis Lynn	1972	188
73 4/8	14 6/8	15 0/8	5 4/8	5 4/8	8 7/8	Fremont County	WY	Joseph "Jay" Carter	1978	188
73 4/8	13 7/8	14 1/8	6 2/8	6 3/8	8 6/8	Sweetwater County	WY	Jim Dougherty	1983	188
73 4/8	15 2/8	15 1/8	6 0/8	6 0/8	8 6/8	Moffat County	CO	Dave Accashian	1997	188
73 4/8	13 4/8	13 4/8	6 1/8	6 1/8	12 4/8	Albany County	WY	Mike Yancey	2005	188
73 4/8	14 3/8	14 3/8	5 6/8	5 7/8	9 6/8	Converse County	WY	Sam Morello	2008	188
73 2/8	13 0/8	13 1/8	6 6/8	6 4/8	11 3/8	Fremont County	WY	Chuck Kroll	1952	197
73 2/8	11 3/8	11 7/8	6 3/8	6 3/8	11 1/8	Carter County	MT	Benny F. Padden	1960	197
73 2/8	14 2/8	14 2/8	5 2/8	5 2/8	7 5/8	Fergus County	MT	Wayne Miller	1962	197
73 2/8	14 6/8	14 3/8	6 0/8	6 0/8	9 2/8	Garfield County	MT	Paul Brunner	1976	197
73 2/8	14 3/8	14 3/8	5 7/8	5 7/8	8 2/8	Converse County	WY	Ron Rockwell	1988	197
73 2/8	13 6/8	13 6/8	6 1/8	6 1/8	12 0/8	Sweetwater County	WY	Fred R. Trujillo	1992	197
73 2/8	16 0/8	15 5/8	5 7/8	5 7/8	10 4/8	Sioux County	NE	Orville J. DeVoss	1992	197
73 2/8	15 4/8	14 7/8	6 1/8	6 1/8	11 2/8	Lake County	OR	Jeffrey A. Eder	1994	197
73 2/8	13 4/8	13 5/8	6 1/8	6 2/8	7 3/8	Twin Falls County	ID	Darrell Nunez	1995	197
73 2/8	14 0/8	13 7/8	6 1/8	6 2/8	10 2/8	Moffat County	CO	Bud Boker	1996	197
73 2/8	13 5/8	13 6/8	6 1/8	5 7/8	10 0/8	Moffat County	CO	Bob Gulliksen	1997	197
73 2/8	14 4/8	14 5/8	6 0/8	5 7/8	8 6/8	Las Animas County	CO	Andrew Bair	2001	197
73 2/8	13 4/8	13 6/8	6 5/8	6 4/8	8 7/8	Gallatin County	MT	Bob Morton	2001	197
73 2/8	13 3/8	13 4/8	6 2/8	6 2/8	14 0/8	Washakie County	WY	Randy Burtis	2003	197
73 2/8	14 2/8	14 2/8	5 7/8	5 6/8	11 2/8	Las Animas County	CO	Barry J. Smith	2008	197
73 2/8	13 6/8	13 4/8	5 4/8	5 4/8	10 1/8	Moffat County	CO	Mark Turner	2010	197
73 0/8	13 5/8	13 5/8	5 3/8	5 3/8	9 6/8	Harding County	SD	Ted G. Carter	1961	213
73 0/8	12 7/8	12 6/8	6 2/8	6 3/8	11 7/8	Stark County	ND	Ronald D. Hauck	1970	213
73 0/8	13 3/8	13 5/8	6 4/8	6 5/8	9 5/8	Fremont County	WY	Gary Laya	1990	213
73 0/8	14 1/8	14 1/8	6 2/8	6 0/8	12 6/8	Sweetwater County	WY	Mike Barrett	1993	213
73 0/8	14 6/8	14 3/8	5 6/8	5 6/8	12 7/8	Lincoln County	WY	Greg Munther	2000	213
73 0/8	14 2/8	14 2/8	5 6/8	5 6/8	8 7/8	Converse County	WY	Leonard Scarborough	2003	213
73 0/8	14 3/8	14 3/8	5 7/8	6 0/8	10 2/8	Albany County	WY	Mike Yancey	2009	213
72 6/8	12 1/8	12 1/8	6 6/8	6 6/8	10 6/8	Natrona County	WY	Dennis Spawn	1974	220
72 6/8	15 1/8	14 6/8	5 6/8	5 7/8	9 2/8	Carbon County	WY	I. C. Benjamin	1976	220
72 6/8	14 0/8	13 6/8	6 0/8	6 0/8	7 0/8	Moffat County	CO	Jim Dougherty	1983	220
72 6/8	14 0/8	14 0/8	5 4/8	5 3/8	8 4/8	Wheatland County	MT	Bob Radocy	1988	220
72 6/8	13 7/8	14 0/8	6 0/8	6 1/8	11 7/8	Moffat County	CO	Ron Rockwell	1990	220
72 6/8	13 6/8	13 6/8	6 0/8	6 0/8	8 0/8	Sheridan County	WY	Michael Briganti	1991	220
72 6/8	14 0/8	14 4/8	5 6/8	5 6/8	10 5/8	Moffat County	CO	Daniel G. Willems	1993	220
72 6/8	14 2/8	14 3/8	6 3/8	6 2/8	11 1/8	Petroleum County	MT	Mark D. Hughes	1998	220
72 6/8	13 2/8	14 2/8	6 0/8	6 0/8	11 3/8	Rich County	UT	Matthew Thomson	2008	220
72 6/8	13 6/8	13 3/8	5 6/8	5 6/8	9 0/8	Converse County	WY	Steve Osminski	2008	220
72 6/8	13 5/8	13 3/8	6 2/8	6 5/8	11 1/8	Park County	WY	Patrick Allen	2010	220
72 4/8	14 6/8	14 3/8	5 7/8	5 7/8	8 2/8	Sweetwater County	WY	Kirby Warnock	1986	231
72 4/8	12 7/8	12 3/8	6 4/8	6 2/8	10 5/8	Lemhi County	ID	Matt March, Jr.	1986	231
72 4/8	13 5/8	13 5/8	6 3/8	6 2/8	7 5/8	Moffat County	CO	Terry Weimer	1987	231
72 4/8	14 0/8	13 6/8	5 7/8	5 6/8	11 0/8	Sweetwater County	WY	Bryan Radakovich	1990	231
72 4/8	14 2/8	14 0/8	5 7/8	5 7/8	11 4/8	Campbell County	WY	Jim Reints	1991	231
72 4/8	14 5/8	15 0/8	5 7/8	5 7/8	10 0/8	Rosebud County	MT	Gene A. Welle	1991	231
72 4/8	15 2/8	14 7/8	6 1/8	6 0/8	8 6/8	Coconino County	AZ	Dan Turner	1993	231
72 4/8	13 3/8	13 3/8	6 0/8	6 0/8	9 2/8	Moffat County	CO	Roger Stewart	1994	231
72 4/8	13 0/8	13 0/8	6 2/8	6 0/8	10 3/8	Campbell County	WY	Ritch A. Stolpe	1994	231
72 4/8	14 3/8	14 3/8	5 5/8	5 5/8	10 0/8	Converse County	WY	Dale A. Storey	1994	231
72 4/8	13 2/8	13 1/8	6 3/8	6 2/8	11 2/8	Sweetwater County	WY	Roger Wintle	1995	231
72 4/8	13 1/8	13 1/8	6 4/8	6 3/8	8 6/8	Moffat County	CO	Clifford R. Neville, Sr.	1997	231
72 4/8	12 6/8	12 6/8	5 7/8	5 7/8	9 4/8	Converse County	WY	Thomas Morang	2008	231
72 4/8	13 6/8	14 2/8	5 5/8	5 5/8	9 2/8	Cibola County	NM	Bill Van Buskirk	2009	231
72 4/8	13 2/8	13 5/8	5 6/8	5 6/8	7 4/8	Moffat County	CO	Todd A. Brickel	2009	231
72 4/8	15 1/8	14 7/8	5 5/8	5 6/8	12 2/8	Sioux County	NE	Brad Thayer	2011	231
72 2/8	13 5/8	13 5/8	6 2/8	6 1/8	11 1/8	Carbon County	WY	William Scoggin	1953	247
72 2/8	14 0/8	14 0/8	6 0/8	5 7/8	10 4/8	Campbell County	WY	K. K. Knickerbocker	1954	247
72 2/8	13 1/8	13 2/8	6 1/8	6 1/8	7 3/8	Campbell County	WY	Carol Wert	1966	247
72 2/8	14 3/8	14 3/8	6 0/8	5 7/8	9 5/8	Valley County	MT	Wayne Anderson	1975	247
72 2/8	14 1/8	14 0/8	6 2/8	6 3/8	8 5/8	Moffat County	CO	Glenn Pritchard	1977	247
72 2/8	14 3/8	14 4/8	6 1/8	6 0/8	12 3/8	Moffat County	CO	Glenn Pritchard	1979	247

PRONGHORN

Minimum Score 67 *Continued*

Score	Length of R Horn	L	Circumference R of Base	L	Inside Spread	Area	State/ Province	Hunter's Name	Date	Rank
72 2/8	11 3/8	11 3/8	6 3/8	6 3/8	10 4/8	Humboldt County	NV	Verlyn Owens	1981	247
72 2/8	14 0/8	13 7/8	6 0/8	6 0/8	7 6/8	Morgan County	CO	Barry Smith	1983	247
72 2/8	14 3/8	14 1/8	6 4/8	6 1/8	10 4/8	Lemhi County	ID	Peter Cintorino	1986	247
72 2/8	13 4/8	13 5/8	6 3/8	6 5/8	11 0/8	Moffat County	CO	Wayne Depperschmidt	1992	247
72 2/8	14 2/8	14 0/8	6 4/8	6 3/8	10 3/8	Moffat County	CO	J. Keith Chastain	1992	247
72 2/8	13 1/8	12 4/8	6 0/8	5 7/8	12 1/8	Garfield County	MT	Christopher Downs	1992	247
72 2/8	14 1/8	14 2/8	5 4/8	5 4/8	10 6/8	Moffat County	CO	Steve Murphy	1993	247
72 2/8	13 6/8	13 6/8	5 6/8	5 6/8	10 4/8	Albany County	WY	Chris Krueger	1999	247
72 2/8	14 7/8	14 6/8	5 5/8	5 5/8	9 4/8	Albany County	WY	Jerry Bowen	2005	247
72 2/8	14 0/8	13 7/8	6 0/8	5 6/8	16 5/8	Converse County	WY	John Hoinowski	2005	247
72 2/8	14 4/8	14 0/8	6 0/8	6 0/8	10 7/8	Weld County	CO	Michael McArtor	2005	247
72 2/8	13 3/8	13 4/8	6 2/8	6 3/8	9 4/8	Converse County	WY	Forrest Boone	2006	247
72 2/8	14 3/8	14 2/8	5 6/8	5 6/8	10 3/8	Lassen County	CA	Blake Marsters	2009	247
72 2/8	14 0/8	13 7/8	5 6/8	5 7/8	9 3/8	Converse County	WY	Ronald Rockwell	2011	247
72 0/8	13 3/8	13 3/8	5 6/8	6 0/8	11 2/8	Sioux County	NE	Richard Koons	1974	267
72 0/8	14 0/8	14 0/8	5 3/8	5 3/8	11 4/8	Converse County	WY	Roy W. Mackey	1990	267
72 0/8	13 2/8	13 2/8	5 6/8	6 0/8	8 6/8	Park County	WY	Craig Childress	1991	267
72 0/8	13 4/8	13 7/8	5 6/8	5 6/8	9 3/8	Saguache County	CO	Mark Wuerthele	1992	267
72 0/8	13 3/8	12 6/8	6 2/8	6 1/8	15 4/8	Albany County	WY	Vince DiMiceli	1992	267
72 0/8	13 2/8	13 0/8	6 6/8	6 6/8	11 6/8	Converse County	WY	Dave Canfield	1993	267
72 0/8	14 2/8	13 6/8	6 1/8	6 1/8	10 4/8	Campbell County	WY	Kevin Brady	1994	267
72 0/8	13 5/8	13 3/8	6 4/8	6 3/8	11 1/8	Converse County	WY	Steve Thomas, Jr.	1995	267
72 0/8	15 1/8	14 6/8	5 5/8	5 5/8	11 1/8	Converse County	WY	David L. MacDonald	1996	267
72 0/8	14 1/8	14 0/8	6 0/8	5 7/8	10 3/8	Weld County	CO	Matt Dickerson	1997	267
72 0/8	15 0/8	14 7/8	5 6/8	5 6/8	12 5/8	Campbell County	WY	Bret Hamm	2004	267
72 0/8	14 3/8	14 3/8	6 1/8	6 1/8	9 5/8	Moffat County	CO	Troy A. Cunningham	2005	267
72 0/8	14 2/8	14 4/8	5 1/8	5 1/8	9 0/8	Converse County	WY	Ron Rockwell	2007	267
72 0/8	13 4/8	13 2/8	6 2/8	6 1/8	9 6/8	Colfax County	NM	Ronald Rockwell	2012	267
71 6/8	14 3/8	14 5/8	5 7/8	5 7/8	11 6/8	Fremont County	WY	Jim Walters	1983	281
71 6/8	13 5/8	14 0/8	6 0/8	6 0/8	8 2/8	Converse County	WY	Frank Moore	1987	281
71 6/8	14 0/8	14 2/8	5 5/8	5 6/8	10 2/8	Moffat County	CO	David Greenwalt	1988	281
71 6/8	13 0/8	12 3/8	6 3/8	6 2/8	8 3/8	Rosebud County	MT	Kent Kaufman	1989	281
71 6/8	15 5/8	15 7/8	5 1/8	5 2/8	7 4/8	Golden Valley County	ND	Wayne Streitz	1990	281
71 6/8	14 6/8	14 4/8	5 4/8	5 4/8	12 3/8	Orion	ALB	Gunter Lemke	1990	281
71 6/8	13 6/8	13 7/8	6 3/8	6 3/8	11 2/8	Moffat County	CO	R. Tim Reed	1991	281
71 6/8	13 5/8	13 4/8	6 3/8	6 4/8	8 2/8	Custer County	ID	Matt March	1991	281
71 6/8	13 2/8	13 1/8	6 4/8	6 1/8	12 2/8	Moffat County	CO	Bruce Eggenberger	1992	281
71 6/8	13 7/8	14 0/8	5 7/8	6 0/8	10 6/8	Moffat County	CO	Dave Parri	1992	281
71 6/8	12 2/8	12 1/8	6 0/8	6 0/8	8 2/8	Converse County	WY	Joe Cronin	1992	281
71 6/8	13 2/8	13 0/8	5 7/8	5 6/8	11 1/8	Moffat County	CO	Kevin C. Massaro	1992	281
71 6/8	14 4/8	14 0/8	6 0/8	5 5/8	7 2/8	Powder River County	MT	John Martinson	1995	281
71 6/8	12 7/8	13 2/8	6 1/8	6 0/8	9 7/8	Rosebud County	MT	Gene A. Welle	1996	281
71 6/8	13 4/8	13 0/8	7 0/8	6 7/8	10 3/8	Weld County	CO	Kevin Ayers	1997	281
71 6/8	15 1/8	14 7/8	5 5/8	5 4/8	8 1/8	Converse County	WY	Duane R. Mabry	1999	281
71 6/8	13 3/8	13 2/8	6 1/8	6 0/8	7 2/8	Converse County	WY	Jim Ellis	2001	281
71 6/8	13 0/8	13 2/8	6 2/8	6 4/8	10 0/8	Natrona County	WY	Stephen P. Turay	2007	281
71 6/8	13 3/8	13 3/8	5 6/8	5 6/8	8 7/8	Gallatin County	MT	Bob Morton	2010	281
71 6/8	13 3/8	13 5/8	6 3/8	6 3/8	11 4/8	Carbon County	WY	Joe Bradley	2011	281
71 4/8	13 6/8	13 4/8	6 4/8	6 2/8	10 0/8	Carbon County	WY	Bill Scoggin	1957	301
71 4/8	14 7/8	14 7/8	6 0/8	6 2/8	13 5/8	Moffat County	CO	Henry Wichers	1959	301
71 4/8	13 4/8	13 3/8	6 0/8	6 0/8	8 2/8	Tripp County	SD	Dan Smith	1965	301
71 4/8	13 6/8	13 6/8	6 0/8	6 0/8	6 6/8	Sheridan County	WY	David Shoop	1980	301
71 4/8	13 0/8	13 0/8	6 2/8	6 0/8	9 1/8	Moffat County	CO	Terry Weimer	1986	301
71 4/8	13 5/8	13 4/8	6 4/8	6 4/8	11 5/8	Converse County	WY	Steve Gorr	1991	301
71 4/8	14 1/8	14 3/8	5 6/8	5 7/8	6 7/8	Sweetwater County	WY	Thad W. Sullivan	1993	301
71 4/8	13 2/8	13 4/8	5 6/8	5 7/8	11 4/8	Custer County	ID	Matt March	1994	301
71 4/8	13 7/8	14 4/8	6 7/8	6 4/8	10 3/8	Campbell County	WY	Roy M. Wible, Jr.	1994	301
71 4/8	14 2/8	14 4/8	6 0/8	6 1/8	12 3/8	Weld County	CO	Mark Johnson	1996	301
71 4/8	14 2/8	13 5/8	6 0/8	5 7/8	9 7/8	Converse County	WY	David H. Boland	1998	301
71 4/8	14 0/8	13 3/8	6 0/8	5 6/8	9 1/8	Moffat County	CO	Bob Beckwith	1999	301
71 4/8	13 3/8	13 6/8	6 1/8	5 7/8	10 7/8	Natrona County	WY	Steve Turay	2008	301
71 4/8	13 6/8	13 3/8	6 2/8	6 1/8	8 7/8	Powder River County	MT	Mike Barrett	2009	301
71 2/8	14 4/8	14 2/8	5 7/8	5 7/8	8 1/8	Musselshell County	MT	A. A. Anderson	1960	315

PRONGHORN

Minimum Score 67 *Continued*

Score	Length of R Horn	L	Circumference R of Base	L	Inside Spread	Area	State/ Province	Hunter's Name	Date	Rank
71 2/8	14 3/8	14 5/8	5 4/8	5 4/8	13 1/8	Perkins County	SD	Elwood Patterson	1961	315
71 2/8	15 0/8	15 3/8	5 3/8	5 2/8	9 7/8	Natrona County	WY	Doug Pope	1976	315
71 2/8	13 0/8	13 2/8	6 2/8	6 2/8	10 7/8	Converse County	WY	Jack M. Conner	1983	315
71 2/8	12 2/8	12 1/8	6 0/8	6 0/8	9 5/8	Converse County	WY	Jack M. Conner	1984	315
71 2/8	14 1/8	14 4/8	5 5/8	5 6/8	6 5/8	Sweetwater County	WY	Larry Dickerson	1991	315
71 2/8	12 6/8	13 0/8	5 6/8	5 6/8	9 1/8	Moffat County	CO	Scott George	1993	315
71 2/8	14 0/8	14 0/8	5 6/8	5 6/8	12 2/8	Yellowstone County	MT	Thomas J. Madden	1994	315
71 2/8	14 3/8	14 3/8	5 7/8	5 7/8	12 7/8	Juab County	UT	Kirk W. Reese	1996	315
71 2/8	14 6/8	14 6/8	5 4/8	5 3/8	11 4/8	McKenzie County	ND	Brad Blanchard	1996	315
71 2/8	13 6/8	13 6/8	6 1/8	5 7/8	11 3/8	Fergus County	MT	Donny Roy	1998	315
71 2/8	15 3/8	15 5/8	5 0/8	5 0/8	10 5/8	Yavapai County	AZ	Charles Fisher	2001	315
71 2/8	14 2/8	14 2/8	5 4/8	5 4/8	9 4/8	Garfield County	MT	Roger Stewart	2001	315
71 2/8	13 3/8	13 4/8	5 6/8	5 6/8	10 4/8	Fremont County	WY	Gary Laya	2005	315
71 2/8	14 4/8	14 4/8	6 0/8	5 7/8	12 2/8	Gunnison County	CO	Matthew Reilly	2006	315
71 2/8	14 1/8	14 3/8	6 1/8	6 1/8	11 1/8	Johnson County	WY	Don Owen	2006	315
71 2/8	14 5/8	14 4/8	5 4/8	5 2/8	11 3/8	Converse County	WY	Greg Dearth	2007	315
71 0/8	14 0/8	13 6/8	6 2/8	5 7/8	10 4/8	Sioux County	NE	Roger F. Rehborg	1975	332
71 0/8	13 3/8	13 3/8	5 7/8	5 7/8	8 1/8	Natrona County	WY	Wade L. Carstens	1983	332
71 0/8	14 0/8	13 5/8	5 6/8	5 6/8	13 1/8	Converse County	WY	Roy G. Burton	1988	332
71 0/8	13 2/8	13 2/8	5 4/8	5 5/8	9 5/8	Moffat County	CO	Dennis Newton	1990	332
71 0/8	12 7/8	12 7/8	6 2/8	6 1/8	11 3/8	Converse County	WY	J. Todd Payne	1991	332
71 0/8	14 2/8	14 2/8	5 7/8	5 5/8	12 6/8	Converse County	WY	Barry J. Smith	1991	332
71 0/8	13 5/8	13 0/8	6 3/8	6 2/8	11 4/8	Moffat County	CO	Troy Cunningham	1994	332
71 0/8	13 3/8	13 4/8	5 3/8	5 3/8	9 1/8	Moffat County	CO	Michael P. McCarty	1995	332
71 0/8	13 6/8	13 0/8	6 0/8	6 0/8	13 6/8	Moffat County	CO	Ron Miller	1996	332
71 0/8	14 6/8	15 1/8	5 6/8	5 7/8	12 2/8	Sweetwater County	WY	Mike Barrett	1998	332
71 0/8	13 4/8	13 4/8	6 0/8	5 7/8	11 0/8	Grand County	CO	Bryan Holley	2004	332
71 0/8	13 3/8	13 2/8	5 7/8	5 6/8	11 6/8	Platte County	WY	Jerry Bowen	2006	332
71 0/8	14 2/8	14 3/8	6 3/8	6 1/8	10 2/8	Natrona County	WY	Steve Turay	2009	332
71 0/8	13 1/8	13 3/8	6 3/8	6 3/8	9 2/8	Sheridan County	WY	Bruce Burr	2009	332
71 0/8	13 6/8	14 2/8	5 4/8	5 3/8	8 5/8	Natrona County	WY	Michael J. Clevinger	2010	332
71 0/8	13 2/8	13 3/8	5 6/8	5 6/8	9 3/8	Johnson County	WY	John Romans III	2012	332
70 6/8	14 7/8	15 2/8	5 5/8	5 4/8	10 3/8	Washoe County	NV	Frank M. Davis	1967	348
70 6/8	12 5/8	12 4/8	5 5/8	5 4/8	9 0/8	Natrona County	WY	Bernard R. Giacoletto	1973	348
70 6/8	14 0/8	14 0/8	5 4/8	5 4/8	12 6/8	Converse County	WY	Frank Moore	1980	348
70 6/8	13 6/8	14 0/8	5 4/8	5 3/8	12 1/8	Converse County	WY	Russell Hull	1980	348
70 6/8	14 3/8	14 1/8	5 4/8	5 3/8	12 5/8	Tillard Ranch	WY	Charles A. Myers	1981	348
70 6/8	13 7/8	13 7/8	5 3/8	5 2/8	9 3/8	Moffat County	CO	Darryl Quidort	1984	348
70 6/8	14 0/8	13 6/8	5 7/8	5 7/8	7 0/8	Moffat County	CO	Mike Ottenbacher	1986	348
70 6/8	14 0/8	14 0/8	5 5/8	5 5/8	11 0/8	Converse County	WY	David M. Ackland, Jr.	1988	348
70 6/8	14 0/8	13 6/8	5 7/8	5 6/8	10 5/8	Moffat County	CO	Wayne Depperschmidt	1989	348
70 6/8	13 1/8	13 1/8	6 2/8	6 1/8	7 3/8	Billings County	ND	Al Zeller	1989	348
70 6/8	12 2/8	12 4/8	6 2/8	6 4/8	10 2/8	Garfield County	MT	Bob Morton	1989	348
70 6/8	14 2/8	14 2/8	5 7/8	5 7/8	8 4/8	Moffat County	CO	Marvin Cochran	1990	348
70 6/8	13 7/8	13 7/8	5 7/8	6 0/8	8 5/8	Fergus County	MT	Kelly Norskog	1992	348
70 6/8	13 3/8	13 5/8	6 3/8	6 2/8	10 3/8	Sweetwater County	WY	Mark L. Preston	1994	348
70 6/8	12 6/8	12 6/8	6 1/8	6 1/8	10 6/8	Weld County	CO	James L. Tatro	1994	348
70 6/8	13 7/8	14 0/8	5 5/8	5 6/8	11 0/8	Tooele County	UT	Chad J. Hall	1998	348
70 6/8	14 1/8	14 3/8	5 3/8	5 2/8	8 7/8	Larimer County	CO	Richard E. DeLia	1998	348
70 6/8	13 7/8	13 7/8	5 7/8	5 7/8	9 5/8	Musselshell County	MT	James Domaskin	2004	348
70 6/8	13 2/8	13 1/8	7 0/8	6 3/8	7 2/8	Gallatin County	MT	Bob Morton	2006	348
70 6/8	13 3/8	13 3/8	5 7/8	6 0/8	11 0/8	Converse County	WY	Jay F. Deones	2006	348
70 6/8	13 7/8	14 0/8	6 0/8	6 0/8	8 6/8	Rosebud County	MT	Scott Rehak	2008	348
70 6/8	13 1/8	13 4/8	6 0/8	6 0/8	7 7/8	Sweetwater County	WY	Brad Jones	2011	348
70 4/8	12 3/8	12 3/8	5 7/8	5 6/8	7 5/8	Ward County	ND	Bob Torgerson	1964	370
70 4/8	14 1/8	14 2/8	5 6/8	5 5/8	13 6/8	Custer County	MT	Bob Torgerson	1964	370
70 4/8	12 5/8	12 4/8	6 0/8	6 0/8	10 5/8	Sweetwater County	WY	Vern A. Butler	1973	370
70 4/8	13 1/8	13 0/8	5 7/8	6 0/8	8 3/8	Moffat County	CO	H. R. 'Rusty' Neely	1984	370
70 4/8	13 5/8	13 6/8	5 3/8	5 3/8	11 6/8	Moffat County	CO	Glenn Pritchard	1986	370
70 4/8	13 1/8	13 2/8	6 0/8	6 0/8	8 3/8	Brewster County	TX	Michael M. Reamy	1986	370
70 4/8	14 7/8	15 0/8	5 3/8	5 3/8	12 5/8	Sweetwater County	WY	Don Waechtler	1988	370
70 4/8	12 7/8	12 7/8	6 4/8	6 4/8	8 7/8	Natrona County	WY	Kevin Davis	1990	370
70 4/8	13 5/8	13 5/8	5 7/8	5 6/8	12 2/8	Park County	WY	Charles R. Durm	1991	370

PRONGHORN

Minimum Score 67 *Continued*

Score	Length of R Horn	L	Circumference R of Base	L	Inside Spread	Area	State/ Province	Hunter's Name	Date	Rank
70 4/8	14 1/8	14 0/8	5 4/8	5 2/8	14 0/8	Divide County	ND	Robert M. Brunner	1991	370
70 4/8	12 4/8	12 4/8	5 6/8	5 7/8	8 7/8	Golden Valley County	ND	Terry Buechler	1991	370
70 4/8	14 3/8	14 3/8	5 5/8	5 6/8	6 4/8	Moffat County	CO	Ken Hoehn	1992	370
70 4/8	13 0/8	13 0/8	6 2/8	6 1/8	8 1/8	Moffat County	CO	Elmer R. Luce, Jr.	1993	370
70 4/8	14 3/8	14 5/8	5 6/8	5 6/8	7 5/8	Thomas County	NE	Phil Chvala	1994	370
70 4/8	13 6/8	13 5/8	5 3/8	5 4/8	6 4/8	Saguache County	CO	Burton R. Thompson, Jr.	1995	370
70 4/8	13 6/8	13 6/8	5 4/8	5 4/8	7 3/8	Natrona County	WY	Alfred J. Gemrich	1998	370
70 4/8	14 2/8	14 2/8	5 4/8	5 5/8	9 2/8	Converse County	WY	William G. Mason	1999	370
70 4/8	13 6/8	13 6/8	5 4/8	5 4/8	9 3/8	Moffat County	CO	Andy Bair	2000	370
70 4/8	14 0/8	13 7/8	6 0/8	6 0/8	7 1/8	Lemhi County	ID	Ben L. Fahnholz	2000	370
70 4/8	13 6/8	13 5/8	6 1/8	6 1/8	8 1/8	Converse County	WY	W. Tennyson Krugh	2001	370
70 4/8	14 0/8	14 1/8	5 4/8	5 5/8	8 5/8	Moffat County	CO	Dru Harms	2006	370
70 4/8	14 3/8	14 4/8	5 7/8	5 5/8	10 4/8	Harding County	SD	Howard A. Bork	2007	370
70 2/8	15 2/8	15 4/8	5 4/8	5 3/8	11 5/8	Natrona County	WY	Larry J. Colombo	1970	392
70 2/8	12 7/8	12 7/8	6 1/8	6 1/8	12 5/8	Converse County	WY	Edward Coy	1972	392
70 2/8	12 6/8	12 7/8	5 7/8	5 6/8	8 7/8	Meade County	SD	Howard Jess Hollenbeck, Jr.	1978	392
70 2/8	13 6/8	13 6/8	6 6/8	6 4/8	10 6/8	Sweetwater County	WY	Vaughn Cross	1980	392
70 2/8	13 3/8	13 3/8	5 6/8	5 5/8	9 2/8	Carbon County	WY	Duane Hicks	1987	392
70 2/8	13 2/8	13 3/8	6 0/8	5 7/8	9 5/8	Converse County	WY	James H. Miller	1987	392
70 2/8	14 7/8	14 7/8	5 4/8	5 3/8	16 2/8	Converse County	WY	Fred Wallace	1989	392
70 2/8	14 0/8	13 6/8	6 0/8	6 0/8	9 1/8	Converse County	WY	Joel M. Riotto	1989	392
70 2/8	13 1/8	13 2/8	6 0/8	5 7/8	10 7/8	Carbon County	WY	Donald P. Peel	1989	392
70 2/8	14 1/8	14 1/8	5 4/8	5 4/8	11 3/8	Petroleum County	MT	Clark Jenner	1989	392
70 2/8	13 7/8	14 2/8	5 7/8	5 5/8	7 6/8	Moffat County	CO	Conrad Anderson	1992	392
70 2/8	13 2/8	13 0/8	6 2/8	6 1/8	9 5/8	San Juan County	UT	Ronald Kirk	1992	392
70 2/8	13 2/8	13 1/8	5 7/8	6 0/8	9 6/8	Moffat County	CO	Rudy Meyers	1994	392
70 2/8	13 2/8	13 4/8	6 0/8	6 2/8	12 0/8	Moffat County	CO	Ken Hoehn	1995	392
70 2/8	14 2/8	14 0/8	5 7/8	5 6/8	10 5/8	Moffat County	CO	Mark Montgomery	1995	392
70 2/8	14 1/8	13 7/8	5 7/8	5 7/8	10 1/8	Rosebud County	MT	David Burrows	1995	392
70 2/8	13 3/8	13 2/8	6 3/8	6 0/8	8 3/8	Moffat County	CO	Connie Renfro	1996	392
70 2/8	13 1/8	13 1/8	6 0/8	5 6/8	9 6/8	Twin Falls County	ID	Darrell Nunez	1997	392
70 2/8	14 2/8	12 5/8	6 7/8	6 6/8	11 6/8	Converse County	WY	Larry Comer	1998	392
70 2/8	13 0/8	13 5/8	6 2/8	6 1/8	9 7/8	Sublette County	WY	Bennett McMillian	2001	392
70 2/8	13 5/8	13 5/8	5 7/8	5 7/8	8 5/8	Gallatin County	MT	Bob Morton	2003	392
70 2/8	13 4/8	13 4/8	5 5/8	5 5/8	10 5/8	Harding County	SD	Jerry Notch	2005	392
70 2/8	13 0/8	13 1/8	5 6/8	5 5/8	11 4/8	Las Animas County	CO	Michele Eichler	2006	392
70 2/8	13 7/8	13 5/8	5 6/8	5 6/8	7 2/8	Converse County	WY	John E. Wencley	2008	392
70 2/8	14 3/8	14 1/8	5 7/8	5 7/8	10 6/8	Converse County	WY	Ronald Rockwell	2009	392
70 2/8	13 1/8	13 1/8	5 4/8	5 4/8	7 7/8	Campbell County	WY	Jerry Notch	2011	392
70 0/8	13 1/8	13 1/8	6 1/8	6 1/8	13 1/8	Butte County	SD	Wayne Wanhanen	1961	418
70 0/8	13 6/8	13 7/8	5 6/8	5 6/8	11 4/8	Garfield County	MT	Paul M. Ramsey	1963	418
70 0/8	14 0/8	14 0/8	5 3/8	5 3/8	9 5/8	Harding County	SD	Ira Hilburn	1964	418
70 0/8	13 4/8	13 2/8	6 1/8	6 0/8	12 5/8	Pennington County	SD	Thomas Huitfeldt	1974	418
70 0/8	13 1/8	13 2/8	5 7/8	5 6/8	8 2/8	Dawes County	NE	Allan Mintken	1974	418
70 0/8	13 0/8	13 0/8	5 7/8	5 6/8	12 4/8	Converse County	WY	Steven A. Wolff	1982	418
70 0/8	13 6/8	14 1/8	5 7/8	5 7/8	8 1/8	Carbon County	WY	Ron Breitsprecher	1982	418
70 0/8	14 5/8	14 0/8	5 6/8	5 6/8	5 7/8	Moffat County	CO	Dale Drilling	1984	418
70 0/8	13 4/8	13 4/8	5 5/8	5 4/8	9 4/8	Moffat County	CO	Todd Clyncke	1984	418
70 0/8	14 3/8	14 3/8	5 3/8	5 1/8	11 2/8	Moffat County	CO	Glenn Pritchard	1988	418
70 0/8	15 3/8	14 6/8	5 7/8	5 5/8	8 6/8	Custer County	ID	Matt March, Jr.	1989	418
70 0/8	13 3/8	13 4/8	5 5/8	5 4/8	8 6/8	Butte County	SD	Bryce Lambley	1991	418
70 0/8	12 6/8	12 6/8	6 2/8	6 4/8	9 1/8	Campbell County	WY	Robert A. Carman	1991	418
70 0/8	14 6/8	15 1/8	5 5/8	5 4/8	7 6/8	Moffat County	CO	Steve Barnhill	1992	418
70 0/8	14 2/8	13 5/8	5 6/8	5 6/8	11 3/8	Torrance County	NM	Dennis A. Muirhead	1993	418
70 0/8	13 3/8	13 4/8	6 1/8	6 1/8	10 5/8	Sioux County	NE	Michael Judas	1994	418
70 0/8	13 1/8	12 5/8	6 1/8	6 1/8	12 4/8	Weld County	CO	Timothy Bradley	1995	418
70 0/8	13 6/8	13 6/8	5 2/8	5 2/8	8 4/8	Moffat County	CO	Thane Anderson	1995	418
70 0/8	14 2/8	14 2/8	5 3/8	5 4/8	9 3/8	Crook County	WY	Kevin Vail	1999	418
70 0/8	12 7/8	13 0/8	6 0/8	5 7/8	8 1/8	Converse County	WY	Pat Lefemine	2000	418
70 0/8	14 1/8	14 0/8	5 4/8	5 4/8	8 1/8	Petroleum County	MT	John H. Holt	2007	418
70 0/8	14 0/8	13 7/8	6 0/8	5 7/8	9 4/8	Natrona County	WY	Donna Turay	2008	418
70 0/8	13 6/8	14 0/8	5 6/8	5 6/8	11 4/8	Campbell County	WY	Irvin Prough	2008	418
70 0/8	13 5/8	13 5/8	5 7/8	5 4/8	10 0/8	Converse County	WY	Jerry Becay	2009	418

PRONGHORN

Minimum Score 67 *Continued*

Score	Length of R Horn	L	Circumference R of Base	L	Inside Spread	Area	State/ Province	Hunter's Name	Date	Rank
69 6/8	12 6/8	12 6/8	6 6/8	6 6/8	11 4/8	Custer County	ID	Dr. Richard Hagerman	1966	442
69 6/8	13 4/8	13 4/8	6 2/8	6 2/8	10 5/8	Carbon County	WY	Maurice Savora	1972	442
69 6/8	13 6/8	13 5/8	5 6/8	5 6/8	9 0/8	Converse County	WY	Edward Pitchkites	1973	442
69 6/8	14 6/8	14 6/8	5 2/8	5 2/8	10 2/8	Perkins County	SD	Marvin R. Bohnet	1974	442
69 6/8	13 5/8	13 6/8	6 0/8	6 0/8	15 4/8	Sweetwater County	WY	Bill Clink	1985	442
69 6/8	14 1/8	14 4/8	5 6/8	5 4/8	13 7/8	Fergus County	MT	Kelly Norskog	1990	442
69 6/8	14 0/8	14 0/8	5 5/8	5 3/8	13 6/8	Converse County	WY	Phil Perry	1994	442
69 6/8	13 5/8	13 4/8	6 0/8	5 7/8	10 2/8	Converse County	WY	Ronald D. Stoynoff	1994	442
69 6/8	12 7/8	12 5/8	6 2/8	6 2/8	9 0/8	Converse County	WY	Ray Maudsley	1995	442
69 6/8	12 6/8	12 6/8	5 5/8	5 4/8	13 6/8	Natrona County	WY	James Mathias	1996	442
69 6/8	11 7/8	12 1/8	5 6/8	5 5/8	10 4/8	Converse County	WY	Eddie Wynne	1997	442
69 6/8	13 3/8	13 0/8	6 0/8	5 7/8	11 0/8	Converse County	WY	Jeff Fitts	2000	442
69 6/8	13 2/8	13 0/8	5 5/8	5 5/8	9 7/8	Powder River County	MT	Gary D. Sumwalt	2004	442
69 6/8	14 2/8	14 3/8	5 6/8	5 5/8	8 5/8	Lemhi County	ID	Ben L. Fahnholz	2006	442
69 6/8	13 5/8	13 3/8	5 6/8	5 6/8	7 2/8	Powder River County	MT	Martin E. Cain	2007	442
69 6/8	12 7/8	13 1/8	5 5/8	5 3/8	6 6/8	Converse County	WY	Ronald Rockwell	2008	442
69 6/8	13 7/8	14 2/8	5 3/8	5 2/8	9 3/8	Johnson County	WY	Allen Clark	2012	442
69 4/8	15 0/8	13 6/8	5 4/8	5 4/8	5 6/8	Rosebud County	MT	Glenn Gibson	1958	459
69 4/8	14 4/8	14 2/8	5 2/8	5 2/8	10 6/8	Grant County	NM	Harold W. Groves	1960	459
69 4/8	13 2/8	13 0/8	6 0/8	6 0/8	10 6/8	Natrona County	WY	John Benetti	1973	459
69 4/8	14 3/8	14 4/8	5 5/8	5 3/8	8 4/8	Park County	CO	Ed Zehner	1974	459
69 4/8	13 5/8	14 3/8	6 1/8	5 7/8	11 5/8	Converse County	WY	Dr. James R. Scott	1981	459
69 4/8	13 6/8	14 7/8	5 7/8	5 7/8	11 6/8	Sweetwater County	WY	Glenn E. Hisey	1985	459
69 4/8	13 2/8	13 2/8	5 7/8	5 6/8	11 4/8	Converse County	WY	John A. Driver	1987	459
69 4/8	12 7/8	12 6/8	6 4/8	6 2/8	6 0/8	Moffat County	CO	Dale Drilling	1988	459
69 4/8	12 5/8	12 7/8	6 0/8	5 7/8	11 0/8	Albany County	WY	William Zahradka	1988	459
69 4/8	12 2/8	12 1/8	6 5/8	6 4/8	11 6/8	Natrona County	WY	Steve Turck	1988	459
69 4/8	13 0/8	13 0/8	5 7/8	5 6/8	11 3/8	Rosebud County	MT	Gene Welle	1989	459
69 4/8	13 4/8	13 2/8	5 6/8	5 5/8	10 4/8	Moffat County	CO	Eugene Ray, Sr.	1990	459
69 4/8	13 3/8	13 2/8	5 7/8	5 6/8	11 3/8	Converse County	WY	Rodney L. Hamann	1992	459
69 4/8	12 5/8	12 5/8	6 4/8	6 4/8	8 5/8	Moffat County	CO	Herbert M. Groetsch	1993	459
69 4/8	13 4/8	13 5/8	6 1/8	6 0/8	15 4/8	Natrona County	WY	Jim Gaffney	1993	459
69 4/8	13 2/8	13 0/8	5 4/8	5 4/8	9 6/8	Weld County	CO	Douglas P. Douthit	1994	459
69 4/8	13 7/8	13 6/8	5 3/8	5 3/8	8 7/8	Moffat County	CO	Todd J. Rider	1994	459
69 4/8	14 2/8	14 2/8	5 4/8	5 5/8	11 2/8	Moffat County	CO	Reggie Spiegelberg	1994	459
69 4/8	12 5/8	13 3/8	6 3/8	6 3/8	8 4/8	Albany County	WY	T. J. Conrads	1996	459
69 4/8	13 1/8	13 3/8	5 1/8	5 2/8	12 2/8	Carbon County	WY	Kevin Anderson	1996	459
69 4/8	13 5/8	13 6/8	5 5/8	5 4/8	9 5/8	Weld County	CO	Dwight D. Pochant	2000	459
69 4/8	14 1/8	14 2/8	5 3/8	5 1/8	10 0/8	Musselshell County	MT	Craig Richardson	2000	459
69 4/8	13 4/8	13 4/8	5 5/8	5 5/8	8 0/8	Converse County	WY	Gary Bauer	2001	459
69 4/8	13 7/8	13 6/8	5 4/8	5 4/8	7 2/8	Moffat County	CO	Tom Hentrick	2002	459
69 4/8	14 1/8	13 1/8	5 7/8	5 5/8	9 2/8	Converse County	WY	William G. Mason	2004	459
69 4/8	12 7/8	13 0/8	5 7/8	6 0/8	8 7/8	Converse County	WY	William G. Mason	2006	459
69 4/8	13 4/8	13 5/8	5 4/8	5 4/8	8 5/8	Albany County	WY	Mike Yancey	2007	459
69 4/8	12 6/8	12 6/8	6 0/8	5 7/8	8 1/8	Converse County	WY	Ed Hendricks	2007	459
69 4/8	13 0/8	12 7/8	6 3/8	6 3/8	8 7/8	Carter County	MT	Brock C. Atwood	2009	459
69 4/8	14 2/8	14 2/8	5 3/8	5 2/8	9 2/8	Natrona County	WY	Alfred J. Gemrich	2009	459
69 4/8	12 6/8	12 4/8	6 0/8	6 0/8	9 0/8	Converse County	WY	Ronald Rockwell	2012	459
69 2/8	14 2/8	13 7/8	5 4/8	5 4/8	11 6/8	Custer County	ID	Jack Edwards	1960	490
69 2/8	13 5/8	13 4/8	5 3/8	5 2/8	6 2/8	Haakon County	SD	Floyd Hauk	1966	490
69 2/8	12 7/8	12 7/8	5 3/8	5 5/8	11 7/8	Converse County	WY	Dr. James L. Emerson	1980	490
69 2/8	14 3/8	13 7/8	5 7/8	5 6/8	10 4/8	Valencia County	NM	Reggie Spiegelberg	1981	490
69 2/8	14 2/8	13 7/8	6 0/8	6 0/8	8 1/8	Natrona County	WY	Joe M. Skipp	1983	490
69 2/8	14 5/8	15 0/8	6 4/8	6 1/8	12 2/8	Sweetwater County	WY	Mike Barrett	1989	490
69 2/8	14 1/8	14 1/8	6 0/8	5 7/8	7 2/8	Converse County	WY	Wayne Radley	1991	490
69 2/8	14 0/8	14 1/8	5 7/8	6 1/8	8 2/8	Lassen County	CA	Randy Jarvis	1992	490
69 2/8	12 7/8	13 1/8	6 6/8	6 6/8	13 2/8	Sweetwater County	WY	Bill Clink	1992	490
69 2/8	13 7/8	13 7/8	5 2/8	5 2/8	12 7/8	Billings County	ND	Douglas A. Baumiller	1994	490
69 2/8	14 6/8	14 5/8	5 1/8	5 1/8	9 1/8	Rosebud County	MT	John Richardson	1995	490
69 2/8	14 2/8	13 5/8	5 4/8	5 4/8	9 5/8	Sweetwater County	WY	Larry Cross	1997	490
69 2/8	13 1/8	13 1/8	5 4/8	5 5/8	8 5/8	Weld County	CO	Joseph Bradley	2002	490
69 2/8	13 6/8	13 6/8	5 4/8	5 4/8	9 2/8	Gallatin County	MT	Bob Morton	2005	490
69 2/8	13 0/8	13 0/8	5 5/8	5 5/8	11 3/8	Rosebud County	MT	Mike Murphey	2007	490

PRONGHORN

Minimum Score 67 | *Continued*

Score	Length of R Horn	L	Circumference R of Base	L	Inside Spread	Area	State/ Province	Hunter's Name	Date	Rank
69 2/8	14 0/8	14 4/8	5 5/8	5 4/8	12 1/8	Converse County	WY	Ed Defibaugh	2008	490
69 2/8	14 1/8	13 5/8	5 3/8	5 3/8	9 0/8	Custer County	MT	Dennis Dunn	2008	490
69 2/8	13 3/8	13 2/8	5 3/8	5 3/8	11 1/8	Niobrara County	WY	Charles A. Myers	2009	490
69 2/8	14 1/8	14 5/8	5 5/8	5 5/8	8 5/8	Natrona County	WY	Steve Turay	2010	490
69 0/8	12 7/8	13 0/8	5 6/8	5 7/8	8 4/8	Fall River County	SD	Francis R. Tovar	1968	509
69 0/8	14 0/8	14 0/8	5 3/8	5 4/8	7 1/8	Natrona County	WY	Ricky A. Wall	1987	509
69 0/8	12 7/8	12 5/8	5 6/8	5 6/8	9 0/8	Albany County	WY	Gene Welle	1989	509
69 0/8	13 3/8	13 1/8	5 4/8	5 3/8	9 1/8	Moffat County	CO	John Morris	1990	509
69 0/8	13 5/8	13 3/8	6 0/8	6 0/8	11 1/8	Natrona County	WY	Tony Zirkelbach	1990	509
69 0/8	14 4/8	14 2/8	5 5/8	5 5/8	9 1/8	Petroleum County	MT	Mark D. Hughes	1991	509
69 0/8	13 5/8	13 7/8	5 5/8	5 5/8	10 7/8	Petroleum County	MT	Mark D. Hughes	1992	509
69 0/8	12 0/8	12 0/8	6 0/8	6 0/8	12 1/8	Converse County	WY	G. William Buxton	1993	509
69 0/8	12 7/8	13 0/8	6 0/8	6 0/8	7 7/8	Moffat County	CO	Jim Shanks	1994	509
69 0/8	12 7/8	12 6/8	6 2/8	6 3/8	9 4/8	Moffat County	CO	Dave Culter	1995	509
69 0/8	12 7/8	12 5/8	5 7/8	6 0/8	12 2/8	Big Horn County	WY	Dave Moss	1995	509
69 0/8	13 4/8	13 4/8	5 4/8	5 4/8	7 5/8	Rosebud County	MT	Kent Kaufman	1995	509
69 0/8	13 6/8	13 6/8	5 6/8	5 4/8	10 2/8	Madison County	MT	James J. Kilmer	2007	509
68 6/8	14 7/8	14 7/8	5 4/8	5 4/8	12 0/8	Sweetwater County	WY	Dr. Fred Mack	1960	522
68 6/8	13 7/8	14 2/8	5 6/8	5 4/8	7 7/8	Haakon County	SD	William Nankivel	1970	522
68 6/8	13 2/8	13 4/8	6 1/8	6 1/8	7 4/8	Gallatin County	MT	Robert Savage	1971	522
68 6/8	12 3/8	12 4/8	6 4/8	6 2/8	11 0/8	Sweetwater County	WY	Keith Dana	1978	522
68 6/8	14 0/8	13 6/8	6 3/8	6 2/8	14 6/8	Converse County	WY	Frank Moore	1985	522
68 6/8	13 5/8	14 0/8	5 5/8	5 4/8	10 1/8	Buffalo	ALB	Orest Popil	1988	522
68 6/8	13 2/8	13 1/8	5 4/8	5 5/8	7 2/8	Natrona County	WY	R. Ray Wix	1990	522
68 6/8	12 4/8	12 3/8	5 5/8	5 5/8	13 0/8	Natrona County	WY	James V. Siebels	1990	522
68 6/8	13 1/8	13 0/8	5 7/8	5 7/8	9 1/8	Converse County	WY	Phil Perry	1991	522
68 6/8	13 3/8	13 7/8	6 0/8	6 1/8	11 2/8	Fergus County	MT	Chris G. Sanford	1991	522
68 6/8	13 6/8	13 5/8	5 6/8	5 5/8	8 0/8	Washoe County	NV	Anthony L. Mudd	1992	522
68 6/8	13 1/8	13 0/8	5 1/8	5 1/8	8 1/8	Moffat County	CO	David Joyce	1992	522
68 6/8	14 4/8	14 2/8	5 4/8	5 4/8	8 3/8	Moffat County	CO	J. Keith Chastain	1994	522
68 6/8	11 4/8	11 4/8	6 4/8	6 6/8	11 6/8	Yavapai County	AZ	Don Parks, Jr.	1995	522
68 6/8	12 5/8	13 2/8	5 4/8	5 4/8	10 0/8	Natrona County	WY	Ray Wix	1996	522
68 6/8	13 3/8	13 5/8	5 4/8	5 5/8	8 5/8	Converse County	WY	William G. Mason	1996	522
68 6/8	12 7/8	12 4/8	5 7/8	5 7/8	10 6/8	Elbert County	CO	Jerry Becay	1996	522
68 6/8	13 7/8	13 7/8	5 2/8	5 2/8	9 3/8	Converse County	WY	Greg Dearth	1999	522
68 6/8	13 6/8	13 3/8	6 6/8	6 2/8	9 3/8	Converse County	WY	John Croom	1999	522
68 6/8	12 2/8	12 1/8	5 6/8	5 6/8	9 3/8	Converse County	WY	Robert Sihlanick	2005	522
68 6/8	14 5/8	14 5/8	5 1/8	5 2/8	13 0/8	Rosebud County	MT	Lynn Walker	2005	522
68 6/8	12 0/8	11 7/8	6 1/8	6 2/8	6 4/8	Alamosa County	CO	Joe Handorf	2008	522
68 4/8	13 4/8	13 0/8	5 4/8	5 5/8	8 6/8	Tripp County	SD	Spike Jorgensen	1965	544
68 4/8	14 1/8	14 1/8	5 2/8	5 1/8	12 3/8	Lincoln County	NM	James H. Simmons	1966	544
68 4/8	12 6/8	12 2/8	6 5/8	6 5/8	12 1/8	McLean County	ND	Tom O'Connell	1970	544
68 4/8	12 6/8	12 7/8	6 0/8	6 0/8	9 4/8	Billings County	ND	Dean Nevland	1975	544
68 4/8	13 4/8	13 5/8	5 3/8	5 3/8	13 3/8	Converse County	WY	Kent Brigham	1982	544
68 4/8	13 6/8	13 6/8	5 6/8	5 4/8	11 2/8	Converse County	WY	Marty Horn	1984	544
68 4/8	12 0/8	11 5/8	5 5/8	5 5/8	9 6/8	Converse County	WY	Steve Woodman	1985	544
68 4/8	13 0/8	13 0/8	5 6/8	5 7/8	10 7/8	Campbell County	WY	Donald Ace Morgan	1986	544
68 4/8	13 5/8	13 6/8	5 5/8	5 5/8	11 1/8	Converse County	WY	Mick Cochrane	1988	544
68 4/8	12 7/8	12 7/8	5 5/8	5 5/8	9 4/8	Campbell County	WY	Dave Hinton	1993	544
68 4/8	12 7/8	12 6/8	6 0/8	6 0/8	9 6/8	Converse County	WY	V. Gene Mathias	1996	544
68 4/8	13 3/8	13 2/8	5 5/8	5 4/8	9 1/8	Owyhee County	ID	Michael W. Schlegel	1999	544
68 4/8	13 1/8	13 2/8	5 2/8	5 2/8	9 0/8	Uintah County	UT	Darrell Aplanalp	2000	544
68 4/8	12 5/8	12 4/8	5 7/8	5 5/8	10 5/8	Converse County	WY	Bob Homberger	2001	544
68 4/8	12 6/8	12 6/8	5 6/8	5 5/8	8 5/8	Campbell County	WY	William H. Peterson	2004	544
68 4/8	12 1/8	12 3/8	6 3/8	6 3/8	9 3/8	Moffat County	CO	Gary Whitten	2008	544
68 2/8	12 4/8	12 4/8	6 2/8	6 0/8	11 6/8	Moffat County	CO	Burl Duckworth	1958	560
68 2/8	12 6/8	12 6/8	5 7/8	5 5/8	10 5/8	Harding County	SD	Chet Wohlhueter	1963	560
68 2/8	13 5/8	13 3/8	5 3/8	5 2/8	8 7/8	Morton County	ND	Fred F. Heer	1973	560
68 2/8	13 2/8	13 2/8	5 6/8	6 0/8	9 1/8	Sweetwater County	WY	Charles Bartlett	1978	560
68 2/8	13 6/8	14 2/8	5 5/8	5 4/8	7 2/8	Lemhi County	ID	Richard Dewey	1978	560
68 2/8	12 6/8	12 6/8	5 4/8	5 5/8	10 0/8	Natrona County	WY	R. G. Williams	1981	560
68 2/8	12 3/8	12 4/8	5 6/8	5 6/8	8 0/8	Moffat County	CO	Barry J. Smith	1981	560
68 2/8	13 1/8	13 3/8	5 6/8	5 6/8	11 3/8	Carbon County	WY	Ron Breitsprecher	1984	560

Minimum Score 67 *Continued*

Score	Length of R Horn	L	Circumference R of Base	L	Inside Spread	Area	State/ Province	Hunter's Name	Date	Rank
68 2/8	13 1/8	13 2/8	5 7/8	5 7/8	8 4/8	Jenner	ALB	Jack Kempf	1987	560
68 2/8	13 3/8	13 4/8	5 4/8	5 4/8	10 3/8	Dawson County	MT	Dave Athas	1987	560
68 2/8	12 7/8	12 7/8	5 5/8	5 6/8	14 5/8	McKenzie County	ND	Wayne R. Streitz	1989	560
68 2/8	13 4/8	13 5/8	5 3/8	5 2/8	7 7/8	Converse County	WY	Ken Mamazzo	1993	560
68 2/8	14 1/8	14 0/8	5 3/8	5 4/8	10 0/8	Washoe County	NV	Clayton D. Johnson	1994	560
68 2/8	13 6/8	14 1/8	5 6/8	5 5/8	8 2/8	Petroleum County	MT	Craig Richardson	1994	560
68 2/8	14 2/8	14 6/8	5 4/8	5 3/8	8 1/8	Montrose County	CO	Dave Emken	1995	560
68 2/8	12 3/8	12 4/8	6 0/8	6 0/8	8 4/8	Moffat County	CO	Eric A. Sawyer	1996	560
68 2/8	12 5/8	12 5/8	5 5/8	5 6/8	8 3/8	Powder River County	MT	John Martinson	1997	560
68 2/8	12 7/8	12 7/8	5 7/8	5 7/8	9 7/8	Weld County	CO	Barry J. Smith	1998	560
68 2/8	12 6/8	12 2/8	6 0/8	5 7/8	9 4/8	Converse County	WY	Michael Travis	1998	560
68 0/8	14 1/8	13 0/8	5 4/8	5 6/8	7 7/8	Campbell County	WY	Reinhold L. Lind	1961	579
68 0/8	13 4/8	13 4/8	5 4/8	5 4/8	7 5/8	Carbon County	WY	Bill Cunningham	1963	579
68 0/8	14 0/8	13 6/8	5 2/8	5 3/8	8 0/8	Morton County	ND	Roy D. Russell, Jr.	1967	579
68 0/8	13 0/8	13 0/8	5 7/8	5 7/8	9 6/8	Saguache County	CO	David Scheid	1974	579
68 0/8	11 4/8	11 3/8	5 7/8	5 6/8	7 3/8	Burke County	ND	Allen L. Nelson	1974	579
68 0/8	12 3/8	12 5/8	5 4/8	5 4/8	11 5/8	Natrona County	WY	David Bouchard	1985	579
68 0/8	12 6/8	12 6/8	6 1/8	6 1/8	8 6/8	Converse County	WY	Dan Bertalan	1987	579
68 0/8	13 7/8	13 6/8	5 5/8	5 3/8	11 0/8	Washoe County	NV	Bill Fuller	1988	579
68 0/8	12 7/8	12 7/8	5 6/8	5 5/8	9 2/8	Garfield County	MT	Bruce Balerud	1989	579
68 0/8	12 3/8	12 3/8	5 6/8	5 6/8	10 2/8	Moffat County	CO	Mike Lamade	1990	579
68 0/8	12 3/8	12 5/8	6 0/8	6 0/8	9 1/8	Moffat County	CO	Ronald Dinger	1991	579
68 0/8	13 3/8	13 5/8	5 6/8	5 6/8	8 4/8	Bowman County	ND	Kendall Bauer	1991	579
68 0/8	13 1/8	13 0/8	6 0/8	5 6/8	9 3/8	Sioux County	NE	Alton Schroeder	1991	579
68 0/8	12 4/8	12 2/8	6 2/8	6 3/8	9 2/8	Converse County	WY	Jeff Reynolds	1991	579
68 0/8	13 1/8	13 2/8	5 4/8	5 4/8	8 0/8	Albany County	WY	Orvie Linsin	1994	579
68 0/8	13 7/8	13 7/8	5 2/8	5 3/8	9 6/8	Converse County	WY	Mark Hockenberry	1994	579
68 0/8	13 3/8	13 2/8	5 3/8	5 4/8	11 5/8	Owyhee County	ID	Anthony L. Mudd	1996	579
68 0/8	12 5/8	12 5/8	5 7/8	5 7/8	11 1/8	Campbell County	WY	Ritch A. Stolpe	1996	579
68 0/8	13 3/8	13 2/8	5 7/8	5 7/8	13 1/8	Lake County	OR	Jeffrey A. Eder	2001	579
68 0/8	12 3/8	12 3/8	6 6/8	7 0/8	12 3/8	Converse County	WY	Brian Tessmann	2003	579
68 0/8	12 5/8	12 7/8	5 7/8	5 6/8	9 6/8	Converse County	WY	David Wildenstein	2004	579
68 0/8	12 1/8	12 1/8	6 2/8	6 2/8	9 3/8	Powder River County	MT	John Martinson	2005	579
68 0/8	12 6/8	12 6/8	5 4/8	5 4/8	9 5/8	Converse County	WY	Mike Sohm	2006	579
68 0/8	13 0/8	13 1/8	5 5/8	5 4/8	8 7/8	Converse County	WY	Doug Baily	2006	579
68 0/8	13 3/8	13 2/8	6 1/8	6 1/8	10 7/8	Rosebud County	MT	Sjon Zunich	2007	579
68 0/8	12 4/8	12 1/8	6 3/8	6 1/8	7 2/8	Converse County	WY	Weston Giesey	2008	579
68 0/8	13 0/8	12 3/8	6 2/8	6 0/8	8 2/8	Weld County	CO	Michael Casey	2008	579
68 0/8	13 7/8	13 5/8	5 4/8	5 3/8	9 2/8	Broadwater County	MT	Jaden Zambon	2009	579
67 6/8	12 3/8	12 4/8	5 6/8	5 5/8	9 0/8	McHenry County	ND	Darryl Ablestad	1967	607
67 6/8	13 4/8	13 4/8	5 5/8	5 5/8	11 2/8	Logan County	CO	Loren Johnston	1968	607
67 6/8	12 6/8	12 4/8	6 2/8	6 0/8	11 3/8	McHenry County	ND	Jim Budeau	1968	607
67 6/8	13 3/8	13 2/8	5 2/8	5 5/8	10 1/8	Fremont County	WY	Doris Clark	1970	607
67 6/8	14 0/8	13 1/8	5 0/8	5 0/8	13 4/8	Butte County	ID	Craig L. Hansen	1974	607
67 6/8	14 0/8	14 0/8	5 2/8	5 2/8	11 1/8	Park County	WY	Fred W. Achilles	1978	607
67 6/8	13 5/8	13 5/8	5 3/8	5 2/8	8 0/8	Beaverhead County	MT	L. C. Trimber	1982	607
67 6/8	13 6/8	13 2/8	5 4/8	5 5/8	9 1/8	Moffat County	CO	Carol Ashurst	1988	607
67 6/8	13 1/8	13 1/8	6 0/8	6 0/8	9 2/8	Bowman County	ND	Gene Welle	1992	607
67 6/8	12 6/8	12 5/8	5 5/8	5 4/8	7 6/8	Converse County	WY	Dr. Andy Jones	1994	607
67 6/8	12 4/8	12 2/8	5 6/8	5 6/8	10 6/8	Moffat County	CO	Nathan Andersohn	1994	607
67 6/8	12 7/8	12 7/8	6 2/8	6 2/8	11 6/8	Rosebud County	MT	Gene Welle	1995	607
67 6/8	12 1/8	12 2/8	5 5/8	6 0/8	10 4/8	Rosebud County	MT	Bill Dollar	1995	607
67 6/8	13 0/8	13 0/8	5 6/8	5 5/8	6 7/8	Hays	ALB	Marty Belisle	1998	607
67 6/8	14 7/8	14 3/8	5 5/8	5 5/8	9 1/8	Rosebud County	MT	William E. Lee, Jr.	1998	607
67 6/8	12 0/8	12 1/8	6 2/8	6 2/8	11 1/8	Converse County	WY	Randy E. Doyle	1999	607
67 6/8	13 7/8	13 1/8	5 5/8	5 3/8	9 0/8	Weld County	CO	Kevin Waller	2002	607
67 6/8	14 1/8	13 3/8	6 1/8	6 1/8	12 1/8	Converse County	WY	Dan J. Ryle, Jr.	2005	607
67 4/8	13 4/8	13 5/8	5 3/8	5 2/8	13 3/8	Cherry County	NE	Marlin Wells	1967	625
67 4/8	13 3/8	13 2/8	5 4/8	5 4/8	9 0/8	Meade County	SD	Jim Bohls	1974	625
67 4/8	13 5/8	13 4/8	5 4/8	5 4/8	10 1/8	Converse County	WY	Ron Breitsprecher	1981	625
67 4/8	12 3/8	12 4/8	6 0/8	5 7/8	9 0/8	Sweetwater County	WY	Craig Richardson	1984	625
67 4/8	12 7/8	12 6/8	5 3/8	5 2/8	9 0/8	Converse County	WY	Burt Thompson, Jr.	1985	625
67 4/8	13 3/8	13 2/8	5 2/8	5 2/8	8 4/8	Moffat County	CO	E. Damon Handley	1988	625

PRONGHORN

Minimum Score 67 *Continued*

Score	Length of R Horn	L	Circumference R of Base	L	Inside Spread	Area	State/ Province	Hunter's Name	Date	Rank
67 4/8	12 2/8	12 2/8	5 6/8	5 7/8	7 5/8	Converse County	WY	J. G. "Rusty" Watson	1988	625
67 4/8	13 1/8	13 1/8	6 0/8	6 0/8	8 5/8	Johnson County	WY	Charles B. Cureton	1993	625
67 4/8	11 7/8	12 0/8	5 6/8	5 7/8	13 2/8	Moffat County	CO	Dave Greenwalt	1994	625
67 4/8	13 4/8	14 4/8	5 6/8	5 6/8	10 2/8	McKenzie County	ND	Pat Weigel	1994	625
67 4/8	11 6/8	11 7/8	6 1/8	6 0/8	10 4/8	Weld County	CO	Justin Wenthe	1998	625
67 4/8	13 3/8	13 5/8	5 4/8	5 3/8	11 7/8	Converse County	WY	William R. Horn, Jr.	1998	625
67 4/8	13 3/8	13 1/8	5 5/8	5 4/8	10 3/8	Rosebud County	MT	Michael Babcock	2000	625
67 4/8	12 2/8	12 3/8	5 1/8	5 1/8	8 5/8	Moffat County	CO	William Abshagen	2001	625
67 4/8	12 0/8	12 1/8	6 0/8	6 0/8	8 5/8	Musselshell County	MT	Don Davidson	2006	625
67 4/8	13 5/8	13 5/8	5 4/8	5 3/8	9 7/8	Powder River County	MT	Martin E. Cain	2008	625
67 4/8	11 7/8	12 2/8	6 0/8	6 0/8	10 3/8	Converse County	WY	Brad Harriman	2010	625
67 4/8	12 3/8	12 5/8	5 5/8	5 5/8	7 2/8	Converse County	WY	Dave Canfield	2012	625
67 4/8	12 6/8	11 7/8	5 6/8	5 7/8	11 1/8	Campbell County	WY	Douglas B. Strecker	2012	625
67 2/8	12 1/8	12 1/8	6 0/8	5 5/8	11 6/8	Carbon County	WY	Dennis Behn	1974	644
67 2/8	12 7/8	13 0/8	5 2/8	5 2/8	6 4/8	Musselshell County	MT	Scott L. Koelzer	1976	644
67 2/8	12 5/8	12 3/8	5 3/8	5 4/8	11 7/8	Meade County	SD	Steve D. Krier	1983	644
67 2/8	13 1/8	13 1/8	4 5/8	4 7/8	11 7/8	Moffat County	CO	Glenn Pritchard	1984	644
67 2/8	12 0/8	12 1/8	5 6/8	5 5/8	7 7/8	Converse County	WY	Ron Rockwell	1984	644
67 2/8	13 3/8	12 6/8	5 4/8	5 4/8	7 6/8	Moffat County	CO	Kurt Keskimaki	1985	644
67 2/8	13 7/8	13 7/8	5 6/8	5 5/8	11 1/8	Converse County	WY	David Jerome	1985	644
67 2/8	12 7/8	13 0/8	6 2/8	6 2/8	7 7/8	Moffat County	CO	Ron Rockwell	1987	644
67 2/8	13 2/8	13 1/8	5 7/8	6 0/8	10 6/8	Petroleum County	MT	Leamon D. Ferrell	1988	644
67 2/8	13 1/8	13 4/8	5 6/8	5 5/8	7 2/8	Campbell County	WY	Robin D. Johnson	1989	644
67 2/8	12 0/8	12 7/8	6 2/8	6 1/8	7 6/8	Converse County	WY	Ed Defibaugh	1991	644
67 2/8	11 5/8	11 7/8	5 7/8	5 6/8	8 7/8	McKenzie County	ND	Mark D. Hughes	1991	644
67 2/8	12 3/8	12 1/8	5 6/8	5 6/8	7 2/8	Converse County	WY	Gene Mathias	1992	644
67 2/8	13 3/8	13 0/8	6 0/8	5 7/8	12 1/8	Converse County	WY	Randy A. Reeves	2001	644
67 2/8	13 2/8	13 2/8	5 7/8	5 3/8	8 5/8	Converse County	WY	William G. Mason	2002	644
67 2/8	13 2/8	13 4/8	6 0/8	6 0/8	7 0/8	Campbell County	WY	William Peterson	2003	644
67 2/8	12 5/8	12 6/8	5 5/8	5 5/8	11 4/8	Las Animas County	CO	Nancy L. Palmer	2004	644
67 2/8	12 4/8	13 2/8	5 6/8	5 6/8	11 2/8	Platte County	WY	Terry Krahn	2004	644
67 2/8	12 7/8	13 0/8	5 3/8	5 3/8	10 6/8	Las Animas County	CO	Lance Guerin	2004	644
67 2/8	12 1/8	12 0/8	5 6/8	6 0/8	9 2/8	Campbell County	WY	Gene A. Hall	2004	644
67 2/8	13 5/8	12 7/8	6 1/8	6 0/8	10 3/8	Washoe County	NV	Lew Webb	2005	644
67 2/8	11 2/8	11 1/8	6 1/8	6 0/8	11 0/8	Converse County	WY	Rick Liebel	2005	644
67 2/8	12 5/8	12 3/8	5 5/8	5 5/8	8 7/8	Converse County	WY	Robert W. Markowski	2010	644
67 0/8	12 7/8	12 7/8	5 6/8	5 4/8	10 2/8	Butte County	SD	David Lind	1961	667
67 0/8	13 0/8	12 4/8	6 0/8	6 0/8	10 4/8	Campbell County	WY	Gerald L. Egbert	1975	667
67 0/8	13 4/8	13 4/8	5 5/8	5 4/8	9 2/8	Moffat County	CO	Terry Weimer	1988	667
67 0/8	14 2/8	14 4/8	5 2/8	5 1/8	6 5/8	Converse County	WY	Jay Deones	1988	667
67 0/8	13 5/8	13 4/8	5 5/8	5 6/8	8 5/8	Eddy County	NM	Jess Stuart	1988	667
67 0/8	13 7/8	13 7/8	5 4/8	5 3/8	10 7/8	Moffat County	CO	Bob Radocy	1990	667
67 0/8	14 7/8	14 4/8	5 7/8	5 6/8	10 7/8	Rich County	UT	Robert K. Paulson	1990	667
67 0/8	12 2/8	12 1/8	5 6/8	5 7/8	9 1/8	Converse County	WY	Michael T. Wheeler	1994	667
67 0/8	11 2/8	11 0/8	6 3/8	6 2/8	10 7/8	Albany County	WY	Joseph Osvath	1996	667
67 0/8	13 3/8	13 2/8	5 6/8	5 6/8	10 5/8	Moffat County	CO	Todd Szmania	1996	667
67 0/8	13 1/8	13 6/8	5 5/8	5 5/8	11 6/8	Lake County	OR	Jeffrey A. Eder	1997	667
67 0/8	13 4/8	14 0/8	5 2/8	5 3/8	9 4/8	Converse County	WY	Ray McFarland	2000	667
67 0/8	12 5/8	12 5/8	5 6/8	5 5/8	9 2/8	Converse County	WY	Christopher C. Hill	2001	667
67 0/8	11 1/8	11 5/8	6 0/8	6 0/8	6 4/8	Converse County	WY	Steven R. Davis	2004	667
67 0/8	13 0/8	11 6/8	6 0/8	6 0/8	9 3/8	Converse County	WY	Michael J. Clevinger	2006	667
67 0/8	14 1/8	14 2/8	4 7/8	5 0/8	6 6/8	Converse County	WY	Phil Layman	2008	667
67 0/8	13 4/8	13 3/8	5 3/8	5 2/8	8 3/8	Converse County	WY	Ralph Fleegle	2010	667
67 0/8	13 2/8	13 1/8	5 2/8	5 3/8	9 4/8	Converse County	WY	Samuel Morello	2011	667

ROCKY MOUNTAIN GOAT

Minimum Score 40 *Oreamnos americanus americanus* and related subspecies

Score	Length of R Horn	L	Circumference R of Base	L	Greatest Spread	Sex	Area	State/ Province	Hunter's Name	Date	Rank
52 4/8	11 3/8	11 2/8	5 7/8	5 7/8	8 1/8	M	Park County	CO	Lyle K. Willmarth	1988	1
51 6/8	10 5/8	10 5/8	5 3/8	5 4/8	6 0/8	M	Cleveland Peninsula	AK	Kurt Kuehl	2000	2
50 2/8	10 7/8	10 6/8	5 3/8	5 3/8	5 6/8	M	Ketchikan	AK	Kurt Kuehl	1989	3
50 0/8	9 7/8	9 6/8	5 5/8	5 5/8	7 1/8	M	Terrace	BC	John H. Holt	2008	4
49 6/8	9 1/8	9 1/8	5 6/8	5 6/8	7 6/8	M	Tesla Lake	BC	Peter Halbig	1970	5
49 6/8	9 5/8	9 6/8	5 5/8	5 5/8	6 2/8	M	Gallatin County	MT	Clark Kelly III	1990	5
49 4/8	9 7/8	10 0/8	5 6/8	5 6/8	7 1/8	M	Baker Mtn.	BC	David A. Shupp	2006	7
49 4/8	9 5/8	10 1/8	5 6/8	5 5/8	7 0/8	M	Weber County	UT	James Eeckhout	2012	7
49 2/8	9 7/8	9 6/8	5 5/8	5 4/8	7 6/8	M	King County	WA	Jerry Solie	1978	9
49 2/8	10 3/8	10 1/8	5 3/8	5 3/8	6 7/8	M	Clear Creek County	CO	Gayle Lippold	1994	9
49 0/8	9 0/8	9 0/8	5 6/8	5 6/8	6 7/8	M	Bennett Lake	BC	Jack Stephen	1982	11
49 0/8	10 2/8	10 3/8	5 3/8	5 3/8	6 0/8	M	Cleveland Peninsula	AK	Kurt Kuehl	1990	11
49 0/8	10 3/8	10 3/8	5 4/8	5 4/8	6 2/8	M	Nakusp	BC	Harvey J. Surina	1992	11
48 6/8	10 2/8	10 0/8	5 4/8	5 4/8	8 5/8	M	Taku River	BC	Andrew Gamatko	2001	14
48 4/8	9 5/8	9 5/8	5 5/8	5 5/8	7 3/8	M	Clallam County	WA	Dr. Charles F. Raab	1967	15
48 2/8	9 3/8	9 4/8	5 6/8	5 5/8	6 6/8	M	Seebe	ALB	Chris Kroll	1962	16
48 2/8	9 3/8	9 2/8	5 6/8	5 5/8	7 2/8	M	Crown Mtn.	AK	Harold W. Jacobson	1973	16
48 2/8	9 7/8	9 2/8	5 7/8	5 7/8	6 4/8	M	Chaffee County	CO	Marvin Clyncke	1978	16
48 2/8	9 6/8	9 6/8	5 3/8	5 3/8	5 6/8	M	Chaffee County	CO	Daniel G. Willems	1999	16
47 6/8	9 2/8	9 2/8	5 5/8	5 4/8	7 0/8	M	Terminus Mtn.	BC	Paul P. Schafer	1975	20
47 4/8	9 4/8	9 4/8	5 2/8	5 2/8	6 7/8	M	Dease Lake	BC	Don Davidson	1993	21
47 2/8	9 4/8	9 3/8	5 3/8	5 4/8	6 2/8	M	Kennedy Springs	MT	Don Leondorf	1964	22
47 2/8	8 6/8	8 7/8	5 3/8	5 4/8	6 2/8	M	Kittitas County	WA	Arnold L. Deckwa	1969	22
47 2/8	9 2/8	9 2/8	6 2/8	6 2/8	7 5/8	M	Kenai Peninsula	AK	John Moline	1971	22
47 2/8	9 5/8	9 5/8	5 2/8	5 2/8	7 0/8	M	Alsek River	AK	F. Wyatt Cook	1976	22
47 0/8	9 3/8	9 3/8	5 3/8	5 4/8	7 2/8	M	Olympic Peninsula	WA	William V. Mishler	1968	26
47 0/8	9 5/8	9 0/8	5 6/8	5 6/8	6 6/8	M	La Plata County	CO	John L. Gardner	1999	26
46 6/8	9 2/8	9 2/8	5 3/8	5 4/8	6 1/8	M	Haines	AK	Lowell Marylin	1962	28
46 6/8	9 7/8	9 4/8	5 2/8	5 2/8	6 4/8	M	Bonneville County	ID	Coby Tigert	1989	28
46 6/8	8 3/8	8 4/8	5 3/8	5 3/8	6 4/8	M	Todagin Mtn.	BC	Len Cardinale	1990	28
46 4/8	9 1/8	8 7/8	5 3/8	5 3/8	7 1/8	M	Clallam County	WA	Thos. J. Smith	1969	31
46 4/8	9 6/8	9 0/8	5 4/8	5 4/8	7 2/8	M	Day Harbor	AK	William L. Ruby	1970	31
46 4/8	10 0/8	9 7/8	5 2/8	5 1/8	7 4/8	M	Thuodadi Lake	BC	Gary Petee	1979	31
46 4/8	9 3/8	9 4/8	5 2/8	5 3/8	6 6/8	M	Olympic Peninsula	WA	Gerald Egbert	1980	31
46 4/8	8 3/8	8 2/8	5 5/8	5 5/8	6 0/8	M	Chaffee County	CO	Todd Clyncke	1987	31
46 4/8	9 0/8	9 2/8	5 5/8	5 5/8	6 2/8	M	Chaffee County	CO	John R. Olson	1996	31
46 2/8	9 2/8	9 5/8	5 2/8	5 3/8	6 5/8	M	Lewis & Clark County	MT	W. J. Fuller	1958	37
46 2/8	9 1/8	9 1/8	5 2/8	5 3/8	6 2/8	M	Lemhi County	ID	Ray Torrey	1967	37
46 2/8	9 4/8	9 3/8	5 1/8	5 1/8	5 5/8	M	Kittitas County	WA	Keith E. Anyan	1978	37
46 2/8	9 1/8	9 2/8	5 3/8	5 3/8	6 6/8	M	Kodiak Island	AK	Troy A. Graziadei	1993	37
46 0/8	9 3/8	9 4/8	5 2/8	5 2/8	6 6/8	M	Whidbey Bay	AK	George Moerlein	1964	41
46 0/8	8 7/8	9 0/8	5 4/8	5 4/8	6 5/8	M	Kechika Range	BC	Roger Stewart	1980	41
46 0/8	9 1/8	9 0/8	5 1/8	5 2/8	5 6/8	M	Chaffee County	CO	Joe B. Farmer	1997	41
46 0/8	9 2/8	9 0/8	5 3/8	5 3/8	6 4/8	M	Grand County	CO	Scott George	2000	41
45 6/8	8 5/8	8 5/8	5 2/8	5 2/8	6 4/8	M	Jefferson County	WA	John Lund	1978	45
45 6/8	8 6/8	9 0/8	5 1/8	5 2/8	6 7/8	M	Cordova	AK	Gary A. Twigg	1980	45
45 6/8	9 6/8	9 6/8	5 0/8	5 1/8	6 2/8	M	Clearwater County	ID	Mike VonLindern	1984	45
45 6/8	8 4/8	9 0/8	5 2/8	5 3/8	7 1/8	M	Stephens Passage	AK	Steve Hohensee	2002	45
45 4/8	9 2/8	9 3/8	5 3/8	5 3/8	6 0/8	M	Lemhi County	ID	A. LaVerne Hokanson	1968	49
45 4/8	8 7/8	8 7/8	5 2/8	5 1/8	8 1/8	M	Kenai Peninsula	AK	Dean Lust	1969	49
45 4/8	8 7/8	8 6/8	5 2/8	5 2/8	6 6/8	M	McCarthy Glacier	AK	John F. Sumrall	1974	49
45 4/8	9 4/8	9 3/8	5 2/8	5 2/8	6 1/8	M	Chaffee County	CO	Troy Cunningham	1993	49
45 2/8	8 6/8	8 7/8	5 5/8	5 4/8	7 4/8	M	Telegraph Creek	BC	Troy M. Miller	1968	53
45 2/8	9 3/8	9 0/8	5 2/8	5 2/8	6 5/8	M	Kenai Mtn.	AK	Robert Borland	1970	53
45 2/8	9 5/8	9 4/8	5 0/8	5 0/8	7 1/8	M	Lemhi County	ID	D. Kittredge/R. Torrey	1973	53
45 2/8	8 3/8	8 5/8	5 1/8	5 2/8	6 2/8	M	Kechika River	BC	Paul Brunner	1974	53
45 2/8	8 7/8	8 7/8	5 3/8	5 2/8	7 0/8	M	Valdez	AK	Rickie D. Snell	1990	53
45 2/8	9 2/8	9 1/8	5 2/8	5 2/8	6 1/8	M	Clark County	ID	Kenneth Ruzicka	1994	53
45 2/8	8 7/8	8 5/8	5 4/8	5 4/8	7 2/8	M	Eagle County	CO	Connie Renfro	1995	53
45 2/8	9 0/8	9 3/8	5 3/8	5 3/8	6 2/8	M	Gunnison County	CO	K. Craig Vaughn	1995	53
45 0/8	9 3/8	9 1/8	5 3/8	5 3/8	5 5/8	M	Flathead County	MT	Jack Whitney	1962	61
45 0/8	9 0/8	9 1/8	5 0/8	5 0/8	6 3/8	M	Boise County	ID	Bradley H. Jolley	1972	61

ROCKY MOUNTAIN GOAT

Minimum Score 40 *Continued*

Score	Length of R Horn	L	Circumference R of Base	L	Greatest Spread	Sex	Area	State/ Province	Hunter's Name	Date	Rank
45 0/8	8 7/8	8 7/8	4 7/8	5 0/8	5 6/8	M	Jefferson County	WA	Edward H. Boyle	1974	61
44 6/8	8 6/8	8 5/8	5 1/8	5 1/8	0 0/8	M	Kittitas County	WA	Les Turner	1967	64
44 6/8	8 7/8	8 6/8	5 1/8	5 1/8	6 1/8	M	Fox River	AK	John F. Sumrall	1977	64
44 6/8	8 7/8	8 6/8	5 2/8	5 3/8	6 7/8	M	Park County	WY	Scott Steere	1982	64
44 6/8	9 1/8	9 1/8	5 1/8	5 1/8	6 6/8	M	Valdez	AK	Rick D. Snell	1991	64
44 6/8	9 0/8	9 0/8	5 1/8	5 0/8	6 6/8	M	Park County	CO	Mark Turner	2006	64
44 4/8	9 7/8	9 7/8	4 3/8	4 3/8	8 5/8	F	Cold Fish Lake	BC	K. K. Knickerbocker	1957	69
44 4/8	8 7/8	9 0/8	5 1/8	5 1/8	6 4/8	M	Kenai Peninsula	AK	Larry Jones	1969	69
44 4/8	8 5/8	8 5/8	5 1/8	5 2/8	6 1/8	M	Kittitas County	WA	David L. Smartt	1972	69
44 4/8	9 2/8	9 3/8	4 6/8	4 5/8	7 7/8	M	Chugach Mtns.	AK	Darryl Quidort	1985	69
44 4/8	8 5/8	8 5/8	5 1/8	5 1/8	6 4/8	M	Chugach Mtns.	AK	Gary White	1986	69
44 4/8	9 2/8	8 6/8	5 3/8	5 3/8	5 6/8	M	Custer County	SD	Vilas Schoenfelder	1988	69
44 4/8	9 2/8	9 4/8	5 0/8	5 0/8	6 7/8	M	Madison County	MT	Steve Rhodes	1989	69
44 4/8	8 7/8	8 5/8	5 2/8	5 2/8	5 7/8	M	Custer County	SD	R. Craig Oberle	1990	69
44 4/8	8 6/8	8 5/8	5 1/8	5 1/8	5 5/8	M	Clear Creek County	CO	Barry J. Smith	1994	69
44 4/8	9 0/8	9 0/8	5 1/8	5 1/8	7 0/8	M	Clear Creek County	CO	E. Damon Handley	1997	69
44 4/8	8 6/8	8 6/8	5 1/8	5 1/8	6 3/8	M	Clear Creek County	CO	Steve Smith	1997	69
44 4/8	8 6/8	9 6/8	5 0/8	5 0/8	7 3/8	M	Lincoln County	MT	Robert Kehn	2003	69
44 4/8	9 2/8	9 2/8	5 1/8	5 1/8	6 4/8	M	La Plata County	CO	Eric A. Sawyer	2007	69
44 2/8	8 5/8	8 5/8	5 1/8	5 0/8	5 6/8	M	Kenai Lake	AK	James R. Carr	1973	82
44 2/8	9 0/8	8 7/8	5 1/8	5 2/8	6 7/8	M	Custer County	SD	Kent D. Keenlyne	1981	82
44 2/8	9 2/8	9 4/8	5 0/8	5 0/8	6 2/8	M	Clear Creek County	CO	Elmer R. Luce, Jr.	1990	82
44 2/8	8 3/8	8 2/8	5 2/8	5 2/8	6 1/8	M	Stephens Passage	AK	Steve Hohensee	2005	82
44 0/8	7 4/8	9 0/8	5 4/8	5 4/8	6 0/8	M	Clallam County	WA	Virgil T. Cole, Jr.	1973	86
44 0/8	8 5/8	8 4/8	5 0/8	5 0/8	5 3/8	M	Chelan County	WA	Steve Gorr	1975	86
44 0/8	8 4/8	8 3/8	4 6/8	4 6/8	4 0/8	M	Snohomish County	WA	Steve Gorr	1978	86
44 0/8	9 3/8	9 4/8	4 6/8	4 6/8	6 4/8	M	Valley County	ID	Jack Barrett	1980	86
44 0/8	8 7/8	9 0/8	4 7/8	4 7/8	5 7/8	M	Snohomish County	WA	Stan Hansen	1985	86
44 0/8	9 0/8	9 1/8	4 7/8	5 0/8	6 0/8	M	Chaffee County	CO	Mark Montgomery	1994	86
44 0/8	9 2/8	9 2/8	4 6/8	4 6/8	6 5/8	M	Park County	WY	Terry Fieseler	1996	86
44 0/8	9 0/8	9 0/8	5 0/8	5 0/8	6 0/8	M	Clear Creek County	CO	Rick Duggan	1997	86
43 6/8	8 4/8	8 4/8	5 1/8	5 0/8	5 4/8	M	Kittitas County	WA	Joe Walker	1967	94
43 6/8	8 2/8	8 3/8	5 1/8	5 2/8	6 3/8	M	Park County	WY	Pat McAteer	1979	94
43 6/8	8 5/8	8 6/8	5 1/8	5 1/8	5 3/8	M	Chaffee County	CO	Doug Beck	1988	94
43 6/8	9 1/8	8 5/8	5 1/8	5 1/8	6 4/8	M	Clear Creek County	CO	Jeff Lampe	2005	94
43 6/8	8 5/8	8 6/8	5 1/8	5 2/8	6 5/8	M	Park County	MT	Brian Stremcha	2006	94
43 4/8	8 7/8	8 7/8	4 7/8	4 7/8	6 0/8	M	Lake County	MT	Jack J. Whitney	1969	99
43 4/8	10 4/8	10 1/8	4 3/8	4 3/8	6 2/8	F	Clear Creek County	CO	Donald Ace Morgan	1989	99
43 4/8	8 5/8	8 6/8	5 0/8	5 0/8	6 4/8	M	Clear Creek County	CO	Dave Parri	1999	99
43 4/8	8 7/8	8 5/8	5 0/8	5 0/8	6 1/8	M	Tucha Lake	BC	Gregg Bekelja	2010	99
43 2/8	10 0/8	10 0/8	4 2/8	4 2/8	8 0/8	F	Tutaday Lake	BC	Larry Alma	1982	103
43 2/8	9 0/8	9 0/8	5 1/8	5 0/8	5 4/8	M	Ewilka Peak	BC	Larry D. Jones	1990	103
43 2/8	9 0/8	8 5/8	5 1/8	5 1/8	5 6/8	M	Bull River	BC	David C. Manca	2002	103
43 0/8	10 5/8	8 7/8	5 1/8	5 1/8	4 6/8	M		BC	Vic Clarkson	1960	106
43 0/8	10 0/8	10 0/8	4 4/8	4 4/8	6 2/8	F	Kittitas County	WA	Richard L. Thrasher	1968	106
42 6/8	9 7/8	9 7/8	4 3/8	4 3/8	7 6/8	F	Lake County	MT	Jack Whitney	1965	108
42 6/8	8 2/8	8 7/8	5 0/8	5 0/8	6 0/8	M	Chaffee County	CO	Chuck Hutton	1979	108
42 6/8	9 0/8	7 3/8	5 5/8	5 4/8	6 3/8	M	Chaffee County	CO	Duke Prentup	1979	108
42 6/8	9 7/8	9 7/8	4 0/8	4 0/8	9 2/8	F	Knik Glacier	AK	Gary G. Wall	1985	108
42 6/8	9 4/8	9 3/8	4 3/8	4 3/8	6 7/8	F	Park County	CO	Corey Clyncke	1989	108
42 6/8	8 4/8	8 0/8	5 1/8	5 1/8	6 2/8	M	Summit County	CO	Stan Rauch	1994	108
42 4/8	8 4/8	9 0/8	5 2/8	5 1/8	5 0/8	M	Chelan County	WA	G. H. Malinoski	1964	114
42 4/8	8 3/8	8 1/8	5 1/8	5 1/8	5 7/8	M	Eagle County	CO	Jim Willems	1999	114
42 2/8	7 6/8	8 2/8	5 0/8	5 0/8	5 4/8	M	Kleena Kleene	BC	William P. Mastrangel	1956	116
42 2/8	8 5/8	9 0/8	4 7/8	4 7/8	5 7/8	M	Terminus Mtn.	BC	Paul P. Schafer	1976	116
42 2/8	8 3/8	8 5/8	4 7/8	5 0/8	5 2/8	M	Snohomish County	WA	Eric A. Olson	1979	116
42 2/8	8 4/8	8 4/8	4 7/8	4 6/8	5 6/8	M	Park County	CO	Marvin Clyncke	1989	116
42 2/8	8 3/8	8 3/8	5 0/8	5 0/8	6 3/8	M	Caribou Mtns.	BC	Roger Wintle	2002	116
42 0/8	7 2/8	8 2/8	5 1/8	5 1/8	7 0/8	M	Taylor Lake	BC	Bill Brown	1957	121
42 0/8	8 4/8	8 5/8	5 0/8	5 0/8	5 7/8	F	Swan Range	MT	Jack Whitney	1960	121
42 0/8	8 4/8	8 4/8	4 5/8	4 5/8	6 1/8	M	Lemhi County	ID	Frank N. Hough	1968	121
42 0/8	9 4/8	9 4/8	4 0/8	4 0/8	6 1/8	F	Snohomish County	WA	Joseph R. St. Charles	1980	121

ROCKY MOUNTAIN GOAT

Minimum Score 40 *Continued*

Score	Length of R Horn	L	Circumference R of Base	L	Greatest Spread	Sex	Area	State/ Province	Hunter's Name	Date	Rank
42 0/8	9 0/8	9 1/8	4 2/8	4 2/8	6 4/8	F	Okanogan County	WA	T. J. Conrads	1995	121
42 0/8	8 1/8	8 1/8	4 7/8	5 0/8	5 3/8	M	Bonneville County	ID	Brian Frickey	1998	121
42 0/8	8 2/8	8 6/8	5 0/8	5 0/8	7 1/8	M	Gallatin County	MT	Rex Rogers	1999	121
41 6/8	8 1/8	8 0/8	4 7/8	4 7/8	6 3/8	F	Lord River	BC	Dr. R. Congdon	1958	128
41 6/8	7 3/8	7 6/8	5 1/8	5 1/8	5 0/8	M	Snohomish County	WA	Jack Williams	1984	128
41 6/8	8 0/8	8 7/8	5 0/8	5 0/8	5 7/8	M	Chaffee County	CO	Ron Sniff	1992	128
41 4/8	10 2/8	10 2/8	4 0/8	4 0/8	10 1/8	F	Takia River	BC	William L. Sullivan	1966	131
41 4/8	8 1/8	8 1/8	4 6/8	4 7/8	5 4/8	M	Summit County	CO	Wayne Depperschmidt	1979	131
41 4/8	8 2/8	8 2/8	4 7/8	5 0/8	6 0/8	M	La Plata County	CO	Dale Struble	1990	131
41 4/8	9 2/8	8 5/8	4 2/8	4 2/8	6 4/8	F	Chouteau County	MT	Donny Roy	1994	131
41 4/8	8 2/8	8 3/8	4 7/8	4 7/8	5 6/8	M	Gallatin County	MT	William M. Elfland	2011	131
41 2/8	9 7/8	9 6/8	4 2/8	4 2/8	6 4/8	F	Crazy Mtns.	MT	Glenn Gibson	1957	136
41 2/8	8 3/8	8 3/8	4 3/8	4 3/8	5 1/8	F	Kittitas County	WA	Dennis Dunn	1973	136
41 2/8	9 2/8	9 2/8	4 2/8	4 2/8	6 1/8	F	Rusty Creek	BC	Ronald Montross	1984	136
41 2/8	7 7/8	7 7/8	4 7/8	4 7/8	5 3/8	M	Snohomish County	WA	Dale Drilling	1988	136
41 2/8	8 4/8	8 4/8	4 6/8	4 5/8	6 0/8	M	Chaffee County	CO	Larick Spencer	1990	136
41 2/8	9 0/8	9 1/8	4 4/8	4 5/8	7 1/8	M	La Plata County	CO	Scot L. Hamilton	2002	136
41 0/8	8 6/8	8 5/8	4 2/8	4 2/8	7 0/8	M	Penticton	BC	Bill Brown	1958	142
41 0/8	8 7/8	9 0/8	4 4/8	4 4/8	5 5/8	F	Snohomish County	WA	Bud Peck	1960	142
41 0/8	10 0/8	10 0/8	3 6/8	3 7/8	6 2/8	F	Holly Creek	BC	Jim Jackson	1964	142
41 0/8	8 3/8	8 2/8	4 5/8	4 5/8	6 1/8	M	Park County	CO	Scott George	1990	142
41 0/8	8 6/8	7 2/8	5 1/8	5 1/8	6 1/8	M	Clear Creek County	CO	Cary Laman	2001	142
41 0/8	9 1/8	9 0/8	4 3/8	4 3/8	6 7/8	F	Grand County	CO	Matt Archuleta	2002	142
40 6/8	8 3/8	8 3/8	4 6/8	4 6/8	5 2/8	M	Boise County	ID	Ronald L. Sherer	1970	148
40 6/8	8 0/8	7 7/8	4 5/8	4 5/8	6 0/8	M	Kenai Lake	AK	Dennis Lattery	1973	148
40 6/8	8 6/8	8 5/8	4 6/8	4 6/8	5 5/8	M	Park County	WY	Jeff Umphlett	1979	148
40 6/8	9 1/8	9 1/8	4 1/8	4 1/8	5 4/8	F	Kittitas County	WA	Roger Pitman	1980	148
40 6/8	9 2/8	9 1/8	4 1/8	4 1/8	5 6/8	F	Kittitas County	WA	L. James Bailey	1984	148
40 6/8	9 0/8	8 7/8	4 2/8	4 2/8	6 6/8	F	Mt. Hunter	BC	Dennis Kamstra	1990	148
40 6/8	8 2/8	8 4/8	4 5/8	5 0/8	5 1/8	M	Gunnison County	CO	Shawn Kingery	2010	148
40 4/8	9 0/8	8 3/8	4 1/8	4 1/8	7 5/8	F	Little Johnstone Bay	AK	Ray Uhl	1965	155
40 4/8	8 7/8	8 6/8	4 1/8	4 1/8	6 4/8	F	Smithers	BC	Chris VanderHorst	1974	155
40 4/8	8 4/8	8 4/8	4 2/8	4 2/8	6 7/8	F	Clear Creek County	CO	Reggie Spiegelberg	1991	155
40 4/8	8 5/8	8 6/8	4 5/8	4 5/8	5 6/8	M	Park County	MT	Dave Yeats	2008	155
40 2/8	7 4/8	8 7/8	4 6/8	4 7/8	7 0/8	M	Gallatin County	MT	Jerry Karsky	1995	159
40 2/8	9 2/8	9 2/8	4 2/8	4 2/8	7 3/8	F	Idaho County	ID	Brian Holbrook	2000	159
40 0/8	8 3/8	8 2/8	4 5/8	4 6/8	6 1/8	M	Lemhi County	ID	Marvin Tye	1973	161
40 0/8	8 7/8	8 7/8	4 2/8	4 2/8	6 6/8	F	Chaffee County	CO	Wayne Spencer	1973	161
40 0/8	7 5/8	7 5/8	4 6/8	4 6/8	5 0/8	M	Cordova	AK	Ray P. Noregaard	1975	161
40 0/8	9 3/8	9 2/8	4 1/8	4 1/8	6 0/8	F	La Plata County	CO	Daniel Willems	1988	161
40 0/8	8 7/8	9 2/8	4 1/8	4 1/8	6 4/8	F	La Plata County	CO	Tony Casagrande	1992	161
40 0/8	8 4/8	8 3/8	4 4/8	4 4/8	5 0/8	M	Flathead County	MT	Eric Kress	1993	161
40 0/8	9 2/8	9 3/8	4 0/8	4 1/8	6 5/8	F	La Plata County	CO	Todd A. Brickel	2008	161

BIGHORN SHEEP

Minimum Score 140 — *Ovis canadensis canadensis* and certain related subspecies

Score	Length of R Horn	L	Circumference R of Base	L	Greatest Spread	Area	State/ Province	Hunter's Name	Date	Rank
190 0/8	38 7/8	39 7/8	16 2/8	16 1/8	23 7/8	Granite County	MT	Richard King	2001	1
187 1/8	40 2/8	39 1/8	15 5/8	15 4/8	23 0/8	Sanders County	MT	Rex Rogers	2000	2
186 5/8	38 5/8	39 4/8	15 4/8	15 5/8	21 4/8	Granite County	MT	Dave Yeats	1999	3
184 6/8	39 1/8	37 5/8	16 5/8	16 5/8	20 2/8	Ravalli County	MT	Bill A. Richichi	1996	4
183 7/8	38 3/8	38 6/8	15 4/8	15 5/8	24 5/8	Deer Lodge County	MT	Jerry Parsons	1986	5
183 7/8	36 6/8	37 1/8	15 0/8	15 0/8	22 0/8	Clear Creek County	CO	Ray Alt	1988	5
183 4/8	40 3/8	41 1/8	15 5/8	15 6/8	23 3/8	Lincoln County	MT	Paul Schafer	1983	7
182 2/8	36 0/8	36 0/8	15 7/8	15 7/8	23 2/8	Blaine County	MT	Gary Troester	2005	8
182 0/8	39 6/8	39 6/8	15 0/8	15 0/8	22 4/8	Sanders County	MT	John T. Beyer	1990	9
180 7/8	37 3/8	38 4/8	15 4/8	15 4/8	22 1/8	Granite County	MT	Alden Gregory Beard	1992	10
179 4/8	39 3/8	37 5/8	14 5/8	14 5/8	21 6/8	Fergus County	MT	Stan Colton	2006	11
179 1/8	38 1/8	36 4/8	15 0/8	15 1/8	22 1/8	Larimer County	CO	Joe Bradley	2006	12
179 1/8	37 1/8	37 6/8	14 6/8	14 7/8	20 2/8	Ravalli County	MT	Charlie Johnson	2007	12
177 2/8	32 5/8	34 3/8	15 3/8	15 3/8	22 3/8	Larimer County	CO	Mark Montgomery	1992	14
177 2/8	39 3/8	37 1/8	14 4/8	14 4/8	23 2/8	Granite County	MT	Neil L. Jacobson	1998	14
176 3/8	37 3/8	39 0/8	14 6/8	14 4/8	22 0/8	Sweet Grass County	MT	Ray Alt	1968	16
176 1/8	39 4/8	38 1/8	14 7/8	15 2/8	22 2/8	Phillips County	MT	Ty Milne	1991	17
175 1/8	39 5/8	39 6/8	15 2/8	15 2/8	19 4/8	Canmore	ALB	Dave Addie	1985	18
174 2/8	35 6/8	36 0/8	14 2/8	14 2/8	19 6/8	Saguache County	CO	David "Jake" Powell	1995	19
174 1/8	38 1/8	37 6/8	15 1/8	15 2/8	21 5/8	Canmore	ALB	Todd Zeuske	1993	20
173 7/8	35 1/8	37 0/8	15 1/8	15 3/8	20 7/8	Sanders County	MT	Bart Schleyer	1987	21
173 5/8	34 4/8	35 5/8	14 4/8	14 3/8	21 7/8	Clear Creek County	CO	Ronald Rockwell	1995	22
172 7/8	36 0/8	35 5/8	13 6/8	13 7/8	21 4/8	Boulder County	CO	Scott George	1995	23
171 5/8	35 4/8	34 5/8	15 3/8	15 3/8	21 4/8	Granite County	MT	Marlon Clapham	1994	24
170 0/8	35 6/8	36 2/8	14 2/8	14 3/8	22 2/8	Clear Creek County	CO	Troy Cunningham	1988	25
169 6/8	35 2/8	34 6/8	14 4/8	14 4/8	23 0/8	Clear Creek County	CO	Janet George	1989	26
169 2/8	33 2/8	33 6/8	14 5/8	14 6/8	21 4/8	Clear Creek County	CO	Barry J. Smith	1992	27
169 0/8	33 2/8	35 2/8	14 3/8	14 2/8	20 2/8	Cougar Canyon	ALB	Paul Schwengler	1980	28
168 2/8	32 0/8	35 2/8	14 3/8	14 3/8	21 6/8	Clear Creek County	CO	Mark Turner	2002	29
167 6/8	36 3/8	36 1/8	14 6/8	14 6/8	21 1/8	N. Saskatchewan River	ALB	Larry Jones	1962	30
167 4/8	31 4/8	32 0/8	15 4/8	15 3/8	22 2/8	Clear Creek County	CO	Reggie Spiegelberg	1988	31
166 5/8	35 4/8	32 7/8	15 5/8	15 5/8	20 7/8	Sanders County	MT	Kurt D. Fisher	2002	32
166 2/8	31 2/8	35 0/8	15 3/8	15 0/8	22 6/8	Las Animas County	CO	Gary Troester	2004	33
166 1/8	34 6/8	34 1/8	14 0/8	14 1/8	19 6/8	Saguache County	CO	Jeff Puls	1997	34
166 0/8	33 2/8	32 2/8	14 7/8	14 7/8	19 7/8	Jefferson County	CO	Robert Sorrell	1977	35
165 7/8	31 4/8	32 7/8	14 2/8	14 3/8	21 4/8	Mineral County	CO	Daniel Willems	2012	36
165 6/8	35 2/8	34 0/8	14 2/8	14 2/8	22 2/8	Clear Creek County	CO	Lee Beckwith	1994	37
165 4/8	34 6/8	34 4/8	13 3/8	13 5/8	19 2/8	Boulder County	CO	Frank Piacentino	1999	38
165 3/8	33 7/8	34 4/8	13 7/8	13 6/8	19 4/8	Saguache County	CO	Tom Sieverding	1987	39
164 7/8	38 0/8	37 3/8	14 3/8	14 4/8	20 1/8	Yakima County	WA	Albert Rinaldi, Jr.	1973	40
164 6/8	32 4/8	34 0/8	14 1/8	13 7/8	22 5/8	Saguache County	CO	Rick Duggan	1995	41
164 2/8	34 5/8	34 5/8	14 0/8	14 0/8	21 0/8	Clear Creek County	CO	Troy A. Graziadei	2000	42
164 1/8	32 5/8	32 2/8	14 4/8	14 4/8	22 0/8	Park County	CO	Jim Shanks	2003	43
163 7/8	34 3/8	32 4/8	14 7/8	14 7/8	22 6/8	Teller County	CO	Dale Struble	1987	44
163 7/8	32 5/8	32 2/8	15 5/8	15 5/8	20 7/8	Teller County	CO	William Elfland	1994	44
163 6/8	35 3/8	35 3/8	15 2/8	15 3/8	23 1/8	Lincoln County	MT	Paul Brunner	1983	46
163 5/8	31 2/8	31 1/8	15 6/8	15 6/8	21 0/8	Las Animas County	CO	Carroll Holl	1995	47
163 4/8	30 6/8	31 0/8	15 1/8	15 2/8	21 1/8	Graham County	AZ	David Baker	2006	48
163 0/8	36 3/8	36 3/8	13 7/8	13 7/8	22 0/8	Lake County	MT	Steve Gorr	1973	49
163 0/8	32 5/8	33 3/8	14 3/8	14 3/8	20 3/8	Clear Creek County	CO	Tony Archuleta	2003	49
162 2/8	33 2/8	31 6/8	13 7/8	13 7/8	22 4/8	Clear Creek County	CO	Joe Brown	2000	51
162 2/8	33 7/8	33 5/8	13 0/8	13 0/8	17 4/8	Mora County	NM	Bill J. VanBuskirk	2001	51
162 1/8	30 6/8	33 5/8	15 0/8	15 1/8	23 7/8	Park County	CO	Wayne Depperschmidt	1977	53
162 0/8	35 3/8	31 7/8	14 7/8	15 0/8	21 4/8	Albany County	WY	Jerry Bowen	2001	54
161 3/8	30 1/8	29 4/8	14 4/8	14 4/8	22 2/8	Clear Creek County	CO	Connie Renfro	2002	55
160 6/8	33 3/8	30 7/8	13 7/8	13 6/8	22 0/8	Eagle County	CO	Ken Wolter	2000	56
160 3/8	33 4/8	33 1/8	14 4/8	14 5/8	24 3/8	Owyhee County	ID	Mike J. McCollum	1995	57
159 7/8	36 2/8	33 7/8	13 7/8	13 7/8	23 5/8	Grant County	OR	Christopher Tipton	2011	58
159 2/8	31 0/8	31 2/8	13 7/8	13 7/8	20 3/8	Park County	WY	Thomas M. Taylor	2005	59
159 0/8	31 4/8	31 4/8	13 6/8	13 7/8	24 4/8	Park County	CO	Ron King	2001	60
158 5/8	30 1/8	30 0/8	15 4/8	15 3/8	21 5/8	Park County	CO	Marvin Clyncke	1977	61
158 3/8	30 2/8	27 3/8	14 4/8	14 4/8	20 4/8	Fremont County	CO	John Quick	1974	62

BIGHORN SHEEP

Minimum Score 140 *Continued*

Score	Length of R Horn	Length of L Horn	Circumference of Base R	Circumference of Base L	Greatest Spread	Area	State/ Province	Hunter's Name	Date	Rank
158 3/8	34 4/8	34 5/8	15 0/8	15 0/8	21 1/8	Park County	WY	Terry Constable	1988	62
157 5/8	32 0/8	29 5/8	14 5/8	14 4/8	21 1/8	Larimer County	CO	Glen Vlass	1991	64
157 2/8	33 0/8	31 0/8	13 3/8	13 2/8	19 0/8	Fremont County	CO	Steve Gorr	1973	65
156 5/8	31 1/8	34 2/8	14 5/8	14 5/8	23 4/8	El Paso County	CO	Barry J. Smith	1983	66
156 4/8	32 6/8	32 0/8	12 7/8	12 7/8	19 5/8	Saguache County	CO	John Duggan	1995	67
156 2/8	32 4/8	32 4/8	15 0/8	15 2/8	19 6/8	Las Animas County	CO	Pat Powell	1997	68
155 2/8	30 0/8	30 6/8	14 5/8	14 5/8	21 3/8	Clear Creek County	CO	Garret Decker	1990	69
154 7/8	32 0/8	31 7/8	15 0/8	15 0/8	21 1/8	El Paso County	CO	Lee Kline	1976	70
154 4/8	30 6/8	31 2/8	14 6/8	14 6/8	20 5/8	Chaffee County	CO	Ray Nelson	1986	71
154 0/8	33 4/8	32 0/8	14 3/8	14 5/8	20 1/8	Jefferson County	CO	Dennis Behn	1974	72
154 0/8	30 6/8	29 2/8	15 4/8	15 7/8	20 3/8	Teller County	CO	Shawn Kingery	1996	72
153 4/8	29 2/8	30 0/8	13 0/8	13 0/8	19 4/8	Clear Creek County	CO	Jeff Lampe	1994	74
153 0/8	31 4/8	30 6/8	14 0/8	13 3/8	22 4/8	Saguache County	CO	Larick F. Spencer	1996	75
150 6/8	29 2/8	29 4/8	14 6/8	14 6/8	19 3/8	Saguache County	CO	David 'Jake' Powell	1985	76
150 6/8	30 6/8	31 0/8	15 6/8	15 6/8	21 6/8	Clear Creek County	CO	Todd Bandemer	1999	76
149 4/8	28 6/8	29 4/8	15 1/8	15 0/8	20 0/8	Fremont County	CO	David E. Smith	1996	78
147 6/8	30 0/8	29 2/8	15 5/8	15 5/8	20 1/8	El Paso County	CO	Glenn R. Kuklick	1988	79
147 4/8	27 6/8	24 2/8	15 6/8	15 7/8	20 2/8	Saguache County	CO	Charles Grumley	1988	80
147 1/8	29 7/8	30 0/8	13 6/8	13 7/8	19 6/8	Lake County	CO	G. Fred Asbell	1979	81
144 5/8	29 3/8	29 0/8	14 5/8	14 4/8	20 4/8	Clear Creek County	CO	Rob Firth	1988	82
142 6/8	26 1/8	28 7/8	14 0/8	14 0/8	19 2/8	Teller County	CO	Steve Barnhill	1990	83
141 6/8	25 6/8	27 4/8	14 4/8	14 7/8	19 2/8	Chaffee County	CO	Ron Breitsprecher	1978	84
141 1/8	25 5/8	27 6/8	14 5/8	14 4/8	20 1/8	Pueblo County	CO	Gary Bohochik	1994	85
141 1/8	26 7/8	26 6/8	14 6/8	14 5/8	19 5/8	Saguache County	CO	John L. Gardner	1998	85
140 7/8	27 6/8	33 3/8	14 1/8	14 0/8	19 6/8	Kittitas County	WA	Stan Hansen	1984	87
140 4/8	28 6/8	28 6/8	14 1/8	14 1/8	19 1/8	Valley County	ID	Michael Schlegel	1987	88

Hunter: Richard A. Iano
Species: Typical Whitetail Deer
Score: 138 1/8
Date: 11/15/11
Location: Sherburne County, MN

DALL'S SHEEP

Minimum Score 120 — *Ovis dalli dalli* and *Ovis dalli kenaiensis*

Score	Length of R Horn	L	Circumference R of Base	L	Greatest Spread	Area	State/ Province	Hunter's Name	Date	Rank
164 5/8	40 6/8	41 3/8	13 3/8	13 5/8	28 4/8	Nahanni Butte	NWT	Gary Laya	1986	1
162 3/8	38 7/8	39 0/8	12 4/8	12 4/8	22 6/8	Delta River	AK	Dr. Russell Congdon	1960	2
161 5/8	41 2/8	40 3/8	13 0/8	13 0/8	24 3/8	Liard Range	NWT	Pete Iacavazzi	1994	3
160 6/8	35 3/8	38 3/8	12 4/8	12 4/8	22 3/8	Sag River	AK	Troy Graziadei	2012	4
159 3/8	38 4/8	38 3/8	13 2/8	13 3/8	20 5/8	Atigun Pass	AK	Troy A. Graziadei	2008	5
157 5/8	35 2/8	35 3/8	13 0/8	13 0/8	20 0/8	Roberts Fork	AK	Jeff Falkner	2010	6
157 0/8	35 4/8	35 4/8	13 6/8	14 0/8	21 2/8	Chitina Glacier	AK	Roger Morris	1973	7
155 5/8	37 0/8	37 7/8	13 7/8	13 5/8	27 4/8	Keele River	NWT	Robert J. "Grub" Matthews	2007	8
155 1/8	37 1/8	37 2/8	13 4/8	13 4/8	25 7/8	Rams Head Mtn.	NWT	Ron Breitsprecher	1981	9
155 0/8	36 6/8	36 6/8	13 5/8	13 4/8	23 7/8	Nahanni Butte	NWT	Tom Vanasche	1994	10
154 7/8	35 2/8	34 7/8	14 1/8	14 1/8	21 4/8	Mackenzie Mtns.	NWT	Mike Barrett	1983	11
154 5/8	37 3/8	40 2/8	12 6/8	12 7/8	27 4/8	Dillinger River	AK	John L. Gardner	2004	12
154 5/8	35 1/8	35 6/8	13 1/8	13 1/8	23 6/8	Atigan River	AK	Michael J. Lettis	2012	12
153 5/8	35 1/8	36 6/8	13 3/8	13 3/8	21 6/8	Nahanni Butte	NWT	E. Damon Handley	1989	14
152 7/8	39 0/8	38 7/8	13 3/8	13 4/8	28 3/8	Nahanni Butte	NWT	Monty Moravec	1993	15
152 6/8	37 1/8	36 5/8	13 1/8	13 1/8	30 0/8	Wrangell Mtns.	AK	Ray Torrey	1973	16
152 2/8	37 0/8	36 2/8	12 5/8	12 5/8	23 4/8	Kuskokwim Mtn.	AK	Kenneth R. Wallenberg	1978	17
152 0/8	37 0/8	37 0/8	12 4/8	12 4/8	27 4/8	Wrangell Mtns.	AK	Dr. Rex Hancock	1962	18
152 0/8	35 6/8	35 4/8	13 4/8	13 4/8	26 1/8	Boulder Creek	AK	Rick D. Snell	1993	18
151 2/8	36 6/8	36 6/8	12 6/8	12 6/8	26 4/8	Johnson River	AK	Larry Jones	1963	20
151 2/8	36 4/8	36 4/8	13 7/8	13 6/8	23 4/8	Tonsona Creek	AK	Bruce Stephens	1974	20
151 0/8	35 4/8	34 4/8	13 3/8	13 5/8	27 3/8	Mackenzie Mtns.	NWT	Mike Barrett	1986	22
148 6/8	37 6/8	38 0/8	12 2/8	12 2/8	22 5/8	Nahanni Butte	NWT	Bill Grammer	1986	23
148 5/8	34 6/8	35 1/8	13 6/8	14 0/8	20 7/8	Chugach Mtns.	AK	Ken Wolter	2004	24
147 5/8	31 0/8	33 5/8	12 4/8	12 3/8	17 7/8	Jackfish Mtn.	NWT	Daniel C. Hurd	1996	25
147 4/8	34 6/8	35 4/8	12 0/8	12 0/8	21 5/8	Mackenzie Mtns.	NWT	Paul Brunner	1982	26
147 2/8	32 3/8	32 3/8	12 6/8	12 7/8	20 4/8	Talkeetna Mtns.	AK	Jeffrey S. Stevens	1997	27
146 5/8	36 3/8	35 0/8	11 6/8	11 5/8	21 0/8	Liard Range	NWT	Ron Rockwell	1983	28
146 4/8	33 4/8	34 2/8	12 0/8	12 2/8	25 7/8	Delta River	AK	Elisha Gray	1958	29
146 2/8	33 4/8	32 6/8	11 5/8	11 4/8	17 6/8	Nahanni Butte	NWT	Barry J. Smith	1989	30
145 1/8	32 2/8	33 1/8	13 4/8	13 7/8	23 5/8	Wood River	AK	Art Young	1923	31
145 0/8	33 5/8	33 5/8	13 0/8	13 0/8	25 5/8	Mt. Ibex	YUK	Martin Hanson	1957	32
144 5/8	34 5/8	35 2/8	13 0/8	13 0/8	23 1/8	Mt. Hayes	AK	Keith R. Clemmons	1962	33
144 5/8	34 5/8	35 0/8	12 4/8	12 3/8	22 0/8	Frita River	NWT	Tavis Rogers	2012	33
144 2/8	33 7/8	33 7/8	12 6/8	13 0/8	24 0/8	Yellow Jacket Pass	AK	G. Fred Asbell	1993	35
143 3/8	35 0/8	34 5/8	12 0/8	12 2/8	19 4/8	Endicott Mtns.	AK	Dwane J. Sykes	1968	36
143 2/8	31 0/8	35 2/8	12 4/8	12 4/8	23 6/8	Mackenzie Mtns.	NWT	Paul Schafer	1983	37
143 1/8	32 0/8	32 5/8	13 3/8	13 3/8	21 4/8	Chugach Mtns.	AK	Glenn R. L. Schmidt	1977	38
143 0/8	35 2/8	31 4/8	12 3/8	12 4/8	23 1/8	Nahanni Butte	NWT	T. Sanford Roberts	1993	39
142 1/8	35 7/8	35 6/8	11 6/8	11 6/8	23 7/8	Nahanni Butte	NWT	Todd Szmania	1995	40
142 1/8	33 7/8	33 4/8	12 3/8	12 2/8	22 3/8	June Lake	NWT	Fred Eichler	2005	40
140 5/8	32 4/8	31 5/8	12 4/8	12 4/8	23 7/8	Liard River	NWT	Greg Munther	1984	42
139 1/8	34 2/8	34 7/8	12 3/8	12 3/8	27 7/8	Tok	AK	Gardner Rowell	1991	43
137 2/8	32 4/8	33 0/8	11 7/8	11 7/8	22 4/8	Nahanni Butte	NWT	Lew Webb	2001	44
137 0/8	33 4/8	33 4/8	12 0/8	12 0/8	18 6/8	Liard Mtns.	NWT	Nathan Andersohn	1994	45
136 3/8	32 2/8	32 7/8	12 2/8	12 1/8	22 6/8	Liard Mtns.	NWT	Gregg Bekelja	2012	46
135 0/8	31 2/8	31 2/8	12 0/8	11 7/8	23 6/8	Liard Range	NWT	Mark Checki	1988	47
133 2/8	32 0/8	31 6/8	11 7/8	11 6/8	21 2/8	Nabesna River	AK	George A. Moerlein	1983	48
133 0/8	30 6/8	30 4/8	12 4/8	12 5/8	27 1/8	Wrangell Mtns.	AK	George A. Moerlein	1971	49
132 6/8	30 2/8	30 4/8	12 5/8	12 5/8	18 2/8	Brooks Range	AK	Robert Warpack	1986	50
130 7/8	32 6/8	28 5/8	12 4/8	12 4/8	20 6/8	Talkeetna Mtns.	AK	John L. Wozniak	1984	51
129 6/8	31 2/8	31 6/8	11 1/8	11 0/8	21 2/8	Liard Range	NWT	Rick Duggan	1998	52
127 0/8	29 7/8	29 3/8	11 2/8	11 1/8	21 2/8	Sheep Creek	AK	Ray Uhl, Jr.	1968	53
125 7/8	31 6/8	31 3/8	10 4/8	10 4/8	21 1/8	Alaska Range	AK	Larry Jones	1969	54
125 0/8	29 1/8	29 1/8	11 2/8	11 2/8	17 4/8	Liard Range	NWT	Todd Henck	1988	55
124 0/8	30 1/8	30 5/8	11 2/8	11 0/8	17 5/8	Kuskokwin River	AK	Marvin Clyncke	1997	56

DESERT BIGHORN SHEEP

Minimum Score 120 — *Ovis canadensis nelsoni* and certain related subspecies

Score	Length of R Horn	L	Circumference R of Base	L	Greatest Spread	Area	State/ Province	Hunter's Name	Date	Rank
169 3/8	36 3/8	36 0/8	14 2/8	14 2/8	23 1/8	Nye County	NV	Rick Duggan	2009	1
156 2/8	34 2/8	34 2/8	14 5/8	14 5/8	17 2/8	Carmen Island	MEX	Thomas M. Taylor	2006	2
154 2/8	31 4/8	31 0/8	13 0/8	13 2/8	19 1/8	Carmen Island	MEX	Nathan Andersohn	2007	3
153 6/8	34 0/8	31 6/8	13 5/8	12 5/8	19 2/8	Yuma County	AZ	Jeffery Stevens	1990	4
146 0/8	28 6/8	28 6/8	13 0/8	13 3/8	19 4/8	Yuma County	AZ	Mark F. Vancas	1990	5
128 7/8	28 0/8	28 7/8	12 6/8	12 7/8	23 6/8	Delta County	CO	Wayne Depperschmidt	2004	6

Hunter: Scott Anspaugh
Species: Alaskan Brown Bear
Score: 21 10/16
Date: 10/3/11
Location: Ursus Cove, AK

STONE'S SHEEP

Minimum Score 120 | *Ovis dalli stonei*

Score	Length of R Horn	Length of L Horn	Circumference R of Base	Circumference L of Base	Greatest Spread	Area	State/ Province	Hunter's Name	Date	Rank
174 2/8	41 1/8	42 1/8	12 4/8	12 4/8	22 0/8	Tetsa River	BC	Stanley Walchuk, Jr.	1992	1
158 1/8	40 4/8	38 3/8	12 4/8	13 1/8	27 0/8	Cold Fish Lake	BC	Fred Bear	1957	2
155 2/8	38 1/8	37 1/8	13 1/8	13 2/8	26 6/8	Todagin Mtn.	BC	Reggie Spiegelberg	1984	3
154 2/8	35 2/8	38 2/8	12 2/8	12 3/8	22 6/8	Gathto Creek	BC	Rick Duggan	2000	4
153 7/8	36 3/8	34 6/8	13 1/8	13 0/8	24 6/8	Cassiar Mtns.	BC	Marvin Clyncke	1994	5
153 5/8	38 0/8	39 2/8	14 0/8	14 0/8	24 1/8	Kechika River	BC	Paul Brunner	1974	6
153 0/8	37 2/8	36 2/8	13 3/8	13 3/8	25 5/8	Pelly Mtns.	YUK	Todd R. Zeuske	1997	7
150 3/8	28 0/8	36 5/8	13 4/8	13 3/8	23 6/8	Todagin Mtn.	BC	Bill Nickerson	1987	8
149 1/8	35 4/8	35 3/8	13 2/8	13 1/8	21 6/8	Terminus Mtn.	BC	Paul P. Schafer	1975	9
145 7/8	34 4/8	34 7/8	13 0/8	13 0/8	22 1/8	Todagin Mtn.	BC	Lee Kline	1983	10
144 4/8	37 0/8	33 6/8	11 4/8	11 6/8	22 0/8	Two Peaks	YUK	Fred Eichler	2007	11
144 0/8	34 4/8	34 4/8	12 3/8	12 3/8	23 3/8	Todagin Mtn.	BC	Eugene Arndt	2002	12
142 2/8	33 6/8	34 2/8	12 2/8	12 1/8	22 1/8	Todagin Mtn.	BC	David Hooper	1977	13
130 3/8	30 6/8	30 3/8	11 7/8	11 7/8	22 7/8	Kinaskan Lake	BC	Rick Duggan	1999	14

Hunter: Patrick Allen
Species: Typical American Elk
Score: 328 3/8
Date: 8/30/11
Location: White Pine County, NV

Hunter: Ron Rockwell **Species:** Alaska Brown Bear **Score:** 21 8/16 **Date:** 2/14/12 **Location:** Kiklukh River, Alaska

Hunter: Michael Feldermann Species: Non-typical Whitetail Deer Score: 161 6/8 Date: 11/15/10 Location: Jo Daviess County, Illinois

Hunter: M.R. James Species: Cougar Score: 14 5/16 Date: 1/2/70 Location: Carbon County, Utah

Hunter: Brian Koelzer Species: Typical Mule Deer Score: 174 5/8 Date: 9/11/12 Location: Sheridan County, Wyoming

Hunter: James Carlson Species: Typical American Elk Score: 294 6/8 Date: 9/23/10 Location: Catron County, New Mexico

Hunter: Fred Eichler   Species: Alaska-Yukon Moose   Score: 173 5/8   Date: 9/19/08   Location: Red Creek, Yukon

Hunter: Mark Turner **Species:** Bison **Score:** 100 0/8 **Date:** 10/26/10 **Location:** Pink Mountains, British Columbia

Hunter: Matthew Rehor Species: Typical Whitetail Deer Score: 125 3/8 Date: 10/28/11 Location: Jo Daviess County, Illinois

Hunter: Lee Kline Species: Stone's Sheep Score: 145 7/8 Date: 8/21/83 Todagin Mountain, British Columbia

Hunter: Nathan Andersohn Species: Desert Bighorn Sheep Score: 154 2/8 Date: 11/19/07 Location: Carmen Island, Mexico

Hunter: Dean Derby Species: Non-typical Whitetail Deer Score: 164 5/8 Date: 11/22/08 Jewell County, Kansas